Comprehensive B_{12}

Zenon Schneider · Andrzej Stroiński

Comprehensive B_{12}

Chemistry · Biochemistry · Nutrition Ecology · Medicine

W DE G

Walter de Gruyter · Berlin · New York 1987

Authors
Dr. Zenon Schneider
Institute of Biochemistry
Academy of Agricultural Sciences
35Wolynska
PL-60-637 Poznań

Dr. Andrzej Stroiński
Institute of Applied Botany
Academy of Agricultural Sciences
35 Wolynska
PL-60-637 Poznań

Verlag Walter de Gruyter & Co., Genthiner Str. 13, D-1000 Berlin 30,
Tel.: (030) 26005-0, Telex 184027
Walter de Gruyter, Inc., 200 Saw Mill River Road, Hawthorne, N.Y. 10532,
Tel.: (914) 747-0110, Telex 646677

CIP-Kurztitelaufnahme der Deutschen Bibliothek

Schneider, Zenon:
Comprehensive B_{12} [B]: chemistry, biochemistry, nutrition, ecology, medicine/
Zenon Schneider ; Andrzej Stroiński. - Berlin ; New York : de Gruyter, 1987.
ISBN 3-11-008239-X
NE: Stroiński, Andrzej:

Library of Congress Cataloging-in-Publication Data

Schneider, Zenon, 1934-
Comprehensive B 12.

Includes bibliographies and index.
1. Vitamin B 12. I. Stroiński, Andrzej, 1939-
II. Title. III. Title: Comprehensive B twelve. [DNLM: 1. Vitamin B 12.
QU 194 S359c]
QP772.C9S36 1987 574.19'26 87-8875
ISBN 0-89925-312-1 (U.S.)

 - Typesetting: Georg Appl, Wemding. - Printing: Gerike GmbH, Berlin. - Binding: Lüderitz & Bauer GmbH, Berlin. - Cover design: Hansbernd Lindemann, Berlin. - Printed in Germany

Preface

The biological function of vitamin B_{12} and its complex chemistry have occupied the finest minds among chemists, biochemists and clinical researchers over the last forty years. A wealth of facts obtained by a variety of sophisticated methods has been gathered and it has become increasingly difficult, even for specialists, to become acquainted with the various aspects of this vitamin. The authors address this book to clinicians and students in different areas such as chemists, biochemists, pharmacologists, nutritionists, animal breeders, ecologists and marine biologists who wish to extract readily essential information on particular aspects of vitamin B_{12}. Most of the data presented are accompanied by references to the original literature and are preceded by a brief and comprehensive introduction. Some techniques which at present are of interest to a wider group of researchers, such as immobilization of vitamin B_{12} on supports, methods of B_{12} assay and biological or chemical synthesis of various B_{12} analogues are described in detail.

The authors are indebted to all of those who kindly allowed us to reproduce their results and in particular:
B. M. Babior, R. Bonnet, V. Y. Bykhovsky, I. Chanarin, D. Dolphin, J. M. Elliot, H. C. Friedmann, W. Friedrich, S. Fukui, G. B. J. Glass, R. Graesbeck, D. C. Hodgkin, H. P. C. Hogenkamp, L. Jaenicke, B. C. Johnson, J. W. Leftley, L. Ljungdahl, E. Nexø, J. Pawelkiewicz, F. Pedziwilk, J. M. Poston, J. M. Pratt, J. Retey, G. N. Schrauzer, R. M. Smith, T. Trojanowska, W. Walerych, J. W. Wood, B. Zagalak.

We thank Dr. W. Friedrich for reading the manuscript and for his critical comments and Mrs. Krystyna Baranowska and Danuta Leszczynska for drawing the figures and technical help.

June 1987

Zenon Schneider
Andrzej Stroiński

Contents

1 Historical Outline

The name vitamin B_{12} is synonymous for a group of cobalt containing, closely related compounds which can cure pernicious anaemia, a fatal disease of the red blood cells. Pernicious anaemia was reported already in 1824 [1]. It remained uncurable, however, to the third decade of our century [1–17]. The history of the discovery of vitamin B_{12} and its biological function can be traced back to the early twenties when Minot and Murphy, two American physicians started to cure pernicious anaemia with a liver diet, and Castle observed that the stomach juice contains a protein factor, called by him *intrinsic factor,* which strongly enhances the curing effect of the liver or its extracts administered orally [18, 19] (Table 1-1). Following this discovery and for the next twenty years, liver was the main source of this unknown curing factor, which was prepared, as time passed, in more and more concentrated form [20–22]. Finally in 1948 the isolation of pure crystalline factor was announced by two independent teams from the United States and England [23, 24]. However it is possible to trace several lines of studies of independent origin which were well under way to vitamin B_{12} discovery. These were research on *animal protein factor* (APF) [55, 56] and the *Lactobacillus lactis Dorner factor* (LLDF) essential for growth of this bacteria species [25], or the application of cobalt ions in diets of sheep and cattle to cure the so-called *bush sickness* in Australia [26].

In the course of the years 1950–61 the unique chemical structure of cyanocobalamin and its newly discovered form coenzyme B_{12} was elucidated by a chemical approach and X-ray crystallography [27–29]. Since 1952 a considerable number of vitamin B_{12} analogues has been obtained from natural sources and by guided biosynthesis [30–33] or synthetized chemically from cobyric acid, an incomplete form of vitamin B_{12} [34]. It also became apparent that not the animals but the microorganisms are the primary producers of vitamin B_{12}. Factories making antibiotics by fermentation made use of the *Streptomyces* species for commercial production of vitamin B_{12} at a fraction of the cost of extraction from liver. The introduction of very efficient producers of vitamin B_{12} like *Propionibacterium shermanii,* made the vitamin accessible for broad scientific and medical application. The discovery of the coenzymatic function of vitamin B_{12} in 1958 in Barker's laboratory made a new, fascinating turn in the research of this factor and its molecular function in organisms [35]. During the years 1960–72 more than ten coenzyme B_{12} dependent enzymes or enzyme systems were discovered and characterized [36]. The mechanism of stereospecific hydrogen transfer and methylgroup transfer by these enzymes attracted a considerable number of distinguished investigators of various disciplines.

The biological functions of vitamin B_{12} binding non-enzymatic proteins were simultaneously studied. It became apparent that these proteins play a crucial role in binding the tiny amounts of vitamin B_{12} released from food in the digestion system and in transporting it accross the cell membranes [37]. A great deal of work was done on these proteins mainly at medical laboratories. The efforts directed towards the elucidation of the cause of pernicious anaemia on a molecular level and the neurological disorders often accompanying this disease have been less successful. It is difficult at present to find a direct link of these diseases with the action of two known vitamin B_{12} dependent enzymes: methylmalonyl-CoA mutase and methionine synthetase present in man. Due to efficient medical service and a more appropriate diet, pernicious anaemia has now become a rare disease. The neurological disorders which tend to appear even at a moderate stage of deficiency of vitamin B_{12} are also cured with this vitamin. The during effect of vitamin B_{12} on neurological diseases is not so spectacular, however, as it is in the case of pernicious anaemia [38].

The complex chemical structure of the main part of vitamin B_{12}, the corrin nucleus, presented an ambitious task for biochemists studying its biosynthesis as well as for chemists attempting to synthetize it by chemical means [39]. It was already discovered in the early sixties by the Shemin group that the first steps of biological synthesis of corrinoids are similar or identical to the synthesis of porphyrins [40] (Chapter 4). Further reactions leading to the corrin ring and its side chains are so far only partly elucidated despite intensive research for almost a quarter of a century. The studies are now being accelerated owing to nuclear magnetic resonance techniques. Significant progress has been achieved with regard to the sequence of introduction of methyl groups and the formation of the corrin ring. The synthesis of the nucleotide part of the vitamin B_{12} runs along an independent pathway [41, 42]. Riboflavin has been found as the precursor of the dimethylbenzimidazole base of the nucleotide [42]. The process of joining the nucleotide to the corrin part and the completion of fully active vitamin B_{12} has been found complex as well. Ribosomes, known so far only as machinery of protein synthesis have been found to be involved at this stage [43]. The chemical synthesis of the corrin nucleus (cobyric acid) was successfully already solved in the early seventies by a joint effort of two groups of investigators led by Woodward and Eschenmoser [44, 45]. In the course of the seventies the techniques for synthesis of intermediates were improved and alternative steps for synthesis of the complete vitamin B_{12} were developed.

The corrinoids proved to be interesting models for studies in coordination chemistry. Binding constants of the α- and β-ligands in some analogues were estimated and the subtle electronic structure of the corrin ring was studied by various spectroscopic methods [46]. New vitamin B_{12} analogues have been synthetized for this purpose mostly by substitution of the α- and β-ligands [47]. Studies with some analogues revealed that ligands may undergo chemical rearrangements. These studies, extended also to corrin-like coordination compounds, opened up a new field in coordination chemistry [48, 49, 50, 51]. The isolation of cobalt-free corrinoids from *Chromatium* enabled the introduction of other metals into the corrin ring such as rhodium, iron, zinc or copper thus giving rise to an entirely new type of B_{12} analogues with interesting biological and chemical features [52–54]. The main events from the vitamin B_{12} history are summerized in Table 1-1.

The corrinoids isolated from natural sources or obtained through chemical alterations have been named differently by various investigators. They were also named according to rules proposed by the IUPAC-IUB Commission at various times. The

Table 1-1 Major events in the history of the discovery of vitamin B_{12} and its biological function.

year or year span	event	main contributors or leaders of research groups.	references
1824	First report on pernicious anaemia	J. S. Combe	[1]
1926	Application of liver diet for curing pernicious anaemia	C. R. Minot and W. P. Murphy	[18]
1929	Discovery of a gastric protein which enhances the curing effect of liver	W. B. Castle	[19]
1936	Application of phenol for extraction of vitamin B_{12}	P. LaLand and A. Klem	[22]
1947	Introduction of microbiological assay	M. S. Shorb	[25]
1948	Isolation of crystalline vitamin B_{12} from bovine liver	two research teams led by K. Folkers and E. L. Smith	[23, 24]
1950–1961	Elucidation of chemical structure	Four research teams led by K. Folkers, E. L. Smith, Sir Alexander Todd and D. C. Hodgkin	[27]
1958	Discovery of coenzymatic form of vitamin B_{12} in bacteria, animals and man	H. A. Barker, H. Weissbach, R. D. Smyth, W. Walerych and J. Pawelkiewicz	[28]
1958–1970	*Discovery of coenzymatic function of vitamin B_{12} and isolation of vitamin B_{12} dependent enzymes:*		Chapter 7
1958	– glutamate mutase	H. A. Barker, H. Weissbach, R. D. Smyth	
1960–1962	– methylmalonyl-CoA mutase	R. Swick, E. R. Stadtmann, B. C. Johnson	
1970	– 2-methylene-glutarate mutase	T. C. Stadtmann	
1961	– diol dehydratase	R. H. Abeles, B. Zagalak, S. Fukui	
1962	– glycerol dehydratase	K. L. Smiley, M. Sobolov, Z. Schneider	
1965	– ethanolamine ammonia-lyase	C. Bradbeer, E. R. Stadtmann, B. M. Babior	
1963	– amino mutases	T. C. Stadtmann, R. N. Costilow, H. C. Friedmann	
1965	– ribonucleotide reductase	R. L. Blakley, W. S. Beck, J. Hardy, R. K. Ghambeer	
1967	– methionine synthetase	L. Jaenicke, H. Weissbach, R. T. Taylor, H. Rudiger	
1963–1964	– methane synthetase	T. C. Stadtmann, F. Lynen, J. M. Wood, J. M. Poston	
1964–1966	– acetate synthetase	E. R. Stadtmann, H. G. Wood, J. M. Poston, L. Ljungdahl	
1965–1970	Elucidation of mechanisms of catalytic function	R. H. Abeles, J. Retey, D. Arigoni, B. Zagalak, G. N. Schrauzer	Chapter 7
1953–1970	Isolation and characterization of vitamin B_{12} binding proteins: intrinsic	W. B. Castle, R. F. Schilling, G. B. J. Glass, E. S. Holdsworth,	Chapter 8

Table 1-1

year or year span	event	main contributors or leaders of research groups.	references
	factor, transcobalamin and cobalophilin.	L. Ellenbogen, R. Gräsbeck, D. R. Highley, V. Herbert, I. Chanarin, E. Nexø, C. A. Hall, E. Hippe	
1953–1964	Development of diagnostic techniques	R. F. Schilling, H. L. Rosenthal, H. C. Heinrich	Chapter 8 and 9
	Methods of vitamin B_{12} assay:		
1949–1951	– microbiological methods	S. H. Hutner, B. D. Davis, H. R. Skeggs, K. G. Stahlberg	Chapter 5
1961–1965	– radioisotope dilution method	R. M. Barakat, R. P. Ekins, V. Herbert, K.-S. Lau	Chapter 5
1966	– enzymatic method	Z. Schneider, H. A. Barker, R. H. Abeles	Chapter 5
1970	– neutron activation	Z. Schneider	Chapter 5
	Main achievements in the field of vitamin B_{12} biosynthesis:		
1961–1981	– the synthesis of the porphyrin and the formation of the corrin ring	D. Shemin, K. Bernhauer, A. I. Scott, A. R. Battersby, V. Y. Bykhovsky, G. Müller	Chapter 4
1954–1961	– vitamin B_{12} derivatives	W. Friedrich, J. Pawelkiewicz, D. Perlman, A. DiMarco	Chapters 3, 4, 5 and 6
1965–1981	– the nucleotide moiety	H. C. Friedmann, P. Renz, J. A. Fyfe, Z. Schneider, W. Walerych	Chapters 3 and 4
1961–1966	– formation of the coenzyme B_{12}	H. A. Barker, J. Pawełkiewicz, B. Bartosiński, F. M. Huennekens, H. Weissbach, A. Peterkofsky	Chapter 3 and 4
1965	– isolation of cobalt free corrinoids	I. J. Toohey, V. B. Koppenhagen	Chapter 3
	Chemical modification of vitamin B_{12} and its synthesis:		
1957	– alteration of the corrin ring	K. Bernhauer, A. R. Todd, R. Bonnet, F. Wagner, A. W. Johnson	Chapter 3
1960–1968	– attachment of the 1-amino-2-propanol and derivatives to natural cobyric acid	W. Friedrich, K. Bernhauer	
1971–1979	– synthesis of cobyric acid	Two research groups led by A. Eschenmoser and R. B. Woodward	Chapter 3
1967–1971	– synthesis of coenzyme B_{12} and its derivatives	D. Dolphin, J. D. Brodie, K. Kikugawa, M. Ichino, B. Zagalak, H. P. C. Hogenkamp, S. Fukui	Chapter 3
	– rearrangements at the β-ligand	P. Dowd, G. N. Schrauzer, J. Retey, B. M. Babior, D. Dolphin	Chapter 3
1968–1979	Methylation of heavy metal ions by methylcobalamin	J. M. Wood	Chapter 3
1971–1979	Substitution of cobalt ion in the corrin ring by other ions	V. B. Koppenhagen, J. J. Pfiffner, R. Bieganowski, W. Friedrich	Chapter 3
1963–	Coordination chemistry of corrinoids	J. M. Pratt, G. N. Schrauzer, D. Dolphin, R. A. Firth, H. A. O. Hill	Chapter 3

newest version of nomenclature proposed in 1973 is given in Chapter 2. The names of corrinoids according to these last rules are used in this book wherever possible. For a number of corrinoids the names used prior to 1973 are given in Table 3-1.

References

[1] Combe, J.S., History of a case of anaemia, Trans. Med. Chir. Soc. Edinb. **1**, 194-203, 1824.

[2] Barelay, A.W., Death from anaemia (two cases), Med. Times Gaz. **23**, 480, 1851.

[3] Channing, W., Notes on anhaemia, principally in its connections with the puerperal state and with functional diseases of the uterus: with cases, New Engl. Q. Jl. Med. Surg. **i**, 157, 1842.

[4] Moller, J.O.L., Klinische Bemerkungen über einige weniger bekannte Krankheiten der Zunge, Dt. Klin. **3**, 273, 1851.

[5] Handfield-Jones, C., Pathological and clinical observation respecting morbid conditions of the stomach, London, 1855.

[6] Biermer, A., Über eine Form von progressiver perniciöser Anämie, Schweiz. Ärzte **2**, 15, 1872.

[7] Fenwick, S., On atrophy of the stomach, Lancet **ii**, 78, 1870.

[8] Sorensen, S.T., Taellinger af blodlegemer i 3 Tilfälde af excessiv oligocythaemi Hospitals-Tidende R 2.1. aargang 513, 1874.

[9] Osler, W., and Gardner W., On the changes in marrow in progressive pernicious anaemia, Can. Med. Surg. J. **5**, 385, 1877.

[10] Cahn, A., and Von Mehring, Die Säuren des gesunden und kranken Magens, Dt. Arch. Klin. Med. **39**, 233, 1886.

[11] Lichtheim, L., Zur Kenntniss der perniciösen Anämie, Münch. Med. Wschr. **34**, 300, 1887.

[12]. Russell, J.S.R., Batten, F.E., and Collier, J., Subacute combined degeneration of the spinal cord, Brain **23**, 39, 1900.

[13] Cabot, R.C., Pernicious and secondary anaemia, chlorosis, and leukaemia, in, A System of Medicine, Eds. W. Osler and T. Me-Grae, Frowde, Oxford 1907-1910.

[14] Tempka, T., and Braun, B., Das morphologische Verhalten des Sternum-punktates in verschiedenen Stadien der perniziösen Anämie und seine Wandlungen unter dem Einfluß der Therapie, Folia Haemat. Lpz. **48**, 355, 1932.

[15] Chanarin, I., The Megaloblastic Anaemias, Blackwell Scientific Publications, Oxford 1969, pp. 1-5.

[16] Friedrich, W., in Vitamin B_{12} und Verwandte Corrinoide, Eds. R. Ammon and W. Dirscherl, Georg Thieme Verlag, Stuttgart 1975, pp. 1-4.

[17] Castle, W.B., The history of corrinoids, in Cobalamin, Ed. B.M. Babior, John Wiley, New York 1975, pp. 3-17.

[18] Minot, C.R., and Murphy W.P., Treatment of pernicious anaemia by special diet, J. Am. Med. Ass. **87**, 470-476, 1926.

[19] Castle, W.B., Observations on the etiologic relationship of achylia gastrica to pernicious anaemia. I. The effect of the administration to patients with pernicious anaemia of the contents of the normal human stomach recovered after the ingestion of beef muscle, Am. J. Med. Sci. **178**, 764-777. 1929.

[20] Cohn, E.J., Minot, G.R., Alles, G.A., and Salter, W.T., J. Biol. Chem. **77**, 325-358, 1928.

[21] Dakin, H.D., and West, R., J. Biol. Chem. **109**, 489-522, 1935.

[22] LaLand, P., and Klem, A., Acta Med. Scand. **88**, 620-623, 1936.

[23] Rickes, E.L., Brink, N.G., Koniuszy, F.R., Wood, T.R., and Folkers, K., Crystalline vitamin B_{12}, Science **107**, 396, 1948.

[24] Smith, E.L., and Parker, L.F., Purification of anti-pernicious anaemia factor, Abstract, Biochem. J. **43**, viii, 1948.

[25] Shorb, M.S., Unidentified growth factors for *Lactobacillus lactis* in refined liver extracts. J. Biol. Chem. **169**, 455-456, 1947.

[26] Underwood, E.J., and Filmer, J.F., Aust. Vet. J. II, 84-92, 1935.

[27] Hodgkin, D.C., Pickworth, J., Robertson, J.H., Trueblood, K.N., Prosen, R.J., White, J.G., Bonnett, R., Cannon, J.R., Johnson, A.W., Sutherland, I., Todd, Sir Alexander, and Smith, E.L., Nature **176**, 325-328, 1955.

[28] Bonnet, R., Chem. Rev. **63**, 573, 1963.

[29] Lenhert, P.G., and Hodgkin, D.C., Nature **192**, 937-938, 1961.

[30] Kon, S.K., Biochem. Soc. Symposia No 13: The biochemistry of vitamin B_{12}, Cambridge 1955, p. 17.

[31] Friedrich, W., and Bernhauer, K., in Medizinische Grundlagenforschung vol. II, Ed. K.F. Bauer, Thieme, Stuttgart 1959, p. 661.

[32] Kon, S. K., and Pawełkiewicz, J., Biosynthesis of vitamin B_{12} analogues, Fourth International Congress of Biochemistry Vol XI, Eds. W. W. Umbreit and H. Molitor, pp. 115-149, 1958.

[33] Perlman, D., Barett, J. M., Jackson, P. W., Cobamides synthesized by Propionibacterium species, in Vitamin B_{12} und Intrinsic Factor, 2. Europäisches Symposium, Hamburg 1961, Ed. H. C. Heinrich, Ferdinand Enke, Stuttgart 1962, pp. 58-69.

[34] Friedrich, W., Chemische Partialsynthese kompletter Vitamin B_{12} Arten, ausgehend vom Cobyrinsäure-abcdeg-hexamid, 2. Europäisches Symposion, Hamburg 1961, Ed. H. C. Heinrich, Ferdinand Enke, Stuttgart 1962, pp. 8-28.

[35] Barker, H. A., Weissbach, H., and Smyth, R. D., A coenzyme containing pseudovitamin B_{12}, Proc. Natl. Acad. Sci. USA **44**, 1093-1097, 1958.

[36] The particular vitamin B_{12} enzymes are described in details in Chapter 7.

[37] The vitamin B_{12} binding proteins are described in Chapter 8.

[38] see Chapter 9

[39] For biosynthesis of vitamin B_{12} see Chapter 4, the chemical synthesis is outlined in section 3.1.1.8.

[40] Shemin, D., Corcoran, J. W., Rosenblum, C., Miller, J. M., Science **124**, 272, 1956.

[41] Friedmann, H. C., J. Biol. Chem. **243**, 2065-2075, 1968.

[42] Renz, P., Hörig, J., Wurm, R., On the biosynthesis of the 5,6-dimethylbenzimidazole moiety of vitamin B_{12}, Eds. B. Zagalak, W. Friedrich, Walter de Gruyter, Berlin 1979, pp. 317-322.

[43] Walerych, W., Pezacka, E., Ribosomal proteins share in vitamin B_{12} biosynthesis, Eds. B. Zagalak, W. Friedrich, Walter de Gruyter, Berlin 1979, pp. 345-358.

[44] Eschenmoser, A., 23rd International Congress Pure and Appl. Chem. Bd. II p. 69, 1971.

[45] Woodward, R. B., Pure and Appl. Chem. **33**, 145, 1973.

[46] Pratt, J. M., in Inorganic Chemistry of Vitamin B_{12}, Academic Press, London, 1972.

[47] For introduction of ligands see 3.1.4.2.

[48 a] Atkins, M. P., Golding, B. T., Sellars, P. J., Model systems for adenosylcobalamin dependent enzymic reactions, in Vitamin B_{12}. Eds. B. Zagalak, W. Friedrich, Walter de Gruyter, Berlin 1979, pp. 587-599.

[48 b] Brown, K. L., Vitamin B_{12} in The Chemistry of Enzyme Action, Ed. M. I. Page, Elsevier Science Publishers, B. V., Amsterdam 1984, pp. 433-459

[49 a] Dolphin, D., Banks, A. R., Cullen, W. R., Culter, A. R., and Siverman, R. B., The mechanism of action of coenzyme B_{12}, in Vitamin B_{12}, Eds. B. Zagalak, W. Friedrich, Walter de Gruyter, Berlin 1979, pp. 575-586.

[49 b] Zagalak, B., Vitamin B_{12} als biologische aktive Modellsubstanz, Naturwissenschaften **69**, 63-74, 1982

[50 a] Dowd, P., Nonenzymic models for the enigmatic coenzyme B_{12} dependent carbon-skeleton rearrangements, in Vitamin B_{12}, Eds. B. Zagalank, W. Friedrich, Walter de Gruyter, Berlin 1979, pp. 557-574.

[50 b] Dowd, P. and Trivedi, B. K., On the mechanizm of action of vitamin B_{12}. Model studies directed toward the hydrogen abstraction reaction. J. Org. Chem. **50**, 206-217, 1985

[51] Wood, J. M., Fanchiang, Y. T., Mechanism for B_{12}-dependent methylation, in Vitamin B_{12}, Eds. B. Zagalak, W. Friedrich, Walter de Gruyter, Berlin 1979, pp. 539-556.

[52 a] Koppenhagen, V. B., Elsenhaus, B., and Wagner, F., Methylrhodibalamin and 5'-Deoxyadenosylrhodibalamin, the rhodium analogues of methylcobalamin and cobalamin coenzyme, J. Biol. Chem. **249**, 6533-6540, 1974.

[52 b] Siefert, F., Koppenhagen, V. B., Studies on the vitamin B_{12} auxotrophy of Rhodocydus purpureus, and 2 other vitamin B_{12}-requiring purple nonsulfur bacteria, Arch. Microbiol. **132**, 173-178, 1982.

[53] Koppenhagen, V. B., and Pfiffner, J. J., Currins and zirrins, two new classes of corrin analogues, J. Biol. Chem. **245**, 5865-5873, 1970.

[54] Bieganowski, R., and Friedrich, W., in Vitamin B_{12}, Eds. B. Zagalak, W. Friedrich, Walter de Gruyter, Berlin 1979, pp. 647-648.

[55] Mapson, L. W., Evidence for the existence of a dietary principle stimulating general growth and lactation, Biochem. J. **26**, 970, 1932.

[56] Ott, W. H., Rickes, E. L., and Wood, T. R., Activity of crystalline vitamin B_{12} for chick growth, J. Biol. Chem. **174**, 1047, 1948.

2 Nomenclature of Corrinoids

(1973 Recommendations
by the IUPAC-IUB Commission on Biochemical Nomenclature)

1. The corrinoids are a group of compounds containing four reduced pyrrole rings joined into a macrocyclic ring by links between their α positions; three of these links are formed by a one-carbon unit (methylidyne radicals) and the other by a direct $C\alpha$-$C\alpha$ bond. They include various B_{12} vitamins, factors, and derivatives based upon the skeleton of *corrin*, $C_{19}H_{22}N_4$ (structure I). The atoms are numbered and the rings are lettered as shown in structure I. The numbering is thus the same as that of the porphyrin nucleus, number 20 being omitted to preserve the identity.

Note. The name "corrin" was proposed by those who established its structure because it is the *core* of the vitamin B_{12} molecule; the letters "co" of corrin are *not* derived from the fact that vitamin B_{12} contains cobalt. However, this does not apply to the "cob" terms below, all of which do contain "co" for cobalt.

2. Some important corrinoids that are more unsaturated than corrin itself are derivatives of octadehydrocorrin. This has sometimes been called tetradehydrocorrin because it has four additional double bonds. Although this could be indicated by the prefix "tetrakis(didehydro)," "octadehydro" is preferred.

The octadehydrocorrin system IA has the trivial name *corrole*.

3. Many important corrinoids have a regular pattern of substituents on the methylene carbon atoms of the reduced pyrrole rings and a cobalt atom in the center of the macrocyclic ring. The heptacarboxylic acid II is named *cobyrinic acid*. The carboxyl groups are designated by the locants *a* to *g*, as shown in II. Cobyrinic *a, b, c, d, e, g*-hexaamide, formerly sometimes referred to as *Factor* V_{1a}, is named *cobyric acid*. Substituents on the side chains may be designated by appropriate locants, *e.g.*, 7β-methylcobyrinic acid, if $-CH(CH_3)CO_2H$ replaces $-CH_2CO_2H$ at C-7β of cobyrinic acid.

4. The compound III (R=OH, R'=H), which is the amide formed by combination of cobyrinic acid with D-1-amino-2-propanol at position *f*, is named *cobinic acid;* its hexaamide (III; R=NH_2, R'=H) is named *cobinamide*.

5. The compound III (R=OH, R'=structure V) in which cobinic acid is further substituted at the 2 position of the aminopropanol by an *α-D-ribofuranose 3-phosphate* residue (V) is named *cobamic acid;* its hexaamide (III; R=NH_2, R'=V) is named *cobamide*.

6. Glycosyls and nucleotides (which are *N*-glycosyl derivatives at C-1 of the ribofuranose unit) of cobamides are named by adding the name of the appropriate aglycon radical (ending in "yl") as a prefix to the name of the corrinoid allotted according to **1–5**, *e.g.*, aglyconylcobamide (VI).

7. Most of the important natural products in this series have aglycon radicals containing an imidazole nucleus, one N of the latter being covalently bonded to the ribose while the other is coordinately bonded to what is, by this attachment, defined as the cobalt-α position. *The latter situation (VII) is assumed to exist unless otherwise indicated*. When another ligand occupies the cobalt-α position, it and its locant may be indicated by, *e.g.*, (*Coα*-ligand)-aglyconylcobamide (VIII). The absence of a "*Coα*-ligand" term, as in the cobalamins (see section 9), indicates that the aglycon radical attached to the ribose occupies the cobalt-α position as well.

8. Cobamides bearing a ligand in the cobalt-β position (which implies Co(III)) may be named as follows:

(*Coα*-ligandyl)-(*Coβ*-ligandyl)-(aglyconylcobamide) (IX)

or, if the aglycon is attached to the cobalt-α position, as indicated in section 7

aglyconyl-(*Coβ*-ligandyl)cobamide (Xa)

In a cobalamin (see Section 9), the latter becomes simply

ligandylcobalamin (Xb)

See also section 15.

9. *Cobalamins*. A cobalamin is a cobamide in which 5,6-dimethylbenzimidazole is the aglycon attached by a glycosyl link from its N-1 to the C-1 of the ribose and additionally linked, as stated in Section 7, by a bond between the N-3 and the cobalt (in position α). They may be named as combamides, as above, or according to the pattern:

(ligand in Coβ position, if any)-cobalamin (Xb)

Examples:

Coα-[α-(5,6-Dimethylbenzimidazolyl)]-*Coβ*-cyanocobamide, also known as vitamin B-12, is termed *cyanocobalamin*.

Coα-[α-(5,6-Dimethylbenzimidazolyl)]-*Coβ*-aquacobamide, also known as vitamin B_{12} a, is termed *aquacobalamin*.

Coα-[α-(5,6-Dimethylbenzimidazolyl)]-*Coβ*-hydroxocobamide, also known as vitamin B_{12b}, is termed *hydroxocobalamin*. (Note: aquacobalamin is the conjugate acid of hydroxocobalamin.)

Coα-[α-(5,6-Dimethylbenzimidazolyl)]-*Coβ*-nitritocobamide, also known as vitamin B_{12c}, is termed *nitritocobalamin*.

10. Anion(s) associated with the corrinoids is (are) stated in the usual way after the name of the (cationic) corrinoid, *e.g.*, cobamic dichloride (not dichlorocobamic acid).

11. The state of oxidation of the cobalt may be specified, when necessary, as follows:

vitamin B_{12}	cyanocob(III)alamin
vitamin B_{12r}	cob(II)alamin
vitamin B_{12s}	cob(I)alamin

12. Displacement of the ribosyl-bound aglycon base from its normal coordinate bonding to position α of the cobalt by another ligand (or by water) may be indicated by placing the name and locant of the replacing ligand before the corrinoid name and enclosing the modified corrinoid name (see section 6) in parentheses. (see also section 7).

Table 2-1

Section	Description	Specific Names, in Increasing Order of Complexity	
1	Skeleton (porphyrin nucleus minus C-20)	Corrin (I)	
		Heptaacid	Heptaacid, hexaamide
3	**1**, with standard side chains and with cobalt	Cobyrinic acid (II)	Cobyric acid
4	**3**, with D-1-amino-2-propanol at position *f*	Cobinic acid (III)	Cobinamide
5	**4**, with D-ribofuranose 3-phosphate at position 2 of the aminopropanol	Cobamic acid (III–V)	Cobamide
7	**5**, with heterocyclic base attached by *N*-glycosyl link at position 1 of ribose and attached as an α ligand to cobalt (sections 6, 7)		Aglyconylcobamide (VI)
9	Many "B_{12}" vitamins and derivatives, in which the heterocyclic base is 5,6-dimethylbenzimidazole, are given the trivial name "cobalamin" (section 9)		Cobalamin
15	"B_{12} coenzymes," compounds in which a further organic group (X-yl) is β-ligated to cobalt (sections 9, 15)		X-ylcobalamin; ($Co\alpha$-ligandyl)-($Co\beta$-X-yl)cobamide (X)

Names		Symbols
corrin		Crn
of free acid	of hexaamide	(for the hexaamide)
cobyrinic acid	cobyric acid	Cby
cobinic acid	cobinamide	Cbi
cobamic acid	cobamide	Cba
	cobalamin	Cbl[1]

1 A cobalamin is a cobamide in which 5,6-dimethylbenzimidazole is covalently bonded to the ribose in α-glycosidic linkage; it is thus a dimethylbenzimidazolylcobamide and can be symbolized as such. However, it is often convenient to have a short symbol for this complex, hence Cbl. Cbl is recommended in place of the former B_{12} or B-12 for chemical use.

Example:

Coα-aqua-*Coβ*-methyl(2-methyladenylcobamide), in which the 2-methyladenyl residue is attached to the ribose residue but is not coordinately bound to the cobalt atom, having been displaced by water. Methyl occupies the *Coβ* position.

13. Modified, derived, or related compounds are named systematically from the largest of the compounds I, II, or III that is contained in them

Examples:

cobyrinic acid *a, b, c, d, e, g*-hexaamide *f* 2-hydroxyethylamide

3, 8, 13, 17-tetraethyl-1, 2, 2, 5, 7, 7, 12, 12, 15, 17, 18-undecamethyl-cobalti-corrin dichloride (for the dichloride of fully decarboxylated cobyrinic acid)

12α'-carboxycobyrinic acid (for cobyrinic acid in which the 12α-methyl group has been replaced by CH_2CO_2H)

14. Replacement of the cobalt atom in compounds II or III by another metal or by hydrogen is indicated by replacing "co" in the "cob" part of the name with the name or the root of the name of the replacing metal followed by "o" or "i" according to its valence (*e.g.*, cupri, cupro, zinco). When cobalt is replaced by hydrogen, "hydrogeno" replaces "co".

Examples: ferrobamic acid; hydrogenobamic acid

See note to section 1 concerning corrin. This replacement nomenclature does not apply to corrole (section 2).

15. Cofactor Forms. The coenzymatically active forms of the B_{12} vitamins (section 12) and their analogues possess an organic ligand, either methyl or 5'-deoxy-5'-adenosyl, attached to the β position of the cobalt by a carbon-to-cobalt bond, *i.e.*, in the position of the CN in formula IV. These adducts should be named according to the pattern:

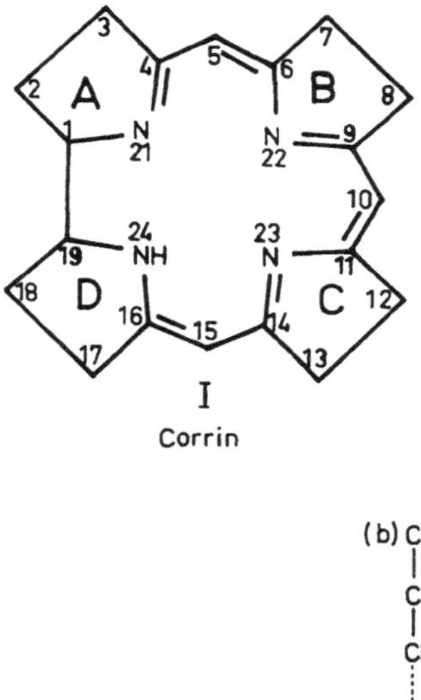

I

Corrin

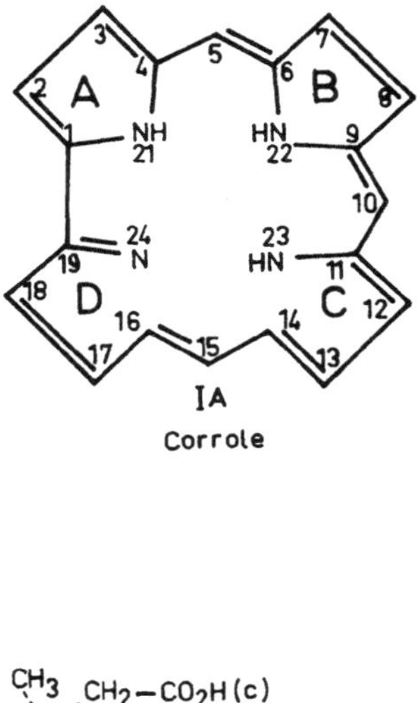

IA

Corrole

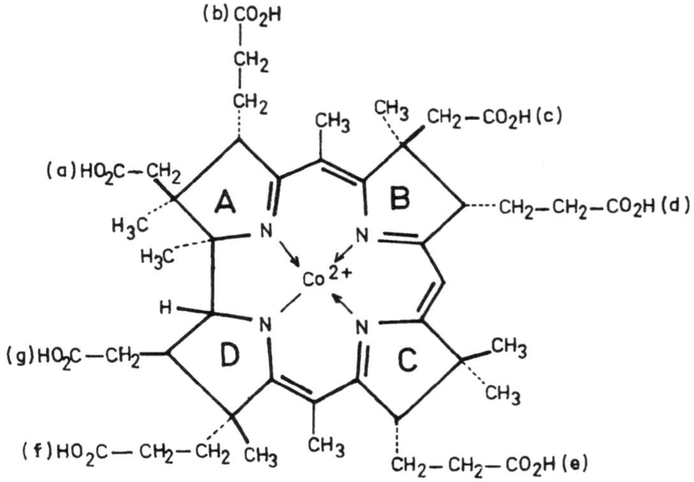

II

Cobyrinic acid

Coα-(radical in *α* position)-*Coβ*-(ligand in *β* position)-(corrinoid name) or (ligand in *β* position)cobalamin, if the radical in the *α* position is dimethylbenzimidazole

Examples:

Coα-[*α*-(5,6-dimethylbenzimidazolyl)]-*Coβ*-adenosylcobamide, or adenosylcobalamin, for the compound formerly known as "coenzyme B_{12}"

Coα-[*α*-(5,6-dimethylbenzimidazolyl)]-*Coβ*methylcobamide or methylcobalamin, for the compound involved in several reactions, including methionine biosynthesis, where a methyl group is ligated to the cobalt in the *β* position

Coα-[*α*-(7-adenyl)]-*Coβ*-adenosylcobamide, the coenzymatically active form of "pseudovitamin B_{12}," capable of replacing adenosylcobalamin in many systems.

16. Summary. The trivial names applied to corrinoids of varying complexity are perhaps confusing to the nonspecialist, and it seems desirable to tabulate (in outline) how they are interrelated (Table 2-1).

Notes on Formulas

(1) In formulas II and III, the corrin nucleus is represented as being roughly in the plane of the paper, *i.e.*, full (heavy) lines are bonds lying *above* the plane of the ring system, while dashed (broken) lines are bonds lying *below* this plane.

(2) Formulas II and III represent the true absolute stereochemical configuration of the structures as determined by X-ray work.

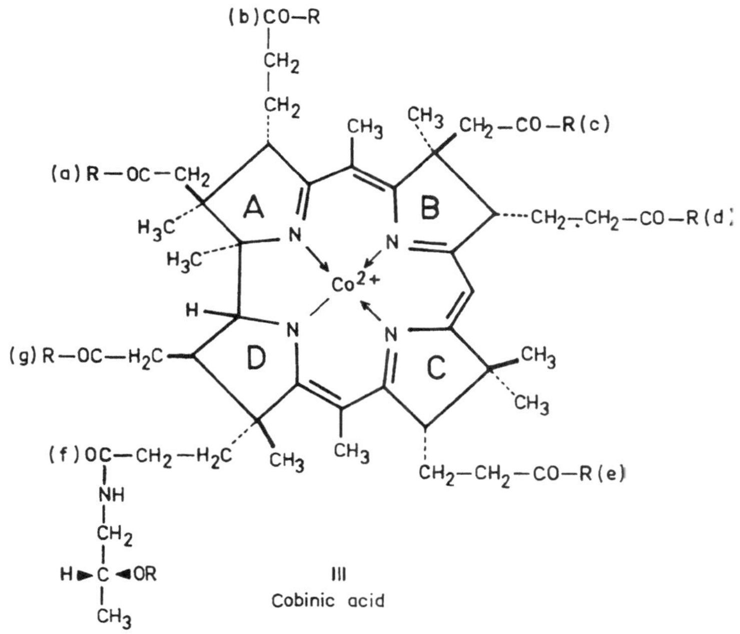

III

Cobinic acid

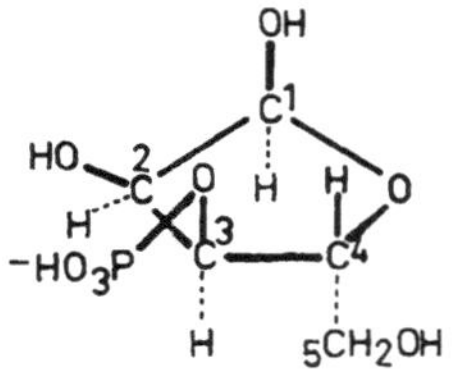

V

α-D-Ribofuranose 3-phosphate residue

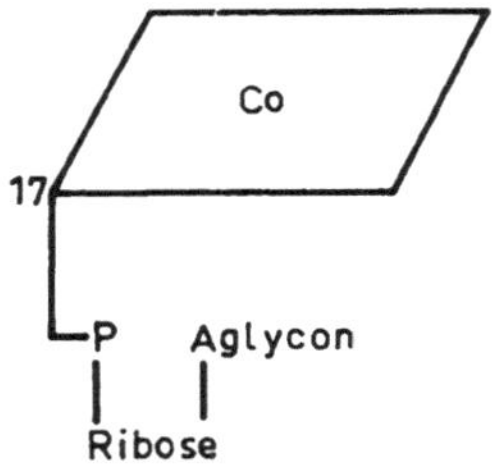

VI

Aglyconylcobamide (III, R' = *P*-Rib-aglycon)

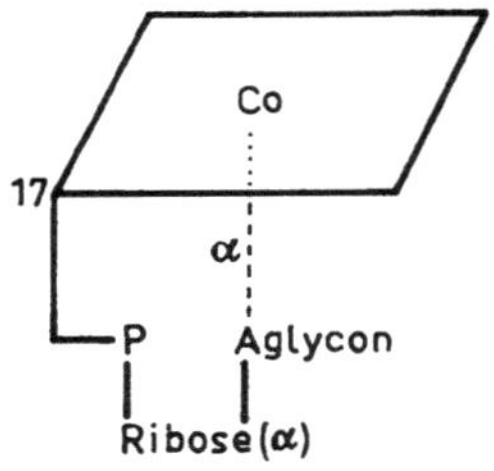

VII

Aglyconylcobamide, with aglyconyl liganded to cobalt

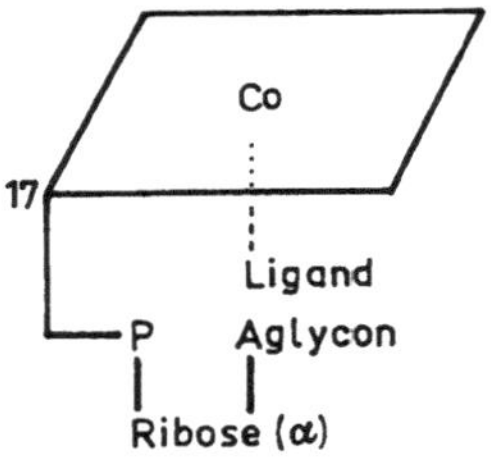

VIII

(*Coα*-Ligandyl)-aglyconyl-cobamide (ligand has "displaced" aglycon of VII)

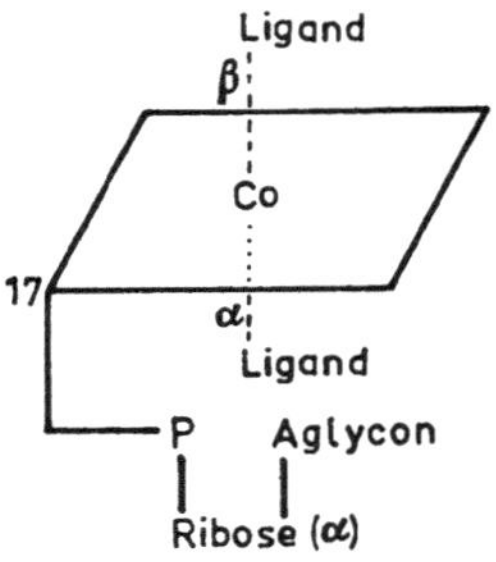

IX

(*Coα*-Ligandyl)-(*Coβ*-ligandyl)aglyconylcobamide (VIII with additional ligand in *Coβ* position)

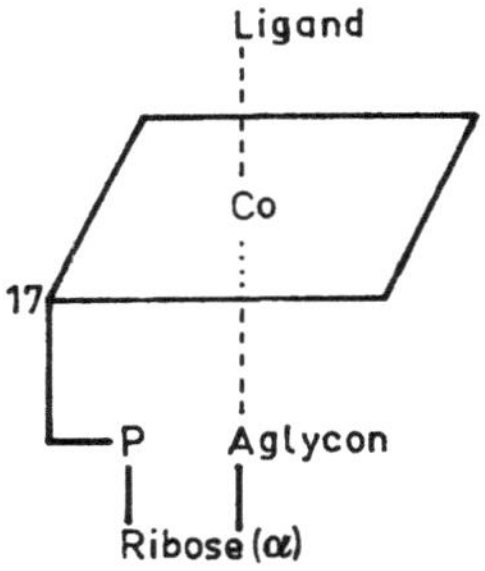

X

Xa, Aglyconyl-(*Coβ*-ligandyl)cobamide (VII with additional ligand in *Coβ* position)
Xb, Ligandylcobalamin (if aglycon is dimethylbenzimidazole, and with CN as *Coβ* ligand)

Appendix. Abbreviations for Corrinoids

I. Names and Symbols

II. Designation of Substituents Attached to Cobalt

Ligands coordinated to the α and β position of the cobalt (below and above the plane of the corrin residue, respectively) are represented by terms that precede the symbol for the corrinoid residue. If the positions of the ligands are *unknown or not specified,* the two terms representing the ligands in the α and β position are enclosed in one set of parentheses and are separated by a *comma.* If the positions are *known and specified,* the α ligand is set apart by parentheses; the β is enclosed (separately) only if its complexity may make it ambiguous. If the ligands are identical, a single term followed by the subscript 2 is used.

A. Anion Substituents. The chemical symbol for the anion is used. Aqua is abbreviated aq. Examples:

(Me)aqCbi	(methyl)aquacobinamide [4] (methyl in α position)
(CN)MeCbi	(cyano)methylcobinamide (methyl in β position)
(CN, aq)Cbi	cyanoaquacobinamide (ligand location unspecified)
(aq, CN)Cbi	

B. Alkyl Substituents.

(1) Primary substituents are designated by naming the alkyl group without denoting the position attached to the cobalt, as it is always 1. Examples:

(aq)EtCbi	(aqua)ethylcobinamide
(CN)(2-OAcBu)Cbi	(cyano)(2-acetoxybutyl)cobinamide

(2) Secondary substituents are named similarly, except that the position attached to cobalt is given by a locant suffixed to the name of the alkyl group (as in the -X-yl name). Examples:

(aq)(Bu-2)Cbi or (aq)Bu^sCbi	(aqua)(*sec*-butyl)cobinamide [5]
(aq)(3-OAcBu-2)Cby	(aqua)(3-acetoxybut-2-yl)-cobyric acid

(3) Alicyclic groups are indicated by a small "c" before the symbol for the alkyl residue. In these compounds, cobalt is always assumed to be substituted in position 1 of the ring. Examples:

(aq)cHxCbi	(aqua)cyclohexylcobinamide [5]
(CN)(2-HOcPe)Cby	(cyano)(2-hydroxycyclopentyl)-cobyric acid

(4) 5'-Deoxy-5'-adenosyl in the β coordination position, as in "coenzyme B-12," is represented by the symbol Ado for "adenosyl" a 2'-deoxyadenosine residue by dAdo [3]. Unusual deoxyadenosyl residues can be indicated by superscripts (*e.g.,* d^3Ado, $d^{2,3}$ Ado). See C below.

C. Cobamides of the Cobalamin Type. As the symbol Cbl designates α-(5,6-dimethylbenzimidazolyl)cobamide [cob(III)-alamin], only those cobamides having this base utilize Cbl. Those containing another base are named as cobamides, utilizing the symbol Cba. Hence, Cbl is preceded by only a single term, the one representing the β-substituent. Examples are given in Table 2-2.

Notes: (i) The hyphenation in the case of secondary alkyl substituents and similar situations of potential confusion may make it necessary to enclose the β-substituent in parentheses, or set it off by a hyphen.

(ii) If the replacing base (in α position) is unspecified, the term (?) is used, *e.g.*, (?) MeCba. The term (OH/base) indicates that the ribose residue is not attached to the *Coα*-linked base.

(iii) If the α substituent (the "base") is displaced from the cobalt by another ligand, but remains attached to the ribosyl residue, the same system is used. Example:

(2-MeAde/aq)MeCba	*Coα*-aqua-*Coβ*-methyl(2-methyl-adenylcobamide)
(Ade/CN)CN-Cba	*Coα*-cyano-*Coβ*-cyano(adenyl-cobamide) or dicyanoadenyl-cobamide

In abbreviating cobalamin derivatives, the base need not be specified. Replacement of the base by another *Coα* ligand is indicated by merely adding to the abbreviation a term corresponding to the replacing ligand. Examples:

(OH)MeCbl	*Coα*-hydroxo-*Coβ*-methylcobalamin or *Coα*-hydroxo-*Coβ*-methyl(dimethylbenzimidazolylcobamide)
$(CN)_2$Cbl	*Coα*-cyano-*Coβ*-cyanocobalamin or dicyanocobalamin

(iv) Cobalt valences of II or I may be indicated by superscripts (*e.g.*, Cbl^{II}).

III. Designation of Alterations and Substitutions on the Corrin Ring

Substituents on the ring itself are represented by symbols following the symbol for the corrinoid, with locants indicating the positions of the substituents. Epimerization is

Table 2-2

CN-Cbl	Cyanocob(III)alamin (vitamin B-12)
AdoCbl	Adenosylcob(III)alamin [6, 7]
PrCbl	*n*-Propylcob(III)alamin; methyl-, etc., similarly [8, 9]
(Ade)(Pr-2)Cba or (Ade)Pr^{i}-Cba[a]	*Coα*-[α-(Aden-9-yl)]-*Coβ*-isopropylcobamide
(Bza)MeCba[b]	*Coα*-(α-Benzimidazolyl)-*Coβ*-methylcobamide
2-(MeOOC)EtCbl	(2-Methoxycarbonylethyl)cob(III)alamin [8]
(Ade-7)AdoCba[a]	*Coα*-[α-(Aden-7-yl)]-*Coβ*-adenosylcobamide [10]
(2-SHAde-7)AdoCba[a]	*Coα*-[α-(2-Thiaaden-7-yl)]-*Coβ*-adenosylcobamide [11]
(5-MeOBza)MeCba	*Coα*-(5-Methoxybenzimidazolyl)-*Coβ*-methylcobamide[d] [12, 13]
(2-MeAde-7)CN-Cba[a]	*Coα*-[α-(2-Methyladen-7-yl)]-*Coβ*-cyanocobamide [11, 14]
(Ade)CN-Cba[a]	*Coα*-[α-(Aden-9-yl)]-*Coβ*-cyanocobamide (pseudovitamin B-12)
(Ade)OH-Cba[a]	*Coα*-[α-(Aden-9-yl)]-*Coβ*-hydroxocobamide (hydroxopseudovitamin B-12)
(Ade)MeCba[a]	*Coα*-[α-(Aden-9-yl)]-*Coβ*-methylcobamide
[4-(Ade-9)Bu]Cbl[c]	[4-(Aden-9-yl)butyl]cob(III)alamin [6]
(6MeSPur)AdoCba	*Coα*-(α-6-Methylthiopurinyl)-*Coβ*-adenosylcobamide [15]

[a] Ade alone represents adenine bonded to the ribosyl moiety through its 7 position (*i.e.*, a 7-α-D-ribofuranosyladenine). Bonding to the cobalt is thus through N-9. When these positions are reversed, Ade-7 and aden-7-yl are used (*i.e.*, the locant specifies the N linked to cobalt). [b] Bza = benzimidazolyl. [c] As this is a cobalamin, the adenine residue is not in the *Coα* position, but is attached (-9-yl) to a but-4-yl residue that is in turn linked to the β position of the cobalt. Named as a cobamide, it would be (Me_2Bza)-[4-(Ade-9)Bu]Cba.
[d] Factor III_m [12, 13].

indicated in a similar manner. Symbols representing replacements on the carboxyl groups at the periphery of the corrin residue follow those that designate substituents directly on the ring. The location of the substituent is indicated by the letter corresponding to the carboxyl group that carries it. Examples:

CN-Cbl(13-epi)	cyano(13-epi)cobalamin
CN-Cbl(13epi-*e*OH)	*Coα*-(*α*-5,6-dimethylbenzimidazolyl)-*Coβ*-cyano-(13-epi)cobamic *a, b, c, d, g*-pentaamide
AdoCbl(10-Cl)	adenosyl-10-chlorocobalamin
(aq)AdoCbi(*e*-PhNH)	*Coα*-aqua-*Coβ*-adenosylcobinic *a, b, c, d, g*-pentaamide *e*-anilide
(CN)ClCby(8-NH-*c*-lactam)	
$(CN)_2Cby(OMe)_7$	dicyanocobyrinic heptamethyl ester [16]

If the location of the carboxyl substituent(s) is unknown, a term of the following structure should be used:

$(a:g\text{-X})_n$

where *a*:*g* indicates a substitution at the periphery of the ring, X is the replacing group, and *n* refers to the number of carboxyl groups substituted. Examples:

(CN, aq)Cby[*a*:*g*-$(NH_2)_5$]	cyanoaquacobyrinic acid pentaamide [17]
$(CN)_2$Cby[10-Cl-*a*:*g*-$(NH_2)_5$]	10-chloro derivative of the above

IV. Replacement of Cobalt by Other Metals [18, 19]

Corrinoids containing metals other than cobalt are symbolized by placing the symbol of the replacing metal in square brackets preceding and attached to the symbol of the corrinoid. Thus, a hydrogenocobamide utilizes [H]Cba, a nickelocobalamin $[Ni]Cbl^{II}$, a zincocobinamide $[Zn]Cbi^{II}$, etc. Phenylcupribamide [19, 20] could be indicated as $(Ph)[Cu]Cba^{II}$. I, II, and III may be added as superscripts when needed.

V. Isotopic Labeling

A labeled position is indicated in the usual fashion [21], *e.g.*

(Bza)Me[^{57}Co]Cba	*Coα*-(*α*-benzimidazolyl)-*Coβ*-methyl-[^{57}Co]cobamide
(Bza)[^{14}C]MeCba	*Coα*-(*α*-benzimidazolyl)-*Coβ*-[^{14}C]-methylcobamide
([4-^{3}H]Bza)MeCba	*Coα*-(*α*-[4-^{3}H]benzimidazolyl)-*Coβ*-methylcobamide

VI. Metallocorrins

As corrin contains no metal (the name "corrin" being derived from "core," not "cobalt"), complexes of metals with corrin require specification of both terms. Example: $Cu^{II}Crn$ for copper(II)corrin.

References

[1] IUPAC-IUB Commission on Biochemical Nomenclature, *Biochemistry* **11**, 1726, 1972, and in other journals.
[2] IUPAC-IUB Commission on Biochemical Nomenclature, *Biochemistry* **11**, 942, 1972, and in other journals.
[3] IUPAC-IUB Commission on Biochemical Nomenclature, *Biochemistry* **9**, 4022, 1970.
[4] W. H. Pailes and H. P. C. Hogenkamp, *Biochemistry* **7**, 4160, 1968.
[5] J. D. Brodie, *Proc. Nat. Acad. Sci. U. S. 62,* 461, 1969.
[6] B. M. Babior, *J. Biol. Chem.* **244**, 2917, 1969.
[7] G. J. Cardinale and R. H. Abeles, *Biochim. Biophys. Acta.* **132**, 517, 1967.
[8] H. P. C. Hogenkamp, J. E. Rush, and C. A. Swenson, *J. Biol. Chem.* **240**, 3641, 1965.
[9] J. Stavrianopoulos and L. Jaenicke, *Eur. J. Biochem.* **3**, 95, 1967.
[10] P. Overath, E. R. Stadtman, G. M. Kellerman, and F. Lynen, *Biochem. Z.* **336**, 77, 1962.
[11] R. Bonnett, *Chem. Rev.* **63**, 573, 1963.
[12] E. Irion and L. Ljungdahl, *Biochemistry* **7**, 2350, 1968.
[13] L. Ljungdahl, E. Irion, and H. G. Wood, *Fed. Proc., Fed. Amer. Soc. Exp. Biol.* **25**, 1642, 1966.
[14] M. Hayashi and T. Kamikub, *FEBS (Fed. Eur. Biochem. Soc.) Lett.* **15**, 213, 1971.
[15] Y. Uschida, M. Hayashi, and T. Kamikubo, *Vitamins (Jap.) 47,* 27, 1973, *Chem. Abstr.* **78**, 94046 e, 1973.
[16] L. Wertheman, Dissertation, E. T. H. Zürich, 1968.
[17] K. Bernhauer, H. Vogelmann, and F. Wagner, *Z. Physiol. Chem.* **349**, 1281, 1968.
[18] J. I. Toohey, *Fed. Proc., Fed. Amer. Soc. Exp. Biol.* **25**, 1628, 1966.
[19] V. B. Koppenhagen and J. J. Pfiffner, *J. Biol. Chem.* **246**, 3075, 1971.
[20] V. B. Koppenhagen and J. J. Pfiffner, *J. Biol. Chem.* **245**, 5865, 1970.
[21] Instructions to Authors, *J. Biol. Chem.* **248**, 4, 1973; and elsewhere.
[22] Nomenclature policy: Generic descriptors and trivial names for vitamins and related compounds. J. Nutr. **109**, 8-15, 1979.

3 Chemistry of Cobalamin and Related Compounds

Z. Schneider

3.1 Chemical alterations of vitamin B_{12}

The vitamin B_{12} is a complex compound with a number of functional groups suscepti-ble to a variety of chemical modifications leading to a tremendeous number of deriva-tives. To make a sensible approach to these various chemical alterations the entire molecule is here divided into four major parts: the corrin ring being the core of vita-min B_{12}, the nucleotide moiety indispensable for biological activity, the aminopropa-nol residue which links the nucleotide to the side chain of the corrin ring and the β ligand linked to the cobalt in the corrin ring by a coordination bond. Fig. 3-1 distin-guishes these four parts by colored background. The chemical reactions described be-low are pertained to modifications within these parts or to substitution of the entire fragment of the molecule by an analogous one. Some of the chemical alterations de-scribed here also occur unintended at various occasions causing inactivation of vi-tamin B_{12} e. g. during extraction, estimation, storage, sterilization, preparation of food and feeding stuffs. Thus, the chemical properties of vitamin B_{12} should be regarded by all who will make use of this compound.

3.1.1 Reactions at the corrin ring

3.1.1.1 Deamidation of the side chains

The corrin ring contains three acetamide and four propionamide side chains which are susceptible to acid hydrolysis. Three of propionamide side chains are faster hy-drolysed than the acetamide ones. The course of hydrolysis is very much dependent on temperature, acid concentration and also on ligands attached to cobalt in the α and β coordination position. The amide groups of the individual side chains hydro-lyse at various velocities. Thus, by controlled time of hydrolysis, acid concentration and temperature, it is possible to obtain partially hydrolysed corrinoids in which one or two among the mono- to heptacarboxylic acids present in the mixture will domi-nate. Monocarboxylic acids of cobalamins are most recently applied for binding of vitamin B_{12} to supports for affinity chromatography [1, 2] (see section 3.1.5). Cyano-cobalamin exposed to mild hydrolysis (0.02 MHCl, 20 °C) yields after 15–30 days mainly monocarboxylic acid [3]. While such harsh conditions as heating cyanocobal-amin in 2 M HCl at 100 °C 0.5–12 hours result in hydrolysis up to five amide residues [3, 4]. The deamidation occurs also on Dowex 50 and P-cellulose columns [5]. Vials and other glass containers may cause the hydrolysis of not buffered solutions of vitamin B_{12} during heat sterilization. The deamidated cobalamins are not active for man and animals, and may suppress the growth of some bacteria species [6].

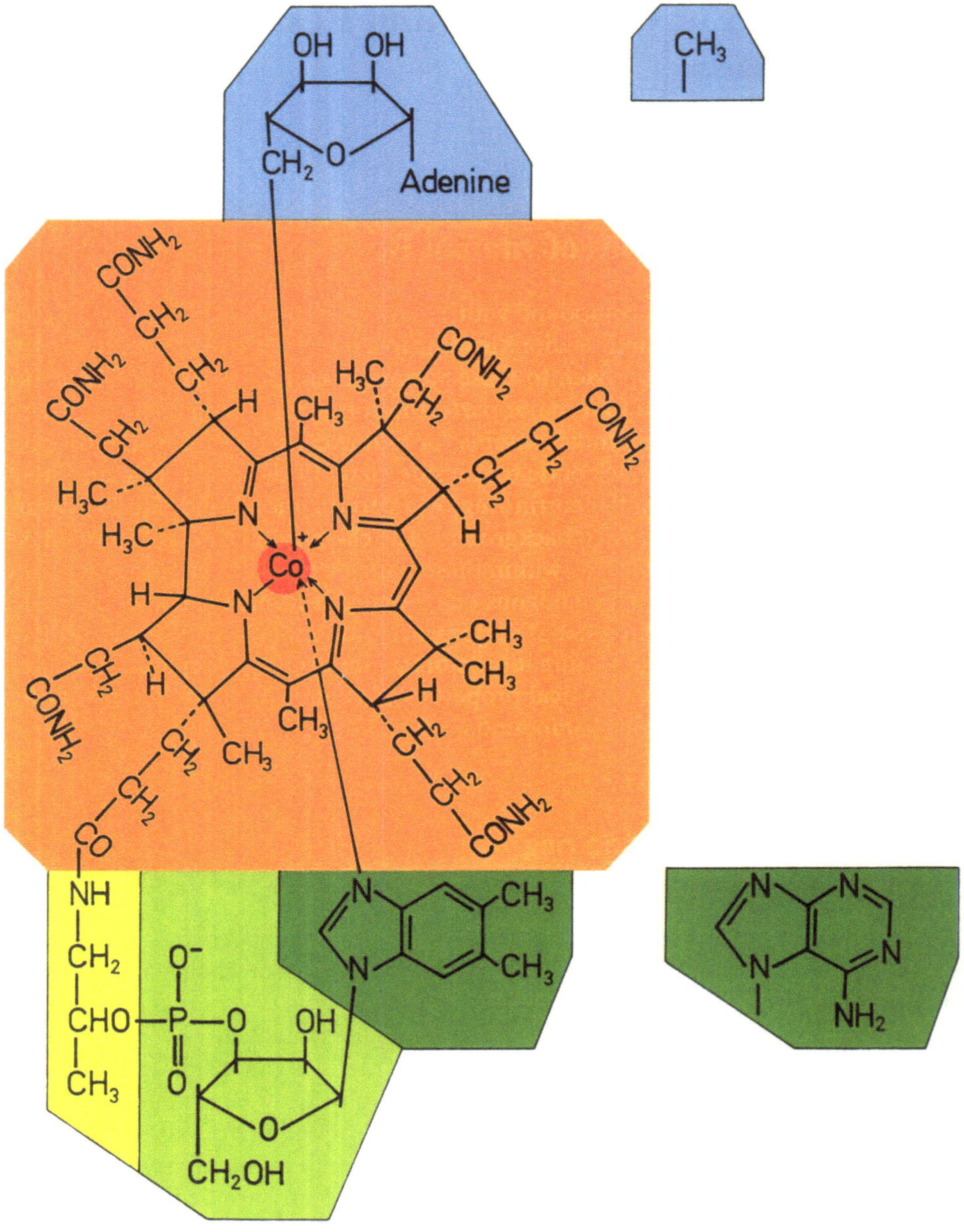

Fig. 3-1 Vitamin B_{12} coenzyme. The figure shows the biological active form of cobalamin (AdoCbl). By substitution of the deoxyadenosyl moiety with the methyl group (blue background) another coenzymatic form, methylcobalamin (MeCbl) is obtained. AdoCbl and MeCbl are distinguished by a unique Co-C bond. This bond plays a crucial role in the course of enzymatic reactions. The essential core of all corrinoids is the corrin ring (red background). Its biological as well as chemical synthesis are so far one of the most complicated processes known. To the side chain of the corrin ring is attached an α-glycosidic bounded benzimidazole nucleotide. Benzimidazole derivatives, substituted in positions B 5 and B 6 (see numbering of atoms in Fig. 3-2), are present in various forms of biologically active vitamin B_{12} derivatives. The corrinoid present in man and ani-

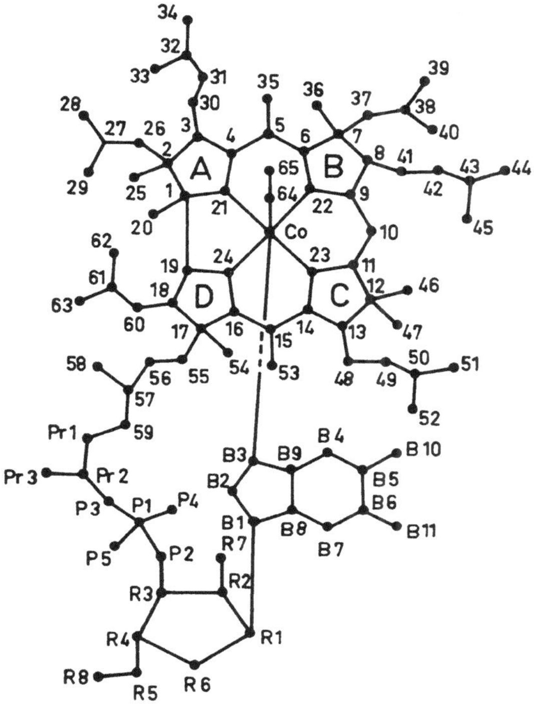

Fig. 3-2 The numbering of atoms in cyanocobalamin (from Moore et al. [159])

3.1.1.2 Amidation of the carboxylic side chains

Carboxylic corrinoids can be amidated by ammonia after being converted to anhydrides with chlorcarbonate esters. By using amines instead of ammonia a number of N-substituted cobalamines were obtained [3, 7, 8]. Some of these derivatives have antivitamine activity [9].

3.1.1.3 Cyclization at the B ring

Lactam Formation. Heating of the cyanocobalamin in 0.1 M NaOH at 100 °C for 10 minutes in the presence of air gives rise to a neutral product which resembles cyanocobalamin in physical properties but is biologically inactive [10–13]. The compound called "dehydrovitamin B_{12}" has been found to be a lactam formed by cyclization of the acetamide side chain of the B ring.

◁ mals, called cobalamin (Cbl) or simply vitamin B_{12} contains 5,6-dimethyl-benzimidazole in the nucleotide moiety (green background). However, various corrinoids differing with the nucleotide moiety like pseudovitamin B_{12} containing adenine (dark green background) are produced by bacteria, only cobalamin is specifically absorbed by man and animals. The parts of the B_{12} molecule may be exposed to various chemical or biological modifications giving rise to many analogues. A number of them are described in Chapter 3.

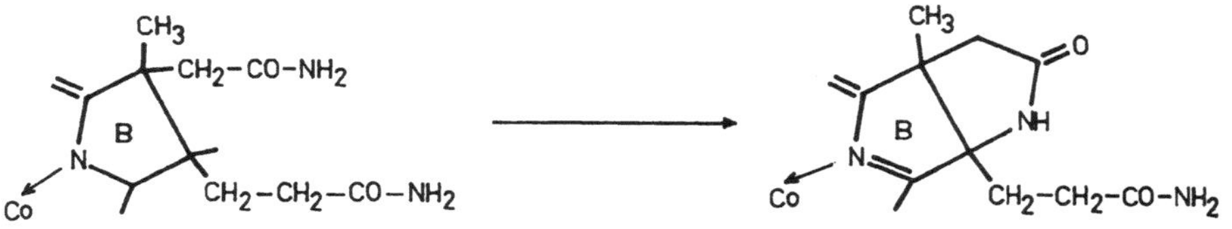

The reaction runs through a reduced intermediate which is subsequently oxidized by air. Reducing agents such as ascorbic acid inhibit the lactam formation. See also CN-Cbl (c-lactam) in Table 3-1.

Lactone Formation. Treatment of cyanocobalamin with an equimolar amount of Chloramine T or bromine water at pH 4 produces a neutral, halogen free compound which has been characterized as [8-hydroxy-α(5,6-dimethylbenzimidazolyl) cobamic acid a, b, d, e, g-pentaamide-c-lactone]. Gentle alkaline hydrolysis leads to hydroxy acid [10, 14, 15].

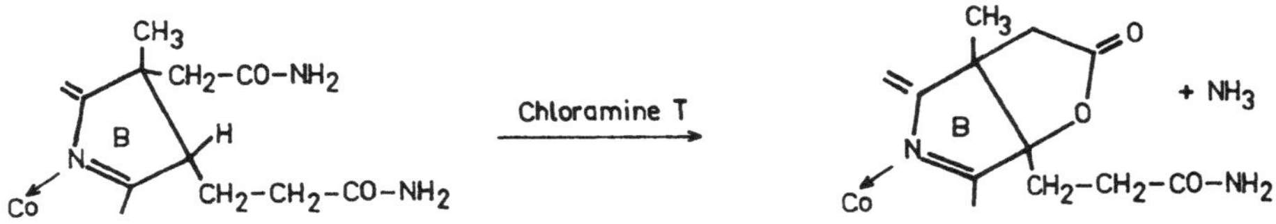

The chlorine released from Chloramine T acts probably in intermediate reactions. Excess of Chloramine T will increase the amount of byproducts such as C-10 chlorinated Cbl (c-lactone) (see below).

3.1.1.4 Substitution at C-10

Halogenation. Methylcobalamin, adenosylcobalamin or sulphitocobalamin but not cyanocobalamin, treated with an equimolar amount of Chloramine T produce C-10 chlorinated derivatives and also lactone as a byproduct. Similar reactions with N-bromosuccinimide lead to the C-10 brominated corrinoids [14, 16, 17]. For structure and properties of MeCbl (C-10-Br) see Table 3-1

Nitration. Methylcobalamin dissolved in acetic acid reacts with 2 equivalents of nitrosyl chloride (NOCl) to give C-10 substituted nitro-cobalamin. It is distinguished from CN-Cbl by a brick red colour. Sodium borohydride reduction yields C-10 substituted aminocobalamin. MeCbl (10-NO_2) and MeCbl (10-NH_2) are the strongest known cobalamin antagonists, shown by *L. leichmannii* [18].

3.1.1.5 Isomerization at C-13

The treatment of cyanocobalamin with CF_3COOH at room temperature for 2 hours gives rise to a new corrinoid with many chemical and physical properties similar to CN-Cbl. It may be distinguished, however, from CN-Cbl by ORD- and CD-spectra (see Fig. 3-22). The X ray data show that the e-propionamide side chain attached to C-13 is projected up instead of down to the plane of the corrin ring. In the light of X-ray studies of this compound and AdoCbl, it was possible to explain the difficulties in synthetizing AdoCbl (13-epi) – the propionamide residue occupies in the 13-epi form the space of the adenosyl ligand [19–22]. For structure and other properties of CN-Cbl (13-epi) see Table 3-1.

3.1.1.6 Removal of methyl groups at C-5 and C-15. Norcorrinoids

The norcorrinoids are obtained from cobalamin by chemical deprivation of the methyl residues at C-5 and C-15 [23]. This is achieved by oxydation of these methyls to carboxyls followed by decarboxylation. The nucleotide moiety damaged by this vigorous oxydative treatment ($KMnO_4$ in dry pyridine) is subsequently removed by hydrolysis with $Ce(OH)_3$ and norcobalamin is restored by chemical or biological attachment of the 5,6-dimethylbenzimidazole nucleotide. Norcobalamins are almost as active biologically as CN-Cbl in the *L. leichmannii* test [24]. For structure see Table 3-1.

3.1.1.7 Conformational changes at the corrin ring - isomeric form of cobalamin [27 a, b]

A new isomeric form resulting from conformational changes at the corrin ring has been recently discovered. The change in conformation is borne out by subtle differences in the proton magnetic resonance on the corrin ring and the base and by very marked variation in the Mössbauer emission spectra [27 a]. The latter is indicative of appreciable changes in the spatial conformation in the immediate vicinity of cobalt. The isomeric form labelled B'_{12} is present in minor quantities in commercial products of cobalamin forms: OHCbl' (ca. 1%), AdoCbl' (ca. 0.2%) and MeCbl' (ca. 0.2%). It may be produced from commercial Cbls by heat treatment. The Cbl' has only 15% of the biological activity of Cbl in the *L. leichmannii* test.

3.1.1.8 Substitution of cobalt

The introduction of other metallic ions to the corrin ring instead of cobalt has been achieved by heating appropriate salt solutions with a cobalt free cobalamin, hydrogenobalamin [25]. Hydrogenobalamin is obtained from *Chromatium,* a photosynthetic microorganism. The removal of cobalt from cobalamin by chemical means, however, possible by ascorbic acid treatment, causes irreversible changes in the corrin ring. Zincobalamin, cupribalamin, rhodibalamin [25] and ferribalamin [26] have been characterized. For structure and properties see Table 3-1 and spectra in Figures 3-12, 3-13, 3-14, 3-15, 3-16, 3-17.

3.1.1.9 Chemical synthesis of cobyric acid [28–36]

The chemical synthesis of vitamin B_{12} was one of the most complicated problems in organic chemistry succesfully solved in the early seventies. It was the synthesis of cobyric acid (the corrin core) which created so many problems. The chemical attach-

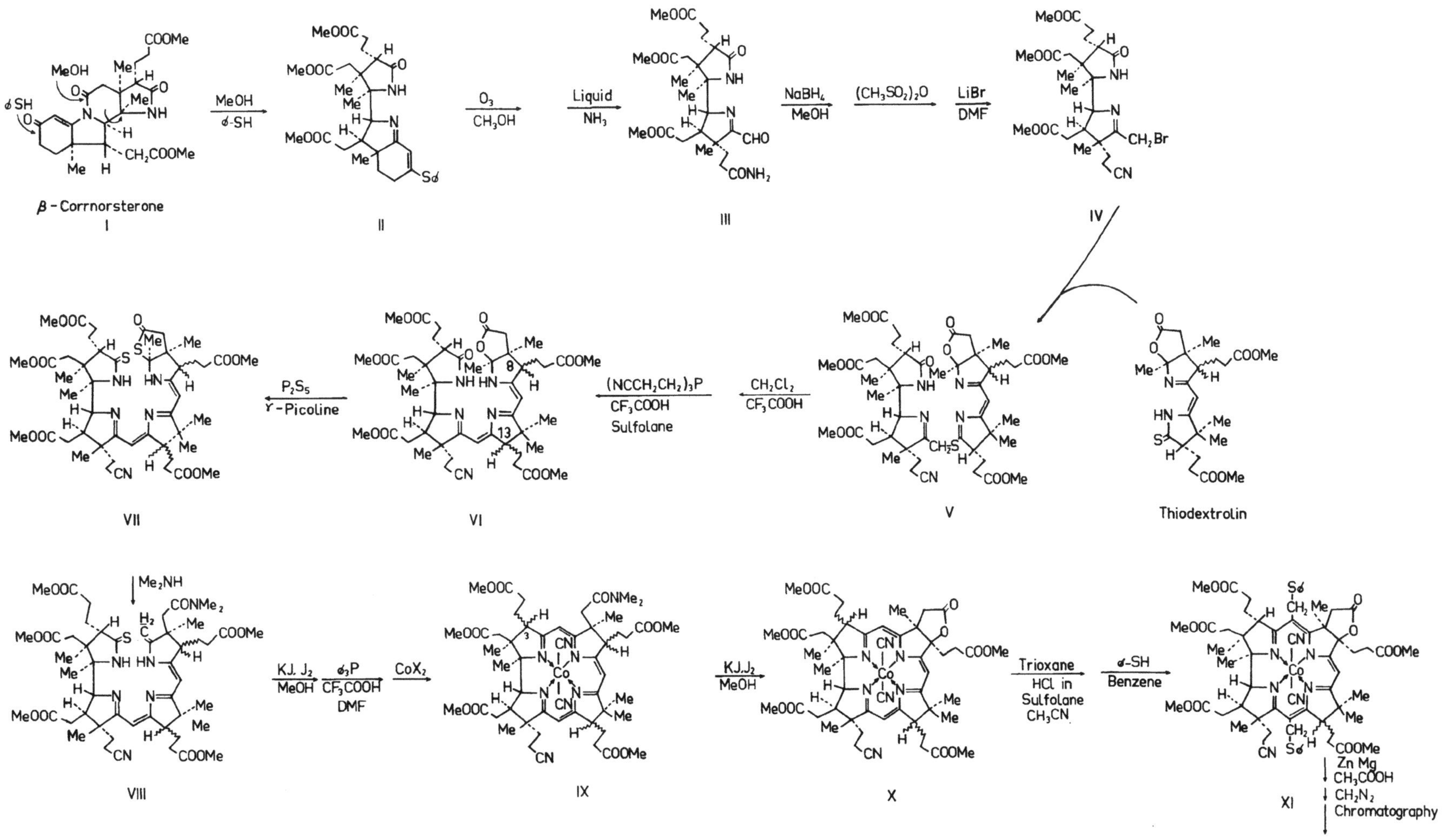
β-Corrnorsterone
I
II
III
IV
V
Thiodextrolin
VI
VII
VIII
IX
X
XI
MeOH
ϕ-SH
O3
CH3OH
Liquid
NH3
NaBH4
MeOH
(CH3SO2)2O
LiBr
DMF
CH2Cl2
CF3COOH
(NCCH2CH2)3P
CF3COOH
Sulfolane
P2S5
γ-Picoline
Me2NH
KJ. J2
MeOH
ϕ3P
CF3COOH
DMF
CoX2
KJ.J2
MeOH
Trioxane
HCl in
Sulfolane
CH3CN
ϕ-SH
Benzene
Zn Mg
CH3COOH
CH2N2
Chromatography

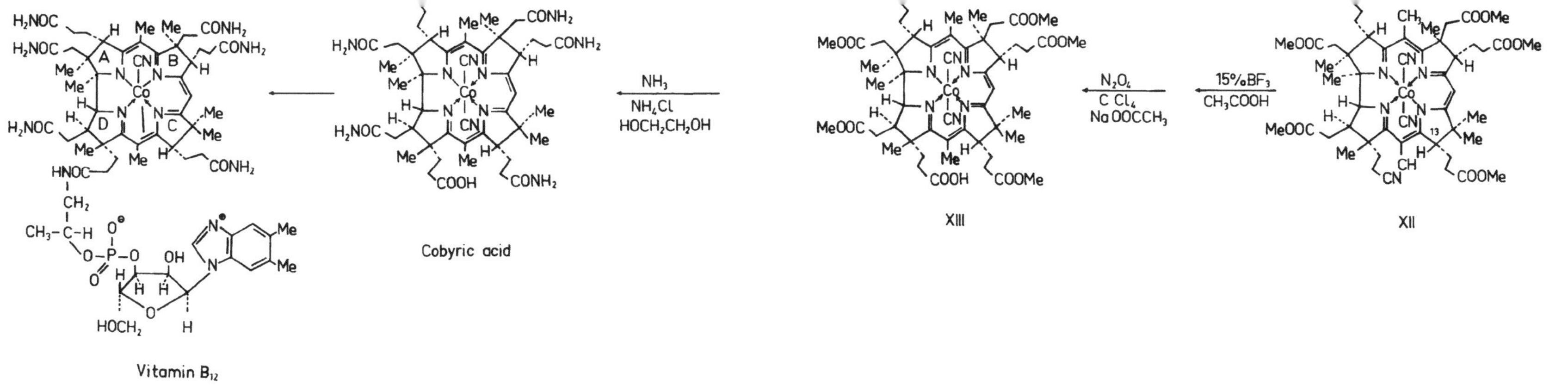

Fig. 3-3

Chemical synthesis of cobyric acid (from Maugh, T. H., Science **179**, 266, 1973. Copyright 1973 by the AAAS) The β-cornorsterone (I) is subjected to methanolysis and conversion to a thioether with thiophenol. Ozonolysis of (II) yields a thioester side chain which is converted to an amide in liquid ammonia. It also produces an aldehyde group (III). The aldehyde at (III) is then reduced to an alcohol with sodium borohydride and esterified with methane sulfonyl anhydride. The anhydride simultaneously dehydrates the amide to a nitrile. Displacement of methane sulfonate with lithium bromide then gives the cyano-bromide (IV). Compound (IV) which provides the A and D rings of the corrinoid is linked to thiodextrolin the source of the B and C ring. The thiodextrolin is linked to (IV) by a base-catalyzed displacement of bromine to form the methylene thioether (V), which holds the two segments in position for formation of a carbon-carbon bond. Sulfur is then extruded from this link with tris(β-cyanoethyl)phospine which also destroys the chirality at C-13 of the corrin ring (VI). The first step in linking the A and B rings is exchange of sulfur for oxygens in the lactam and lactone rings (VII). Aminolysis with dimethylamine then cleaves the thiolactone, producing a dimethylamide side chain and a methylene group in proximity to the thiolactam (VIII). These two groups are held in position by coordination of the four nitrogens with a zinc halide and linked by iodine oxidation. After removal of the zinc, the sulfur is extruded with triphenylphosphine, which also destroys the chirality at C-3. At this point, the product is a mixture of eight diastereomers. Separation of the diastereomers by high pressure liquid chromatography shows that only 36 percent of the mixture has the correct configuration at all three chiral centers. More than 78 percent of the mixture has the proper configuration at C-3, however, and that proves crucial because the bridgehead methyl group at C-5 can be introduced only when the proper configuration exists at C-3. Iodine oxidation of (IX) converts the dimethylamide to a lactone (X) linked to C-8; because the fused lactone is more stable in the cis-configuration, this reaction regenerates the required chirality at that position. Lactone formation also protects the bridgehead C-10 from substitution in the next step, which is introduction of methyl groups at the C-5 and C-15 bridgeheads. Substitution at the bridgeheads is accomplished by hydroxymethylation with trioxane, followed by reaction with thiophenol to form the thioether (XI). Reduction of (XI) with zinc and a catalytic amount of magnesium desulfurizes the molecule and opens the lactone ring, introducing a hydrogen at C-8. The carboxymethyl group thus formed is then esterified with diazomethane. The two resulting substances, differing only in configuration at C-13, can now be separated by high pressure liquid chromatography; one of the resulting products is stereochemically pure (XII). Treatment of (XII) with boron trifluoride in acetic acid converts the nitrile to an amide, which is then deaminated with dinitrogen tetroxide and a catalytic amount of sodium acetate to yield the free acid (XIII). Ammonolysis of (XIII) in ethylene glycol, catalyzed by ammonium ions, converts the six ester side chains to amides, producing cobyric acid. Cobyric acid is used for synthesis of vitamin B_{12} by the method elaborated by Friedrich and coworkers (see 3.1.2.2)

ment of the other parts to natural cobyric acid has already been achieved ten years earlier (see section 3.1.2.2). The main problem lays in the synthesis of the proper stereoisomers of the intermediates, since cobyric acid has nine asymmetric centres. To overcome these difficulties the corrin ring was built from larger segments (βcornorsterone and thiodextrolin) with the proper configuration, similarly to the way done in nature in the course of biological synthesis (see Chapter 4). Since several research groups approached this problem, alternatives at various steps of the chemical synthesis were developed [28–36]. Fig. 3-3 presents the main steps of a chosen path, as described in [36], showing a general approach to this difficult task. Details of the cobyric acid synthesis and the references to original literature may be found in the reviews of Eschenmoser and Woodward [28–31].

3.1.2 The nucleotide moiety

3.1.2.1 Hydrolysis of the phosphate bond

The benzimidazole nucleotide of cobamides may be split off by hydrolysis with 70% $HClO_4$ at 65 °C within 5 minutes [3]. A more elegant method is the hydrolysis of both phosphodiester linkages with $Cr(OH)_3$ [37]. The reaction proceeds at 100 °C for 2 hours in pH 8 to 9. Hydrolysis yields benzimidazole nucleosides, phosphate and cyanoaquacobinamides. The purine-containing cobamides such as (Ade)CN–Cba resemble the purine ribonucleotides in acid lability. The hydrolysis of these cobamides in 0.05 M HCl, at 100 °C for 15 minutes splits off in the first place the purine moiety while hydrolysis in presence of cereous hydroxide suspension and cyanide ions releases the 7-α-D ribofuranosyladenine [37, 38]. Acid hydrolyses of cobamides under rather severe conditions result in the release of ammonia and 1-amino-2-propanol [39]. Preparation of monocarboxylic acid derivative of vitamin B_{12} by acid hydrolysis is described in section 3.1.5.

3.1.2.2 Chemical attachment of the propanolamine and the nucleotide moiety

The chemical binding of these two compounds to the f-side chain of the cobyric acid yields the complete cobalamine or other cobamides. The synthesis proceeds in the following sequence of reactions [40–42]:

(i) Cyclization of the benzimidazolyl-α-D-ribofuranosyl-3′-phosphate or its base substituted analogues to the corresponding 2′,3′-cyclic nucleotides. Dicyclohexyl carbodiimide (DCC) is used for cyclization.

(ii) Condensation of the 2′,3′-cyclic nucleotides with 1-amino-2-propanol in the presence of tert. Na-butylate. A mixture of 2′ and 3′ nucleotide diesters is produced under these conditions. Other alkanolamines may be used for condensation as well.

(iii) Activation of the cobyric acid at the f-sidechain-carboxyl by condensation with chlorocarbonate esters.

(iiii) Condensation of the nucleotide diester with the activated cobyric acid to yield the corresponding cobamides.

The method permits the introduction of a variety of nucleotides into the corrinoid and thus a production of cobamides not found in the nature. The substitution of D-1-amino-2-propanol by various alkanolamines gives rise to a new range of vitamin B_{12} analogues. A number of them have been synthetized and checked for biological activity [9]. Several of these analogues are strong antagonists of vitamin B_{12} as shown by microbiological tests. Table 3-2 indicates the various alkanolamines incorporated into cobalamin in place of D-1-amino-2-propanol and the biological properties of the cobalamin analogues obtained [27].

3.1.2.3 Introduction of heterocycles by biological means

Some bacteria species may take up from the growth medium benzimidazole and its analogues, purine and its analogues and a number of other bases as well. They are subsequently incorporated into corrinoids in place of the 5,6-dimethylbenzimidazole. This approach termed "guided biosynthesis" provides by a simple procedure several ranges of cobamide analogues not present in the nature [43]. Usually bacteria produce the corrin core for the cobamide themselves however, they may take up from the medium also cobinamide and use it as precursor of cobamides [44]. Among the microorganisms most suitable for guided biosynthesis are *Propionibacterium* species, *E. coli* and *Streptomyces griseus*. This technique may also serve for synthesis of α-glycosidic bounded nucleosides. The following heterocycles and their derivatives can be incorporated into the nucleotide side-chain of corrinoids: imidazole, benzimidazole, benztriazole, benzthiazole, purine, adenine, hypoxanthine, guanine, 8-aza-adenine, pyrazine-imidazole, quinoxaline, quinazoline, naphthimidazole, imididazole-benzimidazole and phenazine [45–47] (see Table 4-2 and Fig. 4-3).

3.1.3 Reduction of the cobalt in the corrin ring

The cobalt atom chelated in the corrin ring of corrinoids may be exposed to chemical, photochemical and electrochemical reduction [48–55]. The reduction of cob(III)alamin runs through two stages yielding cob(II)alamin and cob(I)alamin which contain respectively divalent and monovalent cobalt. These compounds termed throughout the literature as B_{12r} and B_{12s} are extremely labile in presence of air being instantly exydized to cob(III)alamin.

$$\underset{\text{red}}{B_{12}} \underset{O_2}{\overset{NaBH_4,\ H^+}{\rightleftharpoons}} \underset{\text{brown}}{B_{12r}} \underset{O_2}{\overset{NaBH_4,\ OH^-}{\rightleftharpoons}} \underset{\text{gray green}}{B_{12s}}$$

While the B_{12r} is stable under anaerobic conditions the B_{12s} decomposes slowly to B_{12r} with the release of hydrogen. B_{12s} is a highly reactive compound so far not obtained as a solid. It is produced quite often for synthesis of Ado-Cbl, Me-Cbl and other cobalt-alkyl vitamin B_{12} analogues [48]. It is also produced by enzyme systems as intermediate in synthesis of coenzymatic forms of corrinoids [56, 57]. The most convenient method for chemical reduction of Cbl^{III} to Cbl^{I} is the treatment with $NaBH_4$ at slight alkaline pH [52, 59]. The reduction also offers a convenient way of the cyanide removal from CN-Cbl. Subsequent oxydation by air of the reduced form yields OH-Cbl. Some reducing agents are destruction to the corrin ring. The treatment of cobalamin by ascorbic acid or by thiols at elevated temperature causes irreversible changes in the corrin ring resulting in inactivation of cobalamin. Nitrous oxide (N_2O) is a specific oxidant of vitamin B_{12s} [60-74].

$$2\,Cbl^{I} + N_2O \rightarrow 2\,Cbl^{II} + N_2$$

Since N_2O has been found to inactivate irreversibly vitamin B_{12} dependent enzymes it is suggested that B_{12s} appears as an enzyme bound intermediate in the catalytic reaction [60-62]. This discovery is of physiological significance since nitrous oxide used as an anaesthetic has been found to cause abnormality of vitamin B_{12} function [63-71]. From clinical observations it is concluded that nitrous oxide interferes with the function of vitamin B_{12} in the homocysteine-methionine reaction [67-74]. For other properties of Cbl^{I} and Cbl^{II} see Table 3-1 and Chapter 4.

3.1.4 The beta ligand

The β-ligand occupies the upper part of the corrin ring (see Fig. 3-1). The coenzymatic forms of vitamin B_{12} contain the methyl- or adenosyl-moiety as β-ligands. These groups are coordinated with cobalt by a cobalt-carbon bond a phenomenon unique in coordination chemistry. The cleavage and reconstitution of that bond is described below.

3.1.4.1 Photolytic cleavage of the Co-C bond

Methylcobalamin and adenosylcobalamin are photolabile compounds. The primary products of photodecomposition are Cbl^{II} and methyl- or adenosylradical. Under aerobic conditions the Cbl^{II} is oxydized to Cbl^{III} and the corresponding radicals are oxydized to formaldehyde and adenosine-5′-carboxaldehyde. The cleavage of MeCbl by anaerobic photolysis in the absence of radical binding agents is reversible:

$$CH_3\text{-}Cbl \xrightleftharpoons{\text{III light}} CH_3^{\prime} + Cbl^{II}$$

The methyl radicals decay slowly yielding ethane and other products. Thus the rate of photolysis of MeCbl depends very much on the removal of the methyl radicals from the medium [49, 75]. The anaerobic photodecomposition of AdoCbl unlike MeCbl proceeds fast and is irreversible. The produced adenosyl radical recombines with the adenine moiety of its own molecule yielding 8,5′-cyclic adenosine [76]. The introduction of thiols, quinones and certain alcohols increases greatly the rate of anaerobic photolysis since these compounds recombine with the radicals and thus eliminate the reverse reactions. Photolysis of methylcobalamin in presence of cysteine or homocysteine yields S-methyl-cysteine and methionine respectively, while S-adenosylhomocysteine is formed when adenosylcobalamin is photolyzed in presence of homocysteine [77, 78]. Other alkylcobalamine are photolyzed as well yielding Cbl^{II} and alkyl radicals as primary products.

From the practical point of view the light sensitivity of MeCbl and AdoCbl is rather undesired since many experiments with these compounds, such as estimation, purification and the work with enzyme systems, have to be done in dark or in dimmed red light. They are far less photolabile as solids. These two compounds are photolyzed by all waves within the visible range with quantum yields ranged between 0.1 and 0.4 [78]. The irradiation of AdoCbl aqueous solution ($3{,}2 \cdot 10^{-5}$M) by a 60 watt bulb at a distance of 25 cm causes 26% decomposition of the vitamin within 5 minutes [79, 80]. Prolonged irradiation by sunlight inactivates irreversibly all forms of cobalamin. Light shifts also the equilibrium of cyanocobalamin in favour of the aquacobalamin.

$$\text{CN-Cbl} \underset{}{\overset{\text{light}}{\rightleftarrows}} \text{aqCbl} + \text{CN}^-$$

No reduced Cbl is produced by irradiation of CN-Cbl. Bubbling air through the irradiated CN-Cbl solution stops the reverse reaction, due to removal of CN^-, and aquacobalamin is obtained. The MeCbl seems to be somewhat less photolabile if bound to proteins. While some enzymes requiring AdoCbl or MeCbl may protect these corrinoids from photodecomposition the non-enzymatic binding proteins are less protective.

3.1.4.2 The preparation of coenzymic forms of vitamin B_{12} and their alkyl- and acyl analogues

The preparation of alkyl cobalamins is based upon reaction of Cbl^{I} B_{12s}) with suitable electrophiles [48, 78-86]. The reaction results in formation of a cobalt-carbon bond. The principal reactions may be summarized by the following scheme:

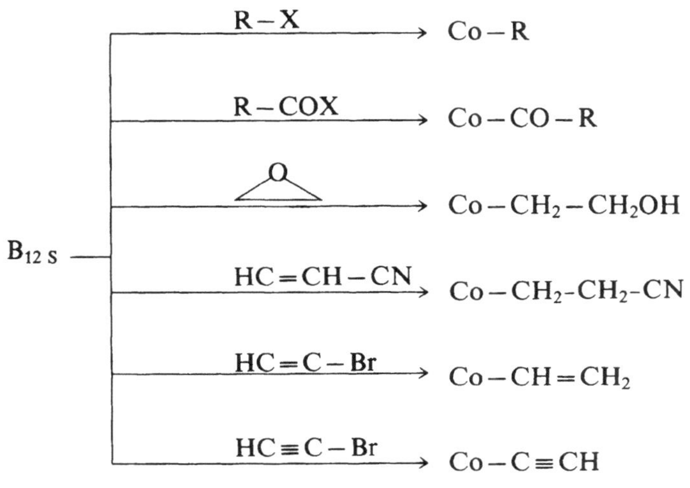

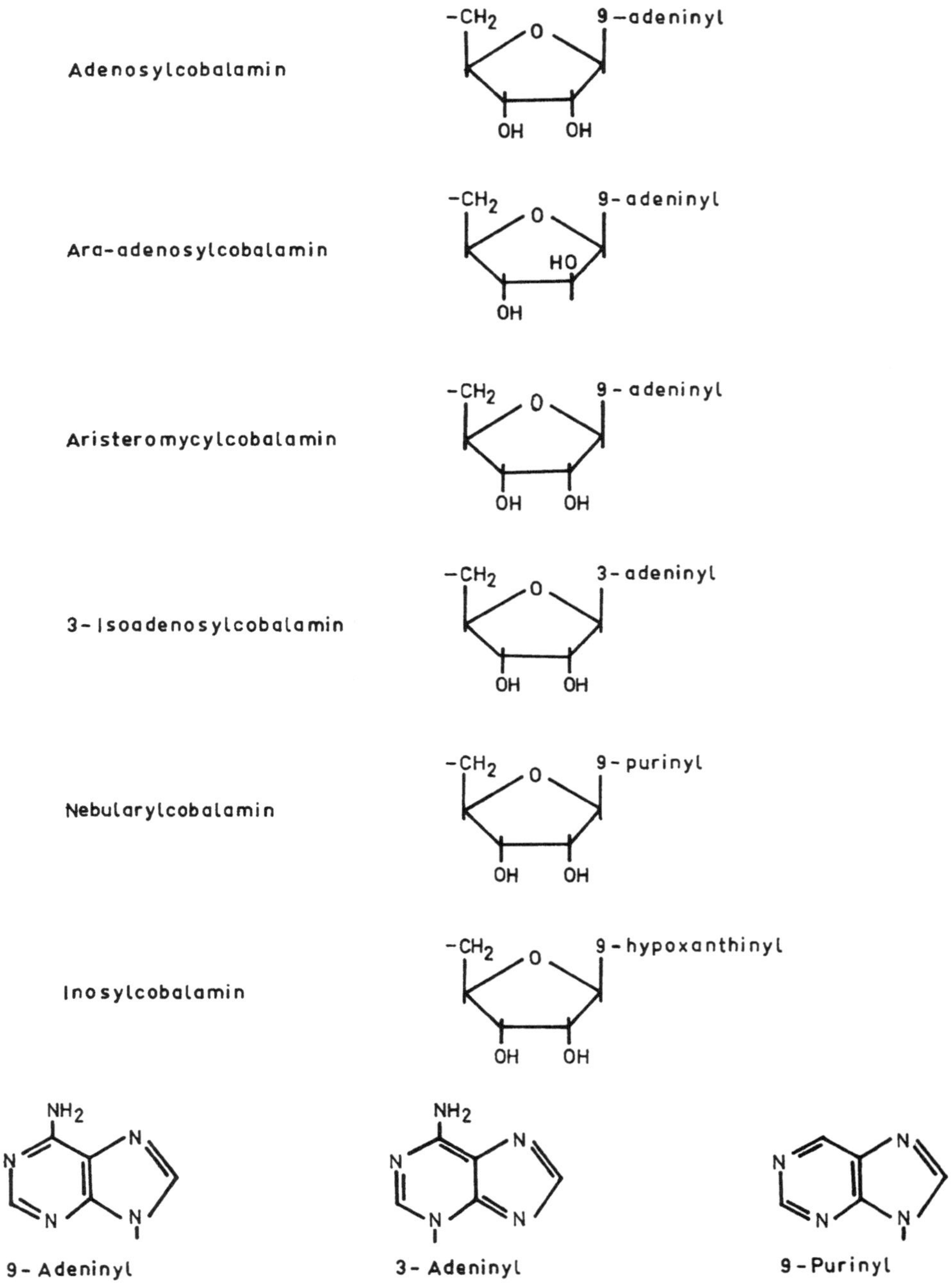

Fig. 3-4 Various nucleosides which may be attached by chemical means to cobalt in the corrinoid at the β position (from Toraya and Fukui [151] see text in 3.1.4.2.)

Cytidylcobalamin

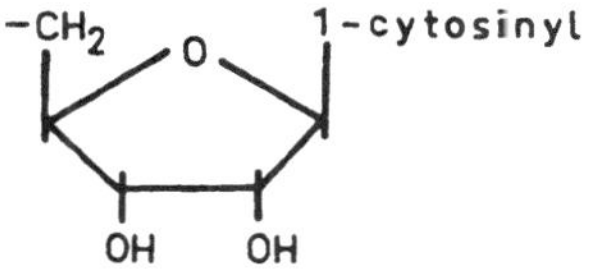

Benzimidazoleribosylcobalamin

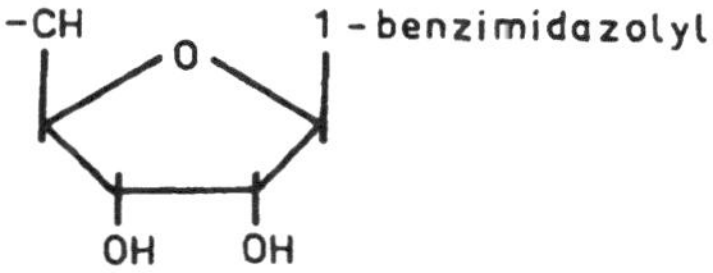

L-Adenosylcobalamin

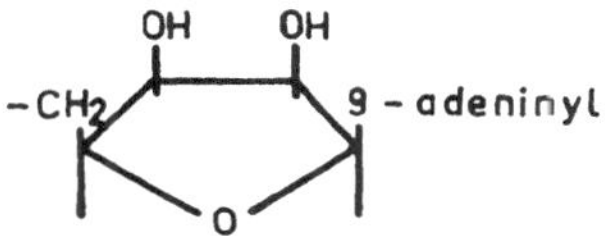

Adenosylethylcobalamin

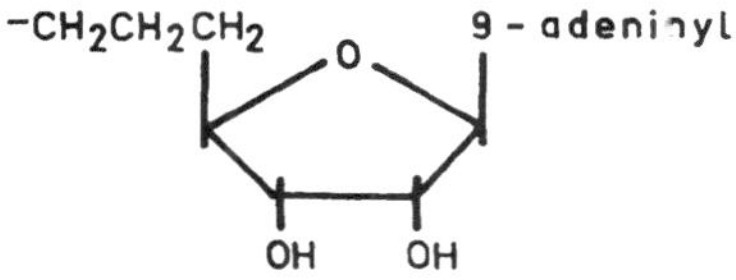

Adeninylhexylcobalamin	$-(CH_2)_6-9-$adeninyl
Adeninylbutylcobalamin	$-(CH_2)_4-9-$adeninyl
Adeninylpropylcobalamin	$-(CH_2)_3-9-$adeninyl
Adeninylethylcobalamin	$-(CH_2)_2-9-$adeninyl

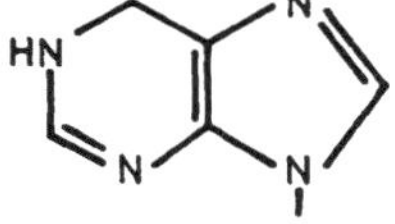

9-Hypoxanthinyl

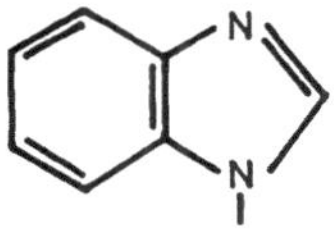

1-Benzimidazolyl

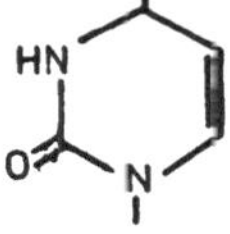

1-Cytosinyl

The coupling to cobalt proceeds very fast, usually within seconds [157]. To obtain the MeCbl, methyliodide or methylchloride is used, while 5'iodo-5'-deoxyadenosine is applied for AdoCbl production [87]. Various AdoCbl derivatives are obtained similarly (Table 3-4 and Fig. 3-4 see also procedures described in section 5.4.3). The Co-C bond weakens the strength of binding of 5,6-dimethylbenzimidazole to cobalt. Unlike CN-Cbl or OH-Cbl, which remain red in a wide range of pH, MeCbl or AdoCbl turn yellow if acidified (see AdoCbl in Table 3-1 and Table 3-6).

The formation of higher alkylcorrinoids is also possible from the Co(II)corrinoids and activated fatty acids. The reaction proceeds under anaerobic conditions in the presence of mild reducing agents such as V^{3+}, Mo^{3+} or Fe^{2+} salts and traces of oxygen or H_2O_2 in moderate acidic pH. The reaction is based on recombination of Co(II)cobalamin (B_{12r}) with an alkyl radical according to the scheme [89 a]:

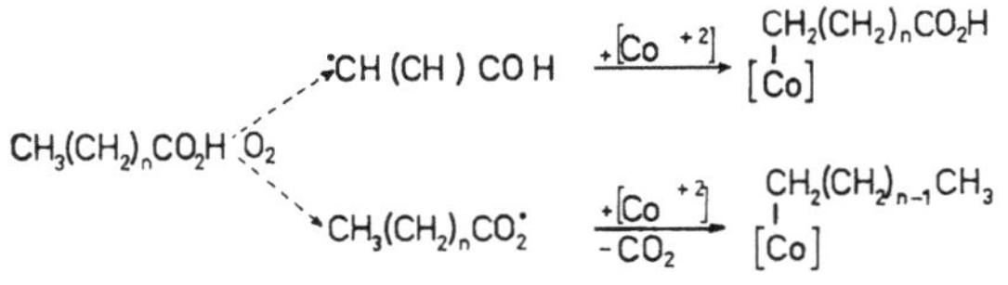

By this technique methyl- and carboxymethyl cobalamin have been obtained from acetate. Below pH 4.5 methylcobalamin is formed preferentially, while above pH 4.5 carboxymethyl cobalamin is produced mainly. Similarly, higher alkylcorrinoids can be obtained. The Co(II)corrinoids seem to be very effective free radical scavenging agents, thus this technique offers a new approach for synthesis of a broader range of corrinoid analogues with various axial β-ligands. Unlike MeCbl the higher alkylcorrinoids are rather unstable. Particularly unstable are secondary alkylcobalamins. They decompose spontaneously yielding corresponding olefins and cob(I)alamin (B_{12s}) under anaerobic conditions. The rate of spontaneous decomposition of secondary alkylcobalamins is affected by the opposite axial ligand (α-ligand) coordinated to cobalt. In case of cobalamin it is the nitrogen of 5,6-dimethylbenzimidazole. As shown in Table 3-3 the secondary alkylcobalamins are much more stable in 1M HCl than at pH 7. At acidic pH the benzimidazole ligand is no more coordinated to cobalt. Its place is taken by H_2O (see also Table 3-7). AdoCbl and other adenosyl cobamides may also be obtained by biological or enzymatic synthesis. *Propionibacterium* species will take up cobamides from the medium and convert them into their coenzymatic forms. Protein extracts from *Propionibacterium* or other bacteria will also catalyze the incorporation of the adenosyl moiety into the corrinoids [56, 90-92]. ATP is the source of the adenosyl residue. Similarly as in chemical synthesis, the Cba^{III} is reduced. A FAD-dependent enzyme system reduces Cbl^{III} to Cbl^{I} prior coupling to adenosine (see Chapter 4).

3.1.4.3 Miscellaneous substitutions

Cyanide treatment. Cyanocobalamin, one of the most widely known forms of vitamin B_{12}, is obtained by cyanide treatment of all other naturally occurring forms. Among the β-ligands known the cyanide has the highest affinity for cobalt. Since vitamin B_{12}

occurs usually in low concentrations, already traces of KCN will convert all the corrinoids into their cyanide forms. The attachment of CN does not require reduction of the cobalt. In the case of AdoCbl the cyanide ion, beside the formation of CN-Cbl, interacts also with the adenosyl moiety [78, 93]. The formation of CN-Cbl from this compound is slower than from OH-Cbl. It takes about 6-8 minutes to decompose in the dark in the presence of 0.1 M KCN half of the AdoCbl [78]. Light increases greatly the CN-Cbl formation due to photodecomposition of AdoCbl. It has to be stressed, however, that corrinoids with an alkyl ligand at the β-position such as MeCbl, EtCbl, PropCbl etc. are not decomposed by KCN in the dark. The substitution of these ligands by the cyanide is possible after photolytic cleavage.

Excess of KCN at high pH produces the dicyanide form, $(CN)_2$-Cbl. This is visualized by a change in color of the vitamin solution from red to purple (see also the remarks on $(CN)_2$-Cbl in Table 3-1).

The formation of analogues with Co-S bonds. The treatment of aquacobalamin with sodium bisulfite yields sulfitocobalamin.

$$[Co]\text{-}H_2O + HSO_3 \rightleftharpoons [Co]\text{-}SO_3H + H_2O$$

In dilute acid solution it changes color from red to yellow indicating that the 5,6-dimethylbenzimidazole moiety is protonated and no longer coordinated to cobalt. Corrinoids with a cobalt-sulfur bond can also be formed by reacting aquacorrinoids with thiols, hydrogen sulfide, sodium thiosulfate and pottassium thiocyanate [94-99]. Cobinamides and by a lower degree also cobalamins catalyze the oxydation of thiols by oxygen [99 a, b]. According to recent investigations [99 b] the reaction runs via a corrinoid-thiol complex which undergoes nucleophylic attack by a thiolate anion to form a disulfide and a reduced corrinoid. The latter is then reoxydized by O_2 which results in hydrogen peroxide formation. For further information see CystCbl and GluthCbl in Table 3-1. A number of other ligands can be introduced with Co-N, Co-Se, Co-P, and Co-As bonds [94, 95].

3.1.4.4 Rearrangements at the β-ligand [100-111]

The ease in which various groups can be attached to cobalt in corrinoids as the sixth ligand resulted in synthesis of a considerable number of cobalamin analogues (see 3.1.4.2.). Some of the β-ligands undergo intramolecular rearrangements. One of such typical interconversions studied by Chemaly and Pratt is the rearrangement of cyclopropylcarbinylcobalamin to but-3-enylcobalamin [106]:

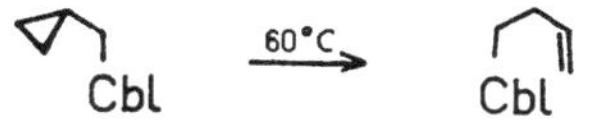

Similar rearrangements were demonstrated in cyclopropylcarbinyl (pyridine) cobaloxime a cobalt organocompound which mimic in many aspects the corrinoids. In studies with the use of the ^{13}C labelled cobaloxime model compound it was demonstrated that the rearrangement is accompanied by a transfer of the label in the ligand group as shown in the scheme [105]:

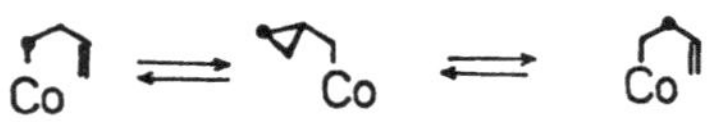

It seems that an identical displacement of the carbon atom takes place in course of rearrangement of cyclopropyl-carbinyl cobalamin. Carbon-skeleton rearrangement has been found also in the carbon bonded methylitaconate-tetrahydropyranyl ester which decomposes spontaneously yielding three compounds:

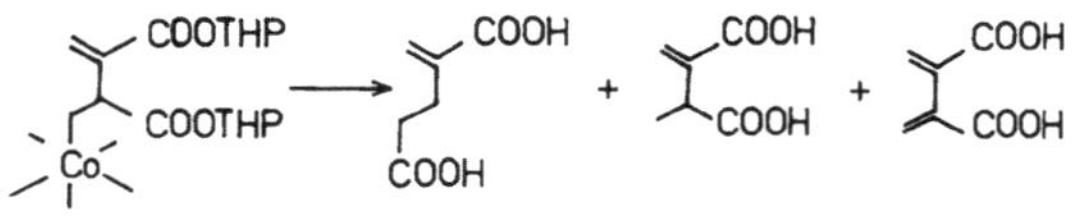

These studies on model compounds extend our knowledge as to the mechanism of action of coenzyme B_{12} in enzyme systems. Similar carbon-skeleton rearrangements of the substrates are observed in enzymatic conversions (for details see Chapter 7).

3.1.4.5 Methylation of metal ions by methylcobalamin [112–136]

Methylcobalamin is capable of transfering by non-enzymatic means its methyl group coordinated to cobalt to heavy metal ions. This property discovered in 1968 by Wood and co-workers [113] is of great scientific and environmental significance. The methylated metals are invariably more toxic than their inorganic substrates. This toxicity is probably due to the nonpolar nature of many organometallic compounds which allows them to diffuse rapidly into and through cell membranes. Some bacteria species produce preferentially methylcobalamin. Their contact with deposits of heavy metals such as e.g. mercury on the bottom of polluted rivers may result in production of very toxic methylmercury which through the foodchain, mainly via fish, reaches the man. The toxic effect of methylmercury is strongly enhanced due to its stability and long retention time. Severe damage of the brain in man by methylmercury have been observed in several countries [132].

Methylcobalamin is a rather stable compound in solution if kept in the dark (see Table 3-3) in contrast to many other alkylcobalamins. The transfer of the methyl group requires its activation, which is done according to Wood and Fanchiang by the metal ion itself or by a system to two metal ions (see schemes below) [112].

Based on both structural and kinetic studies, six different mechanisms have been postulated (Fig. 3-5). Methyl transfer from methylcobalamin requires cleavage of the Co–C bond. This bond can break under different conditions to give a carbanion (CH_3^-), a radical ($CH_3^{\cdot}$), or carbonium ion (CH_3^+). Examples representing some of the mechanisms formulated in Fig. 3-5 are given in the schemes below.

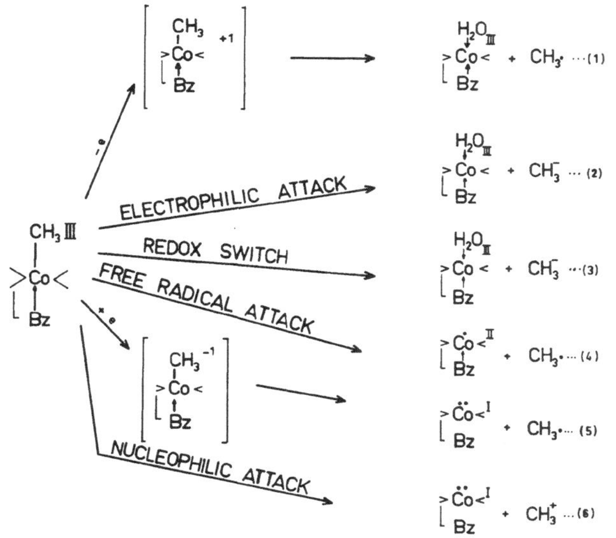

Fig. 3-5 Mechanisms by which the methyl moiety of methylcobalamin can be transferred to various metal ions (from Wood et al. [112] see section 3.1.4.5.)

Electrophilic attack. The carbanion (CH_3^-) is most likely to react with positively charged metal ions. The methylation of mercury ion (Hg^{2+}) is an example for this mechanism.

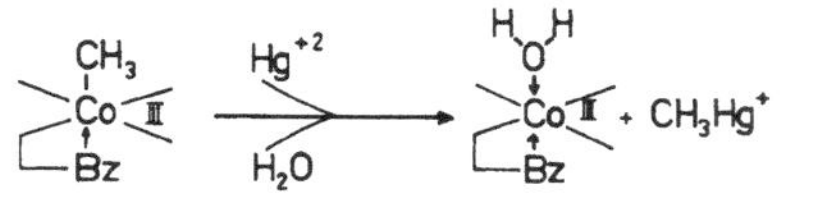

Other metal ions which are known to react with methylcobalamin by a similar mechanism to mercuric ion are lead (Pb^{4+}), thallium (Tl^{3+}) and palladium (Pd^{2+}). These reactions occur under aerobic conditions with rate constants in the order of milliseconds. The transfer of the methyl to mercury proceeds quantitatively so that this reaction can be utilized for the assay of methylcorrinoids (see section 5.3.6).

Free radical attack. Homolytic cleavage of the Co-C bond of methylcobalamin leads to methyl radical ($CH_3^{\cdot}$) and to Cob(II)alamin (B_{12r}). The methyl radical recombines under anaerobic conditions easily with B_{12r} which is a very effective free radical scavenger and a reducing agent (see also section 3.1.4.2). The presence of a transient metal radical such as $\dot{S}n$(III) or $\dot{A}u$(II) produced *in statu nascendi*, enables the recombination of the methyl radical with it. According to this mechanism proceeds the methylation of Sn^{2+} in presence of Fe^{3+} ions:

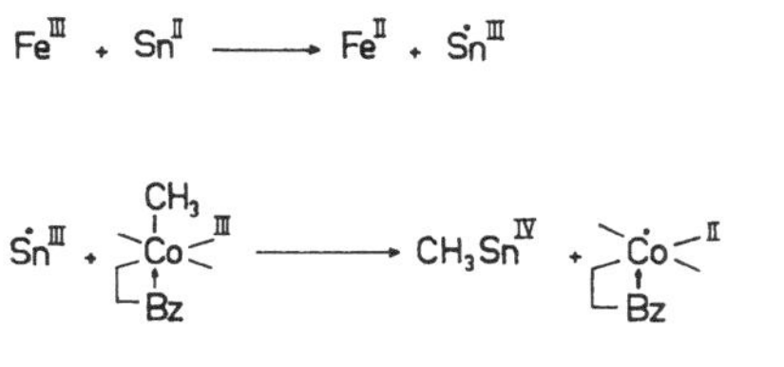

The "Redox Switch" mechanism occurs in case of methylation of Pt. The methylation takes place in presence of both Pt^{2+} and Pt^{4+} salts. When methylcobalamin reacts with a mixture of $Pt^{IV}Cl_6^{2-}$ and $Pt^{II}Cl_4^{2}$ the products were shown to be aquacobalamin and $CH_3PtCl_5^{2}$. The reaction rate is not dependent on the oxygen from air. ^1H NMR studies revealed that $Pt^{II}Cl_4^{2-}$ forms a complex with methylcobalamin.

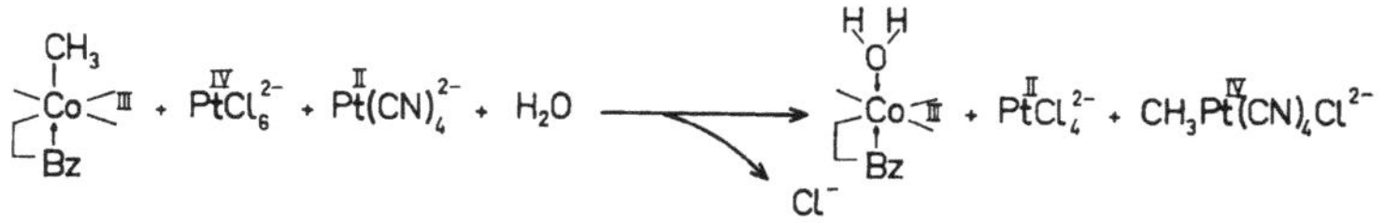

Oxidative cleavage. The demethylation of methylcobalamin is affected by $Ir^{IV}Cl_6^{2}$ which is being oxidized by Ir(IV). It is postulated that the reaction runs through a methylcob(IV)alamin intermediate, which decomposes yielding methyl radical and aquacob(III)alamin. The methyl radical interacts with $Ir^{IV}Cl_6^{2}$ to yield CH_3Cl according to the scheme:

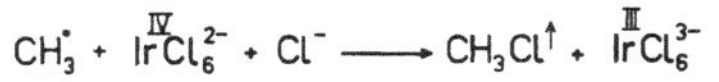

The capability of methylcobalamin to transfer the methyl group by various means points to a great flexibility and mobility of the chemical bonds in the corrin ligand system. It seems that nature worked out a unique compound in terms of its susceptibility to modifications of reactivity by physical and chemical means.

3.1.5 Immobilization of vitamin B_{12} on Sepharose [137–156]

The immobilization of vitamin B_{12} on solid supports is applied recently in affinity techniques for purification of cobalamin dependent enzymes or cobalamin binding proteins and their receptors. The first successful immobilization of vitamin B_{12} on cellulose was achieved already in 1966 [137] followed by a number immobilization techniques mainly on Sepharose [140–152]. Vitamin B_{12} is an organocobalt compound with a considerable number of potential binding sites for chemical attachment. Various immobilized corrinoid derivatives have been obtained and their interaction with the particular binding protein or enzyme were studied. By applying immobilized corrinoids to affinity chromatography a great deal of information as to the particular groups of the vitamin B_{12} involved in binding to the enzymes or binding proteins have been obtained. The most useful immobilized vitamin B_{12} for affinity chromatography revealed to be B_{12}-Sepharose obtained by binding monocarboxyl derivatives of vitamin B_{12} to an extended chain (spacer) with the length of six or more carbon atoms (see formulae I and II in Fig. 3-6) attached with its one end to Sepharose. Less effective for affinity chromatography are B_{12}-Sepharose supports in which vitamin B_{12} is bound to Sepharose through its other sites or those obtained by binding incomplete corrinoids (see formulae IV–IX in Fig. 3-6) [151]. A simple method for preparation of monocarboxylic vitamin B_{12} and its coupling to Sepharose according to Allen et al. [146] is described below:

Preparation of monocarboxylic acid derivatives of vitamin B_{12}. Crystalline vitamin B_{12} (4340 µmoles containing 1 µCi of [^{57}Co]vitamin B_{12}) is dissolved in 500 ml of 0.4 M HCl. After standing in the dark at 22 °C for 64 hours, the deep red solution is chilled to 4 °C and applied to a column (4.5 × 60 cm) of BioRex AG1-X8 (acetate form, 100-200 mesh, Bio Rad Laboratories) that has been washed with 4 liters of water immediately before use. The sample is applied at a flow rate of 200 ml per hour, and the column is eluted subsequently with 25 mM acetic acid. The eluate is collected as the red material approaches the bottom of the column. More than 95% of the applied radioactivity is present in the first 900-1000 ml of collected eluate. The eluate is lyophilized. The residue is dissolved twice in 300 ml of water followed by lyophilization each time. The final chloride-free residue is dissolved in 300 ml of water, and 55 ml of pyridine are added followed by sufficient concentrated NH_4OH(< 1 ml) to adjust the pH to 9.05. The solution, 385 ml, is chilled to 4 °C and applied to a 2.5 × 56 cm column of QAE-Sephadex A-25 (acetate form) that has been washed with 300 ml of 0.2 M pyridine-NH_4OH at pH 9.0. The sample is applied at a flow rate of 100 ml per hour, and 15-ml fractions are collected. After the column is eluted with 600 ml of 0.4 M pyridine, the flow rate is decreased to 50 ml per hour, and a linear acetate gradient is begun in which the mixing chamber contains 1 liter of 0.4 M pyridine and the reservoir contains 1 liter of 0.4 M pyridine – 0.16 m acetic acid. A final elution is performed with 1.0 liter of 0.4 M pyridine – 0.32 M acetic acid. The first major peak eluting after the start of the acetate gradient contains approximately one-third of the starting material in the form of a mixture of monocarboxylic acid derivatives of vitamin B_{12} in which an amino group is missing from one of the propionamide side chains adjacent to rings A, B, and C (Fig. 3-1). This material, approximately 300 ml, is lyophilized and dissolved twice in 400 ml of water followed each time by lyophilization to remove pyridine acetate. The final residue is dissolved in sufficient water such that the vitamin B_{12} derivative concentration is approximately 10-11 µmoles/ml, and the solution is stored at −20 °C.

Comment. It is important that the acid hydrolysis step and all of the lyophilization procedures be performed in the dark since at acid pH there is a light catalyzed conversion of vitamin B_{12} (cyanocobalamin), to hydroxycobalamin which is basic with a pH of approximately of 7-8. The monocarboxylic acid derivatives of hydroxycobalamin are not adsorbed to QAE-Sephadex under the conditions described above.

Coupling of carboxylic acid derivatives of vitamin B_{12} to 3,3′-diamino-dipropylamine-substituted Sepharose [146]. Thirty-four milliliters of water containing 321 µmol of monocarboxylic acid derivatives of vitamin B_{12} are added to 30 ml of packed 3,3′-diamino-dipropylamine-substituted Sepharose and mixed with a magnetic stirrer at 22 °C. The pH is raised from between 3.0 and 4.0 to pH 5.6 with 0.2 M NaOH (about 1.0-1.5 ml) and, while stirring continues, 3 ml of 1-ethyl-3-(3-diethylaminopropyl) carbodiimide, 50 mg/ml, are added in 0.6-ml aliquots at 1-minute intervals. Gentle stirring is continued in the dark for 18 hours. The substituted Sepharose is collected by vacuum filtration on a Büchner funnel and washed at 22 °C successively with 300 ml of water, 600 ml of 0.1 M glycine-NaOH at pH 10.0, 300 ml of water, and 300 ml of 0.1 M potassium phosphate at pH 7.0. The washed substituted Sepharose is deep red and contains from 0.35 to 0.7 µmoles of vitamin B_{12} derivative per milliliter of packed Sepharose, based on measurement of radioactivity. This product is referred to as vitamin B_{12}-Sepharose since the coupling of the single free carboxyl group of the monocarboxylic vitamin B_{12} derivative to the free amino group of the derivatized Sepharose presumably regenerates the amide structure of the vitamin B_{12} molecule attached to Sepharose. The amounts of each of the three monocarboxylic acid isomers that are coupled to Sepharose is not known.

The coupling reaction and subsequent washing of Sepharose substituted with vitamin B_{12} derivatives that lack the nucleotide portion of the vitamin are performed as described above except that the concentration of vitamin B_{12} derivatives in the coupling reaction, prior to the addition of carbodiimide, is 9.2 μmol/ml. The washed Sepharose contains 0.20–0.30 μmol of vitamin B_{12} derivatives per milliliter of packed Sepharose. The amounts of the different carboxylic acid derivatives and their isomers that are coupled to Sepharose are not known.

Comment. Both of the reactions described above give maximal coupling at pH value between 5.0 and 6.0. Experiments have not been performed to determine how the amount of coupling varies as a function of the concentration of the specific vitamin B_{12} derivative.

Coupling of vitamin B_{12} to Sepharose through cobalt [152, 153, 156]. A unique type of chromatography, *called affinity photo-release chromatography,* was developed by at-

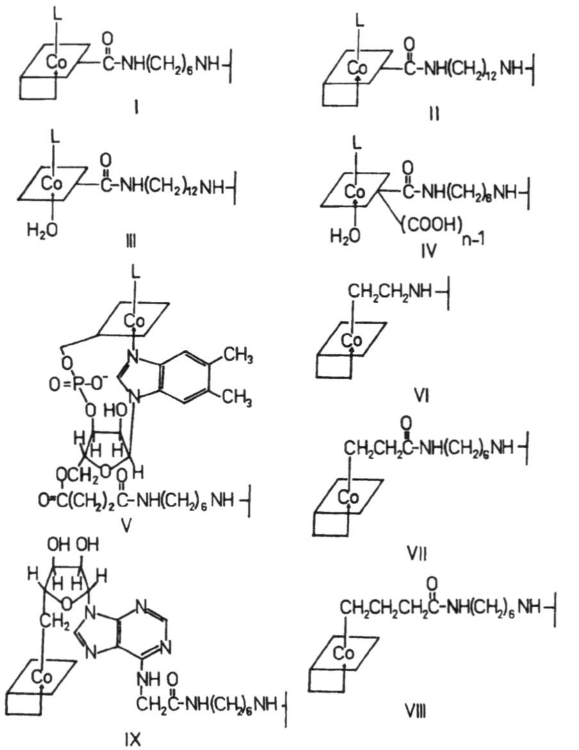

Fig. 3-6 Various chemical attachments of corrinoid to Sepharose (from Toraya and Fukui [151] see text in section 3.1.5.)

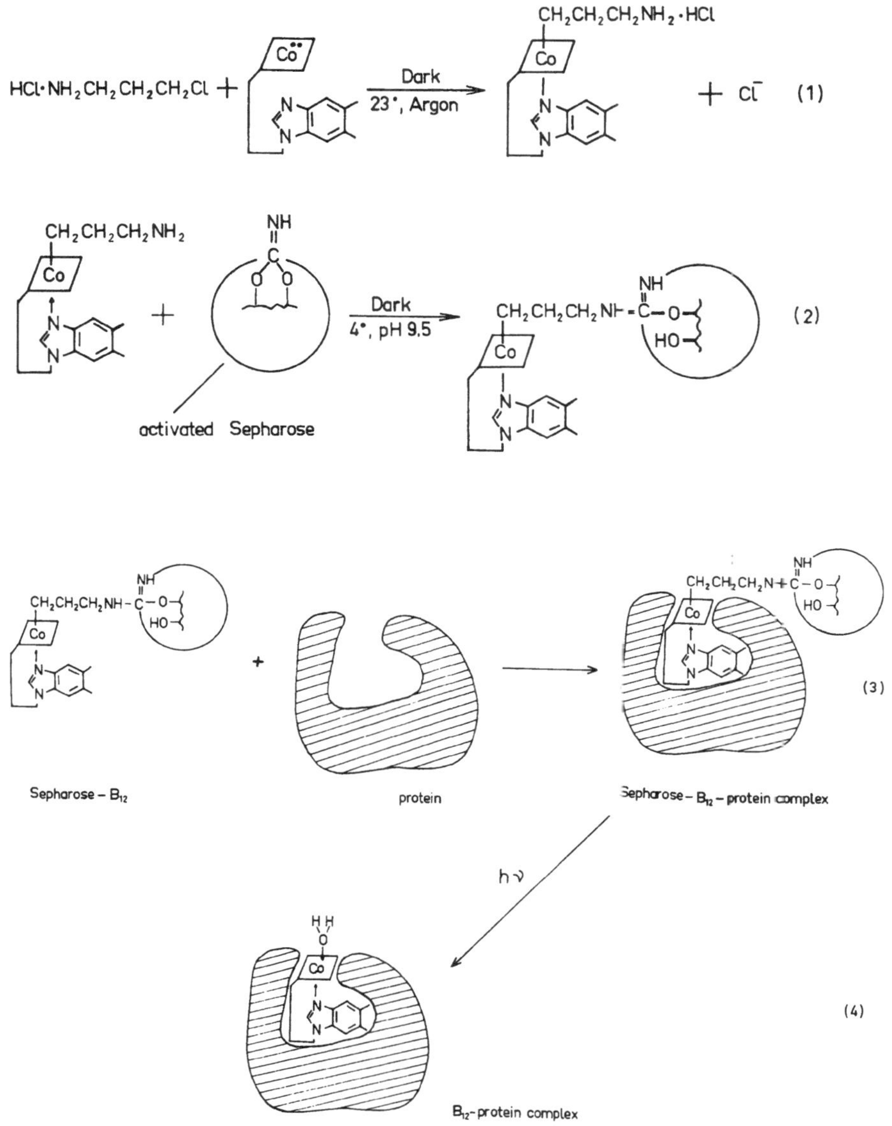

Fig. 3-7 Immobilization of vitamin B_{12} on Sepharose by binding to cobalt and scheme of application in affinity photo-release chromatography (from Jacobsen and Montejano see section 3.1.5. and reference [152] in this Chapter)

tachment of the Sepharose spacer to cobalt of the vitamin B_{12}. The technique of binding to cobalt is as described in section 3.1.4.2. In course of chromatography the specific apoprotein is being complexed with the Sepharose bound vitamin B_{12} similarly as is it done by other affinity techniques. Unlike other affinity methods, however, the release of the protein from the support is effected by light irradiation (see Fig. 3-7). This is possible due to photoability of the Co–C bond (see section 3.1.4.1).

Coupling of vitamin B_{12} to insolubilized 3′, 3′-diaminodipropylamine by a temperature-labile linkage [154, 158 a, b]. The thermolabile immobilization of vitamin B_{12} is utilized in a special type of affinity chromatography for purification of B_{12}-binding proteins. The release of of the protein adsorbed on the column together with bound B_{12} is achieved by heating the column for several hours at 37 °C. Only holo-proteins are obtained by this technique.

References

[1] Allen, R. H., Mehlman, C. S., Isolation of gastric vitamin B_{12}-binding proteins using affinity chromatography, J. Biol. Chem. **248,** 3660-3669, 1973. (see also 3.1.5.)

[2] Christensen, J. M., Hippe, E., Olesen, H., Rye, M., Haber, E., Lee, L., Thomson, J., Purification of human intrinsic factor by affinity chromatography, Biochim. Biophys. Acta **303,** 319-332, 1973.

[3] Armitage, J. B., Cannon, J. R., Johnson, A. W., Parker, L. F. J., Smith, E. L., Stelford, W. H., and Todd, A. R., Chemistry of the vitamin B_{12} group, Part III: The course of hydrolytic degradations, J. Chem. Soc. 3849-3864, 1953.

[4] Bernhauer, K., Vogelman, H., und Wagner, F., Gewinnung und Charakterisierung von Corrinoid-Carbonsäure-Isomeren, Hoppe-Seyler's Z. Physiol. Chem. **349,** 1281-1296, 1968.

[5] Friedrich, W., Ohlms, H., Sandeck, W., Bieganowski, R., Über die a-b-Isomerie der Cyano-aquocorrinoide., Z. Naturforschg. **22 b,** 839-850, 1967.

[6] Mertha-Kelemen, A., Csányi, E., Simon, A., und Eckhardt, S., Untersuchungen über die Monocarbonsäure-Derivate des Vitamin B_{12}, 2. Europäisches Symposion, Hamburg 1961, Ed. Heinrich, H. C., Ferdinand Enke Verlag, Stuttgart 1962, pp. 241-244.

[7] Morley, C. G. D., Blakley, R. L., Hogenkamp, H. P. C., Analogs of deoxyadenosyl cobalamin with alterations in a side-chain of the corrin ring, Biochemistry **7,** 1231-1238, 1968.

[8] Rapp, P., Hildebrand, R., Die Biologische Aktivität von N-alkyl- und N-aryl-substituierten Cobalaminen und Cobinamiden, Hoppe-Seyler's. Z. Physiol. Chem. **353,** 1141-1152, 1972.

[9] Heinrich, H. C., und Gabbe, E. E., Strukturspezifität des Vitamin B_{12}-Stoffwechsels und der Vitamin B_{12}-Bioaktivität, 2. Europäisches Symposion, Hamburg 1961, Ed. Heinrich, H. C., Ferdinand Enke Verlag, Stuttgart 1962, pp. 252-275.

[10] Bonnett, R., Cannon, J. R., Clark, V. M., Johnson, A. W., Parker, L. F. J., Smith, E. L., and Sir Todd, A., Chemistry of the vitamin B_{12} group, Part V: The structure of the chromophoric grouping, J. Chem. Soc. 1158-1168, 1957.

[11] Pratt, J. M., Inorganic Chemistry of Vitamin B_{12}, Academic Press, London, New York 1972, p. 193.

[12] Wagner, F., Bernhauer, K., New aspects of the structure of corrinoid coenzymes, Ann. N. Y. Acad. Sci. **112,** 580, 1964.

[13] Rapp, P., Bozler, G., und Fridrich, E., Positionsermittlung der freien Propionsäuregruppe zweier nucleotidhaltiger Corrinoid-monocarbonsäuren, Hoppe-Seyler's Z. Phys. Chem. **354,** 970-973, 1973.

[14] Wagner, F., Reactions of the cyano and alkyl cobalamines., Proc. Roy. Soc. **A 288,** 344, 1965.

[15] Pratt, J. M., Inorganic Chemistry of Vitamin B_{12}, Academic Press, London, New York 1972, pp. 284-285.

[16] Dolphin, D. M., Johnson, A. W., Shaw, N., Sulphitocobalamin, Nature **199,** 170-171, 1963.

[17] Wagner, F., and Renz P., zur Struktur der Cobamide-Coenzyme, Tetrahedron Letters **4,** 259, 1963.

[18] Friedrich, W., Vitamin B_{12} und verwandte Corrinoide, in Fermente, Hormone, Vitamine, Ammon, R., and Dirscherl, W. Eds., Vol. III/2, Georg Thieme Verlag, Stuttgart 1975, pp. 290-298.

[19] Bonnet, R., Godfrey, J. M., Redman, D. G., Cobyric acid and related compounds from vitamin B_{12}. J. Chem. Soc. **C. 8,** 1163-1166, 1969.

[20] Bonnett, R., Godfrey, J. M., Math, V. B., Edmond, E., Evans, H., and Hodder, O. J. R., Neovitamin B_{12} identified, Nature **229,** 473-476, 1971.

[21] Bonnett, R., Godfrey, J. M., Math, V. B., Cyano-13-epicobalamin (Neovitamin B_{12}) and its relatives, J. Chem. Soc. **C. 22**, 3736-3743, 1971.

[22] Rapp, P., Oltersdorf, V., Darstellung und biologische Eigenschaften von 8 Epicorrinoiden, Hoppe-Seyler's Z. Physiol. Chem. **354**, 32-36, 1973.

[23] Jauering, D., Rapp, P., und Ruoff, G., 5-Nor-, 15-Nor- und 5,15-Dinocorrinoide, Z. Physiol. Chem. **354**, 957-966, 1973.

[24] Rapp, P., Ruoff, G., Die Biologische Aktivität der Norcobinamide, Hoppe-Seyler's Z. Physiol. Chem. **354**, 967-968, 1973.

[25 a] Koppenhagen, V. B., Elsenhaus, B., and Wagner, F., Methylrhodibalamin and 5′-deoxyadenosylrhodibalamin, the rhodium analogues of methylcobalamin and cobalamin coenzyme, J. Biol. Chem. **249**, 6532-6540, 1974.

[25 b] Koppenhagen, V. B., and Pfiffner, J. J., Currins and zirrins, two new classes of corrin analogues., J. Biol. Chem. **245**, 5865-5873, 1970.

[26] Bieganowski, R. and Friedrich, W., in Vitamin B_{12}, Eds. B. Zagalak and W. Friedrich, Walter de Gruyter, Berlin 1979, pp. 647-648.

[27 a] Katada, M., Tyagi, S., Nath, A., Petersen, L. R., Gupta, R. K., A novel form of vitamin B_{12} and its derivatives, Biochim. Biophys. Acta **584**, 149-163, 1979.

[27 b] Mishra, P. K., Gupta, R., Goswami, P. C., Venkatasubramanian, P. N., Nath, A., ^{31}P nuclear relaxation studies of para- and diamagnetic cobalamins in their two isomeric forms, Biochim. Biophys. Acta **668**, 406-412, 1981.

[28] Eschenmoser, A., Quartely Rev. **24**, 366, 1970.

[29] Eschenmoser, A., 23rd International Congress Pure and Appl. Chem. Bd. II p. 69, 1971.

[30] Woodward, R. B., Pure and Appl. Chem. **33**, 145, 1973.

[31] Woodward, R. B., Synthetic vitamin B_{12}, in Vitamin B_{12}, Eds. B. Zagalak, W. Friedrich, Walter de Gruyter, Berlin 1979, pp. 37-87.

[32 a] Eschenmoser, A., Chemical synthesis of corrinoids: current problems and recent advances, In Vitamin B_{12}, Eds. B. Zagalak, W. Friedrich, Walter de Gruyter, Berlin 1979, pp. 89-117.

[32 b] Eschenmoser, A., Wintner, C., Science **196**, 1410, 1977.

[33] Patteuden, G., Specialist periodical report general and synthetic methods, Vol. 1., A review on the literature published during 1976. The Chemical Society: London, England 1978.

[34 a] Stevens, R. V., Recent studies on the synthesis of vitamin B_{12}, in Vitamin B_{12}, Eds. B. Zagalak, and W. Friedrich, Walter de Gruyter, Berlin 1979, pp. 119-135.

[34 b] Stevens, R. V., Studies on the synthesis of corrins and related ligands. An approach to the total synthesis of vitamin B_{12}, Proc. 24th National Organic Symposium, 21-35, 1975.

[34 c] Stevens, R. V., Chang, J. H., Lapalme, R., Schow, S., Schlageter, M. G, Shapiro, R., Weller, H. N., Studies on the synthesis of vitamin B_{12}. J. Am. Chem. Soc. **105**, 7719-7729, 1983.

[35] Inhoffen, H. H., New reactions of the chromophoric system of vitamin B_{12}, in Vitamin B_{12}, Eds. B. Zagalak, and W. Friedrich, Walter de Gruyter, Berlin 1979, pp. 137-139.

[36] Maugh, T. H., Vitamin B_{12}: after 25 years, the first synthesis, Science **179**, 266-267, 1973.

[37] Friedrich, W., and Bernhauer, K., Über den Abbau der Cobalamine mit Ce(III)-hydroxyd. 7. (D-Ribofuranozido)-adenin, ein Abbauprodukt des pseudo vitamin B_{12}, Chem. Ber. **89**, 2507-2512, 1956.

[38] Brink, N. G., Folkers, K., Vitamin B_{12}, XVIII. The degradation of vitamin B_{12} to 1-α-D-ribofuranosyl-5,6-dimethylbenzimidazole, J. Am. Chem. Soc. **74**, 2850-2858, 1952.

[39] Gant, D. E., Smith, Sir. Parker, L. F. J., Removal of nucleotide from vitamins B_{12}, proceedings of the Biochemical Society XXXIV, Biochem. J., p. 56, 1954.

[40] Friedrich, W., Grass, G., Bernhauer, K., and Zeller, P., Helv. Chim. Acta, **43**, 704, 1960.

[41] Friedrich, W., Chemische Partialsynthese kompletter Vitamin-B_{12}-Arten, ausgehend vom Cobyrinsäure-abcdeg-hexamid., 2. Europäisches Symposion, Hamburg 1961, Ed. Heinrich, H. C., Ferdinand Enke Verlag, Stuttgart 1962, pp. 8-28.

[42] Moskophidis, M., and Friedrich, W., Partialsynthese von β-[5,6-Dimethylbenzimidazolyl]-cobamid-cyanid, Z. Naturforsch., **23**, 752-753, 1968.

[43] Perlman, D., Barrett, J. M., and Jackson, P. W., Cobamides synthesized by Propionibacterium species, 2. Europäisches Symposion, Hamburg 1961, Ed. Heinrich, H. C., Ferdinand Enke Verlag, Stuttgart 1962, pp. 58-69.

[44] Pawełkiewicz, J., The methods of the preparation of corrinoids including the analytical methods, 2. Europäisches Symposion, Hamburg 1961, Ed. Heinrich, H.C., Ferdinand Enke Verlag, Stuttgart 1962, pp. 280-294.

[45] Pratt, J.M., Inorganic chemistry of vitamin B_{12}, Academic Press, London, New York 1972, p. 157.

[46] Friedrich, W., Vitamin B_{12} und verwandte Corrinoide, in Fermente, Hormone, Vitamine, Eds. Ammon, R., and Dirscherl, W., Vol. III/2, Georg Thieme Verlag, Stuttgart 1975, 164.

[47] Kon, S.K., and Pawełkiewicz, J., Biosynthesis of vitamin B_{12} analogues, Fourth International Congress of Biochemistry, Vol. XI, 115-149, 1958.

[48] Dolphin, D., Preparation of the reduced forms of vitamin B_{12} and of some analogs of the vitamin B_{12} coenzyme containing a cobalt-carbon bond, Methods in Enzymology, **18 C,** 34-52, 1971.

[49] Yamada, R., Shimizu, S., and Fukui, S., Preparation of solid B_{12r} by anaerobic photolysis of methylcobalamin, in Methods in Enzymology, **18,** 52-54, 1971.

[50] Lexa, D., and Saveant, I.M., Electrochemistry of vitamin B_{12}. Role of the Base-On/Base-Off reaction in the oxidoreduction mechanism of the B_{12}-B_{12s} system, J. Am. Chem. Soc. **98,** 2652-2659, 1976.

[51] Yamada, R., Shimizu, S., Fukui, S., Formation of B_{12s} by anaerobic photolysis of cobalt-alkylcobalamins. Biochim. Biophys. Acta **124,** 195, 1966.

[52] Ljungdahl, L., and Irion, E., Photolytic and reductive cleavage of Co-carboxymethylcobalamin, Biochemistry **5,** 1846-1850, 1966.

[53] Das, P.K., Hill, H.A.O., Pratt, J.M., Williams, R.J.P., The chemistry of vitamin B_{12} Part VIII. Controlled potential reduction of vitamin B_{12a}, J. Chem. Soc. **A,** 1261-1264, 1968.

[54] Beaven, G.R., and Johnson, A.W., The reduction of vitamin B_{12}, Nature **176,** 1264-1266, 1955.

[55] Zagalak, B., and Pawełkiewicz, J., Acta Biochim. Pol. **11,** 49-52, 1964.

[56] Pawełkiewicz, J., Enzymatic preparation of coenzyme B_{12} and some of its analogs, in Methods in Enzymology **18 C,** 99-101, 1971.

[57] Vitols, E., Walker, G.A., and Huennekens, F.M., Enzymatic conversion of vitamin B_{12s} to a cobamide coenzyme, α-(5,6-Dimethylbenzimidazolyl)deoxyadenosylcobamide (Adenosyl-B_{12}), J. Biol. Chem. **241,** 1455-1461, 1966.

[58] Hill, H.A.O., Pratt, J.M., and Williams, The corphyrins, J. Theor. Biol. **3,** 423-425, 1962.

[59] Dolphin, D., Johnson, A.W., Rodrigo, R., and Show, N., The vitamin B_{12} coenzyme, Pure Appl. Chem. **7,** 539-541, 1963.

[60] Schrauzer, G.N., Holland, R.J., Seck, J.A., J. Am. Chem. Soc. **93,** 1503, 1971.

[61] Schrauzer, G.N., Seck, J.A., Holland, R.J., Z. Naturforsch. **28 c,** 1, 1973.

[62] Schrauzer, G.N., Stadlbauer, E.A., Bioinorg. Chem. **4,** 185, 1975.

[63] Amess, J.A.L., Burman, J.F., Cullen, M.H., Mollin, D.L., The in vivo effects of nitrous oxide on human marrow – a morphological biochemical and cell cycle study, in Vitamin B_{12}, Eds. B. Zagalak, W. Friedrich, Walter de Gruyter, Berlin 1979, pp. 1037-1044.

[64] Amess, J.A.L., Burman, J.F., Mollin, D.L., Br. Med. J. **i,** 525, 1976.

[65] Metz, J., Kelly, A., Swett, V.C., Waxman, S., Herbert, V., Br. J. Haemat. **14,** 575-592, 1968.

[66] Wickramasinghe, S.N., Longland, J.E., Br. Med. J. **iii,** 148-150, 1974.

[67] Reed, B., Dinn, J., McCann, S., Wilson, P., O Sullivan, H., Weir, D.G., Scott, J.M., Alterations in mammalian cells induced by inactivation of vitamin B_{12} with nitrous oxide, in Vitamin B_{12}, Eds. B. Zagalak, W. Friedrich, Walter de Gruyter, Berlin, 1979, pp. 1061-1064.

[68 a] Deacon, M., Lumb, M., Muir, M., Perry, J., Chanarin, I., Studies on cobalamin and folate metabolism in rats exposed to nitrous oxide N_2O, in Vitamin B_{12}, Eds. B. Zagalak, W. Friedrich, Walter de Gruyter, Berlin 1979, pp. 1055-1060.

[68 b] McKenna, B., Weir, D.G., Scott, J.M., The induction of functional vitamin B_{12} deficiency in rats by exposure to nitrous oxide. Biochim. Biophys. Acta **628,** 314-321, 1980.

[68 c] Deacon, R., Chanarin, I., Perry, J., and Lumb, B., Impaired deoxyuridine utilization in the vitamin B_{12}-inactivated rat and its correction by folate analogs, Biochem. Biophys. Res. Common, **93,** 516-520, 1980.

[69] Amess, J.A.L., Burman, J.F., Nancekievill, D.G., Mollin, D.L., Lancet, **ii,** 339-342, 1978.

[70] McGing, P., Reed, B., Weir, D.G., Scott, J.M., Biochem. Biophys. Res. Commun. **82,** 540-546, 1978.

[71] Dinn, J.J., McCann, S., Wilson, P., Reed, B., Weir, D., Scott, J., Lancet **ii,** 1154, 1978.

[72] Foster, M.A., Dilworth, M.J., Woods, D.D., Nature. **201,** 39-42, 1964.

[73] Taylor, R.T., Hanna, M.L., Hutton, J.J., Arch. Biochem. Biophys. **165,** 787-795, 1974.

[74] Foster, M.A., B-Vitamine: Klinische und Physiologisch Chemische, Ed. H.F. von Kress and K.U. Blum, Stuttgart 1966, pp. 103-106.

[75] Hogenkamp, H.P.C., The photolysis of methylcobalamin, Biochemistry **5,** 417-422, 1966.

[76] Hogenkamp, H.P.C., J. Biol. Chem. **238,** 477-480, 1963.

[77] Johnson, A.W., Shaw, N., Wagner, F., A chemical synthesis of S-adenosyl-homocysteine and the vitamin B_{12}-coenzyme, Biochim. Biophys. Acta **72,** 107-110, 1963.

[78] Hogenkamp, H.P.C., The chemistry of cobalamins and related compounds, in Cobalamin-Biochemistry and Pathophysiology, Babior, B.M., Ed., John Wiley and Sons, New York 1974, pp. 21-73.

[79] Friedrich, W., Vitamin B_{12} und verwandte Corrinoide, in Fermente, Hormone, Vitamine, Eds. Ammon, R. and Dirscherl, W., Vol. III/2, Georg Thieme Verlag, Stuttgart 1975, pp. 111-115.

[80] Zagalak, B., Pawełkiewicz, J., Synthesis and properties of some analogues of the corrin coenzymes, Acta Biochim. Pol. **11,** 49-59, 1964.

[81] Zagalak, B., and Pawełkiewicz, J., Synthesis and properties of Co-adenine nucleoside analogues of coenzyme B_{12}, Acta Biochim. Pol. 219-228, 1965.

[82] Pratt, J.M., Chemistry of vitamin B_{12} part II. Photochemical reactions, J. Chem. Soc., 5154-5160, 1964.

[83] Johnson, A.W., Mervyn, L., Shaw, N., Smith, E.L., A partial synthesis of the vitamin B_{12} coenzyme and some of its analogues, J. Chem. Soc., 4146-4156, 1963.

[84] Dolphin, D., Johnson, A.W., Rodrigo, R., Reactions of the alkylcobalamins, J. Chem. Soc., 3186-3193, 1964.

[85] Friedrich, W., und Nordmeyer, J.P., Partialsynthese und einige Eigenschaften des Co-Methyl-cobalamins, Z. Naturforschg. **24 b,** 588-596, 1969.

[86] Barnett, R., Hogenkamp, H.P.C., and Abels, R.H., Reactions of the carbon-cobalt bond of alkylcobalamins., J. Biol. Chem. **241,** 1483-1486, 1966.

[87] Kikugawa, K., and Ichino, M., Tetrahedron Lett. **2,** 87-88, 1971.

[88] Schrauzer, G.N., and Deutsch, E., J. Am. Chem. Soc. **91,** 3341, 1960.

[89 a] Schrauzer, G.N., Grate J.H., Hashimoto, M., and Maihub, A., Vitamin B_{12}: current problems and recent advances, in Vitamin B_{12}, Eds. B. Zagalak and W. Friedrich, Walter de Gruyter, Berlin 1979, pp. 511-528.

[89 b] Schrauzer, G.N., and Hashimoto, M., Studies of vitamin B_{12} and related compounds: 49. Direct synthesis of alkylcobalamins and ω-carboxy-alkylcobalamins from vitamin B_{12} and aliphatic carboxylic acids under oxidizing + reducing conditions. J. Am. Chem. Soc. **101,** 4593-4601, 1979.

[90] Brady, R.O., Castanera, E.G., and Barker, H.A., The enzymatic synthesis of cobamide coenzymes., J. Biol. Chem. **237,** 2325-2332, 1962.

[91] Pawełkiewicz, J., Bartosinski, B., Walerych, W., Enzymic synthesis of corrin coenzymes., Ann. N. Y. Acad. Sci. **112,** 638-640, 1964.

[92] Bartosinksi, B., Enzymic synthesis of the coenzyme form of cobinamide (SB_{cp}) Bulletin De L'ACADEMIE Polonaise Des Sciences **10,** 189-194, 1962.

[93] Barker, H.A., Smyth, R.D., Weissbach, J., Toohey, J.I., Ladd, J.N., and Volcani, B.E., J. Biol. Chem. **235,** 480-484, 1961.

[94] Pratt, J.M., Inorganic Chemistry of vitamin B_{12}, Academic Press, London, New York 1972, pp. 252-255.

[95] Friedrich, W., Vitamin B_{12} und verwandte Corrinoide, in Fermente, Hormone, Vitamine, Eds. Ammon, R. and Dirscherl, W., Vol. III/2, Georg Thieme Verlag, Stuttgart 1975, pp. 125-129.

[96] Dolphin, D., and Johnson, A.W., The properties of cobalamins with sulphur containing ligands, J. Chem. Soc. **387,** 2174-2181, 1965.

[97] Peel, J.L., The catalysis of the auto-oxidation of 2-mercaptoethanol and other thiols by vitamin B_{12} derivatives. Biochem. J., **88,** 296-308, 1963.

[98] Hogenkamp, H.P.C., The chemistry of cobalamins and related compounds, in Cobalamin-Biochemistry and Pathophysiology,

Ed. Babior, B.M., John Wiley an Sons, New York 1974, pp.50-52.

[99 a] Aronovitch, J., and Grossowicz, Cobalamin catalyzed oxidation of sulfhydryl groups, Biochem. Biophys. Res. Comm. **8,** 416-420, 1962.

[99 b] Jacobsen, D.W., Troxell, L.S., Brown, K.L., catalysis of thiol oxidation by cobalamins and cobinamides: Reaction products and kinetics, Biochemistry **23,** 2017-2025, 1984.

[99 c] Balt, S., De Bolster, M.W.G., Van Garderen, C.J., Van Herk, A.M., Lammers, K.R. and Van der Velde, E.G., Effects of solvent and ionic medium on the kinetics of axial ligand substitution in vitamin B_{12}. Part III. Reactions of aquacobalamin and aquamethylcobaloxime with sulfur-coordinating ligands, Inorg. Chim. Acta **106,** 43-47, 1985.

[100] Firth, R.A., Hill, H.A.O., Mann, B.E., Thorp, R.G., and Williams, R.J.P., J. Chem. Soc. **(A),** 2419, 1968.

[101] Brodie, J.D., Proc. Nat. Acad. Sci. USA **62,** 461, 1969.

[102] Schrauzer, G.N., Lee, L.P., and Sibert, J.W., J. Am. Chem. Soc. **92,** 2997, 1970.

[103] Schrauzer, G.N., and Holland, R.J., J. Am. Chem. Soc. **93, 4060,** 1971.

[104] Schrauzer, P.G., Proc. Roy. Soc. A, **303,** 43, 1968.

[105] Atkins, M.P., Golding, B.T., Sellars, P.J., Model systems for adenosylcobalamin dependent enzymic reactions, in Vitamin B_{12}, Eds. B.Zagalak, W.Friedrich, Walter de Gruyter, Berlin 1979, pp.587-598.

[106] Chemaly, S., Pratt, J.M., J. Chem. Soc. Chem. Comm. 988-989, 1976.

[107 a] Dolphin, D., Banks, A.R., Cullen, W.R., Cutler, A.R., and Siverman, R.B., The mechanism of action of coenzyme B_{12}, in Vitamin B_{12}, Eds. B.Zagalak, W.Friedrich, Walter de Gruyter, Berlin 1979 ,pp.575-586.

[107 b] Dowd, P. and Trivedi, B.K., On the mechanizm of action of vitamin B_{12}. Model studies directed toward the hydrogen abstraction reaction, J. Org. Chem. **50,** 206-217, 1985.

[108] Dowd, P., Nonenzymic models for the enigmatic coenzyme B_{12} dependent carbon-skeleton rearrangements, in Vitamin B_{12}, Eds. B.Zagalak, W.Friedrich, Walter de Gruyter, Berlin 1979, pp.557-574.

[109 a] Dowd, P., Shapiro, M., Kang, K., J. Am. Chem. Soc. 97, 4754, 1975.

[109 b] Scott, A.I., Kang, J., Dowd, P., and Trivedi, B.K., Vitamin B_{12}: catalyst for a nonenzymic carbon-skeleton model rearrangement. Bioorg. Chem. **9,** 227-230, 1980.

[110] Bidlingmaier, G., Flohr, H., Kempe, M., Krebs, T., and Retey, J., Angew. Chem. **87,** 877, 1975.

[111] Flohr, H., Pannhorst, W., Retey, J., Angew. Chem. **88,** 613, 1976.

[112] Wood, J.M., Fanchiang, Y.T., Mechanisms for B_{12}-dependent methylation, in Vitamin B_{12}, Eds. B.Zagalak, and W.Friedrich, Walter de Gruyter, Berlin 1979, pp. 539-556.

[113] Wood, J.M., Kennedy, F.S., Rosen, C.G., Nature **220,** 173-175, 1968.

[114] Wood, J.M., Science **183,** 1049-1054, 1974.

[115] Ridley, W.P., Dizikes, L.J., Wood, J.M., Science **197,** 329-332, 1977.

[116] Wood, J.M., La Recherche **70,** 711-716, 1976.

[117] Wood, J.M., Naturwissenschaften **62,** 356-367, 1975.

[118] Brinkman, F.E., Bellama, J.M., Organometalloids, occurrence and fate in the environment. ACS Symposium Series **82,** 1978.

[119] DeSimone, R.E., Charbonneau, L., Smith, S.G., Penley, M.W., Wood, J.M., Hill, H.A.O., Pratt, J., Ridsdale, S., Williams, R.J.P., Biochim. Biophys. Acta **304,** 851-863, 1973.

[120] Wood, J.M., Fanchiang, Y.T., Ridley, W.P., Quarterly Reviews of Biophysics **9, 10,** 1979.

[121] Agnes, G., Hill, H.A.O., Pratt, J., Ridsdale, S., Kennedy, F.S., Williams, R.J.P., Biochim. Biophys. Acta **252,** 207-210, 1971.

[122] Scovell, W.H., J. Amer. Chem. Soc. **96,** 3451-3456, 1974.

[123] Dizkes, L.J., Ridley, W.P., Wood, J.M., J. Amer. Chem. Soc. **100,** 1010-1012, 1978.

[124] Espenson, J.H., Seelers, T.D., J. Amer. Chem. Soc. **96,** 94-97, 1974.

[125] Frick, T., Francia, M.D., Wood, J.M., Biochim. Biophys. Acta **428,** 808-818, 1976.

[126] Fritz, H.L., Espenson, J.H., Williams, D.A., Molander, G.A., J. Amer. Chem. Soc. **96,** 2378-2381, 1974.

[127] Agnes, G., Hill, H.A.O., Ridsdale, S., Williams, F.R., Williams, R.J.P., Chem. Communs. 850-851, 1971.

[128] Taylor, R.T., Hanna, M.L., Bio-inorg. Chem. **6,** 181-293, 1976.

[129] Fanchiang, Y.T., Ridley, W.P., Wood, J.M., Methylation of platinum complexes by

methylcobalamin, J. Amer. Chem. Soc. **101,** 1442-1447, 1979.
[130] Hensens, O.D., Hill, H.A.O., Thornton, J., Turner, A.M., Williams, R.J.P., Philos. Trans. R. Xoc. London, Series B. **273,** 353-356, 1976.
[131] Abley, P., Dockal, E.R., Halpern, J., J. Amer. Chem. Soc. **94,** 659-660, 1972.
[132] D Itri, F., Andren, A., Doherty, W., Wood, J.M., STAEP report on Mercury, National Research Council, Environmental Research Assessment Committee, Ed. A. King, U.S. National Academy of Sciences Monograph, 1977.
[133] McBride, B.C., Merilees, H., Cullen, W.R., Pickett, W., Organometals and Organometalloids, ACS Symposium Series **82,** Ed. Brinkman, F., Bellama, J.M., 94-116, 1978.
[134] Huber, F., Schmidt, U., Kirchmann, H., ACS Symposium Series **82,** 65-82, 1978.
[135] Chau, Y.K., Wong, P.T.S., ACS Symposium Series **82,** 39-54, 1978.
[136] Ruediger, H., Europ. J. Biochem. **21,** (2) 264-268, 1971.
[137] Fenrych, W., Schneider, Z., Pawełkiewicz, J., Eine einfache Methode zur Isolierung des Intrinsic Factors mittels B_{12}^{+}-Zelluose, 5. Jahreskongreß der Arbeitsgemeinschaft Klinische Pathologie und Klinische Chemie, Mai 1966.
[138] Olesen, H., Hippe, E., and Haber, E., Biochim. Biophys. Acta **243,** 66-74, 1971.
[139] Yamada, R.H., and Hogenkamp, H.P.C., J. Biol. Chem. **247,** 6266-6270, 1972.
[140] Allen, R.H., and Majerus, P.W., J. Biol. Chem. **247,** 7695-7701, 1972.
[141] Allen, R.H., and Mehlman, C.S., J. Biol. Chem. **248,** 3670-3680, 1973.
[142] Schlecht, M., and Muller, O., Z. Naturforsch. Teil C **28 c,** 351-352, 1973.
[143] Lien, E.L., Ellenbogen, L., Law, P.Y., and Wood, J.M., J. Biol. Chem. **249,** 890-894, 1974.
[144] Sato, K., Hiei, E., Shimizu, S., and Abeles, R.H., FEBS Lett. **85,** 73-76, 1972.
[145] Allen, R.H., and Majerus, P.W., J. Biol. Chem. **247,** 7702-7708, and 7709-7717, 1972.
[146] Allen, R.H., Burger, R.L., Mehlman, C.S., Majerus, P.W., Vitamin B_{12}. Methods in Enzymology, **34,** 305-314, 1974.
[147] Burger, R.L., and Allen, R.H., J. Biol. Chem. **249,** 7220-7227, 1974.
[148] Burger, R.L., Mehlman, C.S., and Allen, R.H., J. Biol. Chem. **250,** 7700-7706, 1975.
[149] Yamada, S., Itaya, H., Nakazawa, O., and Fukuda, M., Biochim. Biophys. Acta **496,** 571-575, 1977.
[150] Seligman, P.A., and Allen, R.H., J. Biol. Chem. **253,** 1766-1772, 1978.
[151] Toraya, T., Fukui, S., The synthesis of several immobilized derivatives of vitamin B_{12} coenzyme and their use as affinity adsorbents for a study of interactions of diol dehydrase with the coenzyme, J. Biol. Chem. **255,** 3520-3525, 1980.
[152] Jacobsen, D., and Montejano, Y., Affinity photo-release purification of cobalamin-binding proteins, in Vitamin B_{12}, Eds. B. Zagalak, W. Friedrich, Walter de Gruyter, Berlin 1979, pp. 949-954.
[153] Gräsbeck, R., Puutula, L., Aro, H., J. Clin. Lab. Invest. **113,** 64, 1970.
[154] Nexø, E., A new principle in biospecific affinity chromatography used for purification of cobalamin binding proteins. Biochim. Biophys. Acta **379,** 189-192, 1975.
[155] Kouvonen, I., Gräsbeck, R., Subunit structure of the pig ileal intrinsic factor receptor, in Vitamin B_{12}, Eds. B. Zagalak and W. Friedrich, Walter de Gruyter, Berlin 1979, pp. 909-916.
[156] Nexø, E., Spectral studies on hog intrinsic factor and hog non-intrinsic factor, in Vitamin B_{12}, Eds. B. Zagalak, and W. Friedrich, Walter de Gruyter, Berlin 1979, pp. 895-904.
[157 a] Halpern, J., Some aspects of organocobalt chemistry related to vitamin B_{12}, Ann. N. Y. Acad. Sci. **239,** 2-21, 1974.
[157 b] Blaser, H.U., Halpern, J., Reactions of vitamin B_{12} with organic halides, J. Am. Chem. Soc. **102,** 1684-1689, 1980.
[158 a] Lindemans, J., van Kapel, J., Abels, J., Purification of human transcobalamin II-cyanocobalamin by affinity chromatography using thermolabile immobilization of cyanocobalamin. Biochim. Biophys. Acta **579,** 40-51, 1979.
[158 b] van Kapel, J., Loef, B.G., Lindemans, J., Abels, J., An improved method for large scale purification of human holo-transcobalamin II, Biochim. Biophys. Acta **676,** 307-313, 1981.
[159] Moore, F.M., Willis, B.T.M., and Hodgkin, D.C., Nature **214,** 130, 1967.

3.2 Spectroscopy of corrinoids

3.2.1 Absorption spectra

The corrinoids are distinguished by beautiful and intensive colours varying from yellow or brown through red to purple or violet. This variation of the spectra – which makes the corrinoids distinguishable if in solution or on chromatograms – together with the high extinction coefficients aide in the elaboration of separation-and-assay methods [1] (see section 5.1) and provide useful techniques for kinetic and equilibrium studies [2, 3]. The light absorption is due to π-π transitions in the corrin ring of which chromophore consists of six conjugated double bonds [4–6]. The spectra are affected particularly by the metal in the corrin ring [7–9], the axial ligands [2, 3, 5, 10–18] and specific proteins [19–24]. The nucleotide moiety of cobamides and the deoxyadenosine (β-ligand) coordinated to cobalt in the coenzymatic forms of corrinoids contribute mainly to the absorption of light in the UV-region [25]. In Figs. 3-18 and 3-19 the spectra of two nucleotides isolated from cobamides are shown. In Figs. 3-8, 3-9, 3-10, 3-11 spectra of those forms of cobalamin and a cobamide are presented which are frequently identified by their spectra [16, 26–28].

Besides the naturally-occurring cobalt-containing- or cobalt-free corrinoids [7] a number of corrinoids have been obtained by introduction of other than cobalt metal ions into the macroring, e.g. ions of rhodium, iron, copper or zinc [7–9]. The spectra of the cobalt-free corrinoids and the corrinoids obtained by insertion of metal ions are presented in Figs. 3-12, 3-13, 3-14, 3-15, 3-16, 3-17. The extinction coefficients and other properties of these corrinoids are given in Table 3-1.

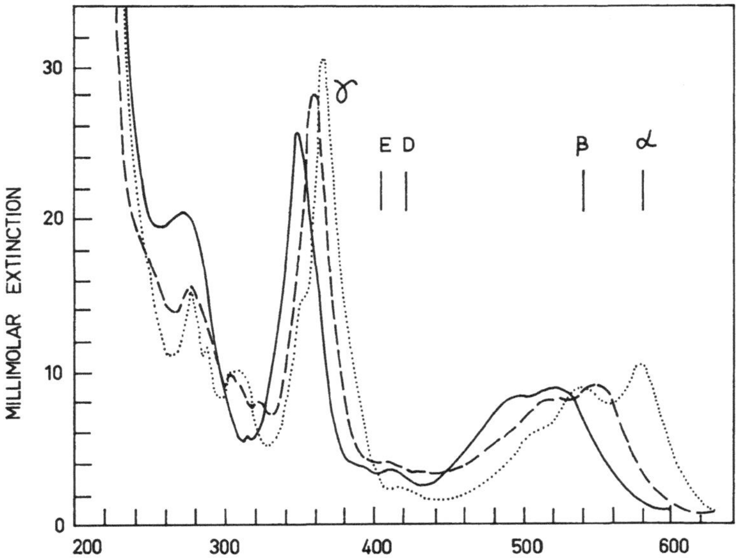

Fig. 3-8 Absorption spectra of cyanocobalamin (- - -), aquacobalamin (——), and dicyanocobalamin (· · · · ·). The letters α, β, D, E and γ refer to the bands of the dicyanocobalamin see also text under section 3.2.1 (from Heinrich [26])

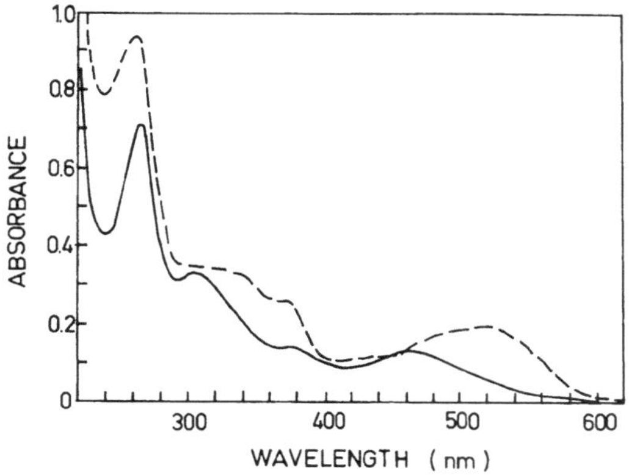

Fig. 3-9 Absorption specta of AdoCbl $2{,}64 \times 10^{-5}$ M and (Ade)AdoCba $1{,}51 \times 10^{-5}$ M in 0.02 M sodium acetate buffer, pH 6,7 (from Weissbach et al. [28])

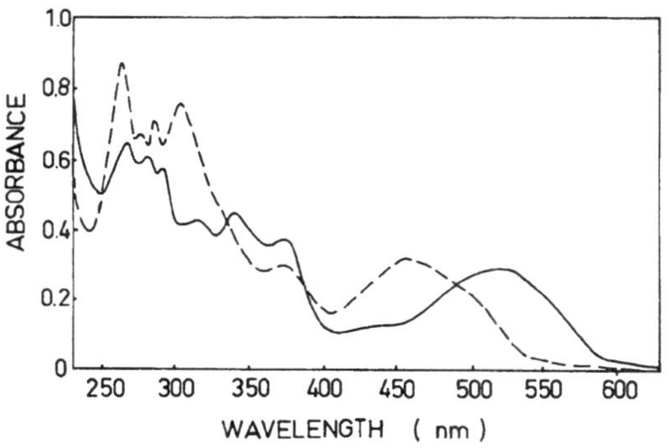

Fig. 3-10 Absorption spectra of methylcobalamin $3{,}58 \times 10^{-5}$ M, in water (——) and in 0.05 M HCl (- - -) (from Müller and Müller [16])

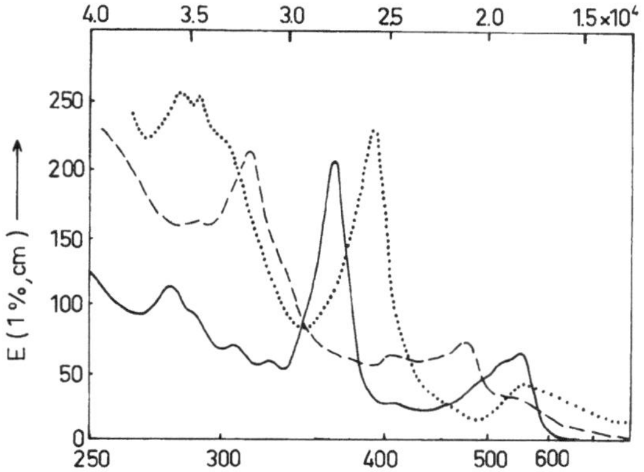

Fig. 3-11 Absorption spectra of cyanocob(III)alamin (——), cob(II)alamin (- - -) and cob(I)alamin (· · · · ·) (from Beaven and Johnson [27])

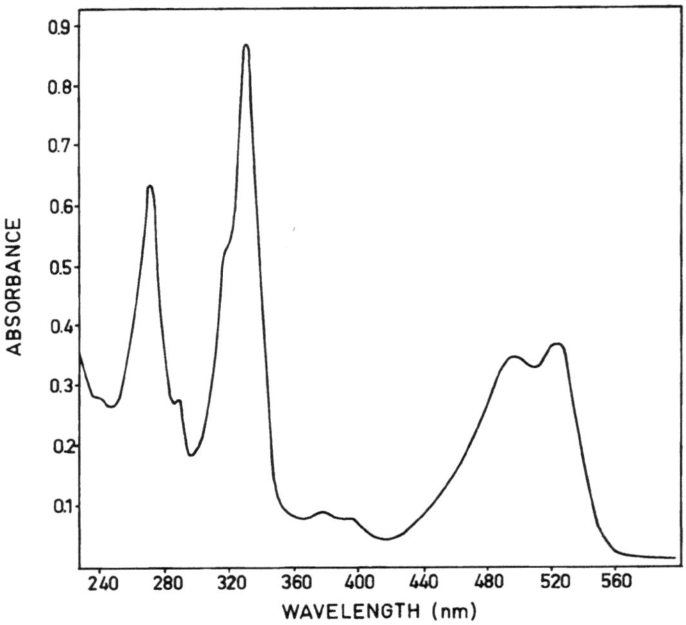

Fig. 3-12 Absorption spectrum of hydrogenobalamin, 2.71×10^{-5} M in water at pH 7 (from Koppenhagen and Pfiffner [7])

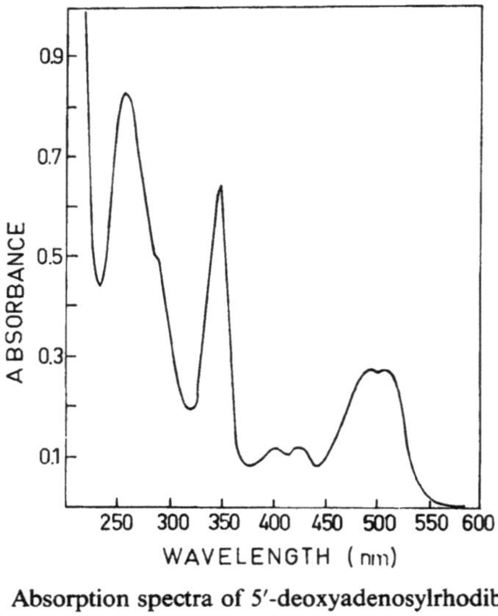

Fig. 3-13 Absorption spectra of 5′-deoxyadenosylrhodibalamin, 2.68×10^{-5} M, in water at pH 7 and 11 (from Koppenhagen et al. [8])

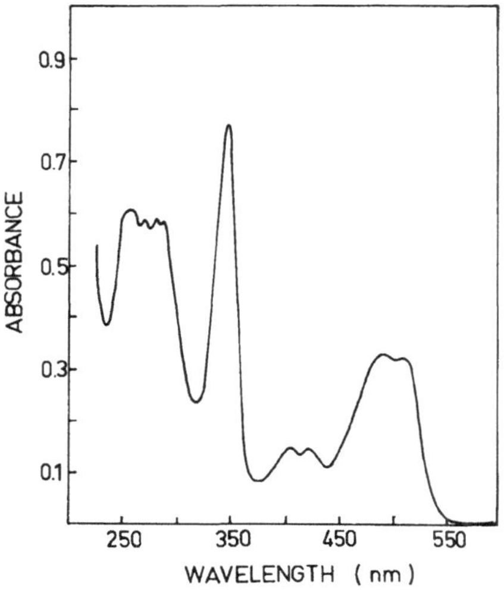

Fig. 3-14 Absorption spectrum of methylrhodibalamin, 3.07×10^{-5} M, in water at pH 2.5, 7 and 11 (from Koppenhagen et al. [8])

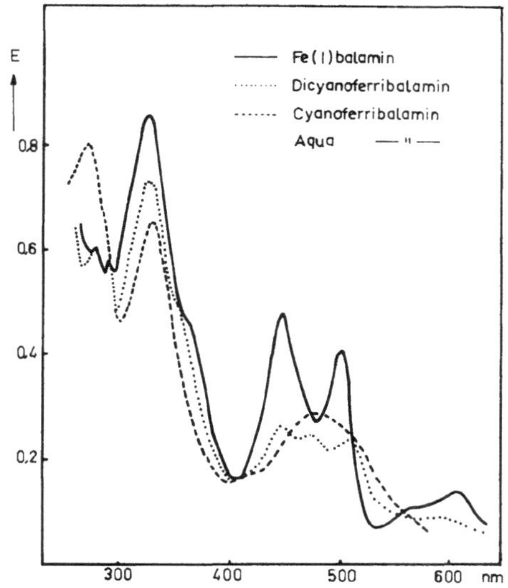

Fig. 3-15 Absorption spectra of various forms of ferribalamin (from Bieganowski et al. [9])

The axial ligands coordinated to the metal are readily substituted by other ligands, in particular by an OH ion or a water molecule in aqueous solution, which brings about a drastic change in the spectrum. Temperature [29-32], pH [15-18], light [33] and the concentration of the potential ligand [3, 34] are the main factors effecting the change of the spectra of corrinoids. This feature is applied for monitoring ligand substitution (Tables 3-6, 3-7) [2, 3, 11]. The coenzymatic forms of cobalamin and other cobamides change their colour from red to yellow if acidified [10-13, 16-18] (Table 3-7). This is

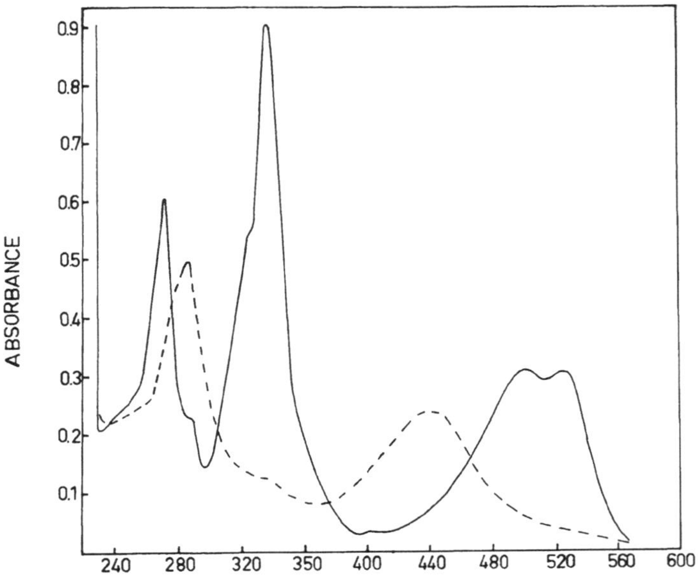

Fig. 3-16 Absorption spectra of zincobalamin, 2.94×10^{-5} M in water at pH 7 or 0.01 M NaOH (——) and zincobalamin 2.94×10^{-5} M in water at pH 7 after exposure to light for 1 min (- - -) (from Koppenhagen and Pfiffner [7])

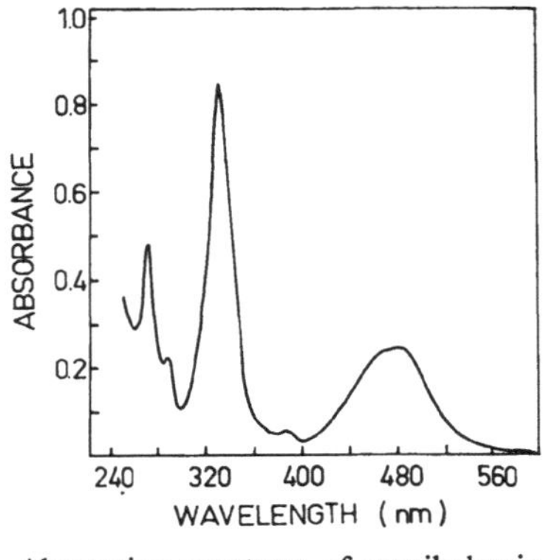

Fig. 3-17 Absorption spectrum of cupribalamin, 2.76×10^{-5} M, in water at pH 7 or in 0.01 M NaOH (from Koppenhagen and Pfiffner [7])

due to reversible substitution of the α-ligand (the benzimidazole base or adenine) by the water molecule. Cobinamides, which do not have the nucleotide moiety such as Ado-Cbi, AdoCbi-P, AdoCbi-GDP (spectra not shown here, but see the absorption bands of these corrinoids in Table 3-1) remain yellow even at alkaline pH [11, 12]. This property is conveniently utilized for rapid differentiation of the coenzymatic forms of cobamides from those of cobinamides.

The bands observed in the spectra of corrinoids are grouped in the following four regions [26] (Fig. 3-8):

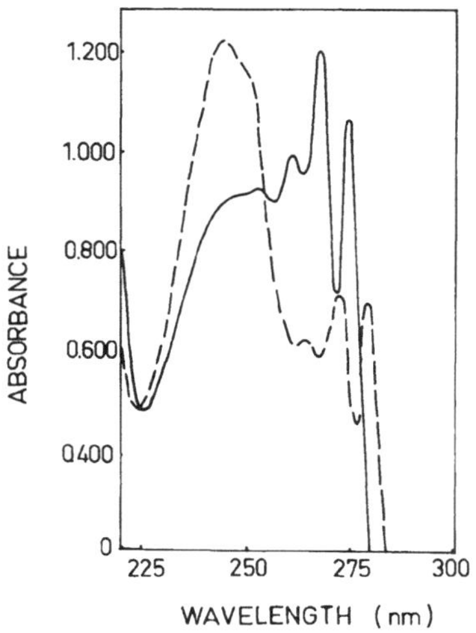

Fig. 3-18 Absorption spectra of benzimidazolyl nucleotide from benzimidazolylcobamide at pH 1 (——) and pH 12 (- - -) (from Bauer, K. F. [25])

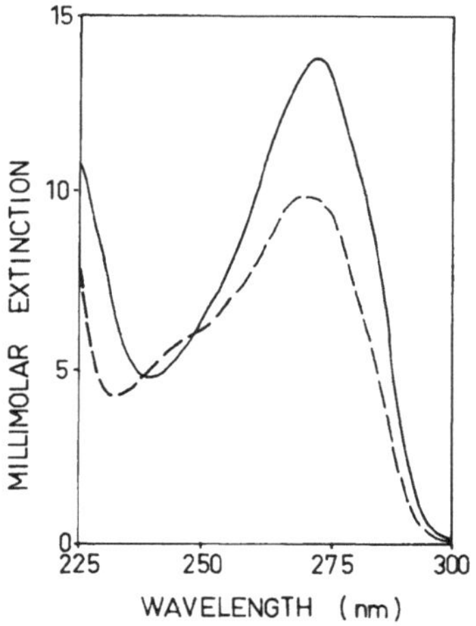

Fig. 3-19 Absorption spectra of nucleoside from pseudovitamin B_{12} at pH 1 (——) and pH 12 (- - -) (from Bauer, K. F. [25])

I $\alpha\beta$ bands. A series of low to medium intensity bands in the 420–600 nm region. The lowest and next lowest in energy are termed the α and β bands respectively.
II D and E bands. Two bands of low intensity in the 390–420 region
III γ band. The most intense band in the 350–370 nm region
IV δ bands. Bands of low to medium intensity in the 300–330 nm region.

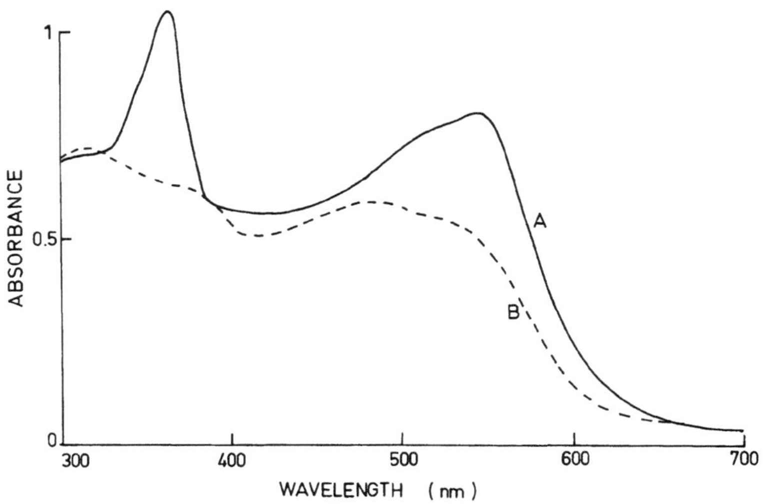

Fig. 3-20 Reflection spectra of crystalline cyanocobalamin (A) and crystalline adenosylcobalamin (B). The crystals were ground with 5-10 volumes of synthetic zeolite as inert diluent (from Hill et al. [6])

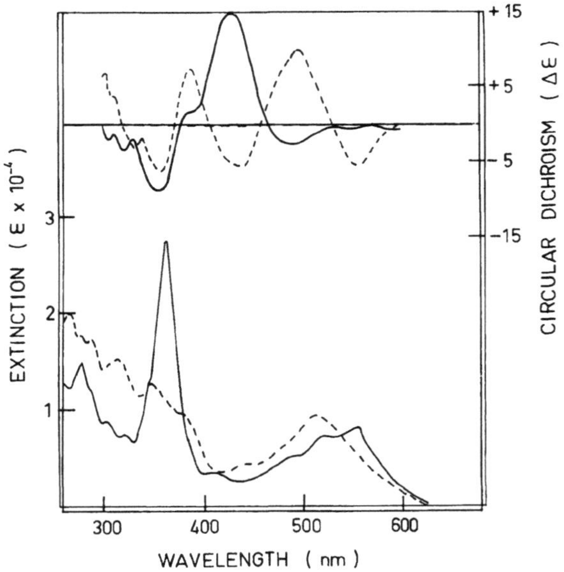

Fig. 3-21 Absorption spectra and circular dichroism of cyanocobalamin (—) and ethylcobalamin (- - -). The solvent was water, room temperature (from Firth et al. [31])

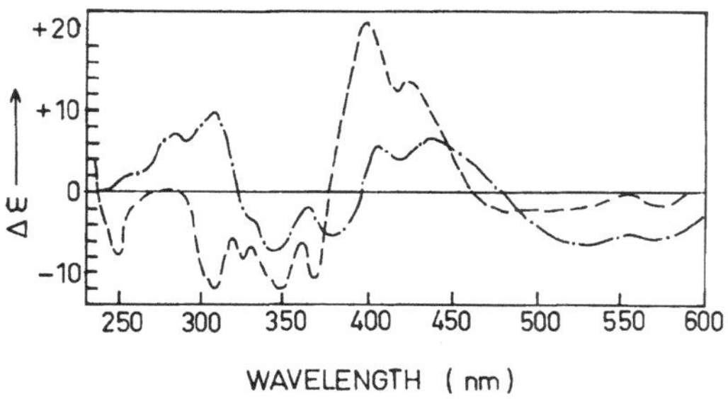

Fig. 3-22 Circular dichroism spectra of dicyanocobalamin (A) and (13-epi) dicyanocobalamin (B) in 0.1 M KCN. These two compounds have similar properties and produce indistinguishable visible absorption spectra. Their CD spectra differ significantly, however. For properties of 13-epicobalamin see Table 3-1 (from Bonnet et al. [35])

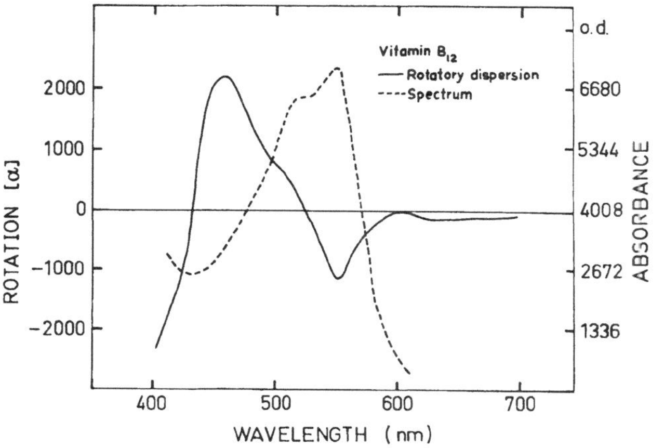

Fig. 3-23 Rotatory dispersion curve (B) and the spectrum (A) of cyanocobalamin. The vitamin B_{12}, for rotatory dispersion measurements, was in 0.05% solution and for the spectrum in 0.01% solution. Ligands at the α and β coordination positions as well as reduction of cobalt remarkably affect the shape of the rotatory dispersion curve (Reprinted with permission from Tetrahedron **13**, Eichhorn, G. L., Copyright 1961, Pergamon Press Ltd)

3.2.2 Other spectra

Various spectroscopical techniques are applied for investigating the structure and function of corrinoids. Circular dichroism- and rotatory dispersion spectroscopy produce information regarding conformational changes obtained by chemical alterations in the corrin ring [31, 35, 36] (Figs. 3-21, 3-22, and 3-23). IR spectroscopy is applied for studies of transeffects of ligands [37] (Fig. 3-24). Reduced cobalt in cob(II)alamin produces distinct EPR spectra of the "base-on" and "base-off" forms (Fig. 3-25) [21, 38-41]. Spin labelled cob(III)alamin permits studies of protein-vitamin B_{12} interactions [39] (see nitroxalkyl-Cbl in Table 3-1 pos. 16). A great deal of work has been done with the use of high resolution ^{13}C and ^{1}H NMR spectroscopy to elucidate the positions of methyl groups in intermediates on the path of the corrin

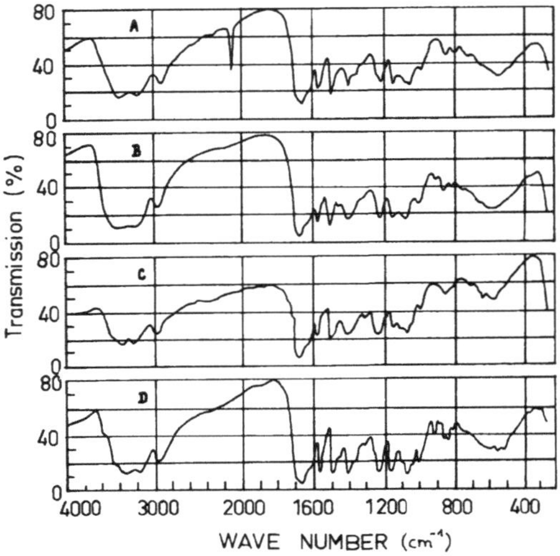

Fig. 3-24 Infrared spectra of cyanocobalamin (A), hydroxocobalamin (B), adenosylcobalamin (C) and methylcobalamin (D) (from Hogenkamp et al. [37])

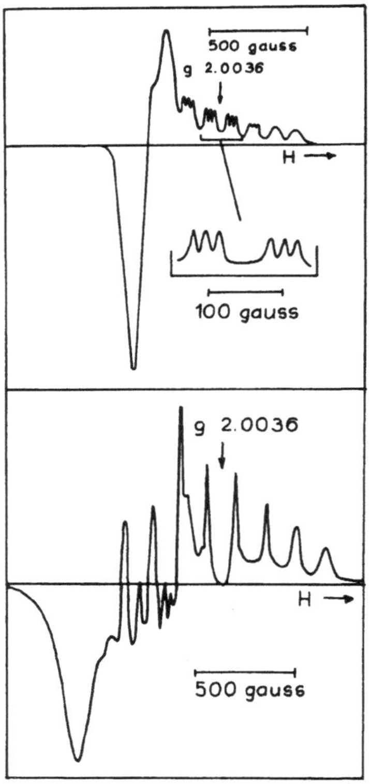

Fig. 3-25 The EPR spectra of "base-on" cob(II)alamin (A) and "base-off" cob(II)alamin (B) in propane-1,2-diol-water 1:4 at pH 7 and pH 0 respectively. The spectra were taken at 100 K using X-band irradiation. The cob(II)alamin was obtained by reduction of cob(III)alamin with H_2/Pt. The substitution of the N-14 ligand atom from 5,6-dimethylbenzimidazole in the "base-on" form by oxygen atom from water in the "base-off" form cause significant changes of the spectrum. The divalent cobalt in cob(II)alamins being paramagnetic produce distinct EPR spectra only. The diamagnetic forms of vitamin B_{12}: cob(III)alamin and cob(I)alamin are not suitable for this purpose. Recently a spin-labelled compound was linked to cobalt in cob(III)alamin. See nitrox-alkyl-cobalamin in Table 3-1 (from Cockle et al. [38])

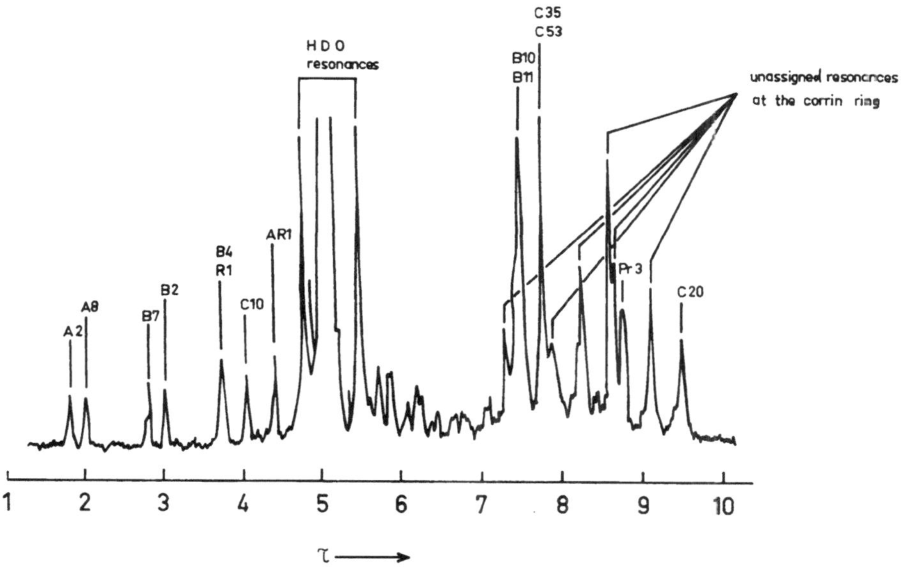

Fig. 3-26 ^1H NMR spectrum of adenosylcobalamin in D_2O at 220 MHz and 18 °C. The letters A, B, C, Pr, AR, R refer to carbon atoms in adenine, benzimidazole, corrin ring, propanolamine, ribose in adenosine and ribose of the nucleotide moiety respectively. The carbon atoms are numbered as indicated in Fig. 3-2 (The spectrum and supplementing data are from Hill et al. [45])

ring biosynthesis (see Chapter 4). The resonance positions of hydrogen bound carbon atoms in the ^1H NMR spectrum have been assigned [42-45] (Fig. 3-26). This also permits studies on the effects of temperature, pH, various ligands and other factors on subtle conformational changes.

References

[1] Shimizu, T., Narahane, H., Kamikubo, T., Application of difference spectra of corrinoids to their determination, Analyt. Biochem. **28**, 85-91, 1969.

[2] Hayward, G.C., Hill, H.A.O., Pratt, J.M., Vanston, N.S., and Williams, R.J.P., The chemistry of vitamin B_{12} part IV. The thermodynamic trans effect, J. Chem. Soc. 6485-6493, 1965.

[3] Pratt, J.M., Inorganic chemistry of vitamin B_{12}, Academic Press, London, New York 1972, pp. 44-67.

[4] Bonnet, R., Cannon, J.R., Clark, V.M., Johnson, A.W., Parker, L.F.J., Smith, E.L., and Sir Todd, A., Chemistry of the vitamin B_{12} group. Part V. The structure of the chromorphic grouping, J. Chem. Soc. **227**, 1158-1168, 1957.

[5] Friedrich, W., Fermente, Hormone, vitamine, Vol III/2, Vitamin B_{12} und Verwandte Corrinoide, Eds Ammon, R., and Dirscherl, W., Georg Thieme Verlag, Stuttgart 1975, pp. 46-78.

[6] Hill, J.A., Pratt, J.M., and Williams, R.J.P., The chemistry of vitamin B_{12}. Part I. The valency and spectrum of the coenzyme, J. Chem. Soc. 5149-5153, 1964.

[7] Koppenhagen, V.B., and Pfiffner, J.J., α-(5,6-Dimethylbenzimidazolyl) hydrogenobamide and its copper and zinc analogues, J. Biol. Chem. **246**, 3075-3077, 1971.

[8] Kooppenhagen, V.B., Elsenhans, B., Wagner, F., Methylrhodibalamin and 5'-deoxyadenosylrhodibalamin, the rhodium analogues of methylcobalamin and cobalamin coenzyme, J. Biol. Chem. **249**, 6532-6540, 1974.

[9] Bieganowski, R., and Friedrich, W., in Vitamin

B_{12}, Eds. B. Zagalak and W. Friedrich, Walter de Gruyter, Berlin 1979, pp. 647-648.

[10] Pawelkiewicz, J., Bartosinski, B., and Walerych, W., Enzymic synthesis of light-sensitive cobalto II-corrins, Bulletin de L'Academie Polonaise Des Sciences, **8,** 131-142, 1961.

[11] Bartosinski, B., Pawelkiewicz, J., Colorimetric method of determination of synthetase activity of the coenzyme form of cobinamide, Bulletin De L'Academie Polonaise Des Sciences, **10,** 121-124, 1962.

[12] Bartosinski, B., Enzymic synthesis of the coenzyme form of cobinamide (SB_{12p}), Bulletin De L'Academie Polonaise Des Sciences **10,** 189-194, 1962.

[13] Barker, H. A., Smyth, R. D., Weissbach, H., Petersen, A., Toohey, J. I., Ladd, J. N., Volcani, B. E., and Wilson, R. M., Assay purification and properties of the adenylcobamide coenzyme, J. Biol. Chem. **235,** 181-190, 1960.

[14] Barker, H. A., Chemistry and biology of the vitamin-B_{12}-coenzymes, 2. Europäisches Symposion Hamburg 1961, Ed. Heinrich, H. C., Ferdinand Enke Verlag, Stuttgart 1962, pp. 82-105.

[15] Ladd, J. N., Hogenkamp, H. P. C., Barker, H. A., Structure of cobamide coenzymes influence of pH on absorption spectra and ionophoretic mobilition, J. Biol. Chem. **236,** 2114-2118, 1961.

[16] Müller, O., Müller, G., Synthesen auf dem Vitamin B_{12}, XIV. Über die Synthese von Corrinoiden mit Kobalt-Kohlenstoff-Bindung, Biochem. Z. **336,** 299-313, 1962.

[17] Smith, E. L., Mervyn, L., Muggleton, P. W., Johnson, A. W., and Shaw, N., Ann. N. Y. Acad. Sci. **112,** 565, 1964.

[18] Hogenkamp, H. P. C., Rush, J. E., and Swenson, C. A., J. Biol. Chem. **240,** 3641, 1965.

[19] Schneider, Z., Larsen, E. G., Jacobson, G., Johnson, B. C., Purification and properties of glycerol dehydrase, J. Biol. Chem. **245,** 3388-3396, 1970.

[20] Babior, B. M., and Kai Li Ting. The binding of cobamides to ethanolamine deaminase, Biochemistry **8,** 154-160, 1969.

[21] Hamilton, J. A., Yamada, R., Blakleay, R. L., Hogenkamp, H. P. C., Looney, F. D., and Winfield, M. E., Cobamides and ribonucleotide reduction, VII. Cob(II)alamin as a sensitive probe for active center of ribonucleotide reductase, Biochemistry **10,** 347-352, 1971.

[22] Weissbach, H., and Taylor, R., Fed. Proc. **25,** 1649, 1966.

[23] Kaplan, B. H., and Stadtman, E. R., J. Biol. Chem. **243,** 1794, 1968.

[24] The spectra of the B_{12}-binding protein complexes are shown in Chapter 8.

[25] Bauer, K. F., Medizinische Grundlagenforschung, Vol. 2, Ed. K. F., Bauer, Thieme, Stuttgart 1959, p. 661.

[26] Heinrich, H. C., Pharm. Ztg. **108,** 1319-1322, 1963.

[27] Beaven, G. R., and Johnson, E. A., Nature **176,** 1264-1268, 1955.

[28] Weissbach, H., Toohey, J. I., Barker, H. A., Proc. Nat. Acad. Sci. **45,** 521, 1959.

[29] Schneider, Z., Friedmann, H. C., Studies on the enzymatic dephoshorylation of vitamin B_{12} 5'-phosphate, Arch. Biochem. Biophys. **152,** 488-495, 1972.

[30] Fox, J., Banninger, R., Draper, R. D., and Ingraham, L. L., Arch. Biochem. Biophys. **125,** 1022, 1968.

[31] Firth, R. A., Hill, H. A. O., Pratt, J. M., Williams, R. J. P., and Jackson, W. R., The circular dichroism and absorption spectra of some vitamin B_{12} derivatives, Biochemistry **6,** 2178-2188, 1967.

[32] Firth, R. A., Hill, H. A. O., Mann, B. E., Pratt, J. M., Thorp, R. G., and Williams, R. J. P., J. Chem. Soc. **A,** 2419, 1968.

[33] Pailes, W. H., and Hogenkamp, H. P. C., The photolability of Co-alkyl-cobinamides, Biochemistry **7,** 4160-4166, 1968.

[34] Hill, H. A. O., Pratt, J. M., and Williams, R. J. P., The chemistry of vitamin B_{12}, Chemistry in Britain **5,** 156-161, 1969.

[35] Bonnett, R., Codfrey, J. M., and Math, V. B., Circular dichroism of some vitamin B_{12} derivatives, J. Chem. Soc. Perkin. I, 252-257, 1973.

[36] Eichhorn, G. L., Tetrahedron **13,** 213, 1961.

[37] Hogenkamp, H. P. C., Rush, M., and Swenson, R., J. Biol. Chem. **240,** 3642-3646, 1965.

[38] Cockle, S. A., Hill, H. A. O., Pratt, J. M., Williams, R. J. P., Electron paramagnetic resonance spectra of some vitamin B_{12} derivatives, Biochim, Biophys. Acta **177,** 686-688, 1969.

[38b] Jorin, E., Schweiger, A., Gunthard, H. H., Single-crystal EPR of the oxygen-17-enriched dioxygen adduct of vit. B_{12}. Reversible oxygen bonding, electronic and geometric structure and molecular dynamics, J. Am. Chem. Soc. **105,** 4277-4286, 1983.

[39] Buckman, T., Kennedy, F. S., and Wood, J. M., Spin labeling of vitamin B_{12}, Biochemistry **8,** 4437-4442, 1969.

[40] Hogenkamp, H. P. C., and Parker, H. A., Mason, H. S., An electron-spin resonance study of

coenzyme B_{12}, Biochemistry and Biophysics **100**, 353-359, 1963.

[41] Schrauzer, G. N., and Lee Lian Pin, Cobaloximes (II) and vitamin B_{12} as oxygen carriers. Evidence for monomeric and dimeric peroxides and superoxides, J. Am. Chem. Soc. **92**, 1551-1557, 1970.

[42] Hill, H. A. O., Pratt, J. M., and Williams, R. J. P., The chemistry of vitamin B_{12}, Part III. The proton magnetic resonance spectra of some cobalamins, J. Chem. Soc. Part 2, 2859-2865, 1965.

[43] Broide, J. D., and Poe, M., Proton magnetic resonance of vitamin B_{12} derivatives functioning of B_{12} coenzymes. Biochemistry **11**, 2534-2542, 1972.

[44] Brodie, J. D., and Poe, M., Proton magnetic resonance studies of vitamin B_{12} proton magnetic resonance spectra of some cobalamins and cobinamides, Biochemistry **10**, 914-921, 1971.

[45] Hill, H. A. O., Pratt J. M., and Williams, R. J. P., Identification and investigation of cobalamins and cobamide coenzymes by nuclear magnetic resonance and electron paramagnetic resonance spectroscopy. Methods in Enzymology **18 C**, 5-32, 1971.

3.3 Basic data of selected corrinoids

The number of corrinoids isolated from natural sources or obtained by guided biosynthesis or chemical alterations is tremendous. Many of them obtained for a particular experiment have not found wider practical applications nor have they been sufficiently characterized. A limited number of corrinoids of practical- or biological significance are characterized in Tables 3-1 to 3-7.

Table 3-1 Basic data of selected cobamides and cobinamides (selected by Z. Schneider)

Systematic name abbreviation	Common names and abbreviatons	Structure[a] and formula[b]	Absorbancy λ_{max} (nm) in parentheses $10^3 \cdot \varepsilon$	Occurrence and biological activity	General remarks
No 1					
Coα-[α-(5,6-dimethylbenzimidazolyl)]-(Coβ-cyano) cobamide CNCbl	cyanocobalamin vitamin B_{12}, Factor II, anti pernicious anaemia factor (A.P.A.F.); extrinsic factor, animal protein factor (A.P.F.); Lactobacillus lactis Dorner factor (L.L.D.F.)	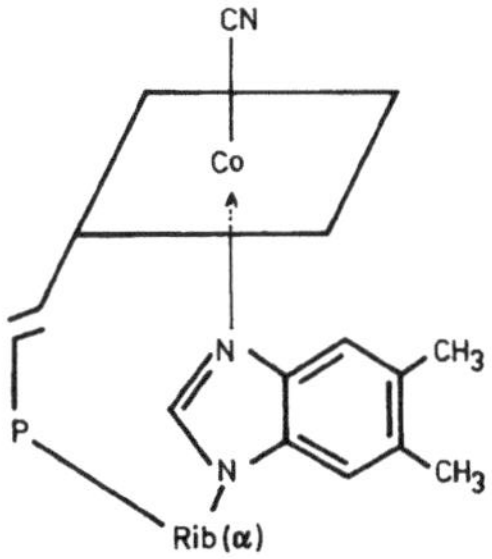 $C_{63}H_{89}O_{14}N_{14}PCo$ M_r 1355.5	278 (15.5) 361 (28.06) 551 (8.74) 361 $E^{1\%}_{1\,cm}=207$ in water [1]	Artificial product obtained from other cobalamins by addition of KCN (see Chapter 6). The absorption by human and animal digestion system is somewhat worse than OH-Cbl. The organisms convert it fast to other forms mainly aqCbl, CH_3-Cbl and AdoCbl [2]. Active in man, animals and microorganisms. Not active in enzyme systems.	Cherry crystaline substance. Heated under reduced pressure at 100 °C for 30 minutes loses 14% of water. No melting point; decomposition at 300 °C. Stable as solid at room temperature for several years. Soluble in water (1,2 g/100 ml), alcohols and phenols. Unsoluble in aceton, ethers, benzene and chloroform. In presence of light is converted into aqCbl; prolonged exposure causes irreversible destruction. In solutions at pH 4–6 may be kept for several years. Solutions may be heated shortly at 120 °C for sterilization. In presence of diluted mineral acids or on ion exchange columns deamidation at the corrin ring occurs. Unstable in strong alkalines, chlorine, $KMnO_4$, H_2O_2, thiamin and thiols. In presence of ascorbic acid complete inactivation at 65 °C, pH 5 with the release of Co^{II} ions [3, 45–49]. For spectrum see Figs. 3-8 and 3-11

[a] Co is for the corrin ring with cobalt and propanolamine attached. For full structure see Fig. 3-1
[b] The formula includes the hydrogen at the phosphate

No 2 Coα-[α-(5,6-dimethylbenzimdazolyl)]-(Coβ-aquo)cobamide aqCbl	aquacobalamin vitamin B_{12}, anti pernicious anaemia factor (A.P.A.F.) extrinsic factor, animal protein factor (A.P.F.) Lactobacillus lactis Dorner factor (L. L. D. F.)	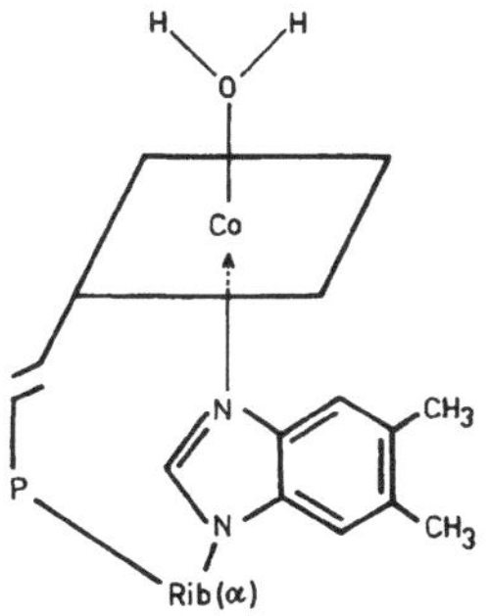 $C_{62}H_{91}O_{15}N_{13}PCo$ M_r 1347.5	274(20.6) 317 (6.1) 351(26.5) 411 (3.7) 499 (8.1) 527 (8.5) in water [6]	Produced by microorganisms. Occurs in man, animals, faeces, sewage sludge, soil and plant roots [10] (Chapter 6).	aq-Cbl and OH-Cbl are dark red crystalline substances containing various amounts of water, solubility similar to CN-Cbl [3, 4]. Much less resistant than CN-Cbl to high temperature and chemical agents. aqCbl and OH-Cbl are interconvertible depending on pH [14, 15] (Table 3-6). $aqCbl \underset{H^+}{\overset{(+)\ OH^-}{\rightleftharpoons}} OH\text{-}Cbl$ For spectrum see Fig. 3-8.
No 3 Coα-[α-(5,6-dimethylbenzimidazolyl)]-(Coβ-hydroxo)cobamide OH-Cbl	hydroxocobalamin vitamin B_{12}, Factor II, anti pernicious anaemia factor (A. P. A. F.) extrinsic factor, animal protein factor (A. P. F.) Lactobacillus lactis Dorner factor (L. L. D. F.) B_{12a}; B_{12b}, B_{12d}.	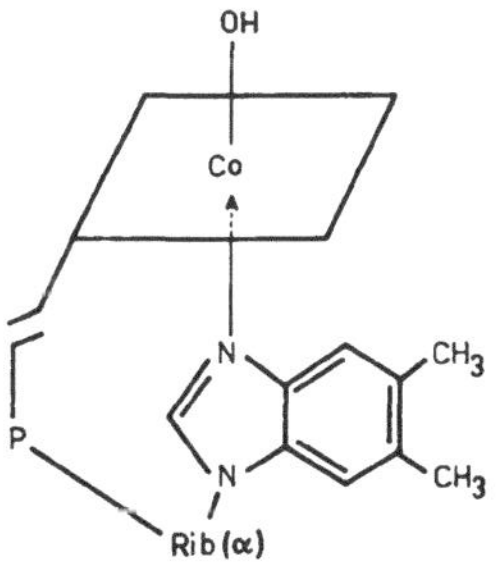 $C_{62}H_{90}O_{15}N_{13}PCo$ M_r 1346.5	279(19.0) 325(11.4) 359(20.6) 421 (4.2) 516 (8.9) 537 (9.5) in water [6]	Is absorbed by human and animal guts mucosa [11]. Active in man, animals and microorganisms [12]. Not active in enzyme systems in vitro [13].	

Table 3-1 continued

Systematic name abbreviation	Common names and abbreviatons	Structure[a] and formula[b]	Absorbancy λ_{max} (nm) in parentheses $10^3 \cdot \varepsilon$	Occurrence and biological activity	General remarks
No 4					
Coα-[α-(5,6-dimethylbenzimidazolyl)]-(Coβ-adenosyl)cobamide AdoCbl	vitamin B_{12} coenzyme, coenzyme B_{12}, DBCC, DBC, DMBC, Co B_{12}	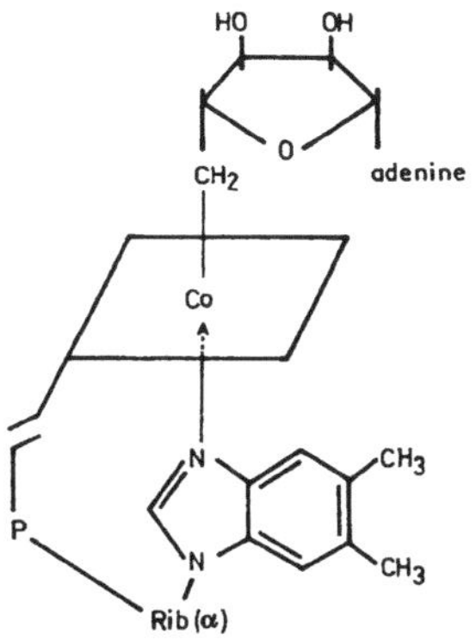 $C_{72}H_{101}O_{17}N_{18}PCo$ M_r 1579.6	233(26.2) 260(34.7) 288(18) 315(13.2) 340(12.3) 375(10.9) 522 (8.0) 0.01 phosphat pH 6,7 262(43.5) 285(25.5) pH 1	Present in man, animals, and microorganisms [10]. Obtained by chemical and enzymatic synthesis from CN-Cbl or aqCbl [16-22] (see 3.1.4.2 and Chapter 4). Absorbed by man somewhat less than OH-Cbl. Active in man, animals and in microorganisms [23, 24]. Cofactor for a number of enzymes acting in hydrogen transfer in bacteria, higher plants, animals and man [12, 34] (see Chapter 7).	Dark-red needles with variable amounts of water. Solubility similar to CN-Cbl. Less stable than CN-Cbl at high temperatures and in presence of mineral acids. Decomposed by light in presence of air to aqCbl^{III} and in absence of oxygen to Cbl^{II}. In dry state is less photolabile and may be stored for several years. Solutions are stable in the dark at pH 4-6. The addition of a few drops of conc. HCl to AdoCbl solution causes a change of color from red to yellow. Light decomposed compound will remain red or pink after acidification. By this simple test one can distinguish AdoCbl from aqCbl, OH-Cbl or CN-Cbl. Excess of KCN will convert it to purple(CN)CN-Cbl. In this form its concentration can be estimated spectrophotometrically. For estimation see (CN)CN-Cbl. For other properties see CN-Cbl [25-34]. Spectrum Fig. 3-9, pk_a value Table 3-7.

No 5 Coα-[α-(5,6-dimethylbenzimidazolyl)]-(Coβ-methyl)cobamide MeCbl	vitamin B_{12}, methylcobalamin	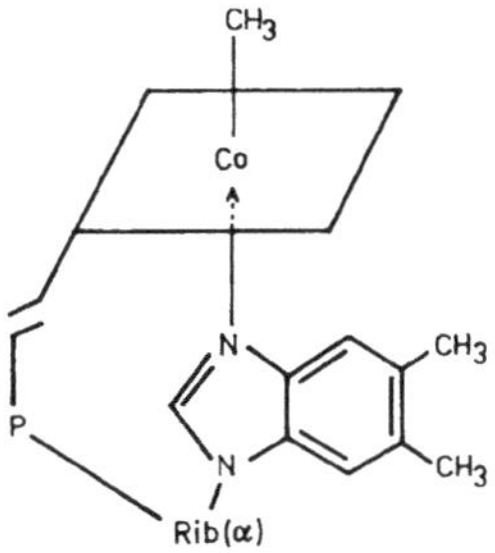 $C_{63}H_{92}O_{14}N_{13}PCo$ M_r 1344.5	340(12.7) 377(10.5) 528 (7.9) base not protonated	Present in man and animals [35]. In microorganisms present in various amounts depending on species and growth conditions [12]. Obtained by chemical synthesis from Cbl^I and CH_3I [16, 36] (see section 3.1. 4.2). Absorbed by man and animals somewhat less than aqCbl. Active in man, animals and microorganisms. Cofactor for enzymes involved in methyl group transfer in man, animals and microorganisms [37–42] (see Chapter 7).	Crystalline substance with variable amounts of water. Solubility like CN-Cbl. Somewhat less stable than CN-Cbl. MeCbl in solution is fast decomposed by light under aerobic conditions to aqCbl. Under anaerobic conditions light produces a methyl radical which may bind to thiols and 1,4-benzoquinone [36]. The cobalt coordinated methyl group may be transferred by chemical means to Hg, Au, Pt and Pd salts [43–48] (see also section 3.1.4.5.). Excess of KCN will convert MeCbl to purple(CN)CN-Cbl in presence of light and air. See also AdoCbl and CN-Cbl. For spectrum see Fig. 3–10, pk_a value Table 3–7.
No 6 Coα-[α-(5,6-dimethylbenzimidazolyl)]-(Coβ-ethyl)cobamide EtCbl	ethylcobalamin	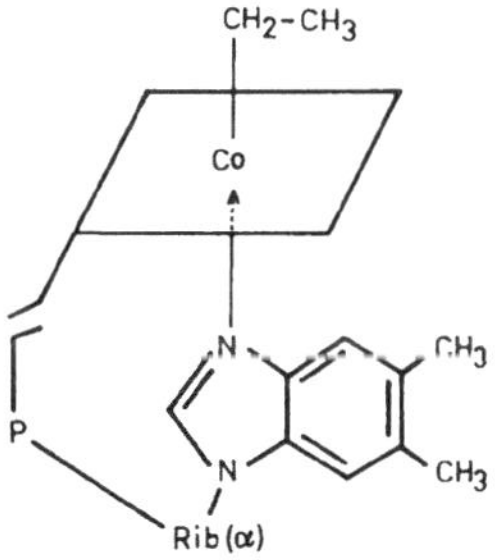 $C_{64}H_{94}O_{14}N_{13}PCo$ M_r 1358.5	264(52) 278(40) 303(30.8) 385 (9.7) 444 (9.0) in water acidified [1]	Artificial product not present in living systems. Obtained by chemical synthesis from Cbl^I and C_2H_5I [16]. Active in man, animals and in microorganisms. Not active in enzyme systems as cofactor.	Physical and chemical properties are similar to Me-Cbl. In contrast to Me-Cbl it does not transfer readily its cobalt coordinated ethyl group to metal ions. Irradiation with light under oxygen free conditions produces a reactive ethyl radical [16].

Table 3-1 continued

Systematic name abbreviation	Common names and abbreviatons	Structure[a] and formula[b]	Absorbancy λ_{max} (nm) in parentheses $10^3 \cdot \varepsilon$	Occurrence and biological activity	General remarks
No 7 Coα-[α-(5,6-dimethylbenzimidazolyl)]-(Coβ-sulfito)cobamide HSO_3-Cbl	sulfitocobalamin	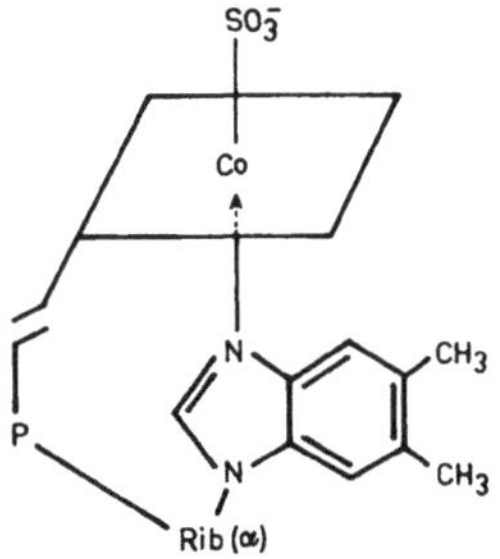 $C_{62}H_{91}O_{17}N_{13}PSCo$ M_r 1411.6	266(19.0) 278(20.0) 315(13.3) 365(14.0) 421 (4.8) 515-540(7.6) in water [1, 52]	Artificial compound, traces occur in man (Chapter 6). Obtained from OH-Cbl or aqCbl by treatment with H_2SO_3 [49-51]. Active in man, animals and microorganisms [1]. Not active in enzyme systems as cofactor.	Light decomposes it in presence of air to aq-Cbl and sulfate. At pH below 2.0 it turns yellow due to dissociation of the α-ligand, 5,6-dimethylbenzimidazole, from cobalt. The compound represents the group of analogues with the Co-S coordination bond [49-52]. A large number of sulfur containing ligands are known which coordinate to cobalt of the corrinoids e.g. thiosulfate $S_2O_3^{2-}$, sulfinates RSO_2^-, sulfoxylate SO_2^{2-}, thiolates RS^-, thiocyanate NCS^- and probably thiourea $SC(NH_2)_2$ [50]. See also Cyst-Cbl. For pk_a value see Table 3-7.
No 8 Coα-[α-(5,6-dimethylbenzimidazolyl)]-(Coβ-vinyl)cobamide ($H_2C{=}CH$)-Cbl vinCbl	vinylcobalamin	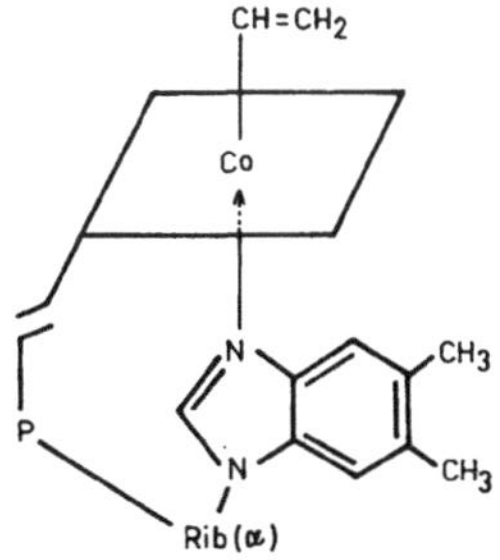 $C_{64}H_{92}O_{14}N_{13}PCo$ M_r 1356.4	260(21.8) 278(21.0) 286(20.8) 325(14.5) 350(10.7) 460 (7.9) diluted mineral acid [1, 16]	Obtained by chemical synthesis from vinyl bromide or acetylene and CBl^{I} [16]. (see 3.1.4.2.). Probably active in man, animals and bacteria. Not active in enzyme systems.	Light splits off the β-ligand yielding Cbl^{III} in presence of air. In presence of mineral acids is decomposed to ethylene and to degradated cobalamin. Forms an additive with Ag^+ [16, 50].

No 9 Coα-[α-5,6-dimethyl-benzimidazolyl)]-(Coβ-ethinyl)cobamide (HC≡C)-Cbl ethinCbl	ethinylcobalamin	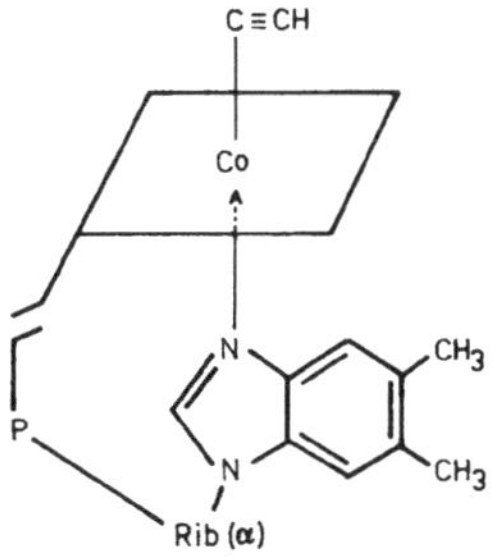 $C_{64}H_{90}O_{14}N_{13}PCo$ M_r 1354.4	266(20.2) 277(17.8) 287(17.3) 320(14.0) 350(10.0) 457 (7.5) diluted mineral acid [1, 16]	Obtained by chemical synthesis from Cbl[I] and Br-C≡CH [16] (see 3.1.4.2.). It is probably biologically active.	Is decomposed by light in presence of air to Cbl[III]. The ligand resembles, in chemical properties, the unsaturated hydrocarbons. Acid treatment produces acetylcobalamin(CH_3CO-Cbl) [50, 53]. Ag^+ ions coordinate with the ethin moiety to form(Ag-C≡C)-Cbl [50].
No 10 Coα-[α-(5,6-dimethylbenzimidazolyl)]-(-Coβ-carboxy-methyl)cobamide HOOC CH_2-Cbl CM-Cbl	carboxymethylcobalamin	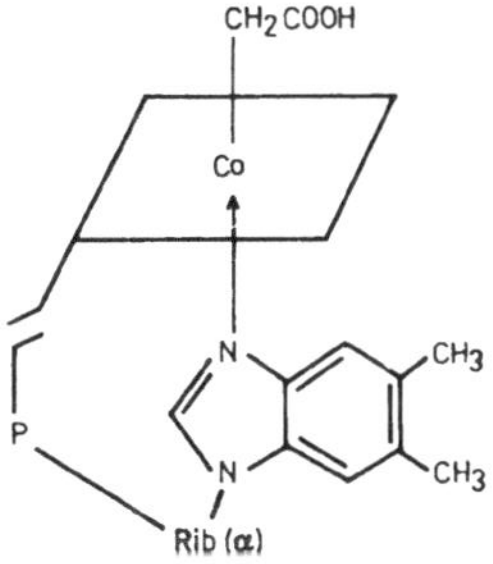 $C_{64}H_{92}O_{16}N_{13}PCo$ M_r 1386.2	331(14.6) 370(13.5) 421 (6.1) 527 (8.5) dimethyl sulphoxide [1, 55]	Obtained by chemical synthesis [16] (see section 3.1.4.2). The postulated cofactor or intermediate in acetate fermentation [54]. Active in man and animals after the removal by the organisms of the carboxymethyl ligand.	Light splits off the cobalt coordinated corboxy-methyl ligand yielding under aerobic conditions aqCbl and acetic acid [54].

Table 3-1 continued

Systematic name abbreviation	Common names and abbreviatons	Structure[a] and formula[b]	Absorbancy λ_{max} (nm) in parentheses $10^3 \cdot \varepsilon$	Occurrence and biological activity	General remarks
No 11 (Coα-cyano)-(Coβ-cyano)-5,6-dimethylbenzimidazolylcobamide (CN)CN-Cbl	dicyanocobalamin	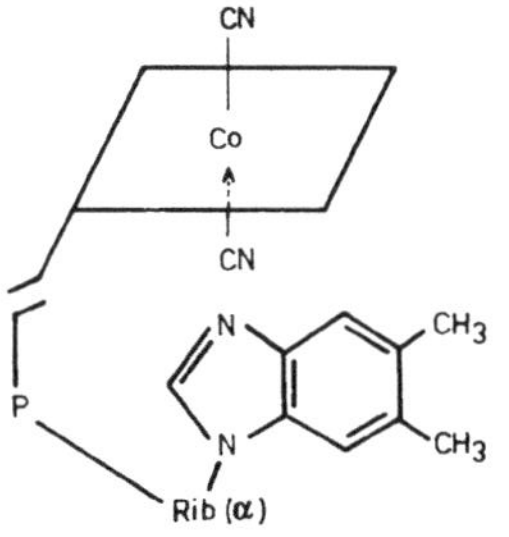 $C_{64}H_{89}O_{14}N_{15}PCo$ M_r 1381.4	279(14.4) 289(10.6) 308 (9.7) 313 (9.8) 369(30.4) 419 (2.2) 543 (8.6) 584 (9.8) excess of KCN, pH 11 [1, 56]	Obtained from other Cbl's in presence of excess of KCN.	Purple solution stable in excess of KCN and at high pH. By acidification is converted to CN-Cbl. In presence of excess of KCN Cby's, Cbi's, and Cba's produce the dicyano form. It is convenient to measure photometrically the concentration of corrinoids in that form since their molar extinction coefficients at 367–369 nm are almost identical and approach the value of 30.4×10^3 [1, 3, 53]. For spectrum see Fig. 3-8 and assay of Cbl by spectroscopic means section 5.3.5.
No 12 Coα-[α-(5,6-dimethylbenzimidazolyl)]-(Coβ-cysteinyl)cobamide CystCbl	Cysteinylcobalamin	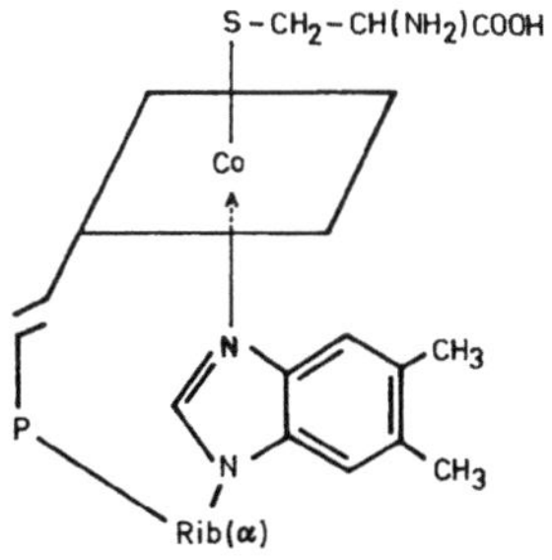 $C_{65}H_{95}O_{16}N_{14}PSCo$ M_r 1449.3	280(19.0) 285(22.0) 290(22.0) 335(14.0) 374(12.5) 430 (4.0) 532 (7.0) 560 (7.0) [1]	Forms from OH-Cbl in presence of excess of cysteine. Active in man, animals and bacteria [4]. Inactive in enzyme systems.	Cyst-Cbl represents a group of thiols derived analogues of the [Co]-S-R type. The binding of thiols to cobalt is of medium strength and varies upon the R. Gluth-Cbl is bound weaker than CystCbl. The spectrum of analogues with a Co-S bond is distinguished from spectra of other Cbl's. There is no direct evidence of the involvement of [Co]SR cobalamins in biological activity. It is suggested, however, that they are formed as enzyme-bound intermediates during catalytic

					action. HS^- and RS^- reduce CBl^{III} to Cbl^{II} in alkaline solution. Cobalamins heated in presence of thiols are being irreversibly inactivated [1, 50, 53, 56–58].
No 13 Coα-[α-(5,6-dimethyiobenzimidazolyl)]-(Coβ-glutathionyl)-cobamide GluthCbl	glutathionylcobalamin	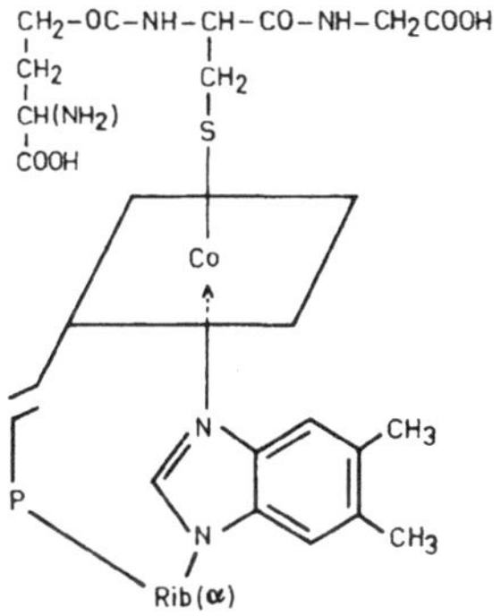 $C_{72}H_{105}O_{20}N_{16}PSCo$ M_r 1635.0	280(18.9) 285(22.3) 291(22.4) 333(14.7) 375(12.7) 408 (3.6) 428 (4.0) 535 (7.1) 562 (7.1) [1]	Obtained from OH-Cbl and excess of gluthation [4, 50, 53]. Active in man, animals and bacteria. Inactive in enzyme systems.	This violet compound does not change its spectrum in pH range from 1 to 10. It is relatively stable to light, but KCN converts it to $(CN)_2$-Cbl. See also CystCbl [4, 50, 56–58]
No 14 Coα-[α-(5,6-dimethylbenzimidazolyl)]-(Coβ-azido)cobamide AzCbl N_3-Cbl	azidocobalamin	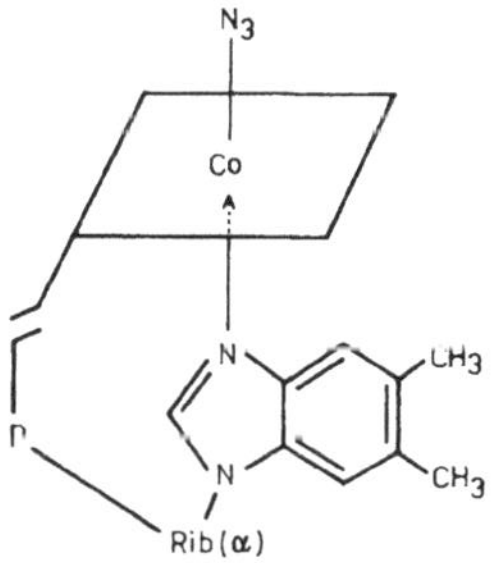 $C_{62}H_{89}O_{14}N_{16}PCo$ M_r 1371.4	358 418 516 540 in water [1, 6]	Obtained through chemical synthesis from OH-Cbl [6]. Active in man, animals and bacteria. Inactive in enzyme systems.	The N_3^- ligand forms a weak coordination bond with cobalt thus may be substituted by stronger binding ligands such as: OH^-, SO_3^{2-} and CN^- [4, 6, 59–60].

Table 3-1 continued

Systematic name abbreviation	Common names and abbreviatons	Structure[a] and formula[b]	Absorbancy λ_{max} (nm) in parentheses $10^3 \cdot \varepsilon$	Occurrence and biological activity	General remarks
No 15 Coα-[α-(5,6-di-methylobenzimidazo-lyl)]-(Coβ-nitrito)cob-amide NO_2-Cbl NitoCbl	nitritocobalamin vitamin B_{12c}	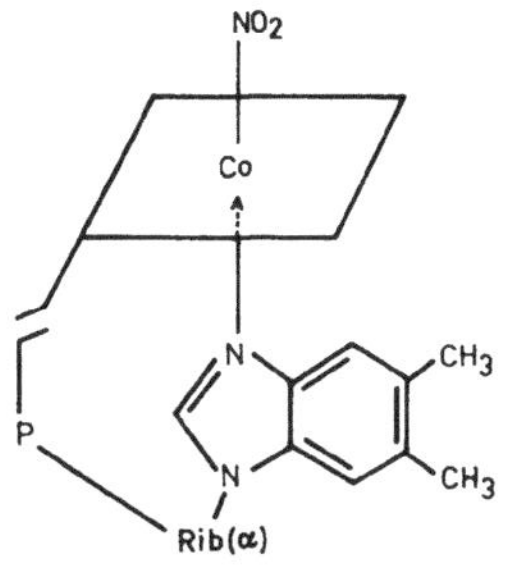 $C_{62}H_{89}O_{16}N_{14}PCo$ M_r 1376.0	263(22.7) 271(21.6) 319 (9.4) 356(21.7) 415 (3.8) 508 (8.3) 534 (8.9) in water [1, 6]	Isolated from fermentation fil-trates of *Streptomyces griseus.* or obtained from aqCbl and HNO_2 [61]. Biologically active. Inactive as cofactor in enzyme systems.	The nitrite ligand coordinates prob-ably with cobalt through the nitrogen atom. Other nitrogen containing com-pounds such as imidazole, pyridine and ammonia, but aminoacids will not form weak coordination bonds with cobalt. The addition of these com-pounds to aqCbl causes a change in the spectrum [61, 62].
No 16 Coα-[α-(5,6-dimeth-ylbenzimidazo-lyl)]-(Coβ-nitroxal-kyl)cobamide Nitroxalkyl-Cbl	nitroxalkylcobal-amin	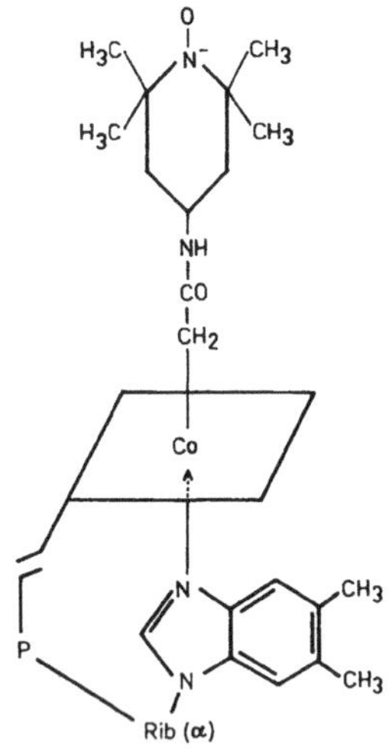 $C_{73}H_{107}O_{16}N_{15}PCo$	329 357 525(9.1) in 0,1 phos-phat pH 7 [63]	Obtained from CBl[I] and 4-brom-acetamino-2,2, 6,6-te-tramethylpiperidin-N-oxyl [63, 64]. Probably active in man, ani-mals and bacteria. May form inactive complexes with en-zymes.	Used as a spin marker for studies of B_{12} dependent enzymes. Is photolabile and decomposes under aerobic condi-tions to aqCbl. Anaerobic photolysis causes a decrease of the EPR signal due to formation of a nitroxalkyl bi-radical [63, 64].

No 17 Coα-[α-(5,6-dimethylbenzimidazolyl)]-[Coβ-(H)]cob(II)amide Cbl^{II}	vitamin B_{12r} cob(II)alamin	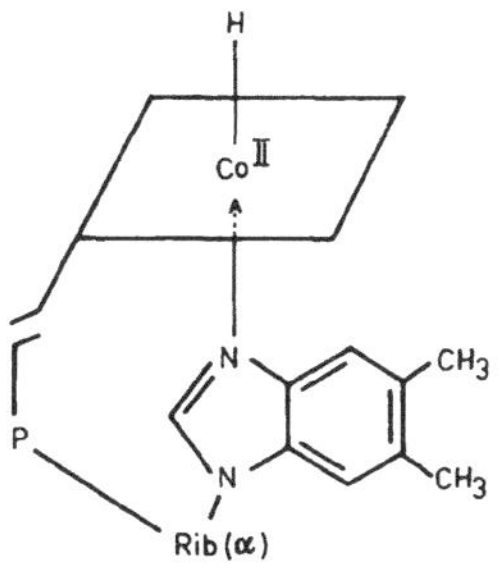 $C_{62}H_{90}O_{14}N_{13}PCo$ M_r 1329.4	267 285 315 467 in water pH 7 [6]	Obtained by chemical, or photochemical reduction of Cbl^{III} [65] (see section 3.1.3 and 3.1.4.1). Intermediate in biological synthesis of MeCbl and AdoCbl (see Chapter 4).	Obtained as a solid by anaerobic photolysis of Me-Cbl [67]. In solution, under aerobic conditions, is oxydized to Cbl^{III}. In presence of oxygen forms an unstable compound Cbl^{II}-O_2. Paramagnetic; produces a distinct EPR spectrum. Cbi^{II} and Cba^{II} are produced under similar conditions [4, 53, 65-69]. Electronic spectrum Fig. 3-11, EPR spectrum Fig. 3-25.
No 18 Coα-[α-(5,6-dimethylbenzimidazolyl)]-cob(I)amide Cbl^{I}	vitamin B_{12s} cob(I)alamin	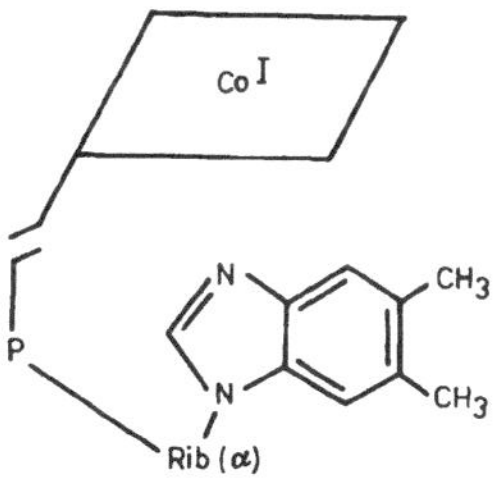 $C_{62}H_{89}O_{14}N_{13}PCo$ M_r 1329.4	280 288 386 455 545 680 800 [6]	Obtained by chemical, or electrochemical reduction of Cbl^{III}or Cbl^{II} [4, 53, 69] (see section 3.1.3). Intermediate in biological synthesis of Me-Cbl and Ado-Cbl in man, animals and microorganisms (see Chapter 4).	Not obtained as a solid so far. Extremely unstable. In presence of air instantly oxydized to Cbl^{III}. Under anaerobic conditions decomposes in aqueous solution within hours yielding Cbl^{II} and hydrogen. Used for preparation of Me-Cbl, Ado-Cbl and other alkyl- and acyl analogues. Cbi^{I} and Cba^{I} are produced under similar conditions [69]. For spectrum see Fig. 3-11, stability Table 5-12 and section 3.1.3.

Table 3-1 continued

Systematic name abbreviation	Common names and abbreviatons	Structure[a] and formula[b]	Absorbancy λ_{max} (nm) in parentheses $10^3 \cdot \varepsilon$	Occurrence and biological activity	General remarks
No 19 Coα-[α-(5,6-dimethylbenzimidazolyl)]-(Coβ-cyano)-(5,15-dinorcobamide) CN-Cbl(5,15-di-nor)	cyanocobalamin-(5,15-di-nor)	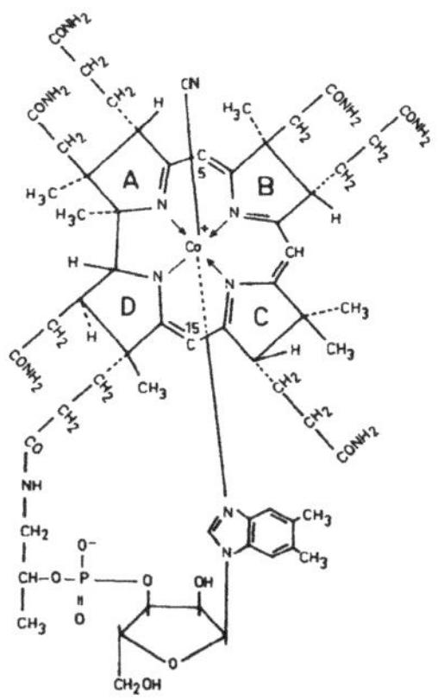 $C_{62}H_{83}O_{14}N_{14}PCo$ M_r 1325.4		Obtained from Cbl by chemical oxydation at the corrin ring and subsequent reconstruction of the damaged nucleotide moiety by biological or chemical synthesis [70] (see section 3.1.1.6.). Exhibits limited activity in *E. coli* and *L. leichmannii* [71].	The structure has not been proved by X-ray crystallography. The absorption spectrum does not differ from CN-Cbl. CN-Cbl-(5-nor) and CN-Cbl-(15-nor) are obtained as byproducts during synthesis of the di-nor form [70–72].
No 20 Coα-[α-(5,6-dimethylbenzimidazolyl)]-(Coβ-cyano)-(c-lactone)-cobamide CN-Cbl-(c-lactone)	cyanocobalamin-(c-lactone)	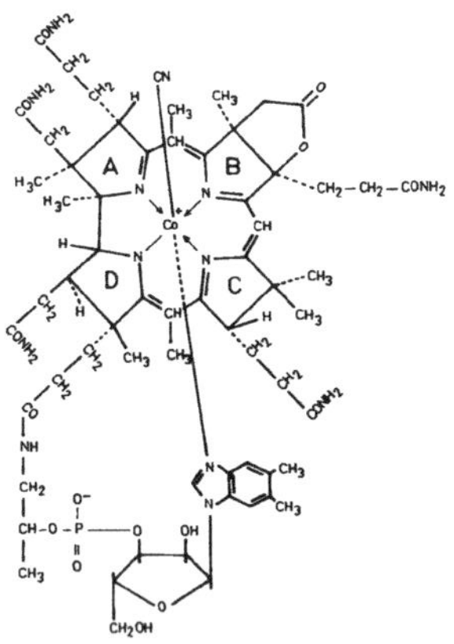 $C_{63}H_{86}O_{15}N_{13}PCo$ M_r 1354.5		Obtained by treatment of CN-Cbl with Ag_2O in a mixture of ethanol and puridine or isolated as byproduct during halogenation of Cbl [73, 74]. Biologically inactive [74].	The formation of the lactone at the B ring has been proved by chemical means. By synthesis of c-lactone, with the use of chlorine or Chloramine T, CN-Cbl-(C-lactone)(10-Cl) is formed as a byproduct [73–76] (see section 3.1.1.3).

No 21 Coα-[α-(5,6-dimethylbenzimidazolyl)]-(Coβ-cyano)-(c-lactam) cobamide CN-Cbl-(c-lactam)	dehydrovitamin B_{12}; cyanocobalamin-(c-lactam)	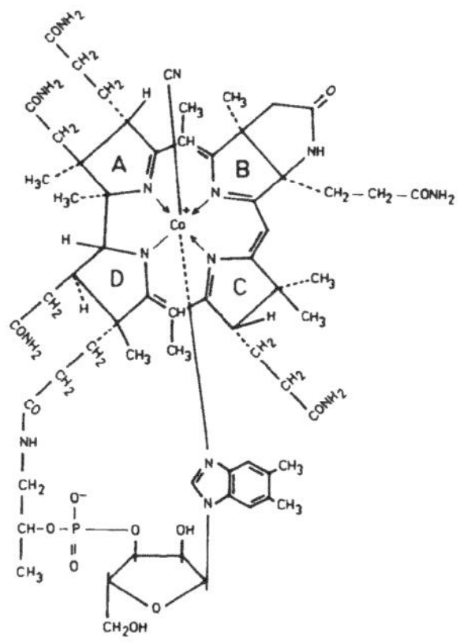 $C_{63}H_{87}O_{14}N_{14}PCo$ M_r 1353.5	359 517 548	Obtained by heating of OH-Cbl at 100 °C in 0.1 N NaOH in presence of air or by treatment with alkaline solution of I_2 in KI [73] (see section 3.1.1.3.). Biologically inactive [76]. AdoCbl-(c-lactam) is in *E. coli* test a non-competitive antagonist [76].	The compound was obtained in crystalline form. Its spectrum is identical with the corresponding spectrum of CN-Cbl. The formation of the lactam from Cbl passes through an internal reduction at the corrin ring as concluded from a change of the spectrum in course of the reaction. The reduced form is subsequently oxydized in presence of air with accompanied lactam formation from the c-acetamide side chain. Incomplete corrinoids,(Cbi), form also lactams under similar conditions [73, 76–78].
No 22 Coα-[α-(5,6-dimethylbenzimidazolyl)]-(Coβ-methyl)-10-bromocobamide MeCbl-(10-Br)	(10-Br)methylcobalamin	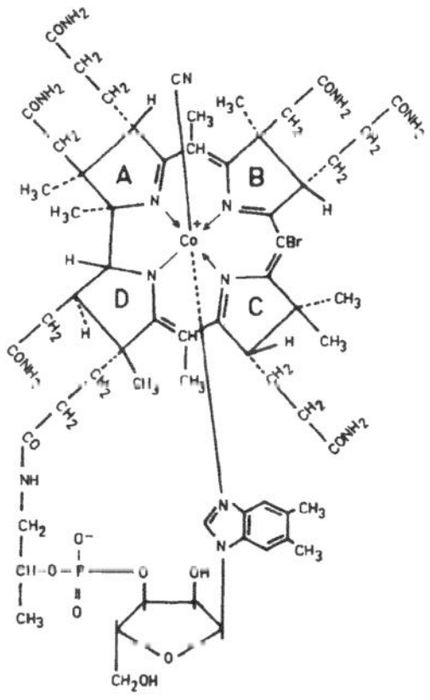 $C_{63}H_{91}O_{14}N_{13}CoBr$ M_r 1424.4	363 550 575	Obtained by chemical synthesis from Me-Cbl or CN-Cbl and N-bromosuccinimide [74] (see section 3.1.1.4.). Growth suppressor of *E. coli*, inhibition index, 9 [81].	This analog represents a group of corrinoids substituted at the C-10 of the corrin ring. Its C-10 Cl counterpart has been obtained by Chloramine T treatment of HSO_3Cbl. The treatment of Me-Cbl with NOCl yields MeCbl-(10-NO_2). The reduction with $NaBH_4$ of the nitro group of MeCbl-(10-NO_2) yields MeCbl-(10-NH_2), one of the most powerful suppressors of *E. coli* growth (inhibition index 4). The β coordinated ligands seem to be activated from substituents at C-10 by a "cis-effect". The HSO_3-Cbl-(10-Cl), in contrast to HSO_3-Cbl, transfers the sulfate without enzyme action to the amino group of β-naphthylamine to give β-naphthyl-sulfamic acid [74, 79, 80, 81].

Table 3-1 continued

Systematic name abbreviation	Common names and abbreviatons	Structure[a] and formula[b]	Absorbancy λ_{max} (nm) in parentheses $10^3 \cdot \varepsilon$	Occurrence and biological activity	General remarks
No 23 Coα-[α-(5,6-dimethylbenzimidazolyl)]-(Coβ-cyano)-(13-epi)cobamide CN-Cbl-(13-epi)	cyano-13-epicobalamin neovitamin B_{12}	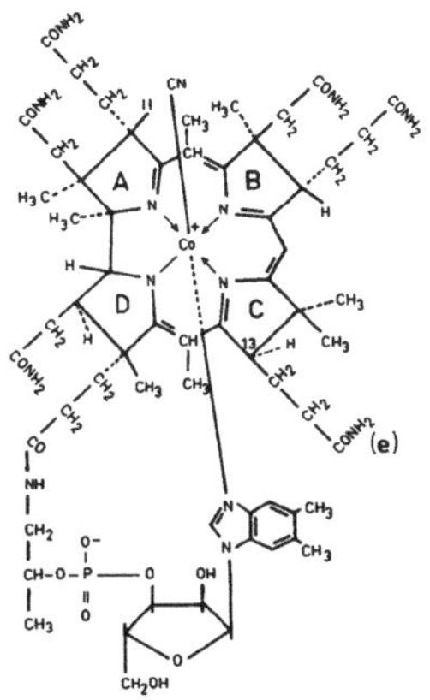 $C_{63}H_{89}O_{14}N_{14}PCo$ M_r 1355.5	278(11.1) 287(10.7) 307 (8.0) 324 (7.4) 360(19.0) 553 (6.1) in water	Not found in nature. Obtained by epimerization of CN-Cbl in CF_3COOH or in conc. HCl [82, 83 a, b] (see section 3.1.1.5). Biologically active in *E. coli* test (10% of CN-Cbl activity).	The structure of this corrinoid has been established by X-ray analysis. It is an epimeric form of CN-Cbl obtained by displacement of the e-propionamide chain at C-13 in the C-ring. Similar epimerization has been achieved at C-8 of the B-ring [85]. Corresponding CN-Cbi-(13-epi) and CN-Cbi-(8-epi) were also obtained. The circular dichroism spectrum of the epimer is distinguished from that of CN-Cbl. The electronic spectrum of the dark-red compound does not differ significantly from CN-Cbl, however the molecular extiction coefficients of the major absorption bands are lower by about 31% both in water and in 0.1 M KCN. The epimerization is reversible [82, 83, 85]. C. D. spectrum, Fig. 3-22.
No 24 α-(5,6-dimethylbenzimidazolyl)-hydrogenobamide [H]Cbl	hydrogenobalamin	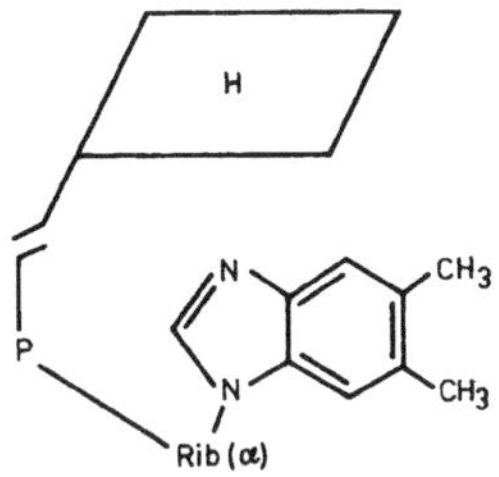 $C_{62}H_{89}O_{14}N_{13}P$ M_r 1270.6	288 329 377 392 497 524 in water [86]	Isolated from photosynthetic bacteria, *Chromatium* (ATCC 17899) grown in presence of 5,6-di-Me-Bza [86, 87]. It suppresses the growth of *L. leichmannii*, inhibition index, 42 [86].	This metal-free corrinoid may be converted to vitamin B_{12}, cupribalamin, zincobalamin and rhodibalamin by insertion of the corresponding metal ions of Co,Cu,Zn and Rh. *Chromatium* produces in absence of dimethylbenzimidazole a cobalt-free corrinoid, phenylhydrogenobamide [86-90]. For spectrum see Fig. 3-12 (see also 3.1.1.8.).

No 25 [Feα-(α-5,6-dimethylbenzimidazolyl)]-(Feβ-cyano)ferribamide CN-[Fe]Cbl	cyanoferribalamin	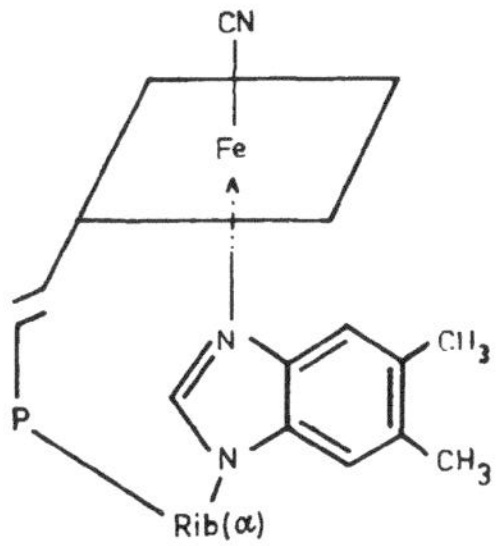 $C_{62}H_{89}O_{14}N_{13}PCu$ M_r 1270.6	271 332 472 in water	obtained from[H]Cbl by insertion of Fe^{3+} ions [84]. Shows approximately one third of the Cbl activity in ochromonas test [84].	Red-brown colour in aqueous solution. In analogy to the true vitamin B_{12} the axial ligands of ferribalamin can be exchanged. It can be reduced to [Fe^{I}]Cbl. The Fe in CN[Fe]Cbl seems to be trivalent and have six ligands like Co(III) in vitamin B_{12}. The spectra of CN[Fe]Cbl, F^{I} OH[Fe]Cbl, $(CN)_2$[Fe]Cbl and [Fe^{I}]Cbl are presented in Fig. 3-15.
No 26 Rhα-[α-(5,6-dimethylbenzimidazolyl)]-(Rh β-adenosyl)rhodibamide Ado[Rh]Cbl	adenosylrhodibalamin	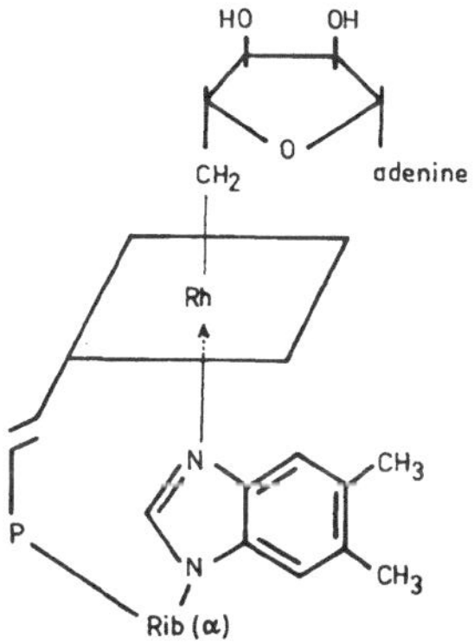 $C_{72}H_{101}O_{17}N_{18}PRh$ M_r 1623.0	350 490 510 in water	Obtained from[H]Cbl by insertion of Rh followed by chemical or enzymatic introduction of the adenosyl ligand [87, 178]. Ado[Rh]Cbl competes with CN-Cbl for binding to human serum trascobalamin and for serum mediated uptake by human blood cells. It is a weak suppressor for *E. coli* and *L. leichmannii* [86, 91].	This red corrinoid is photolabile. In presence of KCN it is converted to its cyanide and dicyanide form. Like Cbl, if reduced by $NaBH_4$, may react with CH_3I to form CH_3[Rh]Cbl [86-91]. For absorption spectrum see Fig. 3-13.

Table 3-1 continued

Systematic name abbreviation	Common names and abbreviatons	Structure[a] and formula[b]	Absorbancy λ_{max} (nm) in parentheses $10^3 \cdot \varepsilon$	Occurrence and biological activity	General remarks
No 27 [Znα-(α-5,6-dimethylbenzimidazolyl)]-zincobamide [Zn]Cbl	zincobalamin	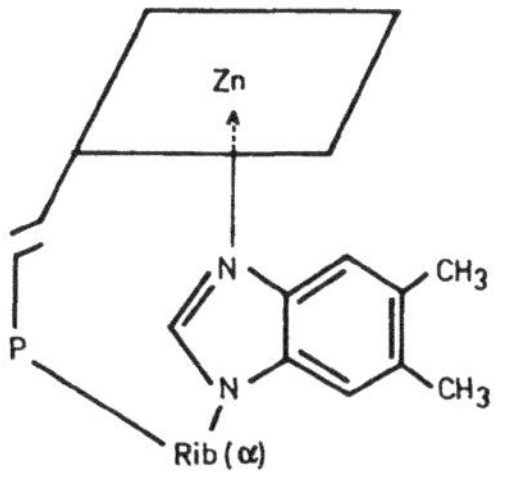 $C_{62}H_{89}O_{14}N_{13}PZn$ M_r 1335.4	338 500 526 in water	Obtained from [H]Cbl by insertion of Zn^{II} [90, 178]. It suppresses the growth of *L. leichmannii*, inhibition index 17 [86].	It is a red compound with an orange-yellow fluorescence. For absorption spectrum see Fig. 3-16. Undergoes irreversible photochemical transformation in presence of oxygen with a marked change in the absorption spectrum. The dimethylbenzimidazole moiety is probably coordinated with Zn [86, 90].
No 28 α-(5,6-dimethylbenzimidazolyl)-cupribamide [Cu]Cbl	cupribalamin	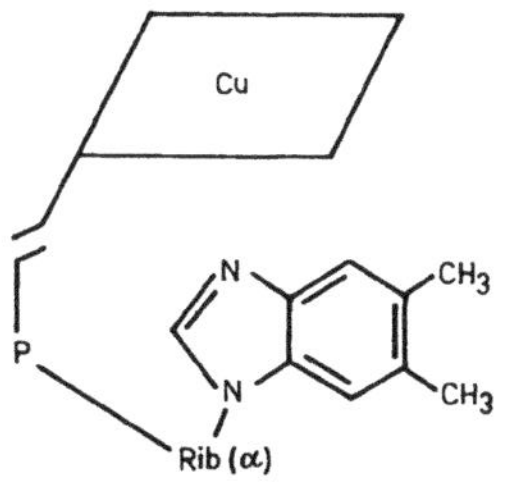 $C_{63}H_{89}O_{14}N_{14}Fe$ M_r 1352	288 330(30.8) 464 484 in water	Obtained from[H]Cbl by insertion of Cu^{II}[90, 178]. It suppresses the growth of *L. leichmannii*, inhibition index 16 [86].	It is a yellow, non-fluorescent compound. For absorption spectrum see Fig. 3-17. [Cu]probably does not coordinate with dimethylbenzimidazole nor does it seem to attach other ligands at the α and β position [86, 90].

No 29 Coα-[α-(benzimidazolyl)]-Coβ-cyano)-cobamide (Bza)CN-Cba	Benzimidazolylcobamide	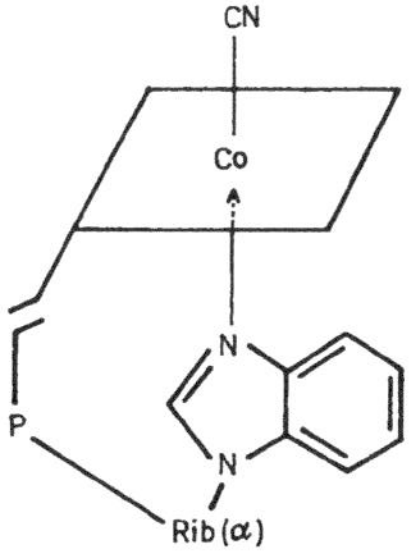 $C_{61}H_{85}O_{14}N_{14}PCo$ M_r 1327.5	274 $E^{1\%}_{1\,cm} = 108$ in water [97]	Obtained by guided biosynthesis in bacteria (see Chapter 4). Active in man, animals and bacteria [96]. Not active in enzyme systems. The coenzyme forms (Bza) AdoCba and (Bza) MeCba are active in enzyme systems derived from animals and bacteria [92–97] (see Chapter 7 for coenzymatic activity).	Its properties are similar to CN-Cbl.
No 30 Coα-[α-(5-methylbenzimidazolyl)]-(Coβ-cyano)cobamide (5-Me-Bza)CN-Cba	5-Methylbenzimidazolylcobamide	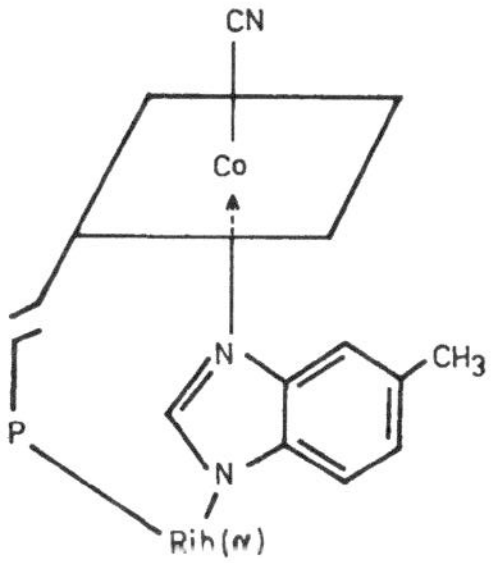 $C_{62}H_{87}O_{14}N_{14}PCo$ M_r 1341.5	278 $E^{1\%}_{1\,cm} = 108$ in water [97]	(5-Me-Bza)CN-Cba and(6-Me-Bza)CN-Cba are obtained by guided biosynthesis in bacteria [96]. (see Chapter 4)	The chemical and biological properties of 5-methylbenzimidazolylcobamide and 6-methylbenzimidazolylcobamide are similar to CN-Cbl.

Table 3-1 continued

Systematic name abbreviation	Common names and abbreviatons	Structure[a] and formula[b]	Absorbancy λ_{max} (nm) in parentheses $10^3 \cdot \varepsilon$	Occurrence and biological activity	General remarks
No 31 Coα-[α-(6-methyl-benzimidazolyl)]-(Coβ-cyano)-cobamide (6-Me-Bza)CN-Cba	6-Methylbenzimidazolylcobamide	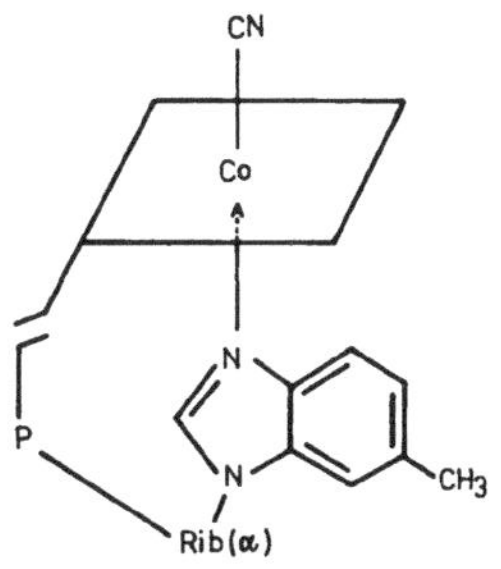 $C_{62}H_{87}O_{14}N_{14}PCo$ M_r 1341.5	278 $E^{1\%}_{1\,cm} = 108$ in water [97]	They are absorbed by animals somewhat less than CN-Cbl. Active in man, animals and microorganisms almost as well as CN-Cbl [96]. Inactive in enzyme systems. The coenzyme forms (see above) are active in enzyme systems [93, 96, 98].	
No 32 Coα-[α-(5-hydroxy-benzimidazolyl)]-(Coβ-cyano)cobamide (5-OH-Bza)CN-Cba	5-hydroxybenzimidazolylcobamide Factor III	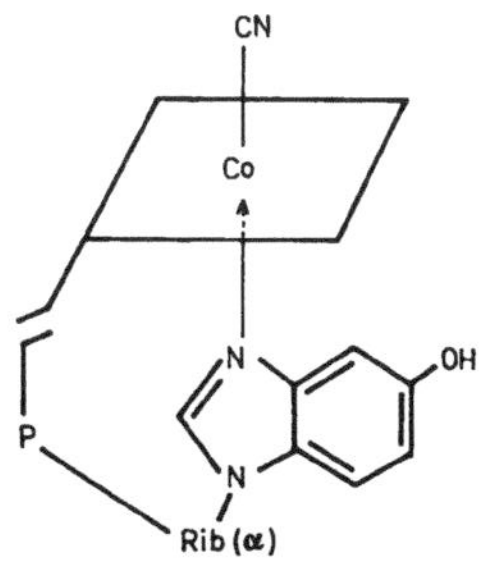 $C_{61}H_{85}O_{15}N_{14}PCo$ M_r 1343.5	295 $E^{1\%}_{1\,cm} = 103$ in water [97]	Isolated from *Methanobacillus omelianskii* or from sewage sludge in presence of KCN. It may be obtained also by guided biosynthesis [99]. Is poorly absorbed by man and animals. Active in man, animals and in microorganisms [101, 102]. Inactive in enzyme systems. The coenzymatic form (5-OH-Bza)AdoCba is active in enzyme systems. This form is also present in some bacteria.	The physical and chemical properties are similar to CN-Cbl except that the phenolic OH group is susceptible to various chemical conversions. It may be methylated by chemical and biological means to form (5-Me-O-Bza)CN-Cba [96, 99–100].

No 33 Coα-[α-(5-methoxy-benzimidazo-lyl)]-(Coβ-cyano)cob-amide (5-Me-O-Bza)CN-Cba	5-methoxybenzimi-dazolylcobamide Factor III_m	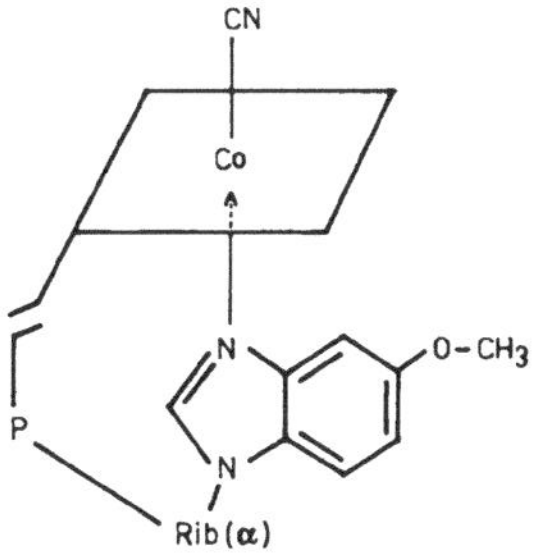 $C_{62}H_{87}O_{15}N_{14}PCo$ M_r 1357.5		Obtained from sewage sludge and *Clostridium thermoaceticum* after cyanide treatment. May be synthetized by chemical methylation of (5-OH-Bza)CN-Cba [100, 103] (see section 6.1). Its Co-methyl or Co-carboxymethyl form is probably active as cofactor in acetate biosynthesis [104]. Active in man, animals and microorganisms [105]. Inactive in enzyme systems. The coenzyme forms (5-Me-O-Bza)AdoCba and (5-Me-O-Bza)MeCba are probably active in some enzyme systems.	The properties are similar to CN-Cbl. A number of other ethers have been obtained through chemical synthesis. Some of them are biologically active. See also (5-OH-Bza)CN-Cba [100–105].
No 34 [Coα-(α-naphtimi-dazolyl)]-(Coβ-cya-no)cobamide (Nza)CN-Cba	Naphtimidazolyl-cobamide	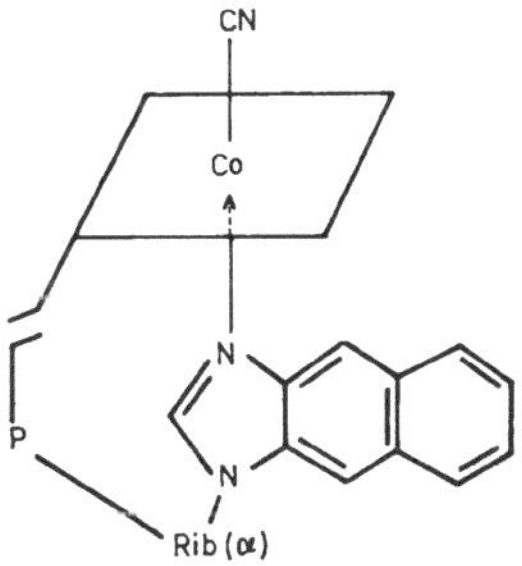 $C_{65}H_{85}O_{14}N_{14}PCo$ M_r 1375.6	279 $E^{1\%}_{1\,cm} = 90$ in water 323 $E^{1\%}_{1\,cm} = 109$	Isolated from sewage sludge after cyanide treatment or obtained through guided biosynthesis in bacteria [105] (see chapters 4 and 6). Active in man, animals and microorganisms. Inactive in enzyme systems. The coenzyme form (Nza)Ado-Cba is active in enzyme systems [105] (see Chapter 7).	Physical and chemical properties are similar to CN-Cbl [105–107].

Table 3-1 continued

Systematic name abbreviation	Common names and abbreviatons	Structure[a] and formula[b]	Absorbancy λ_{max} (nm) in parentheses $10^3 \cdot \varepsilon$	Occurrence and biological activity	General remarks
No 35 Coα-[α-(aden-9-yl)]-(Coβ-cyano)cobamide (Ade)CN-Cba	Pseudovitamin B_{12} ψB_{12} B_{12f} Factor IV	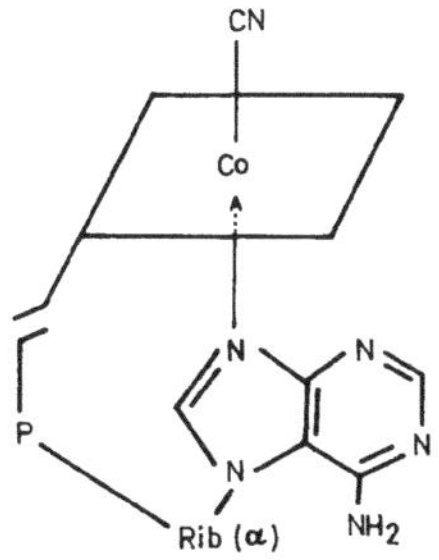 $C_{59}H_{86}O_{14}N_{17}PCo$ M_r 1346.5	360 (27,8) in water 276 $E^{1\%}_{1 cm} = 134$	Artifical product obtained by addition of KCN to other (Ade)X-Cba forms [108]. Active in bacteria, not active in man and animals, not active in enzyme systems. See (Ade)AdoCba [105].	Unlike ATP and NAD this corrinoid contains adenine bound to ribose through its 7 position. Adenine is coordinated with cobalt through N-9 [116] (see Nomenclature). The coordination bond of adenine is rather weak, thus (Ade)CN-Cba, if in solution, occurs mainly in the "base off" form [3, 105, 108–111]: (Ade)CN-Cba$\rightleftarrows$(Ade/aq)CN-Cba. (Ade)CN-Cba is less stable than CN-Cbl toward acids at elevated temperature due to cleavage of the glycosidic bond [105].
No 36 Coα-[α-(aden-9-yl)]-(Coβ-adenosyl)cobamide (Ade)AdoCba (Ade/aq)AdoCba	Pseudovitamin B_{12} coenzyme ψB_{12} coenzyme pseudo-B_{12} coenzyme AC	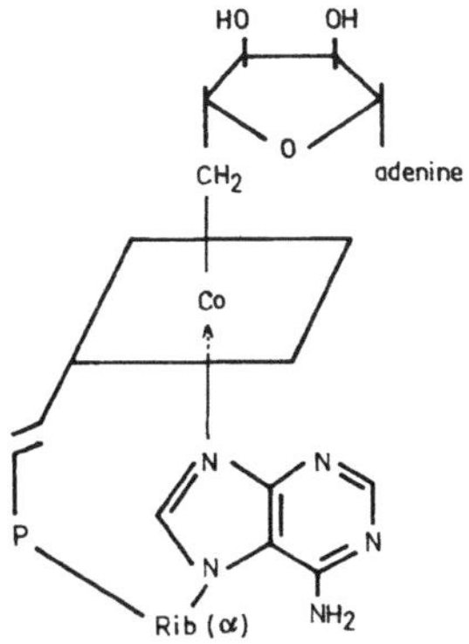 $C_{68}H_{96}O_{17}N_{21}PCo$ M_r 1568.6	239 (26.7) 263 (45.3) 303 (21.6) 375 (9.0) 458 (8.8) 0.01 M phosphate pH 6,7 [94, 105]	Occurs in bacteria, faeces and in sewage sludge [108]. Active in bacteria, not active in man and animals [105, 113]. Cofactor for bacterial enzymes. Not active in enzymes derived from man or animals [112–113] (see Chapter 7).	First corrinoid found to exhibit coenzymatic activity. Though (Ade)-AdoCba and (Ade)MeCba are produced by bacteria in digestion systems in appreciable quantities, they are not absorbed by human and animals. Light decomposes it in presence of air to (Ade)aqCba. For structure and other chemical properties see (Ade)CN-Cba [18, 105, 111–114]. For spectrum see Fig. 3-9.

No 37 Coα[α-(2-methyladen-9-yl)]-(Coβ-cyano)cobamide (2-Me-Ade)CN-Cba	2-methyladenylcobamide Factor A ψB_{12d} B_{12m}	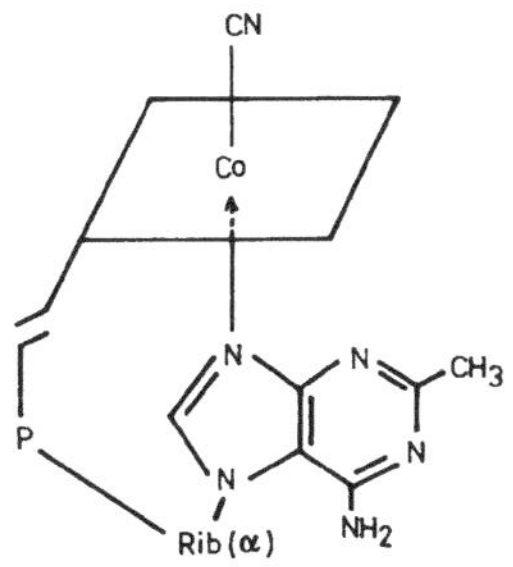 $C_{60}H_{88}O_{14}N_{17}PCo$ M_r 1360.4	277 $E^{1\%}_{1\,cm} = 128$ in water [97, 105]	Isolated from anaerobically grown bacteria, in presence of KCN. Occurs in aquo or coenzymatic forms in faeces, sewage sludge and traces in man and animals [115] (Table 6-1). Active in some bacteria species. Inactive in man and animals [115]. Not active in enzyme systems. The coenzyme form of this corrinoid, (2-Me-Ade)AdoCba, is probably active in some bacterial enzymes.	Its chemical properties are similar to (Ade)CN-Cba [3, 97, 115–118].
No 38 Coα[α-(2-methylmercapto-aden-9-yl)]-(Coβ-cyano)cobamide (2-Me-S-Ade)CN-Cba	2-methylmercaptoadenylcobamide Factor S	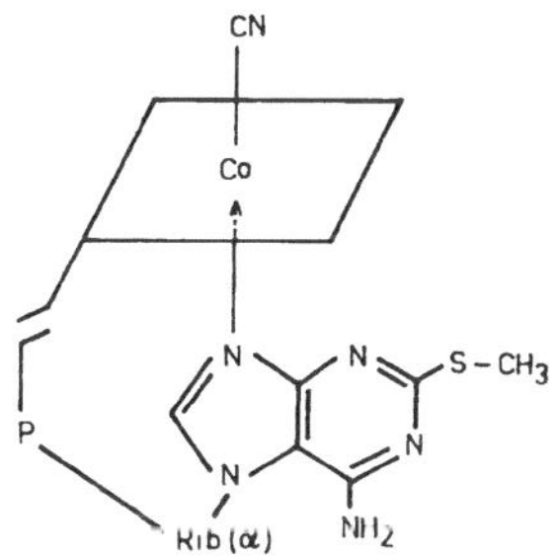 $C_{60}H_{88}O_{14}N_{17}PSCo$ M_r 1392.0	303 $E^{1\%}_{1\,cm} = 98$ in water [97]	Isolated from feaces and sewage sludge in presence of KCN [3, 115]. Active in some bacteria species [3, 115]. Inactive in man and animals [3, 115]. Not active in enzyme systems.	One of the first isolated sulfur containing corrinoids. Its chemical properties are similar to (Ade)CN-Cba [3, 23, 97, 115, 117, 119–122].

Table 3-1 continued

Systematic name abbreviation	Common names and abbreviatons	Structure[a] and formula[b]	Absorbancy λ_{max} (nm) in parentheses $10^3 \cdot \varepsilon$	Occurrence and biological activity	General remarks
No 39 (Coα-aquo)-(Coβ-aquo)-cobin-amide (aq)aqCbi	diaquocobinamide Cobinamide	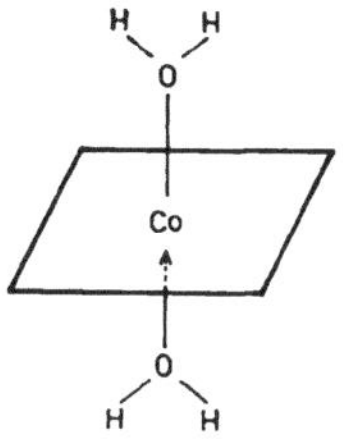 $C_{48}H_{76}O_{10}N_{11}Co$ M_r 1025.4	348 (28) 490 519 [123]	Isolated from *Propionibacterium* and other bacterial species or obtained by chemical degradation of Cba's [124-126]. Growth factor for some bacteria species e.g., *E. coli*[96, 125]. Inactive in man and animals [96]. Inactive in enzyme systems as cofactor.	Cobinamides are called "incomplete corrinoids" due to lack of the nucleotide moiety. The biological attachment of the nucleotide is a slow going process mainly due to slow biosynthesis of dimethylbenzimidazole. Thus young cultures of some bacteria species contain considerable quantities of cobin-amides. The addition of dimethylbenzimidazole to the growth medium will increase the yield of Cbl at the expense of Cbi. Other benzimidazole- and purine analogues may also be incorporated to corrinoids by this way. This process called guided biosynthesis allows for a biosynthetic production of a great variety of cobamides (see Chapter 4). Cobinamides are used also as substrates for chemical synthesis of Cbl and Cba's. The cobalt α and β ligands are interchangeable under special conditions [61-62, 124-129, 130-131].
No 40 (Coα-aquo)-(Coβ-cyano)-cobin-amide (aq)CN-Cbi	cyanoaquocobinamide Cobinamide Factor B Factor Ib B_{12p}	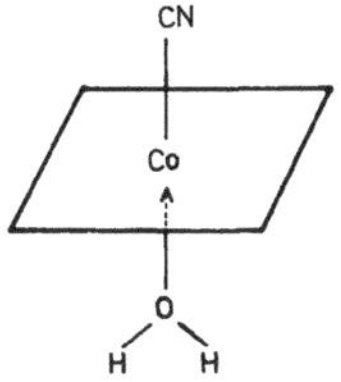 $C_{49}H_{74}O_9N_{12}Co$ M_r 1033.4	275 (11.7) 322 (11.8) 355 (27.3) 404 (5.4) 496 (8.7) 528 (8.1) in water [128]	Isolated from bacteria in presence of traces of KCN or obtained by chemical degradation of Cbl or Cba's [124-126]. See (aq)aqCbi.	

No 41 (Coα-cyano)-(Coβ-cyano)-cobin-amide (CN)CN-Cbi	dicyanocobinamide Factor B Factor Ib B_{12p}	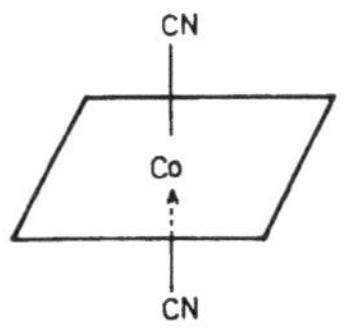 $C_{50}H_{72}O_8N_{13}Co$ M_r 1041.4	266 (9.6) 277 (11.5) 314 (11.6) 368 (30.4) 420 (2.9) 542 (9) 583 (9.9) in water [6]	Isolated from bacteria in presence of excess of KCN or obtained by chemical degradation of Cbl or Cba's [124–126]. See (aq)aqCbi, (aq)CN-Cbi.	
No 42 [Coα(or β)-methyl]-[Coβ(or α)-cyano]-cobamide (CN, Me)Cbi	methylcyanocobinamide	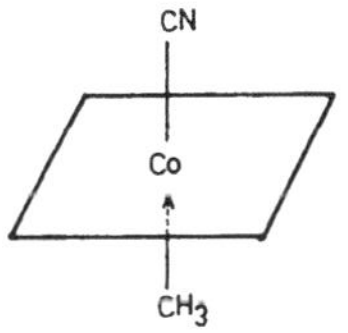 $C_{50}H_{75}O_8N_{12}Co$ M_r 1030.4	390 (12.7) 469 (4.3) 580 (7.5) in water [6]	Obtained by chemical synthesis [16]. See (aq)aqCbi and MeCbl.	Light decomposes it quickly to Cbi^{III} in presence of air. (see MeCbl). The position of the α and β ligands has not been established [129]. For other properties see above [124–132].

Table 3-1 continued

Systematic name abbreviation	Common names and abbreviatons	Structure[a] and formula[b]	Absorbancy λ_{max} (nm) in parentheses $10^3 \cdot \varepsilon$	Occurrence and biological activity	General remarks
No 43 [Coα(orβ)-aquo]-[Coβ(orα)-adenosyl]-cobinamide (aq,Ado)Cbi	cobinamide coenzyme SB_{12p}	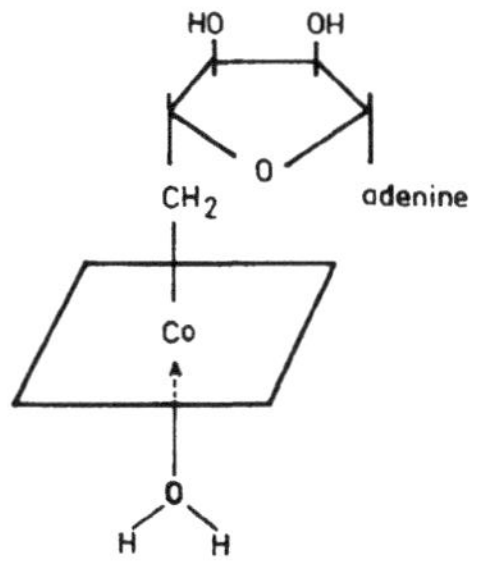 $C_{58}H_{86}O_{12}N_{16}Co$ M_r 1257.5	264 304 375 462 in water [17]	Obtained by chemical and enzymatic synthesis from Cbi [16, 17]. Isolated from *Propionibacterial* species [117]. Not active in man and animals. Growth factor for some bacteria species [95, 126]. Not active in enzyme systems (?).	Light decomposes it to Cbi^{III} in presence of air and to Cbi^{II} under anaerobic conditions. The position of α and β ligands has not been established. For other properties see AdoCbl and (aq)aqCbi [117, 133-137] (see Fig. 4-2).
No 44 [Coα(orβ)-aquo]-[Coβ(orα)-adenosyl]-cobinamide guanosine-di-phosphate (aquo,Ado)Cbi-GDP	Factor C Factor y_1	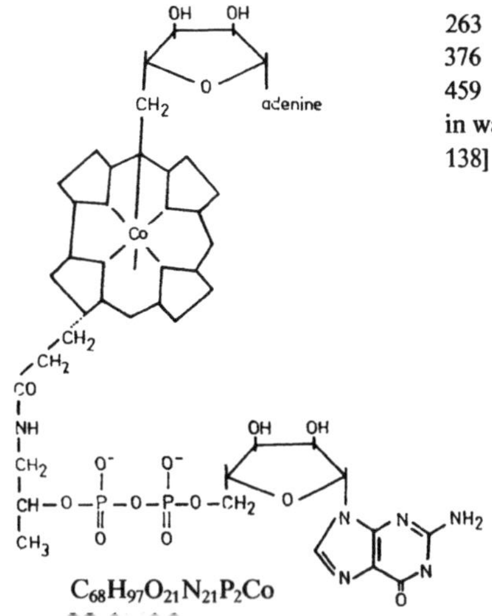 $C_{68}H_{97}O_{21}N_{21}P_2Co$	263 376 459 in water [17, 138]	Isolated from young cultures of *Propionibacterium shermanii* at neutral pH [140-142] (Table 6-1 and section 5.4.2). Growth factor for some bacterial species [96]. Inactive in man and animals [96]. Inactive as cofactor in enzyme systems.	Photolabile in solution. At pH > 8 hydrolyses to Cbi-P derivatives. Postulated precursor in the last steps of Cba biosynthesis. Used as substrate for Cbl biosynthesis by enzyme systems in vitro [138-143] (see Chapter 4).

No 45 (Coα-cyano)-(Coβ-cyano)cobinamide-phosphate (CN)CN-Cbi-P	Factor y_2	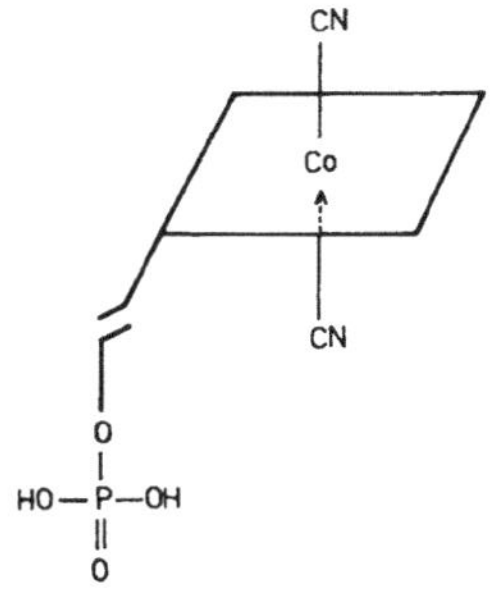 $C_{50}H_{73}O_{11}PN_{13}Co$ M_r 1121.4	Isolated from young cultures of ***Propionibacterium shermanii*** in presence of excess of KCN [144–145] (see section 5.4.4). Growth factor for some bacteria species [138]. Inactive for man and animals. Inactive in enzyme systems.	This compound seems to be an artifact derived from $(CN)_2$Cbi-GDP during extraction of corrinoids in alkaline solutions. Its Ado-form is postulated as precursor in Cbl biosynthesis [141–145] (see Chapter 4).

Table 3-2 Alkanolamines incorporated into cobalamin instead of 1-amino-2-propanol by chemical and biological means and the biological properties of the cobalamin analogues obtained (data compiled by Friedrich see ref. [18] in section 3.1).

Alkanolamine moiety in the corrinoid	*β*ligand	Source chemical synthesis C.S. or biosynthesis B.S.	Biological property: *E. coli* 113-3 biological activity	*E. coli* 113-3 Inhibition index*	*Ochromonas malhamensis* biolog. activity	*Ochromonas malhamensis* Inhib. index*	*L. leichmannii* biolog. activity	*Eug. gracilis* Inhib. index*
(D)$-NH-CH_2-CH(CH_3)-O-$**	CN	C.S.	100 96		100 98		100 100	
(L)$-NH-CH_2-CH(CH_3)-O-$	CN	B.S.	60 50 57.8		100 66.1		73.4	
(DL)$-NH-CH_2-CH(CH_3)-O-$	CN	C.S.	80 78.5		100		80	
$-NH-CH_2-CH_2-O-$	CN	B.S.	83		60			
(L)$-NH-CH(CH_3)-CH_2O-$	CN	C.S.	<0.1	6	4	80		
$-NH-CH_2-CH_2-CH_2O-$	CN	C.S.	<0.01	12	16			
$-NH-C(CH_3)_2-CH_2O-$	CN	C.S.	<0.01	3	<0.01	60	25	500
	CH_3	C.S.	<0.01	5	<0.1	30		
	$CH_2 \cdot CH_3$	C.S.	<0.01	140	<0.1	60		
	Propyl	C.S.	<0.01	180	<0.1	250		
	Aminoethyl	C.S.	<0.01	14	<0.1	580		
	Allyl	C.S.	<0.01	30	<0.1	50		
	Isopropyliden 5′-Deoxy-adenosyl	C.S.	<0.01	7	<0.1	180		
	5′-Deoxy-adenosyl	C.S.	<0.01	4	<0.1	30		

(DL)$-NH-CH_2-CHO-$ 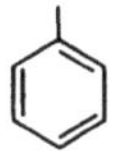	$-CN$	C.S.	<0.01	17	<0.01
$-NH-CH-CH_2O-$ CH_2 	$-CN$	C.S.	<0.1	200	<0.01
$-N-CH_2-CH_2O-$ CH_3	$-CN$	C.S.	10		20
$-N-CH_2-CH_2O-$ CH_2CH_3	$-CN$	C.S.	<0.01	700	0.2
$-N-CH_2-CH_2O-$ CH_2-CH_2OH	$-CN$	C.S.	<0.01		
(DL)$-NH-CH_2-CH-O-$ CH_2F	$-CN$	B.S.	40	8	
(DL)$-NH-CH_2-CH-O$ CHF_2	$-CN$	B.S.	65		
(DL)$-NH-CH_2-CH-O-$ CF_3	$-CN$	B.S.	55		
(DL)$-NH-CH-CH-O$ CH_3 CH_3	$-CN$	B.S.	6		

* inhibition index $= \frac{\text{moles of } B_{12} \text{ antagonist}}{\text{moles of vitamin } B_{12}}$ for 50% of growth inhibition.

** present in true vitamin B_{12}

Table 3-3 First order rate constants for the spontaneous decomposition of alkylcobalamins in acidic and neutral solutions at 20 °C (from Schrauzer et al., see ref. [89] in 3.1.4.2.).

Alkyl Group	Observed rate constants k and $t_{½}$ times of decomposition			
	pH 7		1 M HCl	
	$k(s^{-1})$	$t_{½}$	$k(s^{-1})$	$t_{½}$
$t_{½}$				
CH_3	stable		[a]	
C_2H_5	ca. 10^{-8}	ca. 6 mos	[a]	
$n-C_3H_7$	ca. 10^{-8}	4–5 mos	[a]	
$1-C_3H_7$	4.2×10^{-3}	2.8 min	8.8×10^{-6}	22 h
$2-C_4H_9$	7.3×10^{-3}	1.5 min	2.4×10^{-5}	8 h
$2-C_5H_{11}$	5.6×10^{-3}	2.1 min	2.0×10^{-5}	9.6 h
$3-C_5H_{11}$	2.2×10^{-1}	3.2 s	3.5×10^{-4}	30 min
$3-C_6H_{13}$	5×10^{-1}	1 s	4.7×10^{-4}	25 min
$4-C_7H_{15}$	1.5×10^{-1}	4.5 s	4.4×10^{-4}	26 min
3-methyl-2-butyl	(N. D.)[b]	1 s	1.1×10^{-3}	10 min
$c-C_3H_5$	ca. 10^{-8}	ca. 6 mos	[a]	
$c-C_4H_7$	ca. 10^{-8}	4–6 mos	[a]	
$c-C_5H_9$	3.7×10^{-3}	3.1 min	1 M H_3PO_4: 2.4×10^{-6}	82 h
$c-C_6H_{11}$	2.6×10^{-4}	44 min	1 M H_3PO_4: 1.2×10^{-7}	67 days
$c-C_7H_{13}$	(N. D.)		1.1×10^{-4}	102 min
$c-C_8H_{15}$	(N. D.)		4.3×10^{-4}	27 min

[a] Slow side-reactions prevent measurement of rate. [b] N. D. = not determined.

Table 3-4 Characteristic of vitamin B_{12} containing radioactive cobalt isotopes (data compiled from [148-154]).

Isotope	$t_{½}$	Emission (major emissions only)			Effective radiant energy absorbed by liver per disintergration Σ(bE)	MPC (maximum permissible body concentration) µCi/total body	Relative dose to liver per MPC	Effective half-life in liver days	Radiochemical decomposition in aqueous solution, loss
		radiation	MeV	%					
^{56}Co	77 d	β^+	1.50	18	0.963	10	4	66	
		γ	0.84	100					
			1.24	70					
			0.56	36.2					
			1.75	17.8					
			2.60	15.9					
			1.03	15.7					
			3.25	12.3					
			2.02	11.0					
^{57}Co	270 d	γ	0.123	93	0.0488	200	1	161	
			0.014	93					
			0.137	7					
^{58}Co	71 d	β^+	0.47	14.5	0.293	30	2.3	60	25% after 30 days, 19 µCi/ml, 8.4 µg/ml
		γ	0.81	99.5					
			0.51	29					
^{60}Co	5.27 y	β^-	0.31	100	0.736	10	29	331	11% after 70 days, 10 µCi/ml, 10 µg/ml
		γ	1.17	100					
			1.33	100					

Table 3-5 Polarographic half-wave potentials ($E_{1/2}$). Values are given in volts versus the saturated calomel electrode. They can be converted into potentials versus the standard hydrogen electrode by adding +0.246. The $E_{1/2}$ value can vary considerably with the nature of the other electrolytes present in the solutions and pH. In particular compounds which may substitute the β and α ligands such as: KCN, K_2SO_3, KNO_2, NH_3, glutathion etc. may interfere (from Pratt [167], see also references [166, 168-173]).

Corrinoid	Solution	Range studied (V)	$E_{1/2}$ (V)	Number of electrons	Remarks	References
CN-Cbl	0.1 N K_2SO_4	+0.2 to −2.0	−1.11	2		[155]
CN-Cbl	0.1 M Na-EDTA pH 9.5 or 9.6	−0.7 to −1.3	−1.021			[156]
aqCbl	0.1 N K_2SO_4	+0.2 to −1.8	−0.06 −1.02	1 1		[155]
SftoCbl	0.1 N K_2SO_4		−1.48			[157]
AdoCbl	0.1 N K_2SO_4		−1.52			[157]
AdoCbl	solutions various pH 4.75-11.4	+0.25 to −1.9	−1.2 to −1.6	1.5-2		[158]
MeCbl	0.1 N K_2SO_4		−1.55			[159]
EtCbl	0.1 N K_2SO_4		−1.53			[159]
PrCbl	0.1 N K_2SO_4		−1.53			[159]
BuCbl	0.1 N K_2SO_4		−1.53			[159]
CN-Cbl	0.1 N K_2SO_4		−1.48			[159]
$(CN)_2Cbl$	0.1 N KCN		−1.33	2		[160]
(CN,aq)Cbi	0.1 N K_2SO_4		−0.84			[161]
(Ade)CN-Cba	0.1 N K_2SO_4		−1.03			[161]
(Ado,aq)Cbi	0.1 N K_2SO_4		−1.30			[161]
(Ade)AdoCba	0.1 N K_2SO_4		−1.49			[159]
Cbl^{II} (B_{12r})	0.1 N K_2SO_4	+0.2 to −1.8	−0.04 −0.95 −1.55	1 1 variable	anodic wave, pH 5-10	[155, 162, 163]
Cbl^{I} (B_{12s})	0.1 M Na-EDTA pH 9.5 or 9.6		−0.87		anodic wave	[155]

Table 3-6 Equilibrium- and rate constants for β ligand substitution in aquacob(III)alamin (data selected from Pratt, [15]). The aquacobalamin is particularly suitable for producing various analogues by exchange reactions at the β coordination position:

$$aqCbl + X \underset{k_1}{\overset{k_2}{\rightleftharpoons}} X\text{-}Cbl + H_2O.$$

The reaction is reversible and proceeds without change of valency of the cobalt in the corrin ring. The formation of the X-Cbl analog may be measured spectrophotometrically [164, 165]. The equilibrium constant K is expressed in units of M^{-1} according to equation:

$$K = \frac{[X\text{-}Cbl]}{[aqCbl][X]}.$$

The concentration of free H_2O is neglected.

X (ligand introduced in place of H_2O)	K (M^{-1})	$\log_{10}K$	Rate constants at 25 °C ($M s^{-1}$)	
			k_2	k_1
Cl^-	1.3	0.11		
Br^-	1.9	0.28		
J^-	3.2	1.51		
HO^-	7.8-6.0	7.1-6.2		
ClO_4^-		< -1		
NO_3^-		< -1		
PO_4^{3-}		-		
$CH_3CO_2^-$	4.5	0.65		
$C_6H_5O^-$		2.9		
NH_3		7		
Glycine		5.8		
Imidazole	3.9×10^4	4.6		
NO_2^-	2.3×10^5	5.4		
N_3^-	7.2×10^4	4.9	1.7×10^3	0.03
Histidine		5.8		
CN^-		$\geqslant 12$	$\approx 1.5 \times 10^3$	$< 10^{-9}$
NCO^-	530	2.7	7.3×10^2	0.95
NCS^-	1.2×10^3	3.1	7.1×10^3	1.8
HS^-	+	+		
Cystein		6.0 or 8.3		
Glutathione		4.9		
SO_3^{2-}	2.2×10^7	7.3		
$S_2O_3^{2-}$	7.3×10^3	3.86		
$NCSe^-$	8.3×10^3	3.9		

Table 3-7 pK_a values of some cobalamins* (from Hogenkamp [53]).

Cobalamin	Axial ligand	pK_a
Aquacobalamin	H_2O	−2.4
Cyanocobalamin	CN^-	0.1
Ethinylcobalamin	$HC{\equiv}C-$	0.7
Vinylcobalamin	$H_2C{=}CH-$	2.4
Methylcobalamin	CH_3-	2.72
Adenosylcobalamin	5′-Deoxyadenosyl	3.5
n-Propylcobalamin	$CH_3CH_2CH_2-$	3.8
Cob(II)alamin	H	2.5

* In dilute acid adenosylcobalamin and the alkyl- and acylcobalamins change color from red to yellow. Under these conditions the 5,6-dimethylbenzimidazole base is protonated and no longer coordinated to cobalt. The pK_a value of the 5,6-dimethylbenzimidazole moiety depends on the nature of the trans ligand. In aquacobalamin the coordination bond of the 5,6-dimethylbenzimidazole base is the strongest while in propylcobalamin is the weakest.

References

[1] Friedrich, W., Fermente, Hormone, Vitamine, Band III/2, Vitamin B_{12} und verwandte Corrinoide, Eds. Ammon R. and Dirscherl W., Georg Thieme Verlag, Stuttgart 1975, pp. 47-55.

[2] Ellenbogen, L., Absorption and transport of cobalamins, in Cobalamin-Biochemistry and Pathophysiology, Ed. Babior, B.M., John Wiley and Sons, New York 1974, pp. 219-273.

[3] Smith, E.L., Vitamin B_{12}, Methuen, London 1965.

[4] Friedrich, W., Fermente, Hormone, Vitamine, Bad III/2 Vitamin B_{12} und verwandte Corrinoide, Eds. Ammon R. and Dirscherl W., Georg Thieme Verlag, Stuttgart 1975, pp. 44-46.

[5] Hodgkin, D.C., Fortschr. Chem. Org. Naturst. **15,** 167, 1958.

[6] Firth, R.A., Hill, H.A.O., Pratt, J.M., Williams, R.J.P., Jackson, W.R., Biochemistry **6,** 2178, 1967.

[7] Brink, C., Hodgkin, D.C., Lindsey, J., Pickworth, J., Robertson, J.H., and White, J.G., X-ray crystallographie evidence on the structure of vitamin B_{12}, Nature **174,** 1169-1171, 1954.

[8] Johnson, A.W., The Chemistry of Vitamin B_{12}, Svensk Kemisk Tidskrift **72 (10),** 637-643, 1960.

[9] Wagner, F., Vitamin B_{12} and related compounds, Ann. Rev. Biochem. **35,** 405-434, 1966.

[10] See Chapter 6.

[11] Glass, G.B.J., Skeggs, H.R., Lee, D.H., Jones, E.L., and Hardy, W.W., Applicability of hydroxocobalamin as a long-acting vitamin B_{12} in man, 2. Europäisches Symposion, Hamburg 1961, Ed. Heinrich, H.C., Ferdinand Enke Verlag, Stuttgart 1962, pp. 673-685.

[12] See Chapter 5.2.

[13] See Chapter 7.2.

[14] Smith, E.L., Martin, J.L., Gregory, R.J., and Shaw, W.H.C., Standardisation of hydroxocobalamin., Journal of the Society for Analytical Chemistry **87,** 183-186, 1962.

[15] Pratt, J.M., Inorganic Chemistry of Vitamin B_{12}, Academic Press, London, New York 1972, pp. 140-153, 215-217.

[16] Dolphin D., Preparation of the reduced forms of vitamin B_{12} and of some analogues of the vitamin B_{12} coenzyme containing a Cobalt-carbon bond, Methods in Enzymology **18C,** 34-52, 1971.

[17] Pawelkiewicz, J., Enzymatic preparation of coenzyme B_{12} and some of its analogs, Methods in Enzymology **18C,** 99-101, 1971.

[18] Barker, H.A., Chemistry and biology of the vitamin-B_{12}-coenzymes, in 2. Europäisches Symposion Hamburg 1961, Ed. Heinrich,

H.C., Ferdinand Enke Verlag, Stuttgart 1962, pp. 82-104.
[19] Dolphin, D., Johnson, A.W., Rodrigo, R., and Show, N., The vitamin B_{12} Coenzyme, Pure Appl. Chem. **7**, 539-541, 1963.
[20] Schmidt, R.R., Huennekens, F.M., An improved chemical synthesis of adenosyl-B_{12}, Arch. Biochem. Biophys. **118**, 253-254, 1967.
[21] Kikugawa, K., and Ichino, M., Tetrahedron Lett. **2**, 87-88, 1971.
[22] Fenrych, W., Pawelkiewicz, J., and Magas, S., Conversion of cyanocobalamin, in vivo, into its coenzyme form in the rabbit, Bulletin De L'Academie Polonaise Des Sciences **10**, 117-119, 1962.
[23] Coates, M.E., Doran, B.M., and Harrison, G.F., Studies on vitamin-B_{12}-coenzyme in rat and the chick, in 2. Europäisches Symposion Hamburg 1961, Ed. Heinrich, H.C., Ferdinand Enke Verlag, Stuttgart 1962, pp. 147-155.
[24] Gabbe, E.E., and Heinrich, H.C., Das Stoffwechselverhalten des 5,6-Dimethyl-benzimidazol-aquo-cobamid und des 5,6-Dimethylbenzimidazol-cobamid-Coenzym im Säugetierorganismus, in 2. Europäisches Symposion Hamburg 1961, Ed. Heinrich, H.C., Ferdinand Enke Verlag, Stuttgart 1962, pp. 116-147.
[25] Barker, H.A., Smyth, R.D., Weissbach, H., Toohey, J.I., Cadd, J.N., Volcani, B.E., Isolation and properties of crystalline cobamide coenzymes containing benzimidazole or 5,6-dimethylbenzimidazole, J. Biol. Chem. **235**, 480-488, 1960.
[26] Weissbach, H., Ladd, I.N., Volcani, B.E., Smyth, R.D., Barker, H.A., Structure of the adenylcobamide coenzyme. Degradation by cyanide, acid and light, J. Biol. Chem. **235**, 1462-1473, 1960.
[27] Barker, H.A., Smyth, R.D., Weissbach, H., Munch-Petersen, A., Toohey, J.I., Lodd, J.N., Volcani, B.E., Wilson, R.M., Assay purification and properties of the adenylcobamide coenzyme, J. Biol. Chem. **235**, 181-190, 1960.
[28] Lenhert, P.G., Hodgkin, D.C., Structure of the 5,6-dimethylbenzimidazolylcobamide coenzyme, Nature **192**, 937-938, 1961.
[29] Johnson, A.W., and Shaw, N., Isolation of a crystalline nucleoside from the vitamin B_{12} coenzyme, Proc. Chem. Soc., 447-448, 1961.
[30] Johnson, A.W., and Shaw, N., Some reacams, F.R., Williams, R.J.P., Chem. Comm. 850-852, 1971.
tions of the vitamin B_{12} coenzyme, J. Chem. Soc. **888**, 4608-4614, 1962.
[31] Dolphin, D.H., Johnson, A.W., and Rodrigo, R., Some reactions of the vitamin B_{12} coenzyme and its alkyl analogues., Ann. N.Y. Acad. Sci. **112**, 590, 1964.
[32] Abels, R.H., Myers, C., Smith, T.A., An enzymic assay for the determination of millimicrogram quantities of B_{12}-coenzyme, Analyt. Biochem. **15**, 192-194, 1966.
[33] Schrauzer, G.N., and Sibert, J.W., Coenzyme B_{12} and coenzyme B_{12} model compounds in the catalysis of the dehydration of glycols, J. Am. Chem. Soc. **92**, 1022-1030, 1970.
[34] Stadtman, T.C., Vitamin B_{12}, Science **171**, 859-867, 1971.
[35] Lindstrand, K., Nature **204**, 188-202, 1964.
[36] See 3.1.4.2.
[37] Friedrich, W., Fermente, Hormone, Vitamine, Band III/2, Vitamin B_{12} und verwandte Corrinoide, Eds. Ammon R. and Dirscherl W., Georg Thieme Verlag, Stuttgart 1975, pp. 231-243.
[38] Johnson, B.C., Holdsworth, E.S., Porter, J.W.G., and Kon, S.K., Vitamin B_{12} and methyl-group synthesis, British J. Nutrition **2**, 313-323, 1957.
[39] Guest, J.R., Fredman, S., Dilworth, M.J., and Woods, D.D., Methyl group of methionine, Ann. N.Y. Acad. Sci. **112**, 774-790, 1964.
[40] Stadtman T.C., Synthesis of Adenine-B_{12} coenzyme by *Clostridium sticklandii:* relationship to one-carbon metabolism, J. Bacteriology **79**, 904-905, 1960.
[41] Ljungdahl, L., Irion, E., and Wood, H.G., Total synthesis of acetate from CO_2, I. Co Methylcobyric acid and Co-(Methyl)-5-methoxybenzimidazolylcobamide as intermediates with *Clostridium thermoaceticum,* Biochemistry **4**, 2780-2790, 1965.
[42] Wood, J.M., and Wolfe, R.S., Components reguired for the formation of CH_4 from methylcobalamin by extracts of *Methanobacillus omelianskii,* Journal of Bacteriology **92**, 696-700, 1966.
[43] Wood, J.M., Kennedy, F.S., and Rosen, C.G., Synthesis of methyl-mercury compounds by extracts of a methanogenic bacterium, Nature **220**, 173-174, 1968.
[44] Bertiesson, L., and Neujahr, H.Y., Methylation of mercury compounds by methylcobalamin, Biochemistry **10**, 2805-2808, 1971.
[45] Agnes, G., Bendle, S., Hill, H.A.O., Willi-

[46] Imura, N., Sukegawa, E., Pan, S., Nagao, K., Kim, J., Kwan, T., and Ukita, T., Chemical methylation of inorganic mercury with methylcobalamin, a vitamin B_{12} analog, Science **172,** 1248-1249, 1971.

[47] Schrauzer, G.N., Weber, J.H., Beckham, T.M., and Ho, R.K.Y., Alkyl group transfer from cobalt to mercury: The reaction of alkylcobalamins, alkylcobaloximes and of related compounds with mercuric acetate, Tetrahedron **3,** 275-277, 1971.

[48] Pratt, J.M., Inorganic Chemistry of Vitamin B_{12}, Academic Press, London, New York 1972, pp.250-251.

[49] Dolphin, D.M., Johnson, A.W., Shaw, N., Sulphitocobalamin, Nature **199,** 170-171, 1963.

[50] Pratt, J.M., Inorganic Chemistry of Vitamin B_{12}, Academic Press, London, New York 1972, pp.252-256.

[51] Bernhauer, K., Wagner, F., Biochem. Z. **337,** 366-370, 1963.

[52] Dolphin, D., and Johnson, A.W., The properties of cobalamins with sulphur-containing ligands, J. Chem. Soc. **387,** 2174-2181, 1965.

[53] Hogenkamp, H.P.C., The chemistry of cobalamins and related compounds, in Cobalamin-Biochemistry and Pathophysiology, Ed. Babior, B.M., John Wiley and Sons, New York 1974, pp.21-73.

[54] Ljungdahl, L., and Irion, E., Photolytic and reductive cleavage of Co-carboxymethylcobalamin, Biochemistry **5,** 1846-1850, 1966.

[55] Firth, R.A., Hill, H.A.O., Pratt, J.M., Thorp, R.G., Williams, R.J.P., J. Chem. Soc. (A) 2428, 1968.

[56] Heinrich, H.C., Pharm. Ztg. **108,** 1319-1325, 1963.

[57] Hill, H.A.O., Pratt, J.M., Thorp, R.G., Ward, B., Williams, R.J.P., Biochem. J. **120,** 263-266, 1970.

[58] Beaven, G.H., and Johnson, E.A., Nature **176,** 1264-1266, 1955.

[59] Heim, A., Grassi, R.J., and Wilmarth, W.K., In mechanisms of inorganic reactions, Ed. Gould, R.F., Amer. Chem. Soc., Washington, U.S.A. 1965.

[60] Pratt, J.M., Inorganic chemistry of vitamin B_{12}, Academic Press, London, New York 1972, 216-217.

[61] Pratt, J.M., Inorganic chemistry of vitamin B_{12}, Academic Press, London, New York 1972, 120-156.

[62] Firth, R.A., Hill, H.A.O., Pratt, J.M., Thorp, R.G., and Williams, R.J.P., J. Chem. Soc. (A) 381, 1969.

[63] Buckman, T., Kennedy, F.S., Wood, J.M., Spin-labeling of vitamin B_{12}, Biochemistry **8,** 4437-4442, 1969.

[64] Law, P.Y., Brown, D.G., Lien, E.L., Babior, B.M., and Wood, J.M., Synthesis and catalytic activity of spin-labeled cobinamide coenzymes, Biochem. **10,** 3428, 1971.

[65] Diehl, H., Haar, R.W.V., and Sealock, R.R., The magnetic susceptibility of vitamin B_{12}, J. Am. Chem. Soc. **72,** 5312, 1950.

[66] Lee Pin, and Schrauzer, G.N., The reaction of vitamin B_{12a} and of cobaloximes with carbon monoxide evidence for self-reduction of vitamin B_{12a} in neutral solution, J. Am. Chem. Soc. **90,** 5274-5276, 1968.

[67] Yamada, R., Shimiz, S., Fukui, S., Preparation of solid vitamin B_{12r} by anaerobic photolysis of methylcobalamin, Methods in Enzymol. **18 C,** 52-54, 1971.

[68] Das, P.K., Hill, H.A.O., Pratt, J.M., Williams, R.J.P., The so-called hydridocobalamin, Biochim. Biophys. Acta **141,** 644-646, 1967.

[69] For further references see Me-Cbl and Ado-Cbl in table 3.1., Chapter 3.1.3. and Figures 3-6. and 3-19.

[70] Jauering, D., Rapp, P., and Ruoff, G., 5-Nor-, 15-Nor-, and 5,15-Dinorcorrinoide, Hoppe-Seyler's Z. Physiol. Chem. **354,** 957-966, 1973.

[71] Rapp, P., Ruoff, G., Die biologische Aktivität der Norcobinamide, Hoppe-Seyler's Z. Physiol. Chem. **354,** 967-969, 1973.

[72] See also 3.1.1.6.

[73] Bonnett, R., Cannon, J.R., Clark, V.M., Johnson, A.W., Parker, L.F.J., Smith, E.L., and Todd, A., Chemistry of the vitamin B_{12} group, Part V. The structure of the chromophoric grouping, J. Chem. Soc., 1158-1168, 1957.

[74] Wagner, F., Reactions of the cyano and alkyl cobalamines, Proc. Roy. Soc. **A 288,** 344, 1965.

[75] Pratt, J.M., Inorganic Chemistry of Vitamin B_{12}, Academic Press, London, New York 1972, pp.284-285.

[76] Wagner, F., and Bernhauer, K., New aspects of the corrinoid coenzymes, Ann. N.Y. Acad. Sci. **112,** 280-284, 1964.

[77] Pratt, J.M., Inorganic Chemistry of Vitamin B_{12}, Academic Press, London, New York 1972, pp.193, 280-285.

[78] Rapp, P., Bozler, G., und Friedrich, E., Posi-

tionsermittlung der freien Propionsäuregruppe zweier nucleotidhaltiger Corrinoidmonocarbonsäuren, Hoppe-Seyler's Z. Phys. Chem. **354**, 970-973, 1973.
[79] Dolphin, D. M., Johnson, A. W., Shaw, N., Sulphitocobalamin, Nature **199**, 170-171, 1963.
[80] Wagner, F., and Renz, P., Tetrahedron Letters, 259, 1963.
[81] Friedrich, W., Fermente, Hormone, Vitamine, Vol III/2, Vitamin B_{12} und verwandte Corrinoide, Eds. Ammon, R., and Dirscherl, W., Georg Thieme Verlag, Stuttgart 1975, p. 298.
[82] Bonnet, R., Godfrey, J. M., Redman, D. G., Cobyric acid and related compounds from vitamin B_{12}, J. Chem. Soc. **C.8**, 1163-1166, 1969.
[83 a] Bonnet, R., Godfrey, J. M., Math, V. B., Edmond, E., Evans, H., and Hodder, O. J. R., Neovitamin B_{12} identified, Nature **229**, 473-476, 1971.
[83 b] Bonnett, R., Godfrey, J. M., Math, V. B., Cyano-13-epicobalamin (neovitamin B_{12}) and its relatives, J. Chem. Soc. **C.22**, 3736-3743, 1971.
[84] Bieganawski, R., and Friedrich, W., in Vitamin B_{12}, Eds. B. Zagalah, W. Friedrich, Walter de Gruyter, Berlin 1979, pp. 647-648.
[85] Rapp, P., Oltersdorf, V., Darstellung und biologische Eigenschaften von 8-Epicorrinoiden, Hoppe-Seyler's Z. Physiol. Chem. **354**, 32-36, 1973.
[86] Koppenhagen, V. B., and Pfiffner, J. J., α-[5,6-dimethylbenzimidazolyl]hydrogenobamide and its copper and zinc analogues, J. Biol. Chem. **246**, 3075-3077, 1971.
[87] Koppenhagen, V. B., Elsenhaus, B., and Wagner, F., Methylrhodibalamin and 5′-deoxyadenosylrhodibalamin, the rhodium analogues of methylcobalamin and cobalamin coenzyme, J. Biol. Chem. **249**, 6532-6540, 1974.
[88] Toohey, J. I., A Vitamin B_{12} compound containing no cobalt, Proc. Natl. Acad. Sci. **54**, 934-942, 1965.
[89] Sato, K., Shimizu, S., and Fukui, S., A cobalt-free corrinoid compound in *Streptomyces olivaceus*, Biochem. Biophys. Res. Comm. **39**, 170-174, 1970.
[90] Koppenhagen, V. B., and Pfiffner, J. J., Currins and zirrins, two new classes of corrin analogues, J. Biol. Chem. **245**, 5865-5873, 1970.
[91] Carmel, R., and Koppenhagen, V. B., The effect of rhodium and copper analogs of cobalamin on human cells in vitro, Arch. Biochem. Biophys. **184**, 135-140, 1977.
[92] Bernhauer, K., Friedrich, W., Angew. Chem. **66**, 776-780, 1954.
[93] Ford, J. E., Holdsworth, E. S., and Kon, S. K., Biochem. J. **59**, 86-89, 1955.
[94] Fantes, K. H., and O'Callaghan, C. H., Biochem. J. **59**, 79, 1955.
[95] Pawelkiewicz, J., and Nowakowska, K., Acta Biochim. Polon. **2**, 259-264, 1955
[96] Smith, E. L., Vitamin B_{12}, John Wiley and Sons, Methuen, New York 1960, pp. 81-96.
[97] Friedrich, W., and Bernhauer, K., Medizinische Grundlagenforschung Vol 2, Ed. Bauer, K. F., Thieme, Stuttgart 1959, p. 661.
[98] Pawelkiewicz, J., Acta Biochim Polon. **1**, 313-316, 1954.
[99] Friedrich, W., and Bernhauer, K., Angew. Chem. **65**, 627-631, 1953.
[100] Friedrich, W., and Bernhauer, K., Chem. Berichte **89**, 2030-2035, 1956.
[101] Coates, M. E., Ford, J. E., and Harrison, G. F., Activity for chicks and microorganisms of some analogues of cyanocobalamin, 2. Europäisches Symposion, Hamburg 1961, Ed. Heinrich, H. C., Ferdinand Enke Verlag, Stuttgart 1962, pp. 276-280.
[102] Heinrich, H. C., und Gabbe, E. E., Strukturspezifität des Vitamin B_{12}-Stoffwechsels und der Vitamin B_{12}-Bioaktivität, 2. Europäisches Symposion, Hamburg 1961, Ed. Heinrich, H. C., Ferdinand Enke Verlag, Stuttgart 1962, pp. 252-275.
[103] Irion, E., Ljungdahl, L., Isolation of factor III_m coenzyme and cobyric acid coenzyme plus other B_{12} factors from *Clostridium thermoaceticum*, Biochemistry **4**, 2780-2790, 1965.
[104] Ljungdahl, L., Irion, E., and Wood, M. G., Total synthesis of acetate from CO_2, I. Co-methylcobyric acid and Co-(methyl)-5-methoxybenzimidazolylcobamide as intermediates with *Clostridium thermoaceticum*, Biochemistry **4**, 2771-2780, 1965.
[105] Friedrich, W., Fermente, Hormone, Vitamine, Band III/2, Vitamin B_{12} und verwandte Corrinoide, Eds. Ammon R. and Dirscherl W., Georg Thieme Verlag, Stuttgart 1975, pp. 27-30, 40, 54, 180-187.
[106] Friedrich, W., and Bernhauer, K., Angew. Chem. **71**, 284-290, 1959.
[107] Kurnatowski, J., Szymkowiak, A., Wisniewski, W., Pawelkiewicz, J., Acta Biochim. Polon. **5**, 19-25, 1958.

[108] Pfiffner, J.J., Calkins, D.G., Dion, H.W., Characterization of pseudovitamin B_{12}-pigments, Fed. Proc. **13,** 274-280, 1954.

[109] Kon, S.K., The Biochemistry of Vitamin B_{12}, Biochemical Soc. Symposium, Cambridge 1955, p. 17.

[110] Friedman, H.C., and Fyfe, J.A., Pseudovitamin B_{12} biosynthesis, J. Biol. Chem. **244,** 1667-1671, 1969.

[111] Barker, H.A., Smyth, R.D., Weissbach, H., Munch-Petersen, A., Toohey, J.I., Ladd, J.N., Volcani, B.E., and Wilson, R.M., Assay, purification and properties of the adenylcobamide coenzyme, J. Biol. Chem. **235,** 181-190, 1960.

[112] Barker, H.A., Weissbach, H., and Smyth, R.D., Proc. Natl. Acad. Sci. USA **44,** 1093-1097, 1958.

[113] See section 7.1.

[114] Pawelkiewicz, J., Enzymatic preparation of coenzyme B_{12} and some of its analogs, Methods in Enzymology **18 C,** 99-101, 1971.

[115] Friedrich, W., Fermente, Hormone, Vitamine, Band III/2, Vitamin B_{12} und verwandte Corrinoide, Ammon R. and Dirscherl W., Eds., Georg Thieme Verlag, Stuttgart 1975, pp. 31-32, 49, 174-183, 304.

[116] Kopf, J., von Deuten, K., Bieganowski, R., and Friedrich, W., X-ray crystal structure analysis of factor A (2-methyladeninyl-cyanocobamide) a native B_{12} analog, Z. Naturforsch. Sect. C Biosci. **36,** 506-515, 1981.

[117] Perlman, D., Barrett, J.M., and Jackson, P.W., Cobamides synthetized by *Propionibacterium* species, 2. Europäisches Symposion Hamburg 1961, Ed. Heinrich, H.C., Ferdinand Enke Verlag, Stuttgart 1962, pp. 58-69.

[118] Bernhauer, K., Müller, O., Wagner, F., Zum Mechanismus der Biosynthese des Vitamin-B_{12}-Moleküls, 2. Europäisches Symposion Hamburg 1961, Ed. Heinrich, H.C., Ferdinand Enke Verlag, Stuttgart 1962, pp. 37-58.

[119] Friedrich, W., Bernhauer, K., 2-Methyl Merkaptoadenin Cobalamin Analogion, ein neuer B_{12}-Faktor des Faulschlammes, Chem. Ber. **90,** 1966, 1957.

[120] Ford, J.E., and Porter, J.W.G., Biochem. J. **51,** 5 P, 1952.

[121] Ford, J.E., and Porter, J.W.G., Brit. J. Nutr. **7,** 326, 1953.

[122] Ford, J.E., Holdsworth, E.S., and Kon, S.K., Biochem. J. **59,** 86, 1955.

[123] Hayward, G.C., Hill, H.A.O., Pratt, J.M., Vanston, N.J., and Williams, R.J.P., J. Chem. Soc., 6485, 1965.

[124] Armitage, J.B., Cannon, J.R., Johnson, A.W., Parker, L.F.J., Smith, E.L., Stelford, W.H., and Todd, A.R., Chemistry of the vitamin B_{12} group. Part III. The course of hydrolytic degradations, J. Chem. Soc., 3849-3864, 1953.

[125] Friedrich, W., and Bernhauer, K., Über den Abbau der Cobalamine mit Cl(III)-hydroxyd. 7-D-Ribofuranozido-adenin, ein Abbauprodukt des Pseudovitamin B_{12}, Chem. Ber. **89,** 2507-2512, 1956.

[126] See Chapter 5.

[127] See section 3.1.2.3.

[128] Friedrich, W., Z. Naturforsch. **21 b,** 595, 1966.

[129] Friedrich, W., and Moskophidis, M., Zur Koordinations-Isomerie von Kobaltmethyl-Faktor Ib und Kobalt-methyl-cobalamin, Zeitschrift für Naturforschung **25 b,** 979-983, 1970.

[130] Hayward, G.C., Hill, H.A.O., Pratt, J.M., and Williams, R.J.P., J. Chem. Soc. (A) 196, 1971.

[131] Stefaniak, O., Łubieńska, B., and Janicka, M., Isolation and purification of corrinoids synthesized by *Propionibacterium shermanii,* Acta Microbiologica Polonica **17,** 83-90, 1968.

[132] Menke, K.H., Trennung und Charakterisierung der Vitamin B_{12}-Analogia im Intestinum des Huhns, 2. Europäisches Symposion Hamburg 1961, Ed. Heinrich, H.C., Ferdinand Enke Verlag, Stuttgart 1962, pp. 74-82.

[133] Pawelkiewicz, J., and Bartosinski, B., Bull. Acad. Polon. Sci. **8,** 5-9, 1960.

[134] Pawelkiewicz, J., Bartosinski, B., and Walerych, W., Bull. Acad. Polon. Sci. **8,** 123-126, 1960.

[135] Bartosinski, B., Enzymic synthesis of the coenzyme form of cobinamide (SB_{12p}), Bulletin De L'Academie Polonaise Des Sciences **10,** 189-194, 1962.

[136] Bartosinski, B., Pawelkiewicz, J., Colorimetric method of determination of synthetase activity of the coenzyme form of cobinamide, Bulletin De L'Academie Polonaise Des Sciences **10,** 121-124, 1962.

[137] Pawelkiewicz, J., Bartosinski, B., and Walerych, W., Enzymic synthesis of light-sensitive cobalto II-corrins, Bulletin De L'Academie Polonaise Des Sciences **8,** 131-141, 1961.

[138] Irion, E., Ljungdahl, L., Biochemistry **7,** 2350, 1968.

[139] Walerych, W., Kato, T., and Pawelkiewicz, J., Enzymic synthesis of vitamin B_{12}, Biochem. Biophys. Res. Comm. **31,** 328-332, 1968.

[140] Zodorow, K., Stefaniak, O., Acta Microbiol. Pol. **12,** 267, 1963.

[141] Pawelkiewicz, J., Walerych, W., Bartosinski, B., Acta Biochim. Pol. **6,** 431, 1959.

[142] Renz, P., Some intermediates in the biosynthesis of vitamin B_{12}, in Methods in Enzymology **18 C,** pp. 99-101, 1971.

[143] DiMarco, A., Boretti, G., Migliacci, A., Julita, K., Minghetti, L., Boll. Sci. Ital. Biol. Sper. **33,** 1513, 1957.

[144] Pawelkiewicz, J., Acta Biochim. Pol. **3,** 581, 1956.

[145] Pawelkiewicz, J., Zodrow, K., Acta Biochim. Polon. **4,** 203, 1957.

[146] Pawelkiewicz, J., Zodrow, K., Prekursory biosyntezy nukleotydocyjanokobalamin, Acta Microbiologica Polonica **6,** 9-15, 1957.

[147] Boretti, G., DiMarco, A., Fuocol, [redacted], Marnatt, M. P., Migliacci, A., Spalla, C., Intermediates in the biosynthesis of vitamin B_{12}, Biochim. Biophys. Acta **37,** 379-380, 1960.

[148] Rosenblum, C., Stability of radioactive vitamin B_{12}, 2. Europäisches Symposion Hamburg 1961, Ed. Heinrich, H. C., Ferdinand Enke Verlag, Stuttgart 1962, pp. 294-305.

[149] Rosenblum, C., Radioactive vitamin B_{12} and its application in biochemical research, 2. Europäisches Symposion Hamburg 1961, Ed. Heinrich, H. C., Ferdinand Enke Verlag, Stuttgart 1962, pp. 306-318.

[150] Smith, E. L., Instability of radioactive vitamin B_{12}, Lancet, 387-388, 1959.

[151] Booth, M. A., and Spray, G. H., The excretion and distribution in the tissues of parenterally administered ^{38}Co-Vitamin B_{12} in normal and gastrectomized rats, British J. Haematology **8,** 114-119, 1962.

[152] Barlow, G. H., and Sanderson, N. D., Self-degradation of [^{60}Co]vitamin B_{12}, Biochim. Biophys. Acta **41,** 146-147, 1960.

[153] Rosenthal, H. L., in The Vitamins, Vol 2, Eds. Sebrell and Harris, Academic Press, New York 1968, p. 145.

[154] Mervyn, L., in Comprehensive Biochemistry, Vol 21, Eds. Florkin and Stotz, Elsevier, Amsterdam 1971, p. 153.

[155] Jaselskis, B., and Diehl, H., The polarography of vitamins B_{12r} and B_{12a}, J. Am. Chem. Soc. **76,** 4345-4348, 1954.

[156] Boos, R. N., Carr, J. E., and Conn, J. B., Science **117,** 603, 1953.

[157] Bernhauer, K., Müller, O., and Wagner, F., Adv. Enzymology **26,** 233, 1964.

[158] Krotochvil, B., and Dichl, H., Talanta **13,** 1013, 1966.

[159] Müller, O., and Müller, G., Biochem. Zeit. **336,** 299, 1962.

[160] Dichl, H., Morrison, J. I., Sealock R. R., Experientia **7,** 60, 1951.

[161] Tackett, S. L., Collat, J. W., and Abbott, J. C., Biochemistry **2,** 919, 1963.

[162] Hogenkamp, H. P. C., and Holmes, S., Polarography of cobalamins and cobinamides, Biochemistry **9,** 1886-1892, 1970.

[163] Costa, Puxeddu, Reisenhofer, Collect. Czechoslov. Chem. Comm. **36,** 1065, 1971.

[164] Randall, W. C., and Alberty, R. A., Kinetics of ligand binding to aquocobalamin, Biochemistry **6,** 1520-1525, 1967.

[165] Randall, W. C., and Alberty, R. A. The kinetics of thiocyanate binding to aquocobalamin, Biochemistry **5,** 3189-3193, 1966.

[166] Saveant, J. M., Tacconi, N., Electrochemistry of vitamin B_{12}. Equilibria, kinetics and mechanisms in the B_{12a}, -B_{12r}, -B_{12s} oxido-reduction, in Vitamin B_{12}, Eds. B. Zagalak and W. Friedrich, Walter de Gruyter, Berlin 1979, pp. 203-212.

[167] Pratt, J. M., Inorganic chemistry of vitamin B_{12}, Academic Press, New York 1972, pp. 109-113.

[168] Lexa, D., Lhoste, J. M., Experientia Suppl. **18,** 395-405, 1971.

[169] Abd-El-Nabey, B. A., J. Electronal. Chem. **53,** 317-324, 1974.

[170] Enhyercz, T. H., Mark, H. B., Anal. Lett. **7,** 1-7, 1974.

[171] Swetik, P. G., Brown, D. G., J. Electroanal. Chem. **51,** 433, 1970.

[172] Levitin, I. Ya., Rudacova, I. P., Sigan, A. L., Pospalova, T. A., Yurkevich, A. M., Volpin, M. E., Zh. Obsch. Khim. **45,** 1879, 1975.

[173] Lexa, D., Electrochemistry of vitamin B_{12}. Alkyl cobalamins and cobinamides, in Vitamin B_{12}, Eds. B. Zagalak, W. Friedrich, Walter de Gruyter, Berlin 1979, pp. 213-216.

[174] Schmidt, C. L., and Swofford, H. S., Electrochemical investigation of surface phenomena at a mercury electrode in vitamin B_{12a} solution, Anal. Chem. **51,** 2026-2033, 1979.

[175] Rubinson, K. A., EPR, electron nuclear double resonance, circular dichroism and magnetic circulator dichroism of copper-63-labeled cobalamin, J. Am. Chem. Soc. **101,** 6105-6110, 1979.

[176] Lexa, D., Saveant, J. M., and Zickler, J., Elec-

trochemistry of vitamin B_{12}: 5. Cyanocobalamins, J. Am. Chem. Soc. **102,** 2654-2663, 1980.

[177] De Tacconi, N. R., Lexa, D., and Saveant, J. M., Electrochemistry of vitamin B_{12}: 4. Kinetics and mechanisms in B-12a-B-12r oxido-reduction, J. Am. Chem. Soc. **101,** 467-473, 1979.

[178] Elsenhans, B., and Rosenberg, J. M., Influence of metal substitution on vitamin B_{12} binding to human intrinsic factor and transcobalamin I and II, Biochemistry **23,** 805-808, 1984.

4 Biosynthesis of Vitamin B_{12}

Z. Schneider

The biosynthesis of vitamin B_{12} is one of the most complex chemical operations performed by living systems. Only microorganisms and some algae are equipped with the complete enzyme systems catalyzing the multistep reactions leading to the biologically active vitamin B_{12} while animals and plants are capable of synthetizing a few precursors of the corrin ring. This is mainly because the basic part of the vitamin B_{12} (the corrin ring) shares similar structural features with other important compounds occurring in nature such as heme and chlorophyll (Figs. 4-1, 4-2) [167]. Indeed, the first steps in the biosynthesis of these compounds are similar or identical [1-7]. They originate from the precursor delta-aminolevulinic acid. The common pathway ends with enzymatic condensation of four porphobilinogen molecules to uroporphyrinogen III (uro'gen III). Beyond uro'gen III, already a complex compound, exists a long processing line, involving approximately 30 enzymatic conversions, which lead to the biologically active corrinoid [7-160]. It took a quarter of a century of research by several research groups to elucidate the main steps. Although many enzymatic conversions may be followed in vitro, most of the enzymes involved have not been obtained in pure state. As far as the elucidation of the sequence of reactions and the structure of intermediate compounds is concerned, a major break-through was made recently in several laboratories owing to the application of ^{1}H and ^{13}C NMR spectroscopy and radioactive precursors labelled with ^{14}C and ^{3}H. The details are described in the legend to Fig. 4-2. The early stages of the tetrapyrrole compound biosynthesis are inhibited by heme and the complete form of vitamin B_{12}. There are, however, differences in the mode of feedback action of these compounds. While heme inhibits the formation of delta-aminolevulinic acid and uro'gen III, the complete vitamin B_{12}, but not its intermediates, represses the methylation of uro'gen III [7]. The nucleotide moiety of vitamin B_{12} is synthetized in an independent pathway [100-124]. It is observed quite frequently in a number of microorganisms that incomplete forms of vitamin B_{12}, depleted of the nucleotide are produced in appreciable quantities [36, 37, 125, 143] (see also Chapter 6). In case of the aerotolerant anaerobe *Propionibacterium shermanii*, (the favourite strain for studies of B_{12} biogenesis), the intermediates: cobinamide (Cbi) and cobinamide-GDP occur in young (several days old) cultures grown anaerobically. Subsequent aeration of the medium increases their conversion to vitamin B_{12} (Table 4-1, Table 6-1). This is due to increased biosynthesis of the precursor 5,6-dimethylbenzimidazole under aerobic conditions. Indeed, synthetic dimethylbenzimidazole added to the not aerated medium provides for the conversion of the cobinamides into vitamin B_{12}. Since only the complete forms of corrinoids inhibit the biosynthesis of the corrin ring, high yield of vitamin B_{12} from bacteria is obtained when the dimethylbenzimidazole is not supplemented to the growth medium until before the harvest. This observation, already made two decades ago, has been widely utilized for preparative and industrial production of vitamin B_{12}, although at that time the exact mode of feedback action of vitamin B_{12} was unknown. A number

Table 4-1 Effect of growth medium and aeration on yield of cobalamin production by *Propionibacterium* species (compiled by Perlman et al. [102]).

species	incubation temp.	composition of medium	batch volume	aeration	yield	ref.
P. freudenreichii	30 °C	glucose; casein hydrolyzate; yeast extract; lactate	150 ml	none	3.4 mg/l in 90 h	[145]
P. freudenreichii	30 °C	glucose; peptone; yeast; $CoCl_2$; PO_4 buffer	15 l	low	2.4 mg/l in 96 h	[146]
P. freudenreichii *L. casei*	30 °C	whey powder; casein hydrolyzate; $CaCO_3$; $MnSO_4$; $CoCl_2$	2.5 l	none	2 mg/l in 96 h	[147]
P. shermanii	30 °C	yeast extract; glucose; $CoCl_2$; NaOH (to pH 7)	100 ml	none	2.8 mg/l in 96 h	[148]
P. freudenreichii	30 °C	autolyzed yeast; beet molasses invert; NH_4OH (to pH 7)	10.800 gallons	slight	4 mg/l in 96 h	[149]
P. freudenreichii	30 °C	glucose; yeast extract; $CaCO_3$; $CoCl_2$; NaOH (to pH 7)	4 l	89 h anaerobic; 24 h aerobic	13 mg/l in 113 h	[150]
P. freudenreichii	30 °C	cornsteep liquor; glucose; $CoCl_2$; NH_4OH (to pH 7)	9.500 gallons	82 h anaerobic; 58 h aerobic	24 mg/l in 140 h	[151]
P. shermanii	30 °C	cornsteep liquor; glucose; $CoCl_2$; NH_4OH (to pH 7)	(not stated)	70 h anaerobic; 98 h aerobic	23 mg/l in 168 h	[151]
P. technicum	30 °C	cornsteep liquor; glucose; $CoCl_2$; NH_4OH (to pH 7)	(not stated)	70 h anaerobic; 98 h aerobic	21 mg/l in 168 h	[151]

of microorganisms, in particular the anaerobes, incorporate, instead of dimethylbenzimidazole, other bases into the corrinoids, such as 5-hydroxybenzimidazole, or its derivatives or adenine and its derivatives [125–158]. The latter cobamides are inactive for man and animals but are active as cofactors in bacterial enzyme systems (see chapters 5, 7 and 8). These forms, particularly those containing the adenine nucleotide (pseudovitamin B_{12}) seem to be the dominant corrinoids in nature. The nucleotide moiety of cobalamin may be exchanged by microorganisms for another one. Due to this feature, man and animals with bacterial overgrowth in the intestine, may be depleted of cobalamin in food, since the pseudovitamin B_{12} can be produced by the intestinal flora from cobalamin [144, 159]. Bacteria may also produce a great number of B_{12} analogues by incorporating synthetic benzimidazole- or adenine-like bases into the nucleotide moiety of the corrinoid (see Fig. 4-3 and Table 4-2). This phenomenon

Table 4-2 Vitamin B_{12} analogues obtained by guided biosynthesis (No. 1-35) or isolated from various sources as naturally occuring compounds (No. 36-49) (compiled by Friedrich [36]).

No.	Base in the nucleotide (or aglycon)	Source	Biological activity relative to CN-Cbl (CN-Cbl = 100)				
			E. coli	*Lactobac. leichm.*	*Euglena gracilis*	*Ochromonas malhamens.*	*Pernicious anaemia*
Cobamides obtained by guided biosynthesis							
1.	5,6-Diethylbenzimidazole	*E. coli* *N. rugosa*	100	130		52	active
2.	5,6-Dichlorobenzimidazole	*E. coli* *Str. griseus* *P. shermanii*	50	100		> 100 91	active
3.	5,6-Dinitrobenzimidazole	*P. shermanii* *C. diphteriae*	100		100		
4.	5,6-Dimethoxybenzimidazole	*P. shermanii*	62 10	44		0 H. I. = 550	
5.	5(6)-Chloro-6(5)-methylbenzimidazole	*E. coli*					
6.	5(6)-Bromo-6(5-)-methylbenzimidazole	*E. coli* *N. rugosa*	active active	active active		active active	
7.	5(6)-Nitro-6(5)-methylbenzimidazole	*P. shermanii*	active		100	active	
8.	5(6)-Methyl-6(5)-methoxybenzimidazole	*P. freudenreichii*				50-67	
9.	5-Methoxy-7-bromobenzimidazole	*P. arabinosum*		active		active	
10.	5-Methyl-6-aminobenzimidazole	*N. rugosa*	125	~ 84	~ 83	~ 60	
11.	6-Methyl-5-aminobenzimidazole	*N. rugosa*	125	~ 84	~ 83	~ 60	
12.	5(6)-Ethoxybenzimidazole	*P. shermanii*			100		
13.	5(6)-Nitrobenzimidazole	*E. coli* *P. shermanii* *P. arabinosum*		active	100	active	
14.	5(6)-Trifluoromethylbenzimidazole	*Str. griseus* *P. arabinosum*		active		active	active
15.	5-Aminobenzimidazole	*P. shermanii*	80.3	36.9		1.3	

Table 4-2 continued

No.	Base in the nucleotide (or aglycon)	Source	Biological activity relative to CN-Cbl (CN-Cbl = 100)				
			E. coli	*Lactobac. leichm.*	*Euglena gracilis*	*Ochromonas malhamens.*	*Pernicious anaemia*
16.	6-Aminobenzimidazole	*P. shermanii*	71.5	37.2		0.7	
17.	5-Acetaminobenzimidazole	*P. shermanii*	86	42		1	
18.	6-Acetaminobenzimidazole	*P. shermanii*	13.2	0.0		0.0	
19.	5(6)-Aminobenzimidazole	*E. coli*	active				active
20.	5,7-Dimethylbenzimidazole	*E. coli*					
21.	Benztriazole	*P. shermanii* *P. arabinosum*		active		active	
22.	5(6)-Chlorobenztriazole	*E. coli*	active			active	
23.	5(6)-Ethyl-6(5)-propylbenzimidazole	*N. rugosa*	120	85	84	34	
24.	5(6)-Chlorobenzimidazole	*E. coli*					
25.	5,6-Imidazobenzimidazole	*P. shermanii*	100		100		
26.	5,6-Tetramethylenbenzimidazole	*N. rugosa*	65	25–50	40	50	
27.	2,6-Diaminopurine	*P. arabinosum*					
28.	6-Methylthiopurine	*P. arabinosum*					
29.	6-Chloropurine	*P. arabinosum*					
30.	Purine	*P. arabinosum* *P. petersonii* *P. shermanii*	66	15		0.0	
31.	2-Chloroadenine	*P. arabinosum*	~ 50	~ 50		0.0	
32.	2-Thioadenine	*P. arabinosum*				18	
33.	6′7′-Dimethylnaphtho-(2′,3′)-imidazole4,5	*P. shermanii*	50	100			
34.	4,5-Dimethylimidazole	*P. shermanii*	50	10.2		7.9	

Table 4-2 continued

No.	Base in the nucleotide (or aglycon)	Source	Biological activity relative to CN-Cbl (CN-Cbl = 100)				
			E. coli	*Lactobac. leichm.*	*Euglena gracilis*	*Ochromonas malhamens.*	*Pernicious anaemia*
35.	6′-Methylmercaptopurine	*P. arabinosum*					
Naturally occurring cobamides							
36.	5-Hydroxybenzimidazole (in factor III)	sewage sediments *M. omelianskii* *E. coli 113-3* *P. shermanii*	64	53 50		41 69.2	active
37.	5-Methoxybenzimidazole (in factor III m)	*Cl. thermoaceticum*	85	87		68	active
38.	5-Methylbenzimidazole	sewage sediments *E. coli 113-3* *E. coli* *P. shermanii* *P. arabinosum*	70	96	active	110	active
39.	6-Methylbenzimidazole	sewage sediments *P. shermanii*		92		106	
40.	Benzimidazole	sewage sediments activated sludge *E. coli* *E. coli 113-3* *P. shermanii* *P. arabinosum* *P. olivaceus* *S. griseus*	30	100	100	40	active
41.	Linear Naphthimidazole	*E. coli* *S. griseus* *P. shermanii* activated sludge	70	80		60 88.84	active
42.	Adenine (in 7 α-adenylcobamide)	*E. coli 113-3* A microorganism from rumen	10	50	100	0.0	inactive
43.	Adenine (in 9 α-adenylcobamide)	*P. shermanii*					
44.	Hypoxanthine (in factor G)	faeces		20–100			
45.	2-Methyladenine (in factor A)	faeces *E. coli*	50	40 20	60	0	inactive or little active

Table 4-2 continued

No.	Base in the nucleotide (or aglycon)	Source	Biological activity relative to CN-Cbl (CN-Cbl = 100)				
			E. coli	*Lactobac. leichm.*	*Euglena gracilis*	*Ochromonas malhamens.*	*Pernicious anaemia*
46.	2-Methylhypoxanthine (in factor H)	faeces	40	15–40		0	
47.	2-Methylmercaptoadenine	sewage sediments *E. coli 113-3*	61	51		3.5	
48.	Guanine	sewage sediments	16	17		0.8	
49.	Kresol (in factor I b)	sewage sediments	15	0.0		0.0	

has been applied in a method termed "guided biosynthesis" for production of corrinoids not present in nature (see Table 4-2 and section 5.4). The pollution of soil and water by industrial compounds, which can give rise to biologically inactive corrinoids, is a potential threat to the environment. The concentration of cobalt is a crucial limiting factor in the synthesis of all corrinoids. As shown by Bykhovsky, the biosynthesis of corrinoids stops in the absence of cobalt at the stage of sirohydrochlorine [7]. The microorganisms are very effective in the uptake of cobalt ions from the environment. *Propionibacterium shermanii* take up [^{60}Co] Co^{2+} quantitatively even at a concentration of 10^{-12} M. The uptake of cobalt by *P. arabinosum* is effectively inhibited at the stage of adsorption to cell surface by divalentions such as (the concentration of $I_{0.5}$ inhibition in brackets): Mn^{2+} (4×10^{-5} M), Mg^{2+} (5×10^{-5} M), Zn^{2+} (2×10^{-4} M) and Cd^{2+} (4×10^{-4} M); it is somewhat less inhibited by Ca^{2+} (8×10^{-4} M) and Pb^{2+} (2×10^{-3} M), while Cu^{2+} and Hg^{2+} are stimulatory at conc. 10^{-2} M [169]. Ruminants need cobalt for vitamin B_{12} production in the rumen (see section 6.5) and depend on supply of this element from the plants. Many plants and in particular grasses do not transport readily ionic cobalt into the green parts, although the roots adsorb cobalt as effectively as do bacteria. Some plants, under water deficient conditions, withdraw readily the ionic cobalt from the leaves back to the roots [169]. The growth (bloom) of many algal species in natural waters depends on the availability of cobalt for vitamin B_{12} production (see Tables 6-3 A, 6-3 B and 6-3 C and also the text and references in sections 6.2 and 6.3). The distribution of this ion in soil and natural waters is thus of ecological and agricultural significance.

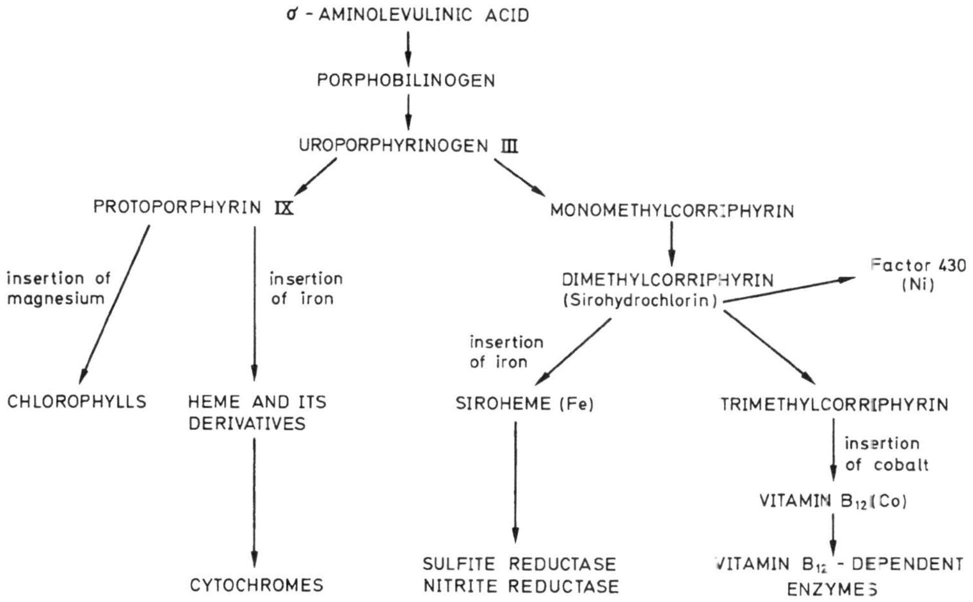

Fig. 4-1 The evolutional and biogenetic regularities of interrelation of tetrapyrrole compounds. From V. Ya. Bykhovsky [7].

Pathway of vitamin B_{12} biosynthesis[a]

(description of the pathway presented in Fig. 4-2). Fig 4-2 can be found as a fold-out in the back cover.

① **Condensation of glycin and succinyl-CoA to delta-aminolevulinic acid (ALA).** The enzyme ALA synthase (EC 2.3.1.37) has been found in bacteria plants and animals [1-6]. It is inhibited by heme [7].

② **Condensation of two ALA molecules to porphobilinogen (PBG)** [8-12]. The enzyme, porphobilinogen synthase (EC 4.2.1.24) is present in bacteria plants and animals. It is

[a] Names of compounds and abbreviations as given in the Figure

1 delta-aminolevulinic acid (ALA)
2 porphobilinogen (PBG)
3 uroporphyrinogen III (Uro'gen III)
4 corriphyrin I (corr. I)
5 corriphyrin II (corr. II)
6 corriphyrin III (corr. III)
7 cobyrinic acid
8 cobyric acid (Cby)
9 cobinamide (Cbi)
10 cobinamide phosphate (Cbi-P)
11 cobinamide guanosinediphosphate (Cbi-GDP)
12 deoxyadenosyl-cobalamin 5'-phosphate (AdoCbl-5'-P)
13 riboflavin
14 5,6-dimethylbenzimidazole (Me_2Bza)
15 1-β-D-ribofuranosyl nicotinic acid (β-N_aMN)
16 1-α-D-ribofuranosyl 5,6-dimethylbenzimidazole-5'-phosphate (α-Me_2Bza-R-5'-P)
17 5'-deoxy-5'-adenosyl-cobalamin (AdoCbl)-vitamin-B_{12}-coenzyme

inhibited by heme, an analog of ALA, 4-keto-5-amino-6-hydroxyhexanoic acid produced by *Bacillus cereus* [163], and by succinylacetone pyrrole [164, 165], but not by corrinoids.

③ **Condensation of four porphobilinogen molecules to uroporphyrinogen III (uro'gen III).** The reaction proceeds in two stages which are catalyzed by two enzymes named porphobilinogen deaminase and uroporphyrinogen cosynthetase (EC 4.3.1.8) [10–19]. The deaminase catalyzes a complex reaction in which four porphobilinogen molecules are deaminated and condensed, in a head-to-tail fashion, into a linear compound which was recently identified as hydroxymethylbilane [14, 15, 19, 166]. The function of the uro'gen III cosynthetase is to rearrange (reverse) the fourth pyrrol ring D in hydroxymethylbilane and close the macroring. In the absence of cosynthetase the labile hydroxymethylbilane forms spontaneously a biologically not utilized macroring isomer uro'gen I [14]. The formation of uro'gen III is inhibited by heme but not vitamin B_{12} [7]. The metabolic pathway common for heme, chlorophyll and the corrinoids ends at this stage.

④ ⑤ **Consecutive reductive methylation of uro'gen III.** [7, 15, 15–35]. The donor of the methyl groups is S-adenosylmethionine (SAM). The methyls are first introduced at C-2 and C-7. The methylation is accompanied by reduction of the A and B pyrrol rings and the formation of at least two double bonds (at C-5 and C-10). As to number of methin bridges at this stage the structural formulae of the two intermediates shown in this figure is the subject of verification. The compound <u>5</u> can be readily oxidized with consecutive formation of further two double bonds in the macroring at C-15 and C-20. From *P. shermanii* culture medium has been isolated also a partially oxidized form of (**5**), which seems to have three methin bridges at C-5, C-10 and C-15 (or C-20) [19]. Which of these forms is the intermediate of the biosynthetic path remains to be solved. The presence of methin bridges has its effect on conformation and size of the entire macroring. They seem to be necessary for introduction of cobalt in the next steps. The methylating enzyme system has been isolated from *Clostridium tetanomorphum* and partially purified [30]. Its molecular weight is 220000 daltons, pH optimum around 7 and the K_m-value 2.17×10^{-4} M for SAM. The enzyme is inhibited by S-adenosylhomocysteine (SAH) and by cobalamin. The stereospecific introduction of the methyl groups has been followed by ^{13}C NMR spectroscopy. The products corriphyrin I (monomethyl corriphyrin) and corriphyrin II (dimethylcorriphyrin) were named also by other authors as factors I and II respectively [7, 30]. The latter isolated independently as metalfree prosthetic group of sulphite reductase from *Desulphovibrio gigas* and related species has been named sirohydrochlorin (see Fig. 4-1). This compound is accumulated also by *P. shermanii* grown in Co^{2+} deficient media [7 a]. Methylation antagonists such as ethionine inhibit in vivo the formation of tetrapyrrole compounds in *P. shermanii* [7 b].

⑥ **Methylation at C-20 and dehydrogenation at C-15 and C-20.** The introduction of the third methyl group at C-20 by SAM and cell-free extract of *C. tetanomorphum* or *P. shermanii* has been followed by ^{14}C and ^{3}H double labelling experiments as well as by ^{13}C NMR spectroscopy [19–28, 30, 31]. The methylation at C-20 was rather unexpected since this carbon joining A and D ring is later excised from the macro-ring. The product is named by various authors as: corriphyrin III, trimethylcorriphyrin, factor III, methylsirohydrochlorin or trimethylisobacteriochlorin [7, 30–35]. At this stage the remaining two methin bridges at C-15 and C-20 seem to be formed. A dehydrogenation at these carbons is required. So far supporting biochemical and structural studies regarding the presence of four methin bridges in corriphyrin III have not been presented.

⑦ **Formation of cobyrinic acid.** Cobyrinic acid 7 is the first known intermediate on the biosynthetic path which contains the typical macroring skeleton for corrinoids called *corrin ring* (see Chapter 2). It is characterized by a direct link of two pyrrole rings A and D. The transformation of 6 to cobyrinic acid involving several steps has not been described so far in terms of sequence of events, intermediates and the particular enzymes engaged except the adenosylating enzyme. The cobyrinic acid has been isolated from many microorganisms at various occasions and it may be obtained from the more accessible cobyric acid 8 by chemical deamidation [3, 7–10, 35–39] (Fig. 3-3). Since all, except one, of the particular enzymes engaged at this stage have not been described yet, the conversions listed below are concluded mainly from the structural formula of cobyrinic acid and a few studies with labelled precursors:

1. Removal of the C-20 carbon and the attached methyl group (see above) and the formation of a direct link between pyrrole rings A and D. In contrast to earlier postulates labelling studies showed that both carbon atoms (C-20 and the attached methyl) are removed in course of the cobyrinic acid formation [25–27].
2. Insertion of cobalt.
3. Decarboxylation of the acetic acid residue at C-12.
4. Methylation at C-1, C-5, C-12, C-15 and C-17.
   SAM is the donor of the methyl groups, but the enzymes have not been isolated. The methylation at C-12 and C-17 is accompanied by reduction of the C and D pyrrole rings. The sequence of methylation is not known.
5. Attachment of 5′-deoxyadenosine to cobalt. The 5′-deoxyadenosine, attached to cobalt with the 5′-carbon, is considered as the sixth ligand or β ligand according to nomenclature (see Chapter 2). The Co-C bond is rather unusual in coordination chemistry (see Chapter 3). Similarly to chemical synthesis the enzymatic adenosylation runs through a reduced corrinoid intermediate (B_{12s}) Thus the reaction consists of two coupled steps: reduction of the corrinoid by two NADP (or NAD) dependent flavoproteins aquacob(III)alamin reductase (EC 1.6.99.8) and cob(II)alamin reductase (EC 1.6.99.9) and adenolysation with ATP by an adenosyl transferase [36, 37, 41–49], Cob(I)alamin adenosyltransferase (EC 2.5.1.17). The enzyme seems to be of low specificity toward corrinoids since it may adenosylate besides cobyrinic acid and cobalamin, all cobalt containing vitamin B_{12} precursors and a great number of B_{12} derivatives including those containing rhodium instead of cobalt in the corrin ring (see rhodibalamin in Chapter 3). It shows also limited specificity toward the nucleoside donor. The purified Coβ-adenosylating enzyme from *C. tetanomorphum* also utilized CTP and somewhat less well ITP, UTP and GTP [41]. The adenosylating enzyme apart from bacteria, is present in animals and man. The presence of this enzyme in animals is necessary due to light sensitivity of the Co–C bond. Congenital defects of the adenosylating enzyme in human limit the coenzymatic function of vitamin B_{12} (see Chapter 9). One can not say with cer-

tainty at present whether the adenosylation occurs in bacteria at the stage of cobyrinic acid as shown in this figure or at a further stage after its partial amidation. Although the cobyrinic acid has been isolated in the adenosylated form, the adenosylation could not be demonstrated *in vitro*, by cell free extracts from *P. shermanii*. Due to seven carboxyl groups the cobyrinic acid tends to attach unspecifically to proteins and thus creates technical problems as substrate for enzymological studies [78].

⑧ ⑨ **Amidation of the cobyrinic acid and incorporation of the isopropanolamine** [39, 50–78]. There are seven carboxyl groups at the side chains in cobyrinic acid 7 designated as *a*, *b*, *c*, *d*, *e*, and *f* which undergo amidation in course of vitamin B_{12} biosynthesis. Six of them are amidated by ammonia while the seventh carboxyl at the side chain *f* is amidated with 1-amino-2-propanol. The sequence of amidation of the side chains and the stage at which the 1-amino-2-propanol is introduced have not been determined definitively. The reactions ⑧ and ⑨ depict simplistically a rather complex multistep process. Assuming that also cobinamide- and cobamide carboxylic acids can serve as substrates for the amidating enzyme(s) no less than 190 different structures can be written between cobyrinic acid and cobamide [51]. The Bernhauer group elaborated a sequence of amidation in *P. shermanii* basing on frequency of particular intermediates in course of amidation [50–55]. According to these studies one propionic acid side chain (probably *e* [56]) is preferentially amidated first followed by amidation of three acetic acid side chains. At last two remaining propionic acid side chains are amidated (probably *d* before *b* according to recent assignments [56]). These findings were verified by other investigators working with the same strain [55, 59]. The Bernhauer group went even further by indicating the particular side chains which are preferentially amidated in this step by step transformation [50]. Unfortunately the assignments of the intermediates isolated in Bernhauers laboratory turned out not to be correct as shown in recent studies on assignments monocarboxylic acid isomers of cobalamin with the ^{13}C nuclear magnetic resonance technique [56]. So far the amidating enzyme(s) has not been isolated; neither the ammonia donor is known. With regard to substitution at the *f* side chain there are no restrictions in *P. shermanii* as to the stage of introduction of the 1-amino-2-propanol in course of amidation of the other carboxyls. Labelling studies on *P. shermanii* revealed cumulation of cobyric acid 8 which seems to suggest that the amidation of the *f* side chain occurs last as shown in this figure [55]. On the other hand the cumulation of an intermediate is the indication of a faulty metabolism. A mutant strain of *Nocardia rugosa* (No. 959) incorporates selectively 1-amino-2-propanol into the *f* side chain before any amidation occurs [78]. Another mutant of this strain (No. 466) is not capable to introduce the isopropanolamine at the side chain *f* while it amidates with ammonia the other side chains [78].

Partially deamidated cobinamide and cobalamin (mono-di- and tricarboxylic acids) are amidated *in vivo* by *N. rugosa* as well. The amidating enzyme system of this strain seems to require the adenosylated form of the corrinoid precursors at the early stage of amidation [78]. These observations provide useful information on the specificity of the amidating enzyme system and suggest alternative pathways of amidation depending on bacteria species and environmental factors. Further studies with purified enzymes might solve this problem. So far only the enzyme amidating the *f* side chain has been studied. L-threonine is the precursor of the R-1-amino-2-isopropanol residue. The enzyme has been partially purified and the mechanism of L-threonine decarboxylation was studied [75-78]. In these studies cobyric acid has been used as substrate for enzymatic attachment of the isopropanol-amine residue.

⑩ **Phosphorylation of the cobinamide.** The cobinamide kinase has been demonstrated by using ATP as substrate and crude extracts of *P. shermanii* [79].

⑪ **Guanosine diphosphate cobinamide formation.** The GDP-cobinamide has been found in a number of anaerobically grown microorganisms (see Chapter 3). Its formation from cobinamide phosphate and GTP was demonstrated by using crude bacterial extracts [79-91].

⑫ **Introduction of the *α*-nucleotide moiety** [92]. The enzymatic reaction between the α-5,6-dimethylbenzimidazole nucleotide and the GDP-cobinamide has been demonstrated by an enzyme system from *E. coli 113-3* mutant [81, 82]. The enzyme system requires for activity ribosomes or the L-18 ribosomal protein. The 5,6-dimethylbenzimidazole is synthetized by *P. shermanii* in a separate, oxygen dependent pathway from riboflavin [100-116, 168]. In anaerobic microorganisms this base is synthetized in a different pathway. Glycine has been found as one of the precursors of the benzimidazole ring [116a], while methionine is the donor of methyl groups [116b]. The 5,6-dimethylbenzimidazole-α-nucleotide, formed in an exchange reaction between nicotinic acid mononucleotide and 5,6-dimethylbenzimidazole by the action of trans-N-glycosidase (nicotinatenucleotide-dimethylbenzimidazole phosphoribosyltransferase EC 2.4.2.21), is the direct substrate to be attached to cobinamide in the reaction depicted in the figure. The α-nucleotide producing enzyme has been isolated from *Clostridium sticklandii, Propionibacterium shermanii* and *Propionibacterium arabinosum* [117-124]. The enzyme from the latter strain has a molecular weight 47000 and a pH optimum 11. It requires high salt concentration for optimal activity. The K_m-values for 5,6-dimethylbenzimidazole and nicotinic acid mononucleotide are 1.4×10^{-4}M and 4.5×10^{-4} M respectively [123]. A considerable number of B_{12} derivatives which differ by the nucleotide moiety are known. Some of them occur naturally, others may be obtained by supplementing benzimidazole- or adenine derivatives into the growth medium of some bacteria species (Table 4-2, Fig. 4-3) [125-158]. This seems to suggest that the N-transglycosidase as well as the enzyme system catalyzing the attachment of the nucleotide to cobinamide is not very specific. New cobamides may be obtained from other cobamides by an exchange of the base in the nucleotide moiety for another one [139, 143, 144]. This conversion has been observed so far *in vivo* only. It could be demonstrated, however, that not the base alone but the nucleoside is being excised [155, 157]. The formation of the new cobamide 5′-phosphate as intermediate is similar to transformations ⑪ and ⑫ [37, 154].

⑬ **Dephosphorylation of vitamin B_{12}-5′-phosphate.** This is the last step of vitamin B_{12} biosynthesis. The dephosphorylation is achieved in vitro by acid or alkaline phosphatase [162]. The specific phosphatase has not been isolated. The coenzyme form of vitamin-B_{12}-5′-phosphate is active as cofactor in B_{12} dependent enzyme [162]. The B_{12}-5′P is accumulated by heat treated *P. shermanii* [156].

Fig. 4.2 als Klapptafel aus Buchdeckel hinten

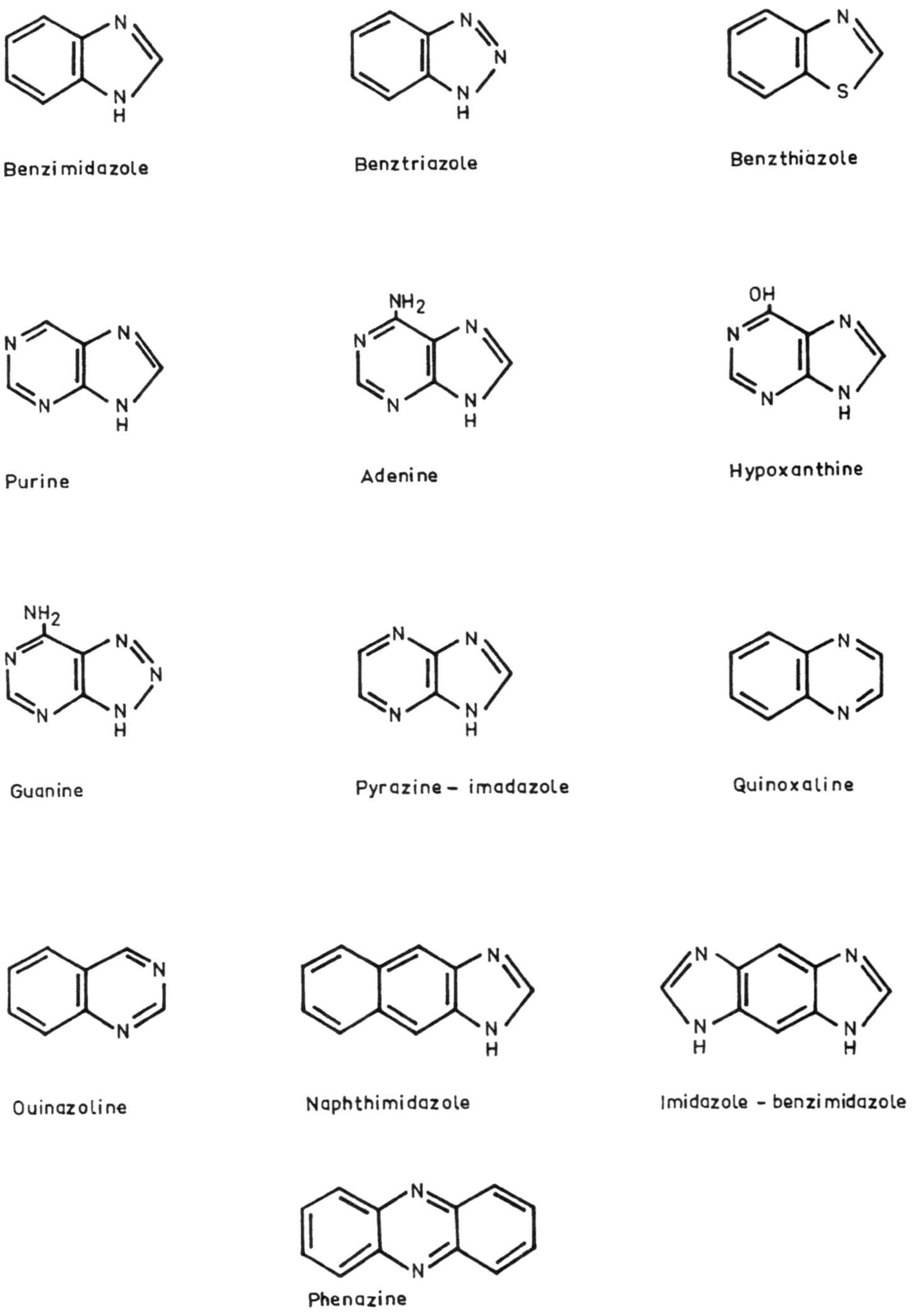

Fig. 4-3 Examples of parent heterocycles and derivatives which can be incorporated into the nucleotide moiety of corrinoids by guided biosynthesis.

References

[1] Shemin, D., Corcoran, J.W., Rosenblum, C., Miller, I.M., On the biosynthesis of the porphyrinlike moiety of vitamin B_{12}, Science **124,** 272, 1956.

[2] Kikuchi, G., Kumar, A., Talmage, and Shemin, D., J. Biol. Chem. **233,** 1214, 1958.

[3] Bernhauer, K., Müller, O., Wagner, F., Zum Mechanismus der Biosynthese des Vitamin B_{12}-Moleküls, in Vitamin B_{12} und Intrinsic Factor 2. Europäisches Symposion, Hamburg 1961, Ed. H.C. Heinrich, Ferdinand Enke Verlag, Stuttgart 1961, pp. 37-58.

[4] Bray, R., Shemin, D., The biosynthesis of the porphyrin-like moiety of vitamin B_{12}, Biochim. Biophys. Acta **30,** 647-648, 1958.

[5] Shemin, D., Bray, R.C., Ann. N.Y. Acad. Sci. **112,** 615, 1964.

[6] Bray, R.C., Shemin, D., On the biosynthesis of vitamin B_{12} (the derivation of the corrin structure from δ-aminolevulinic acid and the methyl group of methionine) J. Biol. Chem. **238,** 1501-1508, 1963.

[7a] Bykhovsky, V.Y., Biogenesis of tetrapyrrole compounds and its regulation, in Vitamin B_{12}, Eds. B. Zagalak and W. Friedrich, Walter de Gruyter, Berlin, 1979, pp. 293-314.

[7b] Bykhovskii, V.Ya., Zaitseva, N.I., Andreeva, N.A. and Yavorskaya, A.N., Effect of methylation antagonists on biosynthesis of tetrapyrrole compounds from *Propionibacterium shermanii,* Prikl. Biokhim. Mikrobiol. **16,** 862-867, 1980.

[8] Corcoran, J.W., Shemin, D., On the biosynthesis of the porphyrin-like moiety of vitamin B_{12}: the mode of utilization of δ-aminolevulinic acid, Biochim. Biophys. Acta **25,** 661, 1957.

[9] Wagner, F., Ann. Rev. Biochem. **35,** 405, 1966.

[10] Friedmann, H.C., Cagen, L.M., Microbial biosynthesis of β_{12}-like compounds, Ann. Rev. Microbiol. **24,** 159-208, 1970.

[11] Shemin, D., Naturwissenschaften **57,** 185, 1970.

[12] Yamasaki, H. and Moriyama, T., Biochim. Biophys. Acta **227,** 698, 1971.

[13] Bogorad, L., The enzymatic synthesis of porphyrins from porphobilinogen, J. Biol. Chem. **233,** 501-510, 1958.

[14] Granick, S., Mauzerall, D., Porphyrin biosynthesis in erythrocytes, J. Biol. Chem. **232,** 1119-1140, 1958.

[15] Scott, A.I., Intermediary metabolism of cobyrinic acid biosynthesis, in Vitamin B_{12}, Eds. B. Zagalak and W. Friedrich. Walter de Gruyter, Berlin 1979, pp. 247-277.

[16] Scott, A.I., Irwin, A.J., Siegel, L.M., Shoolery, J.N., J. Am. Chem. Soc. **100,** 316-318, 1978.

[17] Battersby, A.R., McDonald, E., Bioorg. Chem. **7,** 161-173, 1978.

[18] Kajiwara, M., Ho, K.S., Klein, H., Scott, A.I., Gossauer, A., Engel, J., Neumann, E., Zilch, H., Bioorg. Chem. **6,** 397-402, 1977.

[19] Battersby, A.R., McDonald, E., Hollenstein, R., Ihara, M., Satoh, F., Williams, D.C., J. Chem. Soc. Perkin Trans. **1,** 166-178, 1977.

[20] Pfaltz, A., Buhler, N., Neier, R., Hirai, K., Eschenmoser, A., Helv. Chim Acta **60,** 2653-2672, 1977.

[21] Müller, G., Gneuss, K.D., Kriemler, H.P., Scott, A.I., Irwin, A.J., J. Am. Chem. Soc. **101,** 3655-3657, 1979.

[22] Lewis, N.G., Neier, R., Matcham, G.W.J., McDonald, E., Battersby, A.R, J. Chem. Soc. Chem. Commun. 541-542, 1979.

[23] Battersby, A.R., Matcham, G.W.J., McDonald, E., Neier, R., Thomson, M., Woggon, W.D., Bykhovsky, V.Ya., Morris, H.R., J. Chem. Soc. Chem. Commun. **4,** 185-186, 1979.

[24] Battersby, A.R., McDonald, E., Thompson, M., Bykhovsky, V.Ya., J. Chem. Soc. Chem. Commun. 150-151, 1978.

[25] Nussbaumer, C., Imfeld, M., Worner, G., Muller, G., Arigoni, D., Biosynthesis of vitamin B_{12}: Mode of incorporation of factor III into cobyrinic acid, Proc. Natl. Acad. Sci. **78,** 9-10, 1981.

[26] Mobelli, L., Nussbaumer, C., Weber, H., Müller, G., Arigoni, D., Biosynthesis of vitamin B_{12}: Nature of the volatile fragment generated during formation of the corrin ring system, Proc. Natl. Acad. Sci. **78,** 11-12, 1981.

[27] Battersby, R.A., Bushell, J.M., Jones, C., Lewis, G.N., Pfenninger, A., Biosynthesis of vitamin B_{12}: Identity of fragment extruded during ring contraction to the corrin macrocycle. Proc. Natl. Acad. Sci. **78,** 13-15, 1981.

[28] Deeg, R., Kriemler, H.P., Bergmann, K.H., Müller, G., Hoppe Seyler's Z. Physiol. Chem. **358,** 339-352, 1977.

[29] Dolphin, D., Bioorg. Chem. **2,** 155, 1973.

[30] Müller, G., Deeg, R., Gneu, K.D., Gunzer,

G., On the methylation process in cobyrinic acid biosynthesis, in Vitamin B_{12}, Eds. B. Zagalak and W. Friedrich, Walter de Gruyter, Berlin 1979, pp. 279-291.
[31 a] Battersby, A. R., Recent biosynthetic researches on vitamin B_{12}, in Vitamin B_{12}, Eds. B. Zagalak and W. Friedrich, Walter de Gruyter, Berlin 1979, pp. 217-246.
[31 b] Battersby, A. R., Christopher, C. E., Fookes, J. R., Hook, J. M., Biosynthesis of vitamin B_{12}: Analysis of the proton-NMR and carbon-13 NMR spectra of heptamethyl dicyanocobyrinate cobester, J. Chem. Soc. Perkin. Trans. I, 2265-2278, 1982.
[31 c] Battersby, A. R., and Seo, S., Biosynthesis of porphyrins and related macrocycles: 22. Vitamin B_{12}: Studies on the chlorin, Faktor I, and the detection of 3-epi-Faktor I, J. Chem. Soc. Perkin Trans. I, 3049-3052, 1983.
[31 d] Battersby, A. R., Edington, C., and Fookes, C. J. B., Biosynthesis of vitamin B_{12}: Origin of the hydrogen atoms at C-18 and C-19. J. Chem. Soc. Chem. Commun., 527-530, 1984.
[32 a] Bykhovsky, V. Y., Zaitsera, N. I., Priklad. Biokhim. Microbiol. **12,** 365, 1976.
[32 b] Bykhovskii, V. Ya., Zaitseva, N. S., Microbiological synthesis of tetrapyrrole compounds, Prikl. Biokhim. Mikrobiol. **19,** 163-175, 1983.
[32 c] Bykhovskii, V. Ya., Ilarionor, S. A. and Zaitseva, N. I., Biosynthesis of corrinoids and other tetrapyrrole compounds by the acetogenic Clostridium 99, Prikl. Biokhim. Mikrobiol. **20,** 3-8, 1984.
[32 d] Bykhovskii, V. Ya., Zhilina, T. N., Zaitseva, N. I., Biosynthesis of tetrapyrrole compounds by pure cultures of methanogenic bacteria. Prikl. Biokhim. Mikrobiol. **19,** 579-589, 1983.
[33] Bykhovsky, V. Y., Zaitseva, N. I., Umrikhina, A. V., Yavorskaya, A. N., Priklad. Biokhim. Mikrobiol. **12,** 825, 1976.
[34] Siegel, L. M., Murphy, M. J., Kamin, H., Reduced nicotinamide adenine dinucleotide phosphate-sulfite reductase of *Enterobacteria*. J. Biol. Chem. **248,** 251-264, 1973.
[35] Eschenmoser, A., Chemical synthesis of corrinoids: current problems and recent advances, in Vitamin B_{12}, Eds. B. Zagalak and W. Friedrich, Walter de Gruyter, Berlin 1979, pp. 89-119.
[36] Friedrich, W., Fermente, Hormone, Vitamine, Band III/2, Vitamin B_{12} und verwandte Corrinoide, Eds. R. Ammon und W. Dirscherl, Georg Thieme Verlag. Stuttgart 1975 pp. 181-186.
[37] Friedmann, H. C., in Cobalamin-Biochemistry and Pathophysiology, Ed. Babior, B., John Willey and Sons, New York 1975, pp. 15-111.
[38] Woodward, R. B., Synthetic vitamin B_{12}, in Vitamin B_{12}, Eds. B. Zagalak and W. Friedrich, Walter de Gruyter, Berlin 1979, pp. 37-89.
[39] Hodgkin, D. C., New and old problems in the structure analysis of vitamin B_{12} in Vitamin B_{12}, Eds. B. Zagalak and W. Friedrich, Walter de Gruyter, Berlin 1979, pp. 19-36.
[40] Stevens, R. V., Recent studies on the synthesis of vitamin B_{12}, in Vitamin B_{12}, Eds. B. Zagalak and W. Friedrich, Walter de Gruyter, Berlin 1979, pp. 119-136.
[41] Walker, G. A., Murphy, S., and Huennekens, F. M., Arch. Biochem. Biophys. **134,** 95-102, 1969.
[42] Walker, G. A., Murphy, S., Schmidt, R. R., Huennekens, F. M., Fed. Proc. **26,** 343, 1967.
[43] Brady, R. O., Castanera, E. G., Barker, H. A., The enzymatic synthesis of cobamide coenzymes, J. Biol. Chem. **237,** 2325-2332, 1962.
[44] Peterkofsky, A., Weissbach, H., Ann. N. Y. Acad. Sci. **112,** 622-637, 1963.
[45] Peterkofsky, A., Weissbach, H., Release of inorganic tripolyphosphate from adenosine triphosphate during vitamin B_{12} coenzyme biosynthesis. J. Biol. Chem. **238,** 1491-1497, 1964.
[46] Barker, H. A., Biochem. J. **105,** 1-15, 1967.
[47] Vitols, E., Walker, G. A., Huennekens, F. M., Enzymatic conversion of vitamin B_{12s} to a cobamide coenzyme, α-(5,6-dimethylbenzimidazolyl) deoxyadenosyl cobamide (adenosyl-B_{12}) J. Biol. Chem. **241,** 1455-1461, 1966.
[48] Pawełkiewicz, J., Enzymatic preparation of coenzyme B_{12} and some of its analogs, Methods in Enzymology, **18 C,** pp. 99-101, 1971.
[49] Pawełkiewicz, J., Bartosinski, B., Walerych, W., Enzyme synthesis of light-sensitive cobalto-(II)-corrins. Acta Biochim. Polon. **8,** 131-142, 1961.
[50] Bernhauer, K., Wagner, F., Michna, H., Rapp, P., Vogelmann, H., Z. Physiol. Chem. **349,** 1297-1309, 1968.
[51] Bernhauer, K., Becher, E., Gross, G., Wilharm, G., Biochem. Z. **332,** 562-572, 1960.
[52] Bernhauer, K., Wagner, F., Beisbarth, H., Rietz, P., Vogelmann, H., Zur Chemie und

Biochemie der Corrinoide, Biochem. Z. **344**, 289-309, 1966.

[53] Bernhauer, K., Vogelmann, H., Wagner, F., Z. Physiol. Chem. **349**, 1271-1280, 1968.

[54] Bernhauer, K., Vogelmann, H., Wagner, F., Z. Physiol. Chem. **349**, 1281-1296, 1968.

[55] Bartosiński, B., Zagalak, B., Pawełkiewicz, J., The route of vitamin B_{12} biosynthesis on *Propionibacterium shermanii*. Biochim. Biophys. Acta **136**, 581-584, 1961.

[56] Anton, D. L., Hogenkamp, H. P. C., Walker, T. E. and Nicholas A. Matwiyoff, Carbon-13 nuclear magnetic resonance studies of the monocarboxylic acids of cyanocobalamin. Assignments of the b-, d-, and e-monocarboxylic acids, J. Amer. Chem. Soc. **102**, 2215-2219, 1980.

[57] Friedrich, W., Zur Biochemie der nucleotidfreien Vitamin B_{12}-Carbonsäuren, Biochem. Z. **342**, 143-160, 1965.

[58] Friedrich, W., Sandeck, W., Z. Naturforsch. **20b**, 79-80, 1965.

[59] Rapp, P., Z. Physiol. Chem. **354**, 136-140, 1973.

[60] Bernhauer, K., Wagner, K., Zahn, W., Synthesen auf dem Vitamin B_{12}-Gebiet, Biochem. Z. **342**, 207-216, 1965.

[61] Friedrich, W., Nutr. Dieta (Swiss) **8**, 178-225, 1966.

[62] Krasna, A. J., Rosenblum, C., Sprinson, D. B., The conversion of L-threonine to the Dg-1-amino-2-propanol of vitamin B_{12}, J. Biol. Chem. **225**, 745-750, 1957.

[63] Müller, G., Müller, O., Z. Naturforsch. **21b**, 1159-1164, 1966.

[64] Lowe, D. A., Turner, J. M., J. Gen. Microbiol. **64**, 119-122, 1970.

[65] Neuberger, A., Tait, G. H., Biochim. Biophys. Acta **41**, 164-165, 1960.

[66] Laver, W. G., Neuberger, A., Scott, J. J., α-amino-β-keto-acids, Part II: rates of decarboxylation of the free acids and the behaviour of derivatives on titration, J. Chem. Soc. 1483-1491, 1959.

[67] Elliott, W. H., Aminoacetone formation by *Staphylococcus aureus*. Biochem. J. **74**, 478-485, 1960.

[68] Neuberger, A., Tait, G. H., Production of aminoacetone by *Rhodopseudomonas spheroides*, Biochem. J. **84**, 317-328, 1962.

[69] Turner, J. M., Biochem. J. **99**, 427-433, 1966.

[70] Turner, J. M., Microbial metabolism of amino ketones, Biochem. J. **104**, 112-121, 1967.

[71] Lowe, D. A., Turner, J. M., Enzymic oxidation of D-1-aminopropan-2-ol by diol dehydrogenases of microbial origin, Biochim. Biophys. Acta **170**, 455-456, 1968.

[72] Lowe, P. A., Turner, J. M., J. Gen. Microbiol. **63**, 49-61, 1970.

[73] Dekker, E. E., Swain, R. R., Biochim. Biophys. Acta **158**, 306-307, 1968.

[74] Campbell, R. L., Dekker, E. E., Biochem. Biophys. Res. Commun. **53**, 432-438, 1973.

[75] Friedmann, H. C., Straight approaches to the nucleotide loop, in Vitamin B_{12}, Eds. B. Zagalak, W. Friedrich, Walter de Gruyter, Berlin 1979, pp. 331-343.

[76] Ford, S. H., Friedmann, H. C., Biochem. Biophys. Res. Commun. **72**, 1077-1083, 1976.

[77] Ford, S. H., Friedmann, H. C., Vitamin B_{12} biosynthesis. A model system for isopropanolamine formation by reaction between reduced corrinoid and threonine, Biochim. Biophys. Acta **500**, 217-222, 1977.

[78] DiMarco A, Marnati, M. P., Migliacci, A., Rusconi, A., Spalla, C., Requirements of some *Nocardia rugosa* mutant strains for the biosynthesis of vitamin B_{12}, in Vitamin B_{12} and Intrinsic Factor 2. Europäisches Symposion, Hamburg 1961, Ed. H. C. Heinrich, Ferdinand Enke Verlag, Stuttgart 1961, pp. 69-74.

[79] Renz, P., Biophys. Res. Commun. **30**, 373-378, 1968.

[80] Renz, P., Some intermediates in the biosynthesis of vitamin B_{12}, Methods Enzymol. **18C**, 86-88, 1971.

[81] Walerych, W., Pezacka, E., Ribosomal proteins share in vitamin B_{12} biosynthesis in Vitamin B_{12} Eds. B. Zagalak, W. Friedrich, Walter de Gruyter, Berlin 1979, pp. 345-358.

[82a] Pezacka, E., Walerych, W., The ribosomals proteins L2, L5, L18, and L25 involved in vitamin B_{12} biosynthesis, in Vitamin B_{12}, Eds. B. Zagalak, W. Friedrich, Walter de Gruyter, Berlin, 1979, pp. 359-371.

[82b] Pezacka, E., Walerych, H., Biosynthesis of vitamin B_{12}. Part I role of the ribosomal proteins in vitamin B_{12} biosynthesis, Biochim. Biophys. Acta, **678**, 300-310, 1981.

[83] Ford, J. E., Holdsworth, E. S., Kon, S. K., Biochem. J. **59**, 86-93, 1955.

[84] Dellweg, H., Becher, E., Bernhauer, K., Biosynthesen in der Cobalamin-Reihe, Biochem. Z. **328**, 81-87, 1956.

[85] DiMarco, A., Boretti, G., Ghione, M., Migliacci, A., Sanfilippo A., Ital. J. Biochem. **6**, 259-269, 1957.

[86] Ford, J. E., Porter, J. W. G., Biochem. J. **51**, 5, 1952.

[87] Ford, J.E., Holdsworth, E.S., Kon, S.K., Porter, J.W.G., Nature **171**, 150-151, 1953.

[88] Pawełkiewicz, J., Badania nad biosyntezą związków kobalaminowych, Acta Biochim. Polon. **3**, 581-590, 1956.

[89] Barchielli, R., Boretti, G., Julita, P., Migliacci, A., Minghetti, A., A natural guanine-containing analogue of vitamin B_{12}, Biochim. Biophys. Acta **25**, 452, 1957.

[90] DiMarco A, Boretti, G., Migliacci, A., Mighetti, A., Boll. Soc. Ital. Biol. Sper. **33**, 1513-1516, 1957.

[91] Borchielli, R., Boretti, G., Di Marco, A., Julita, A., Migliacci, A., Minghetti, A., Spalla, C., Biochem. J. **74**, 382-387, 1960.

[92] Pawełkiewicz, J., Walerych, W., Bartosiński, B., Acta Biochim. Polon. **6**, 431-440, 1959.

[93] Irion, E., Ljungdahl, L.G., Isolation of Co-5-deoxyadenosylcobinamide guanosine diphosphate, α-(2-methyladenyl)-Co-5′-deoxyadenosylcobamide and Co-Methylcorrinoids from *Clostridium thermoaceticum*, Biochemistry **7**, 2350-2355, 1968.

[94] Ronzio, R.A., Barker, H.A., Enzymic synthesis of guanosine diphosphate cobinamide by extracts of propionic acid bacteria, Biochemistry **6**, 2344-2354, 1967.

[95] Ford, J.E., Holdsworth, E.S., Biochem. J. **56**, XXXV, 1954.

[96] Renz, P., Febs Lett. **6**, 187-189, 1970.

[97] Lu, S.H., Winkler, M.F., Alworth, W.L., J. Chem. Soc. Chem. Commun. 191-192, 1971.

[98] Alworth, W.L., Lu, S.H., Winkler, M.F., Biosynthesis of 5,6-dimethylbenzimidazole from 6,7-dimethyl-^{14}C-8-ribityllumazine. Biochemistry **10**, 1421-1424, 1971.

[99] Kuhnle, H.F., Renz, P., Z. Naturforsch. **26b**, 1017-1020, 1971.

[100a] Renz, P., Horig, J., Wurin, R., On the biosynthesis of the 5,6-dimethylbenzimidazole moiety of vitamin B_{12}, in Vitamin B_{12}, Eds. B. Zagalak, W. Friedrich, Walter de Gruyter, Berlin 1979, pp. 317-322.

[100b] Hoellriegl, V.L.L., Rowold, J., Hoerig, J., Renz, P., Biosynthesis of vitamin B_{12}: Different pathways in some aerobic and anaerobic microorganisms, Arch. Microbiol. **132**, 155-158, 1982.

[101] Hörig, J., Renz, P., The enzyme system of propionic acid bacteria transforming riboflavin into 5,6-dimethylbenzimidazole, in Vitamin B_{12}, Eds. B. Zagalak, W. Friedrich, Walter de Gruyter, Berlin 1979, pp. 323-329.

[102] Plaut, G.W.E., Smith, C.M., Alworth, W.L., Ann. Rev. Biochem. **43**, 899-908, and 916-922, 1974.

[103] Schlee, D., Biogenetische Beziehungen zwischen Vitamin B_2 und B_{12}, Die Pharmazie **28**, 284-287, 1973.

[104] Speedie, J.D., Hull, G.W., U.S. Patent 2,951,017, 1960.

[105] Rily, P.B., Jackson, P.W., Ross, D., Savage, P.A., Soc. Chem. Ind. London, Monogr. **12**, 127-139, 1961.

[106] Hoerig, J.A., Renz, P., and Heckmann, G., [$5-{}^{15}N$] Riboflavin as precursor in the biosynthesis of the 5,6-dimethylbenzimidazole moiety of vitamin B_{12}: A study by ^1H and ^{15}N magnetic resonance spectroscopy. J. Biol. Chem. **253**, 7410-7414, 1978.

[107] Renz, P., Weyhenmeyer, R., FEBS-Letters **22**, 124-126, 1972.

[108] Perlman, D., Semar, J.B., Biotechnol. Bioeng. **5**, 21-25, 1963.

[109] Renz, P., Reinhold, K., Angew. Chem. Int. Ed. **6**, 1083, 1967.

[110] Plaut, G.W.E., J. Biol. Chem. **211**, 111-116, 1954.

[111] Alworth, W.L., Baker, H.N., Biochem. Biophys. Res. Commun. **30**, 496-501. 1968.

[112] Alworth, W.L., Baker, H.N., Lee, D.A., Martin, B.A., The biosynthesis of the 5,6-dimethylbenzimidazole moiety of vitamin B_{12}. J. Am. Chem. Soc. **91**, 5662-5663, 1969.

[113] Alworth, W.L., Baker, N., Winkler, M.F., Keenan, A.M., Biochem. Biophys. Res. Commun. **40**, 1026-1031, 1970.

[114] Plaut, G.W.E., Studies on the nature of the enzymic conversion of 6,7-dimethyl-8-ribityllumazine to riboflavin, J. Biol. Chem. **238**, 2225-2243, 1963.

[115] Lu, S.H., Alworth, W.L., Biosynthesis of 5,6-dimethylbenzimidazole from [1-^{14}C,5-^{15}N]6,7-dimethyl-8-ribityllumazine, Biochemistry **11**, 608-611, 1972.

[116a] Lamm, L., Heckman, G., and Renz, P., Biosynthesis of vitamin B_{12} in anaerobic bacteria: Mode of incorporation of glycine into the 5,6-dimethylbenzimidazole moiety in *Eubacterium limosum*, Eur. J. Biochem. **122**, 569-572, 1982.

[116b] Hoellriegl, V., Lamm, L., Rowold, J., Hoerig, J. and Renz, P., Biosynthesis of vitamin B_{12}: Different pathways in some aerobic and anaerobic microorganisms, Arch. Microbiol. **132**, 155-158, 1982.

[117] Friedmann, H.C., J. Biol. Chem. **240**, 413-418, 1965.

[118] Fyfe, J.A., Friedmann, H.C., Vitamin B_{12} biosynthesis, J. Biol. Chem. **244,** 1659-1666, 1969.

[119] Fyfe, J.A., Friedmann, H.C., Methods Enzymol. **188 B,** 197-204, 1971.

[120] Friedmann, H.C., Preparation and characteristics of α-D-ribofuranosylodenine 5 C-phosphate, Methods Enzymol. **18 C,** 96-98, 1971.

[121] Friedmann, H.C., Vitamin B_{12} biosynthesis, J. Biol. Chem. **243,** 2065-2075, 1968.

[122] Hayashi, M., and Kamikubo, T. FEBS Lett. **15,** 213-216, 1971.

[123] Schneider, Z. and Maćkowiak, E., not published data.

[124] Schneider, Z., Maćkowiak, E., Walerych, W., A micromethod for rapid estimation of trans-N-Glycosidase with modified cellulose nitrate membranes. Anal. Biochem. **108,** 112-120, 1980.

[125] Kon, S.K., Pawełkiewicz, J., Biosynthesis of vitamin B_{12} analogues, Fourth Int. Congr. Biochem., Symposium XI, W.W. Umbreit and H.Molitor, eds. pp.115-149, 1958.

[126] Perlman, D., U.S. Patent, 2,872,443, 1959.

[127] Janicki, J., Pawełkiewicz, J., Acta Biochim. Polon. **1,** 307, 1954.

[128] Pawełkiewicz, J., Acta Biochim. Polon. **1,** 313, 1954.

[129] Pawełkiewicz, J., Bull. Acad. Polon. **3,** 581, 1956.

[130] Pawełkiewicz, J., Nowakowska, K., Acta Biochim. Polon. **2,** 259, 1955.

[131] Fantes, K.H., O'Callaghan, C.H., in Vitamin B_{12} and Intrinsic Factor, Ed. H.C. Heinrich, Ferdinand Enke, Stuttgart 1957, pp.60-61.

[132] Dellweg, H., Becher, E., Bernhauer, K., Biosynthesen in der Cobalamin-Reihe, Biochem. Z. **328,** 96-100, 1956.

[133] Friedrich, W., Bernhauer, K., Chem. Ber. **91,** 1665, 1958.

[134] Riley, P.B., Jackson, P.W., Ross, D., Savage, P.A., Soc. Chem. Ind. Conference on Continuous Fermentations, 1960.

[135] Perlman, D., Barrett, J.M., Canad. J. Microbiol. **4,** 9, 1958.

[136] Perlman, D., Barrett, J.M., Bact. **78,** 171, 1959.

[137] Perlman, D., U.S. Patent 2,980,591, 1961.

[138] Bernhauer, K., Becher, E., Wilharm, G., Proc. VII. Int. Conf. Microbiol. **400,** 1958.

[139] Bernhauer, K., Becher, E., Wilharm, G., Arch. Biochem. Biophys. **83,** 248, 1959.

[140] Becher, E., Bernhauer, K., Wilharm, G., U.S. Patent 2,976,220, 1961.

[141] Becher, E., Bernhauer, K., Wilharm, G., U.S. Patent 2,976,221, 1961.

[142] Becher, E., Bernhauer, K., Wilharm, G., U.S. Patent, 2,976,222, 1961.

[143] Perlman, D., Barrett, J.M., Jackson, P.W., Cobamides synthesized by *Propionibacterium* species, in Vitamin B_{12} und Intrinsic Factor, 2 Europäisches Symposion, Hamburg 1961, Ed. H.C. Heinrich, Ferdinand Enke Verlag, Stuttgart 1962, pp. 58-69.

[144] Brandt, L.J., Bernstein, L.H., Wagle, A., Production of vitamin B_{12} analogues in patients with small-bowel bacterial overgrowth, Ann. Int. Med. **87,** 546-551, 1977.

[145] Leviton, A., Hargrove, R.E., Ind. Eng. Chem. **44,** 2651, 1952.

[146] Aso, K., Arita, M., Tokue, S., Ito, M., Nippon Nogei - Kasaku Kaishi **28,** 111, 1954.

[147] Leviton, A., U.S. Patent 2,753,289, 1956.

[148] Hinz, C., Bact. Proc. 26, 1957.

[149] Sudarsky, J.M., Fisher, R.A., U.S. Patent, 2,816,856, 1957.

[150] Grant, D.W., U.S. Patent 2,956,932, 1960.

[151] Speedie, J.D., Hull, G.W., U.S. Patent, 2,951,017, 1960.

[152] Perlman, D., Processes for the biosynthesis of cobamides, Methods in Enzymol. **18 C,** 75-82, 1971.

[153] Ford, J.E., Porter, J.W.G., Brit. J. Nutr. **7,** 326-337, 1953.

[154] Ohlenroth, H., and Friedmann, H.C., Biochim. Biophys. Acta **170,** 465, 1968.

[155] Renz, P., Angew. Chem. **4,** 527, 1965.

[156] Friedmann, H.C., Isolation of vitamin B_{12}-5'-phosphate from *Propionibacterium shermanii,* Methods Enzymol. **18 C,** 92-95, 1971.

[157] Friedmann, H.C., Synthesis of vitamin B_{12}5'-phosphate, Methods. Enzymol. **18 C,** 54-57, 1971.

[158] Kamikubo, T., and Hayashi, M., Structures of some vitamin B_{12} analogues and their biological as well as biochemical functions, in Vitamin B_{12}, Eds. B. Zagalak and W. Friedrich, Walter de Gruyter, Berlin 1979, pp.625-633.

[159] Marston, H.R., The requirement of sheep for cobalt or for vitamin B_{12}, Br. J. Nutr. **24,** 615-633, 1970.

[160] Coulter, C.L., Mawkinson, S.W., Friedmann, H.C., Isolation and molecular structure of vitamin B_{12} 5'-phosphate, Biochim. Biophys. Acta **177,** 293-302, 1969.

[161] Hawkinson, S.W., Coulter, C.L., Greaves, M.L., Proc. Rog. Soc. London Scr. **A 318,** 143-167, 1970.

[162] Schneider, Z., Friedmann, H.C., Studies on the enzymatic dephosphorylation of vitamin B_{12}-5′-phosphate, Arch. Biochem. Biophys. **152,** 488-495, 1972.

[163] Perlman, K.L., Schömer, U., Williams, T.H., Perlman, D., Microbial production of vitamin B_{12} antimetabolites. IV. Isolation and identification of 4-keto-5-amino-6-hydroxyhexanoic acid, J. Antibiotics **34,** 483-488, 1981.

[164] Brumm, P. and Friedmann, H.C., Succinylacetone pyrrole a powerful inhibitor of vitamin B_{12} biosynthesis: Effect on δ-aminolevuline acid dehydratase (EC 4.2.1.24) Biochem. Biophys. Res. Commun. **102,** 854-859, 1981.

[165] Brumm, P., Thomas, G.A., and Friedmann, H.C., The role of 4,5-dioxovalerie acid in porphyrin and vitamin B_{12} formation by *clostridia*. Biochem. Biophys. Res. Commun. **104,** 814-822, 1982.

[166] Battersby, A.R., Kishimoto, S., McDonald, E., Satoh, F., and Wurziger, H.K.W., Biosynthesis of porphyrins and related macrocycles: 12. vitamin B_{12}: Studies with 1-methylbilanes. J. Chem. Soc. Perkin. Trans. I. **0 (7),** 1927-1935, 1979.

[167] Scott, A.I., Irwin, A.J., Siegel, L.M., and Shoolery, J.N., Sirohydrochlorin: Prosthetic group of sulfite and nitrite reductases and its role in the biosynthesis of vitamin B_{12}. J. Am. Chem. Soc. **100,** 7987-7994, 1978.

[168] Horig, J.A., and Renz, P., Biosynthesis of vitamin B_{12}: some properties of the 5,6-dimethylbenzimidazole-forming system from *P. shermanii* and *P. freudenreichii,* Eur. J. Biochem. **105,** 587-592, 1980.

[169] Schneider, Z., not published data.

5 Purification and Estimation of Vitamin B_{12}

Z. Schneider

5.1 Extraction of corrinoids by organic solvents

The corrinoids occur in biological material in very low concentrations. The primary task in the course of their isolation is the separation from the bulk of accompanying material. The extraction with organic solvents is the most effective approach. It provides besides high yield of purification, a several-fold concentration of the vitamin. Prior to their extraction, however, the corrinoids must be freed from complexes with proteins. This can be achieved by heat treatment at pH 3–4. Some microorganisms like *Propionibacterium* species release the corrinoids into the medium, if acidified at room temperature. Mineral acids such as HCl, H_2SO_4 and $HClO_4$ are used for acidification. The liberation of corrinoids from protein complexes is a critical manipulation since it might lead to losses of biologically active vitamin B_{12}, e.g. drastic treatment with acids causes deamidations of the amide groups at the corrin ring. Some vitamin B_{12} analogs such as pseudovitamin B_{12} and GDP-cobinamide are even more susceptible to degradation by heat treatment than cobalamin. Most often, however, cobalamin is extracted and the outlines of extraction described here pertain to this corrinoid. In the presence of small amounts of cyanide, daylight and air, all corrinoids are converted into the more stable cyanide form. If coenzymatic forms of cobalamin are to be isolated, cyanide must be excluded from the procedure and the extraction has to be made in the absence of light or in dimmed red light. The extraction is a two step procedure employing two kinds of organic solvents. In the first step the vitamin B_{12} and its derivatives are extracted from the aqueous solution by phenol, benzyl alcohol or any other good solvent of the vitamin which does not combine with water. The addition of salts such as sodium sulfate or ammonium sulfate to the aqueous solution increases the yield of extraction by the organic solvents. In the second step the cobalamin is driven to a small portion of freshly added water with excess of a bad solvent of the cobalamin like chloroform, benzene or diethyl ether. Both steps provide a high degree of purification. Historically, phenol was the first solvent applied for purification of vitamin B_{12} and it is still used at present [1]. Since phenol is solid at room temperature, and thus inconvenient for the technique it is liquified with chloroform or benzene. A mixture of phenol and chloroform in proportion of 2:1 (v/v) is still a good solvent of the vitamin and can be used for extraction (see extraction with phenol sections 5.4.2 and 5.4.3). The addition of a large excess of chloroform to this mixture effects the transfer of the vitamin back to the aqueous phase. The yield of extraction is very high for fair cobalamin concentrations. However, at low concentrations losses of up to 50% are observed [15, 16]. Particularly high losses occur by extraction of hydroxocobalamin from protein suspensions. Beck elaborated an extraction procedure for very low concentrations of cyanocobalamin (down to 10^{-10} M) [10]. This method may be applied for analytical purposes particularly in conjunctions with radiodilution assays. The procedure is as follows:

1) A known volume of liquid sample containing a known weight of solids (5–8% w/v), such as homogenized tissue, bacteria, benthic marine or fresh-water sediments, etc., is boiled for 45 min with sodium nitrite and sodium cyanide (0.5:0.2 g/100 ml) at a pH of 4.0 using hydrochloric acid.
2) Coagulated tissues, organic and inorganic debris are removed from the boiled extract liquor by suction filtration. Persistent solid suspensions in the filtrate may require a zinc defaecation step prior to filtration or especially oily samples may necessitate a petroleum ether extraction after suction filtration.
3) Sodium sulfate (20% w/v) is dissolved in the clarified liquor and it is then extracted three times with one-tenth volume of benzyl alcohol[a]. Combined alcohol extracts are mixed with one-half volume of chloroform and this mixture is extracted three times with one-tenth volume of water. Stubborn emulsions generated during these extraction steps are easily eliminated by centrifugation at 3000 g. For a 100 ml sample, benzyl alcohol extraction yields ≈91% recovery of 4.5 ng of [^{57}Co]-cyanocobalamin.

References

[1] Laland, P., and Klem, A., Acta Med. Scand. **88**, 620–623, 1936.

[2] Wijmenga, H.G., Veer, W.L.C., and Lens, J., Biophys. Biochim. Acta **6**, 229, 1950.

[3] Kaczka, E.A., Wolf, D.E., Kuchl, F.A., Folkers, K., Vitamin B_{12}, XIII. Additional data on Vitamin B_{12}, J. Am. Chem. Soc. **73**, 3569, 1951.

[4] Smith, E.L., and Ball, S., US. Pat. 2, 630, 401, 1952.

[5] Smith, E.L., Ball, S., Ireland, D.M., Biochem. J. **52**, 395, 1952.

[6] Bacher, F.A., Boley, A.E., and Shonk, C.E., Anal. Chem. **26**, 1146, 1954.

[7] Friedrich, W., and Bernhauer, K. Z., Z. Naturforsch. **9b**, 755, 1954.

[8] Pawełkiewicz, J., The Methods of the preparation of corrinoids including the analytical methods in 2. Europäisches Symposion, Hamburg 1961, Ed. Heinrich, H.C., Ferdinand Enke Verlag, Stuttgart 1962, pp. 280–294.

[9] Perlman, D., Processes for the biosynthesis of cobamides, in Methods in Enzymology **18C**, 75–82, 1971.

[10] Beck, R.A., Essential prerequisites for the analysis of cyanocobalamin in biochemically complex samples using radiometric competitive binding assays, in Vitamin B_{12}, Eds. B. Zagalak, W. Friedrich, Walter de Gruyter, Berlin 1979, pp. 675–679.

[11] Rudkin, G.O., Taylor, R.J., Anal. Chem. **24**, 1155, 1952.

[12] Rosenblum, C., Talanta **11**, 255, 1964.

[13] Beck, R.A., Environ. Sci. Technol. **12**, 435, 1978.

[14] Linnell, J.C., Mackenzie, H.M., Wilson, J., Matthews, D.M., J. Clin. Path. **22**, 545–550, 1969.

[15] Gimsing, P., Hippe, E., Nexø, E., Determination of the plasma cobalamins by one-dimensional thin-layer chromatography, in Vitamin B_{12}, Eds. B. Zagalak and W. Friedrich, W. de Gruyter, Berlin 1979, pp. 665–669.

[16] Kamikubo, T., Hayashi, M., Structures of some vitamin B_{12} analogues and their biological as well as biochemical functions, in Vitamin B_{12}, Eds. B. Zagalak and W. Friedrich, Walter de Gruyter, Berlin 1979, pp. 625–633.

[17] Pawełkiewicz, J., Enzymatic preparation of coenzyme B_{12} and some of its analogs, in Methods in Enzymology **18C**, 99–101, 1971.

[18] Fenton, W.A., and Rosenberg, L.E. Improved techniques for the extraction and chromatography of cobalamins, Anal. Bioch. **90**, 119, 1978.

[a] It is recommended to use benzyl alcohol which has not been stored over a longer period of time otherwise it must be distilled. Alkaline pH (about 10) is required for high yield of cobalamin extraction with benzyl alcohol.

5.2 Other purification techniques

The corrinoids purified by means of extraction with organic solvents are quite a pure preparation as compared with the starting crude material. Depending on the source, they contain, however, a variety of B_{12}-like compounds which have to be separated from one another. Microbiological material contains, besides cobalamin a number of cobamides and incomplete corrinoids such as cobinamide, cobinamide phosphate, cobinamide - GDP, etc, which are inactive for man (see Table 3-1 and Chapters 4 and 6). Corrinoids derived from man and animals are easier to separate since they consist mainly of cobalamins in various forms (Ado-Cbl, Me-Cbl, OH-Cbl and CN-Cbl). The separation is accomplished by several methods:

Paper- and thin layer chromatography [1-34]. Paper chromatography is one of the earliest methods applied for separation of corrinoids. It utilizes a variety of solvents based mainly on aliphatic alcohols, acetic acid and ammonia (Table 5-1). The spots of corrinoids may be identified on paper since they differ by shades of yellow and red colors. Low concentrations of cobalamin may be localized on paper by bioautography or autoradiography [18-26]. Sharp and rapid separation of corrinoids is achieved by thin layer chromatography [10-12, 16] (Table 5-1 and Fig. 5-1). The recovery of corrinoids by these techniques is almost complete. Prolonged storage of corrinoids on paper strips is not recommended due to chemical degradation.

Ionophoresis [35-37]. Good separation is obtained by ionophoresis (called also electrophoresis) on paper. Acetic acid (0.5-1.0 M) is the most convenient medium. The relative mobility of several corrinoids is given in Table 5-2. Care must be taken not to contaminate the corrinoids with chlorine and peroxides produced at the electrodes since these compounds effectively inactivate the cobalamin.

Column chromatography [38-71]. Ion exchange resins, charged cellulose or charged Sephadex are the main supports applied in columns. The non-ionic support XAD-2 has been found very effective for separation of some corrinoids [33, 51-54, 77]. Some ionic exchangers have been found to deamidate cobalamin slowly; thus prolonged storage of corrinoids on ion exchange columns is not recommended.

Adsorption on charcoal [67] is the oldest technique applied on preparative scale in conjunction with (or instead of) extraction with organic solvents. The corrinoids ad-

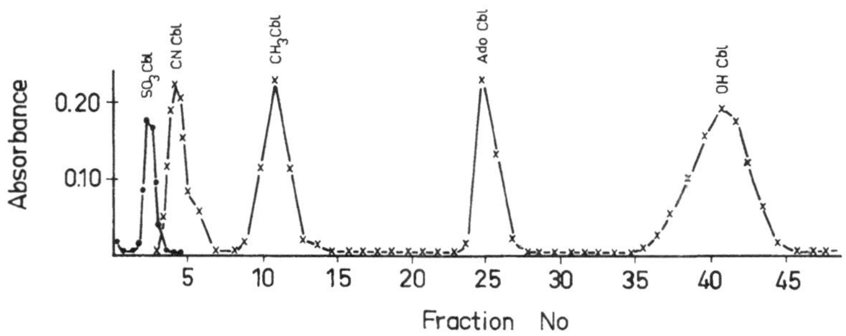

Fig. 5-1 Resolution of cobalamins by SP-Sephadex C-25 cation exchange chromatography (from Begley and Hall [70] see section 5.2).

Table 5-1 Chromatographic separation of corrinoids (data taken mainly from Friedrich [32])

corrinoid	R_f values for paper chromatography (PC) and thin layer chromatography (TLC) in various solvent systems (The references are indicated in parenthesis)					
	a	b	c	d	e	f	g
CN-Cbl	TLC 0.17 [11]	PC* 1.00 [74]	PC 0.34 [1]	TLC 0.3 [11]		PC 0.16 [2]	PC 0.4 [3]
aqCbl	PC* 0.37 [6]		TLC 0.05	TLC 0.05 [15]	PC* 0.45 [6]		
	TLC 0.03 [11]						
AdoCbl	PC* 0.73 [6]		PC 0.22 [1]	TLC 0.55	PC* 0.75 [6]		PC 0.28 [3]
	TLC 0.13 [11]		TLC 0.10 [15]	TLC 0.25 [15]			
MeCbl	PC* 1.6 [6]		PC** 1.5 [8]	PC** 1.66	PC* 1.25 [6]		
	PC** 1.8		0.17 [15]	0.64 [15]	TLC 0.36 [15]		
EtCbl	PC* 1.65 [6]		PC** 1.59 [8]	PC 1.7 [8]	PC* 1.3 [6]		
	PC** 1.91 [8]		TLC 0.20 [15]	TLC 0.64 [15]			
(5-Me Bza)CN-Cba	PC* 0.86		PC* 0.89	PC* 0.98		PC 0.25 [2]	
(6-Me Bza)CN-Cba	PC* 0.92		PC* 0.93	PC* 0.99			
(Bza)CN-Cba	PC* 0.78	PC* 0.90 [74]	PC* 0.95	PC* 0.92			
(Ade)CN-Cba	PC* 0.58	PC* 0.25 [74]		PC* 0.60		PC 0.48 [2]	
(Ade)AdoCba			PC 0.09 [1]				PC 0.13 [3]
(2-Me Ade)CN-Cba	PC* 0.63	PC* 0.45 [74]		PC* 0.65		PC 0.38 [2]	
(CN,aq)Cbi	PC* 1.25	PC* 1.22	PC* 1.00				
(aq,Ado)Cbi	PC** 1.26 [8]		PC** 1.10 [8]	PC** 1.32 [8]			
(aq,Me)Cbi	PC** 1.83 [8]		PC** 1.52 [8]	PC** 1.69 [8]			
(aq)AdoCbi-GDP							PC 0.09 [3]

Chromatographic solvents
a - Sec. butanol sat. with water
b - Sec. butanol: water: 5% KCN = 77:23:0.25 (satur. with $KClO_4$)
c - Sec. butanol sat. with water + 1% acetic acid
d - Sec. butanol: water: NH_3 solution = 50:18:7 (v/v)
e - n-Butanol: isopropanol: water = 10:7:10 (v/v)
f - Water sat. with thyme camphor and with 0.4% pyridin. Paperstrips (whatman 2) 3 cm wide impregnated with thyme camphor.
g - n-Butanol: isopropanol: water: acetic acid = 100:70:1:100 (v/v)

* Mobility relative to CN-Cbl
** Mobility relative to Ado-Cbl

Table 5-2 Ionophoretic mobility of corrinoids on paper in presence of 1 M acetic acid [36, 37]

corrinoid	mobility toward the cathode relative to (CN,aq)Cbi
(CN,aq)Cbi	1.0
CN-Cbl[a]	0.0
aqCbl	1.0
AdoCbl	0.92
(5-OH-Bza)AdoCba	1.24
(5-Me-Bza)AdoCba	1.16
(2-Me-Ade)AdoCba	1.1
(Ade)AdoCba	1.23
(2-Me-Ade)CN-Cba	0.8
(Ade)CN-Cba	0.3
(aq)AdoCbi-GDP	0.24

[a] CN-Cbl moves somewhat toward the cathode due to electroendoosmosis

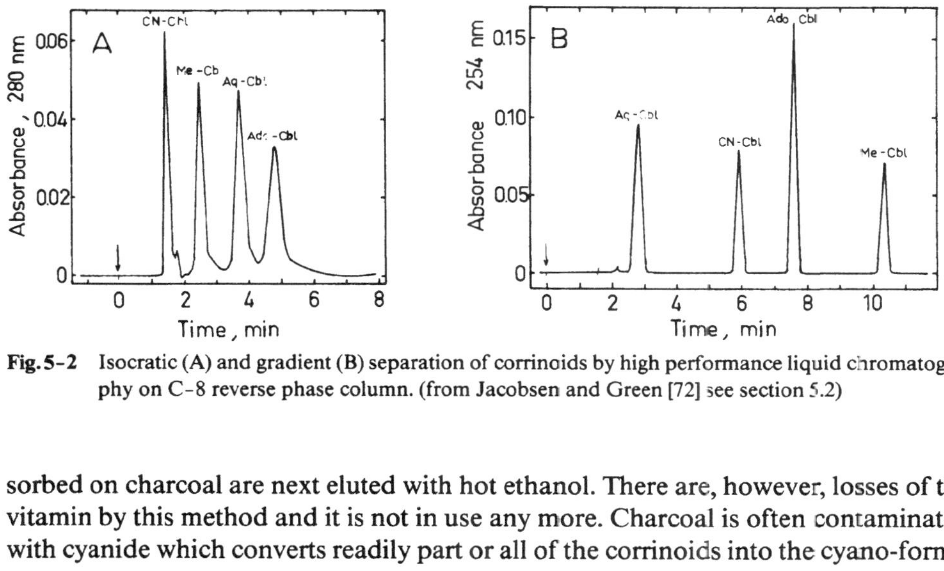

Fig. 5-2 Isocratic (A) and gradient (B) separation of corrinoids by high performance liquid chromatography on C-8 reverse phase column. (from Jacobsen and Green [72] see section 5.2)

sorbed on charcoal are next eluted with hot ethanol. There are, however, losses of the vitamin by this method and it is not in use any more. Charcoal is often contaminated with cyanide which converts readily part or all of the corrinoids into the cyano-forms.

Adsorption on immobilized B_{12}-binding proteins. There is a potential to apply this technique for concentration of very diluted vitamin B_{12}. The most specific binding-proteins, such as the intrinsic factor, will bind cobalamins exclusively [78] (see section 5.3.2).

High pressure liquid chromatography [44, 72, 74, 75]. This method allows a rapid separation of methylcobalamin (Me-Cbl), adenosylcobalamin (Ado-Cbl), aquacobalamin (Aq-Cbl) and cyanocobalamin (CN-Cbl). Both isocratic and gradient separation have been applied successfully (see Fig. 5-2).

References

[1] Barker, H.A., Smyth, R.D., Weissbach, H., Much-Petersen, A., Toohey, J.I., Lodd, J.N., Volcani, B.E., Wilson, R.M., Assay purification and properties of the adenylcobamide coenzyme, J. Biol. Chem. **235,** 181-190, 1960.

[2] Pawełkiewicz, J., and Walerych, W., Fast chromatographic separation of corrinoids (Polish), Acta Biochim. Polon. **6,** 441, 1959.

[3] Pawełkiewicz, J., Bartosinski, B., Walerych, W., Enzymic synthesis of light-sensitive cobalto II-Corrins., Bull. Acad. Pol. Sci. **8,** 131-141, 1961.

[4] Woodruff, H.B., and Foster, J.C., Analysis for vitamin B_{12} and vitamin B_{12a} by paperstrip chromatography., J. Biol. Chem. **183,** 569-576, 1950.

[5] Bayer, J., Eine schnelle papierchromatographische Trennung von Vitamin B_{12}, J. Chromatog. **8,** 123-125, 1962.

[6] Johnson, A.W., Mervyn, L., Shaw, N., Smith, E.L., A partial synthesis of the vitamin B_{12} coenzyme and some of its analogues. J. Chem. Soc. 4146-4156, 1963.

[7] Hilttenrauch, R., und Klotz, L., Die Trennung neutraler und basischer Cobalamine an Ionenaustauscherpapieren, J. Chromatog. **12,** 464-468, 1963.

[8] Müller, O., Müller, G., Biochem. Z. **337,** 179, 1963.

[9] Popova, Ya., Popov, Ch., Illeva, M., Separation of hydroxocobalamin and cyanocobalamin by thin-layer chromatography, J. Chromatog. **24,** 263-264, 1966.

[10] Sasaki, T., Thin-layer chromatography of cyanocobalamin hydroxocobalamin and B_{12} coenzymes, J. Chromatog. **24,** 452-454, 1966.

[11] Morley, C.G.D., Blakley, L.M., Hogenkamp, H.P.C., Biochemistry **7,** 1231, 1968.

[12] Firth, R.A., Hill, H.A.O., Pratt, J.M., Thorp, R.G., Separation and identification of organocobalt derivatives of vitamin B_{12} by thin-layer cellulose, Analyt. Biochem. **23,** 429-432, 1968.

[13] Pawełkiewicz, J., Enzymatic preparation of coenzyme B_{12} and some of its analogs, in Methods in Enzymology **18C,** 99-101, 1971.

[14] Friedrich, W., Gross, G., Bernhauer, K., Mikrochim. Acta **134,** 125, 1956.

[15] Yamada, R., Tamao, Y., Blakley, R.L. Degradation of 5'-deoxyadenosylcobalamin by ribonucleoside triphosphate and binding of degradation product to the active center, Biochemistry **10,** 3959-3968, 1971.

[16] Gimsing, P., and Hippe, E., Determination of the plasma cobalamins by one-dimensional thin-layer chromatography, in Vitamin B_{12}, Eds. B.Zagalak and W.Friedrich, Walter de Gruyter, Berlin 1979, pp.665-669.

[17] Nexø, E., Scand. J. Haematol. **18,** 358-360, 1977.

[18] Linnell, J.C., Mackenzie, H.M., Wilson, J., Matthews, D.M., J. Clin. Path. **22,** 545-550, 1969.

[19] Kanopkaite, S.J., and Brazenas, G.R., in Metal Ions in Biological Systems, vol.10, Marcel Dekker, New York, 1979.

[20] Draper, M.W., North, J.A., Mangum, J.H., Assay for N^5-methyltetrahydrofolatehomocysteine transmethylase activity using instant thin-layer chromatography, Analyt. Biochem. **45,** 653, 1972.

[21] Volcani, B.E., Toohey, J.J., Barker, H.A., Arch. Biochem. Biophys. **92,** 381, 1961.

[22] Linnell, J.C., Hussein, H.A.A., Matthews, D.M., J. Clin. Path. **23,** 820, 1970.

[23] Linnell, J.C., Mackenzie, M.M., Matthews, D.M., J. Clin. Path. **22,** 506, 1969.

[24] Linnell, J.C., Mackenzie, H.M., Wilson, J., and Matthews, D.M., J. Clin. Pathol. **22,** 545, 1969.

[25] Linnell, J.C., Hoffbrand, A.V., Hussein, H.A.-A., Wise, I.J., Matthews, D.M., Clin. Sci. Mol. Med. **46,** 163, 1974.

[26] Matthews, D.M., Distribution of cobalamin in the animal body, in Vitamin B_{12}, Eds. B.Zagalak, W.Friedrich, Walter de Gruyter, Berlin 1979, pp.681-694.

[27] Linnell, J.C., The fate of cobalamin in vivo, in Cobalamin-biochemistry and pathophysiology, Ed. B.M.Babior, John Wiley and Sons, New York 1975, pp.287-334.

[28] Lindstrand, K., and Stahlberg, K.-G., Acta Med. Scand. **174,** 665, 1963.

[29] Lindstrand, K., Nature **204,** 188, 1964.

[30] Stahlberg, K.-G., Scand. J. Haematol. **1,** 220, 1964.

[31] Pawełkiewicz, J., Bartosinski, D., Walerych, W., Enzymic synthesis of light-sensitive cobalto corrins, Bull. Acad. Pol. **8,** 131-141, 1961.

[32] Friedrich, W., Vitamin B_{12} und verwandte Corrinoide, Eds. R.Ammon und W.Dirscherl, Georg Thieme Verlag, Stuttgart, 1975.

[33] Fenton, W.A. and Rosenberg, L.E., Improved techniques for the extraction and chromatography of cobalamins, Anal. Biochem. **90,** 119, 1978.

[34] Bilkus, D.J., Mervyn, L., in The cobalamins, Eds., Arnstein, H.R.V. and Wrighton, R.J., Churchill Livingston, Edinburg and London 1971, pp.17-21.

[35] Armitage, J.B., Cannon, J.R., Johnson, A.W., Parker, L.F.J., Smith, E.L., Stafford, W.H., and Todd, A.R., Chemistry of the vitamin B_{12} group, Part III. The course of hydrolytic degradations. J. Chem. soc., 3849-3864, 1953.

[36] Ford, J.E., Microbiological tests, Nature **171,** 149-150, 1953.

[37] Zagalak, B., Pawełkiewicz, J., Synthesis and properties of Co-adenine nucleoside analogues of coenzyme B_{12}, Acta Biochim. Pol. **12,** 219-228, 1965.

[38] Pawełkiewicz, J., Walerych, W., Bartosinski, B., Identification of y_1, as vitamin B_{12}-guanosinediphosphate (in Polish).

[39] Holdsworth, E.S., Nature **171,** 148, 1953.

[40] Kato, T., Shimizu, S., J. Vitaminol. **9,** 305, 1963.

[41] Bernhauer, K., Irion, E., Biochem. Z. **339,** 530, 1964.

[42] Friedmann, H.C., J. Biol. Chem. **243,** 2065, 1968.

[43] Ertel, R., Brot, N., Taylor, R., Weissbach, H., Arch. Biochem. Biophys. **126,** 353, 1968.

[44] Jacobsen, D.W., Green, R., Quadros, E.V. and Montejano, Y.D. Rapid analysis of cobalamin coenzymes and related corrinoid analogs by High-Performance Liquid chromatography. Anal. Biochem. **120,** 394-403, 1982.

[45] Katz, J., Burtis, J., J. Chromatog. **40,** 270, 1969.

[46] Singhal, C., Cohn, M., Analyt. Biochem. **45,** 585, 1972.

[47] Barker, H.A., Smyth, R.D., Hogenkamp, M.P.C., Biochem. Prepar., **10,** 27, 1963.

[48] Menke, K.H., Kobalt- und Vitamin B_{12}-Stoffwechsel beim Huhn, II. Die Trennung der Cobalamine an Ionenaustauschern, J. Chromat. **7,** 86-97, 1962.

[49] Anton, D.L., and Hogenkamp, H.P.C., Modified adenosylcobalamin: model systems for the active sites of corrinoid-dependent enzymes, in: Vitamin B_{12}, Eds. B.Zagalak and W.Friedrich, Walter de Gruyter, Berlin 1979, pp.605-608.

[50] Riley, W., Taylor, R.T., Analyt. Chim. Acta **46,** 307, 1969.

[51] Vogelmann, H., Wagner, F., Elutionschromatographische Trennungen von Corrinoiden an dem unpolaren Absorbenz Amberlite XAD-2, J. Chromat. **76,** 359-379, 1973.

[52] Koppenhagen, V.B., Warmuth, E., Schlingmann, G., and Dresow, B., Novel metal-free corrinoids and metal analogues, in Vitamin B_{12}, Eds. B.Zagalak and W.Friedrich, Walter de Gruyter, Berlin 1979, pp.635-648.

[53] Zagalak, B., Chemical quantitation of corrinoids, in Vitamin B_{12}, Eds. B.Zagalak and W.Friedrich, Walter de Gruyter, Berlin 1979, pp.671-672.

[54] Bieganowski, R., Friedrich, W., Preparation of ferribalamin, the Fe (III)-analogue of Vitamin B_{12}, in Vitamin B_{12}, Eds. B.Zagalak and W.Friedrich, Walter de Gruyter, Berlin 1979, pp.647-648.

[55] Pawełkiewicz, J., Walerych, W., Über die Anwendung von Zellulose-Ionenaustauschern und Alginsäure zur chromatographischen Reinigung und Trennung der Vitamine der B_{12}-Gruppe, J. Chromatog. **3,** 359-363, 1960.

[56] Friedrich, W., Biochem. Z. **342,** 143, 1965.

[57] Law, P.Y., Brown, D.G., in: E.L.Babior, B., Wood, J.M., Biochemistry **10,** 3428, 1971.

[58] Bernhauer, K., Vogelmann, M., Wagner, F., Hoppe-Seyler's Z. Physiol. Chem. **349,** 1271, 1968.

[59] Gizis, P., Meyer, E.F., Proc. Soc. Exp. Biol. Med. **140,** 326, 1972.

[60] Kamikubo, T., Hayashi, M., Structures of some vitamin B_{12} analoques and their biological as well as biochemical functions, in Vitamin B_{12}, Eds. B.Zagalak and W.Friedrich, Walter de Gruyter, Berlin 1979, pp.625-633.

[61] Menke, K.H., Die Trennung der Cobalamine an Ionenaustauschersäulen, Naturwissenschaften **11,** 263, 1958.

[62] Daisley, K.W., Cell filtration of sea-water separation of free and bound forms of vitamin B_{12}, Nature **191,** 868-869, 1961.

[63] Zagalak, B., and Pawełkiewicz, J., Chromatographic separation on phosphate-cellulose of light-sensitive forms of corrinoids produced by propionic acid bacteria, Acta Biochim. Pol. **9,** 315-319, 1962.

[64] Friedmann, H.C., Isolation of vitamin B_{12} 5'-phosphate from *Propionibacterium shermanii,* Methods in Enzymology **18C,** 92-95, 1971.

[65] Menke, K.H., Über eine neue Methode zur Trennung der Vitamin B_{12} Analoge, Landwirtschaftliche Forschung **12,** 133-138, 1959.

[66] Stefaniak, O., and Muszynska, M., Effect of pH and temperature on the content of corrinoids released from *Propionibacterium shermanii,* Acta Microbiol. Pol. **18,** 25-29, 1969.

[67] Fantes, K.H., Page, J.E., Parker, L.F.J., and Smith, E.L., Crystalline anti-pernicious anaemia factor from liver, Proc. Royal. Soc., **136,** 592-613, 1950.

[68] Inada, M., Kameyama, M., Toyoshima, M.,

The significance of vitamin B_{12} binders in the central nervous system, in Vitamin B_{12}, Eds. B. Zagalak and W. Friedrich, Walter de Gruyter, Berlin 1979, pp. 821-826.
[69] Tortolani, G., Bianchini, P., Mantovani, V., J. Chromatog. **53**, 577, 1970.
[70] Begley, J. A., Hall, C. A., Presence of sulfitocobalamin in cell extracts. Resolution and identification by SP-Sephadex C-25 cation exchange chromatography, in Vitamin B_{12}, Eds. B. Zagalak and W. Friedrich, Walter de Gruyter, Berlin 1979, pp. 971-977.
[71] Kakei, M., and Glass, G. B. J., Separation of bound and free vit. B_{12} on Sephadex G-25 column, Proc. Soc. Expl. Biol. Med. **111**, 270-274, 1962.
[72] Jacobsen, D., Green, R., Savage, M., Rapid determination of corrinoids by high performance liquid chromatography, in Vitamin B_{12}, Eds. B. Zagalak and W. Friedrich, Walter de Gruyter, Berlin 1979, pp. 663-664.
[73] Perlman, D., Processes for the biosynthesis of cobamides, Methods in Enzymology **18C**, 75-82, 1971.
[74] Mattori, T., Asakowa, N., Ueyama, M., Shinoda, A. and Miyake, Y., separation of vitamin B_{12} analogs by high performance liquid chromatography (Japanese), Yakugaku Zasshi **100**, 386-388, 1980.
[75] Binder, M., Kolhouse, J. F., Horne, K. C., Allen, R. M., High-pressure liquid chromatography of cobalamins and cobalamin analogs. Anal. Biochem. **125**, 253-258, 1982.
[76] Gimsing, P. Determination of cobalamins in biological material. Anal. Biochem. **129**, 288-295, 1983.
[77] Gimsing, P., Nexø, E., and Hippe, E. Determination of cobalamins in biological material. Anal. Biochem. **129**, 288-295, 1983.
[78] Kolhouse, J. F., Allen, R. H. Isolation of cobalamin and cobalamin analogs by reverse affinity chromatography. Anal. Biochem. **84**, 486-490, 1978.

5.3 Estimation of corrinoids

5.3.1 Microbiological assay

The microbiological assays of vitamin B_{12} are most suitable for biological material and pharmaceuticals because of their sensitivity and specificity. They are based on the observation that the growth rates of certain microorganisms are proportional to the concentration of vitamin B_{12}. Two techniques are in use: the tube test and the plate test. Both are based on the growth response of the microorganism to tested sample of vitamin B_{12} added to growth medium. In the tube assay the samples containing unknown amounts of B_{12} and the standard are added to the sterile liquid medium in tubes which are subsequently inoculated with the appropriate organism. The plate assay applies medium solidified by agar on Petri dishes or glass plates. Before solidification of the gel, the warm medium is inoculated with microorganism. Paper discs are soaked with samples to be assayed, or with standard B_{12} and placed onto the gel. After appropriate intervals of incubation the growth zone around the discs are measured. The plate method has been also adopted for spotting and assessment of cobalamin on paper after chromatography. The method called bioautography has been found very useful for identification of the various forms of cobalamin like: AdoCbl, MeCbl, OH-Cbl and CN-Cbl [15, 40-45]. The cobalamins are separated by chromatography (see references in section 5.2.) and located bioautographically by overlayering the chromatogram with agar containing a cobalamin sensitive *E. coli* mutant and a tetrazolium growth indicator (see section 5.4.1). The spots are subsequently quantitated by photometric scanning and comparison with standards. The chromato-bioautography technique provides the only means at present available for estimating individual cobalamins in plasma and tissues or in food. For the estimation of vitamin B_{12} in natural products, its freeing from complexing proteins is necessary. Short heating of the homogenated material with the addition of trace of KCN is sufficient in most cases to free the corrinoids. In special cases the treatment with digestive enzymes or acidification for a short time to pH 1-2 is required (see also sections 5.3.1., 5.3.3. and 5.4.1). In presence of cyanide all corrinoids are converted into a more stable cyanide form. The standard vitamin B_{12} is mainly used in the cyanide form (CN-Cbl). Since some material from natural sources may contain, besides cobalamins also other corrinoids, various test organisms may show different growth response to the same samples. Thus if a mixture of corrinoids is estimated the results depend upon the organism used (see Tables 5-3, 5-4, 6-13, 6-14). Among many microorganisms tested, four found the widest application: *Escherichia coli, Lactobacillus leichmannii, Ochromonas malhamensis* and *Euglena gracilis* [1-39]. The most specific strain is *Ochromonas malhamensis*. It produces results closest to the clinical tests. The least specific is *E. coli*, but it is the most convenient in use. In Table 5-5 the growing conditions and the estimation procedures for the above mentioned organisms are specified. For the assay of vitamin B_{12} in the sea water marine algae are suitable [60-71]. Among several species tested, the centric diatom *Thalassiosira pseudonana* clone 3 H has been found useful to assay specifically cobalamin in concentration range 0.2-1.2 ng/l [60, 65, 67]. Similar selective response shows *Amphidinium carteri* [70, 71]. The assay of total corrinoids in sea water can be done with *Thalassiosira pseudonana* clone 13-1 (named previously *Cyclotella nana* 13-1) [67]. This organism shows a linear correlation between uptake of ^{14}C-labelled CO_2 and vitamin B_{12} concentration in the range of 0.2-3.0 ng/l [63, 64]. The feature has been conveniently applied for fast (48 hours) assay of the vitamin. The clone 13-1 of *T. pseudonana* in contrast to the above mentioned clone 3 H is rather unspecific and responds with growth to other corrinoids as well. The following corrinoids have been shown to promote growth of the clone 13-1 (in parantheses percentage of growth relative to cobalamin): BzaCba (43), (5-MeBza)Cba(47), (5-OHBza)Cba(78), AdeCba(60), (2-MeAde)Cba(72), (2-SHAde)Cba(79) and Cbi(65) [64]. Other marine algae which have been applied for B_{12} assay in sea water [60] such as *Bellerochea spinifera* 675-d [65] and *Monochrysis lutherii* [68, 69] show also growth responses to various vitamin B_{12} analogues. A methanolutilizing bacterium has been also applied for bioassay of the vitamin in sea water [72].

Table 5-3 Response of test microorganisms to complete and incomplete corrinoids (from Sourial and Mollin [47])

	Cobamides with benzimidazole nucleotide	Cobamides with purine nucleotide	Cobinamides with no nucleotide
Escherichia coli mutant (bacteria)	+	+	+
Lactobacillus leichmannii (bacteria)	+	+	0
Euglena gracilis (flagellates)	+	+	0
Ochromomas malhamensis (flagellates)	+	0	0

Table 5-4 Biological activity of vitamin B_{12} and its analogues containing various nucleotide moieties instead of 5,6-dimethylbenzimidazole nucleotide (activities expressed in relation to cyanocobalamin, data selected from [6, 13, 16, 24])

Corrinoid	Common name	Nucleotide base	Clinical assay	*E. coli* plate test	*E. coli* tube test	*L. leichmannii* tube test	*Euglena* tube test	*Ochromonas* tube test	Chick assay	
									by mouth	by injection
CN-Cbl	cyanocobalamin	5,6-dimethylbenzimidazole	100	100	100	100	100	100	100	100
(5-MeBza)CN-Cba	–	5-methylbenzimidazole	fully active	100	95	92	active	90	73	74
(5-OHBza)CN-Cba	Factor III	5-hydroxybenzimidazole	active	100–150	50	35	–	50	3.9	5.8
(Bza)CN-Cba	–	benzimidazole	fully active	100	30	100	–	96	23	23
(Ade)CN-Cba	Pseudovitamin B_{12}	adenine	0	100	10	50	100	0	antagonism	0
(2-MeAde)CN-Cba	Factor A	2-methyladenine	slightly active	100	50	40	60	0	<1	<1
(2-MeSAde)CN-Cba	Factor F	2-methylmercaptoadenine	–	135	60	50	60	3	–	–
(aq,CN)Cbi-GDP*	Factor C	guanine in guanosine diphosphate	–	100	20	10	–	0	–	–
(Hpx)CN-Cba	Factor G	hypoxanthine	–	160	–	20–100	–	0	–	–
(2-Me Hpx)CN-Cba	Factor H	2-methylhypoxanthine	–	280	40	15–40	–	0	–	–
AdoCbl	coenzyme B_{12}	5,6-dimethylbenzimidazole	–	115	100	–	–	100	–	–
(Bza)AdoCba	coenzyme B_{12}	benzimidazole	–	–	100	–	–	50	–	–
(Ade)AdeCba	pseudovitamine B_{12} coenzyme	adenine	–	–	100	–	–	0	–	–
(CN,aq)Cbi	Factor B	no nucleotide	0	100–150	20	0	0	0	antagonism	0
(CN,aq)Cbi-P	Cobinamid phosphate	no nucleotide	–	–	20	–	–	–	–	–
(CN-Cbl)15-nor	–	5,6-dimethylbenzimidazole	–	–	13	73	–	–	–	–

* This compound is one of the precursors of cobamides

Table 5-5 Microbiological assay of vitamin B_{12} – basic information (compiled by K. Trojanowska and Z. Schneider)

Test organism	*Escherichia coli mutant 113-3* [1, 2, 4, 7-10] (ATCC 9637 or ATCC 11105)	*Lactobacillus leichmanii 326* [1-4, 17, 18] (ATCC 4797 or ATCC 7830)	*Euglena gracilis* Z-alga [1, 2, 4, 18, 19, 20] ATCC 12716	*Ochromonas malhamensis* [1-3, 18, 21] ATCC 11532
Range of assay of cyanocobalamin (pg/ml)	Tube method: 40.0-250.0 [2] Plate method: 500-50000 [2]	1.0-20.0 [2]	0.3-50.0 [2, 4, 16]	0.5-50.0 [2]
Growth conditions	16-24 hours at 37 °C	16-27 hours at 37 °C	4-7 days at 25 °C-27 °C in light [2, 3]	3-7 days at 24 °C-30 °C [2, 3] Incubation on shaker in presence of light or in the dark [2, 3]
Principles of assay	Turbidimetrically or measurement of diameter of growth zones	Turbidimetrically or titrimetrically [3, 17]	Turbidimetrically or colorimetrically	Turbidimetric or densitometric assay [2, 3]
B_{12} assay medium (basal medium)	Tube method double strength for 1 liter: [2, 4, 7] K_2HPO_4 - 14 g, KH_2PO_4 - 6 g, Na_2 citrate - 1 g, $MgSO_4 \cdot 7H_2O$ - 0,2 g, L-asparagine - 8 g, L-arginine - 0,2 g, L-glutamic acid - 0,2 g, glicine - 0,2 g, L-histidine - 0,2 g, L-proline - 0,2 g, dl-tryptophane - 0,2 g, sodium thioglycolate - 0,2 g, pH 6,8. Plate method [4, 9, 10, 12]: Stock solution A - 54 ml, Stock solution B - 54 ml, Stock solution C - 0,1 ml, 5% (w/v) $MgSO_4 \cdot 7H_2O$ - 2,2 ml, 0,1% $CaCl_2$ - 1 ml L-asparagine-1,62 g, agar- 20 g, and H_2O - 1000 ml, pH = 7,2 Stock sol. A NH_4Cl - 200 g, NH_4NO_3 - 80 g, and distilled water to 2000 ml.	Double strength for 1 liter [3, 17]: Acid hydrolyzed casein - 10 g, adenine - 20 mg, guanine - 20 mg, uracil - 20 mg, L-asparagine - 0,2 g, Tween 80 - 2 g, xanthine - 20 mg, ascorbic acid - 4 g, L-cystine - 0,4 g, dextrose 40 g, sodium acetate - 12 g, dl-tryptophane - 0,4 g, biotin - 4 μg, riboflavin - 2 mg, thiamine HCl - 4 mg, nicotinic acid - 2 mg, p-aminobenzoic acid - 1 mg, pyridoxine HCl - 4 mg, calcium pantothenate - 2 mg, K_2HPO_4 - 1 g, KH_2PO_4 - 1 g, $MgSO_4 \cdot 7H_2O$ - 0,4 g, NaCl - 20 mg, $FeSO_4 \cdot 7H_2O$ - 20 mg, $MnSO_4 \cdot 4H_2O$ - 20 mg, pH 5,8-6,0.	Double strength for 1 liter [2]: KH_2PO_4 - 0,6 g, $MgSO_4 \cdot 7H_2O$ - 0,8 g, L-glutamic acid - 6,0 g, $CaCO_3$ - 0,16 g, NH_4HCO_3 - 0,72 g, sucrose - 30,0 g, dl-asparatic acid - 4,0 g, dl-malic acid - 2,0 g, glicine - 5,0 g, succinic acid - 1,04 g, thiamine HCl - 12 mg, $FeCl_3$ - 60 mg, $ZnSO_4 \cdot 7H_2O$ - 40 mg, $MnSO_4 \cdot H_2O$ - 60 mg, $CuSO_4 \cdot 5H_2O$ - 0,62 mg, $CoSO_4 \cdot 7H_2O$ - 5,0 mg, H_3BO_3 - 1,14 mg, $(NH_4)_6Mo_7O_{24} \cdot 4H_2O$ - 1,34 mg.	Double strength for 1 liter [2]: $(NH_4)_2H$ citrate - 2,4 g, L-glutamic acid - 6,0 g, dl-methionine - 1,2 g, L-histidine HCl - 1,0 g, L-arginine HCl - 1,0 g, hycase-SF - 2,0 g, sucrose - 30 g, $CaCO_3$ - 0,3 g, $MgCO_3$(basic) - 1,0 g, KH_2PO_4 - 0,6 g, NH_4HCO_3 - 0,8 g, $FeCl_3$ - 60 mg, $ZnSO_4 \cdot 7H_2O$ - 40 mg, $MnSO_4 \cdot H_2O$ - 6,0 mg, $CuSO_4 \cdot 5H_2O$ - 0,62 mg, $CoSO_4 \cdot 7H_2O$ - 5,0 mg H_3BO_3 - 1,14 mg, $(NH_4)_6MO_7O_{24} \cdot 4H_2O$ - 1,34 mg, thiamine HCl - 10,0 mg, biotin - 0,02 mg.

Table 5-5 continued

Test organism	*Escherichia coli mutant 113-3* [1, 2, 4, 7-10] (ATCC 9637 or ATCC 11105)	*Lactobacillus leichmanii 326* [1-4, 17, 18] (ATCC 4797 or ATCC 7830)	*Euglena gracilis* Z-alga [1, 2, 4, 18, 19, 20] ATCC 12716	*Ochromonas malhamensis* [1-3, 18, 21] ATCC 11532
	Stock sol. B KH_2PO_4 - 40 g, K_2HPO_4 - 120 g, and distilled water to 2000 ml. Stock sol. C sodium borate - 0,094 g, $CuSO_4 \cdot 5H_2O$ - 0,250 g, $FeSO_4 \cdot 7H_2O$ - 0,540 g, $MnSO_4 \cdot 4H_2O$ - 0,460 g, $ZnSO_4 \cdot 7H_2O$ - 1,9 g, $(NH_4)_6Mo_7O_{24} \cdot 4H_2O$ - 0.9 g and distilled water - 1000 ml.			
Specificity of the organism	*E. coli 113-3* also responds to other cobamides and cobinamides. Some are not active in man. Methionine in concentrations 50000 times greater than vitamin B_{12} may interfere with the assay [2, 4, 5, 13]	*L. leichmannii* may utilize besides cobalamins also other corrinoids inactive for man such as pseudovitamin B_{12} and deamidated cobalamins. Deoxyribosides and deoxynucleotides at concentrations 1000 times greater than vitamin B_{12} may substitute this compound [2, 3, 5, 17]	Less specific than *O. malhamensis*. It responds in certain degree to other cobamides like (Ade)CN-Cba [2, 3] (see Tables 5-3 and 5-4)	The organism is the most specific in its requirements for cobalamins. It responds to all clinically active forms of vitamin B_{12}. Pseudovitamin B_{12} is not utilized by this strain [2, 3] (see Tables 5-3 and 5-4).
Application	For its rapidity, simplicity and low costs is routinely used in many research and industrial laboratories [5, 14]. This method is particularly recommended for assay of vitamin B_{12} in solution eluted from chromatograms and for bioautograms.	Natural products, animal feeds and pharmaceuticals. Is not useful either in bioautography or disk assays. It is less sensitive than *O. malhamensis* and *E. gracilis* and has a narrower assay range compared with the other methods [2, 18]	Applied for vitamin B_{12} assays in natural products, blood, serum and for bioautography [2]. Direct assay of whole blood, milk and other colored or turbidity-producing material is less recommended due to interference with illumination of the alga [4]. The plate meth-	For estimation of cobalamins in natural products such as tissue, blood, gut mucosa or animal feeding stuffs [3, 24] The strain may also be applied for bioautography.

	This organism is not recommended for assays of vitamin B_{12} in natural sources [2, 3]		od may cause such technical difficulties as the length of time required for the assay (7 days) so that all samples must be sterilized before assay and the special procedure which must be followed for preparation of the assay medium [2]	
Sample preparation	In most cases KCN is added to the analysis in order to convert cobalamins to the more stable cyanide form. Because of high affinity of corrinoids to cyanide traces of KCN are sufficient, however usually much greater amounts are added by some authors (up to 1%). High level of cyanide above 10 µg/ml interferes with the estimation of cobalamins by *E. coli 113-3*. The pH of the extracted sample should be in the range of 4.0 to 5,6. The samples are heated for 5-10 minutes at 100-110 °C in order to free the cobalamins from proteins. The sample is next diluted to the range of vitamin B_{12} concentration appropriate to the sensitivity of the assay [4, 5]. In an alternative procedure the proteins are separated from vitamin B_{12} with cold $HC10_4$. To the cyanide treated sample $HC10_4$ is added to obtain the final concentration of about 5% followed by centrifugation of the precipitated proteins and neutralization of the supernatant solution with KOH to pH 5,6. The insoluble $KC10_4$ is removed by decantation. Prolonged treatment of the sample with acid at elevated temperature may damage the vitamin. Some vitamin B_{12}-protein complexes (like CP-B_{12}) may resist the acid treatment (see vitamin B_{12} binding proteins in Chapter 8 and extraction of corrinoids in section 5.1.)			
Comments	Deamidated corrinoids as well as some B_{12} analogues inhibit the growth of this strain.	In presence of excess of vitamin B_{12} a single bacteria may adsorb about 10,500 B_{12} molecules on its cell wall [16]. That quantity is sufficient to grow many generations of this strain.	*Euglena gracilis* has been demonstrated in three variants: varietas bacillans, T strain and Z strain. The Z strain is very sensitive and is being most widely applied [6].	The assay with this strain produces results closest to clinical tests. The cell wall of this strain has a great binding capacity of B_{12} (524.10 B_{12} molecules/cell) [16].

5.3.2 Other biological methods

Biological methods employ chicks, rats and baby pigs as test animals [3, 24-26, 49, 58]. Since microorganisms may respond to a different spectrum of vitamin B_{12} analogs than man, the use of animals for tests, though less convenient and expensive, is justified by their closer response to that of man's (Tables 5-3, 5-4, and 5-5). The chick method uses direct growth response to measure vitamin B_{12} activity [24, 26, 49]. The chicks used for estimation have to be depleted of vitamin B_{12} and kept under restricted conditions. The chicks respond by growth only to benzimidazole containing analogues while other cobamides e.g. containing adenine in the nucleotide moiety are inactive (Table 5-4). Germ free rats depleted of vitamin B_{12} are used for tests. The animals have to be kept and fed under restricted conditions which make the method rather inconvenient. A great deal of fundamental work on estimation of vitamin B_{12} with the use of baby pigs has been done by Johnson [50-53] (Fig. 5-3). By the introduction of dwarf pigs, recently bred, the method may come into use again.

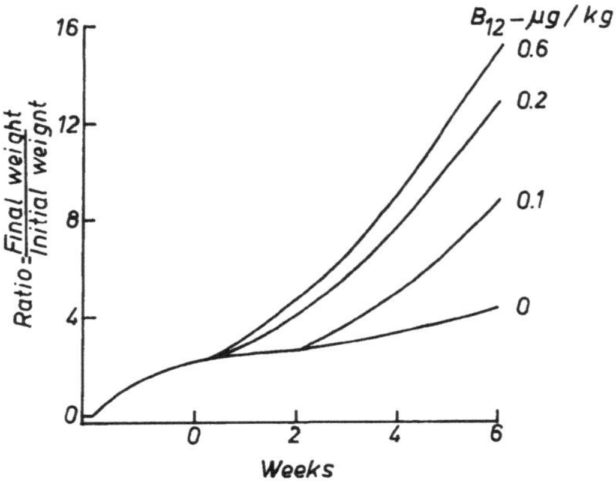

Fig. 5-3 The effect of vitamin B_{12} supplements on the growth of baby pigs while on a vitamin B_{12}-deficient diet (from Johnson, B.C. et al. Am. J. Clin. Nutr. **6,** 34, 1958. Am. J. Clin. Nutr. American Society for Clinical Nutrition)

References

[1] Mücke, D., Einführung in Mikrobiologische Bestimmungsverfahren, 2. Auflage, Georg Thieme, Leipzig 1957, pp. 104-119.

[2] Guttman, H.N., Vitamin B_{12} and congeners, in Analytical Microbiology, Ed. Kavanagh, F., Academic Press, New York, London 1963, pp. 527-550.

[3] The Association of Vitamin Chemists Inc., Methods of vitamin assay, 3rd ed., Interscience Publishers, New York, London, Sydney 1966, pp. 257-270.

[4] Hoff, Jrgensen, E., Microbiological assay of vitamin B_{12}, in Methods of Biochemical Analysis, vol I, Ed. Glick, D., Interscience Publishers, New York 1959, pp. 82-113.

[5] Cook, E.A., Ellis, L.N., Variations in the growth response of four different vitamin B_{12} assay microorganisms to the same tissue and standard preparations, Appl. Microbiol. **16,** 1831-1840, 1968.

[6] Stahlberg, K.G., Studies on methyl-B_{12} in man, Scand. J. Haemat. Suppl. 1, **19,** 7-90, 1967.

[7] Burkholder, P.R., Determination of vitamin B_{12} with a mutant strain of *Escherichia coli*, Science, **114,** 459, 1951.

[8] Myszkowska, K., Tautt, J., Tuszynska, S., Wozniak, W., Mikrobiologiczne metody badania vitamin z grupy B. Wyd. Przemysłu Lekkiego i Spożywczego, Warszawa 1963, pp. 70-79

[9] Lees, K. A., Tootill, J. Rr. P., Microbiological assay on large plates, Part I. General considerations with particular reference to routine assay, Analyst **80**, 95-110, 1955.

[10] Lees, K. A., Tootill, J. Rr. P., Microbiological assay on large plates, Part II. Precise assay, Analyst **80**, 110-115.

[11] Johansson, K. R., Response to and assay of vitamin B_{12} by a mutant of *E. coli*, Proc. Soc. Exptl. Biol. Med. **83**, 448-453, 1953.

[12a] Hamison, G. F., Lees, K. A., Wood, F., The assay of vitamin B_{12}, Part VI. Microbiological estimation with a mutant of *E. coli* by the plate method, Analyst **76**, 696-702, 1951.

[12b] Sourial, N. A., Use of an improved E. coli method for the measurement of cobalamin in serum: comparison with the E. gracilis assay results, J. Clin. Pathol. **34**, 351-357, 1981.

[13] Trojanowska, K., Jaszewski, B., Skupin, J., Some problems concerning the determination of different corrinoids by the plate method with *Escherichia coli 113-3*, Acta Microbiol. Polon. **7**, 103-110, 1975.

[14] Bogucka, J., Iwanowska, J., Kakol, H., Zastosowanie *E. coli* do osznaczania vitaminy B_{12} w ekstraktach watrobowych, Przemysł Chemiczny **IX**, 14-15, 1953.

[15] Volcani, B. E., Toohey, J. I., and Barker, H. A., Detection of cobamide coenzymes in microorganisms by the ionophoretic bioautographic method, Arch. Biochem. Biophys. **92**, 381-391, 1961.

[16] Friedrich, W., Fermente, Hormone, Vitamine, Band III/2, Vitamin B_{12} und verwandte Corrinoide, Georg Thieme, Stuttgart 1975, pp. 190-192 and 304-306.

[17] Skeggs, H. R., *Lactobacillus leichmannii* assay for vitamin B_{12}, in Analytical Microbiology, Ed. Kavanagh, F., Academic Press, New York and London 1963, pp. 551-565.

[18] Strohecker, R., and Henning, H. M., Vitamin-Bestimmungen, E. Merck, Verlag Chemie, Weinheim 1966.

[19] Ross, G. I. M., Hutner, S. H., Back, M. K., An improved Euglena method of vitamin B_{12} assay, in Vitamin B_{12} and Intrinsic Factor, Stuttgart, 1957, pp. 305-310.

[20] Cooper, B. A., Studies with a more rapid method of vitamin B_{12} assay utilizing *Euglena gracilis*, J. Clin. Patho. **12**, 153-156, 1959.

[21] Ford, J. E., The microbiological assay of vitamin B_{12}. The specificity of the requirement of *Ochromonas malhamensis* for cyanocobalamin, Brit. J. Nutr. **7**, 299-306, 1953.

[22] Shaw, W. H. C., and Bessell, Ch. J., The determination of vitamin B_{12}, J. Soc. Anal. Chem. **85**, 389-409, 1960.

[23] Coates, M. E., and Ford, J. E., Methods of measurement of vitamin B_{12}, Biochemical Society Symposia **13**, 36-51, 1955.

[24] Smith, E. L., Vitamin B_{12}, John Willey and Sons, London 1960, pp. 111-123.

[25] Chow, B. F., The B vitamins: B_6, B_{12}, folic acid, pantothenic acid, biotin, in Nutrition, vol. 2, Eds. Beaton, G. H. and McHenry, E., Academic Press, New York, London 1964, pp. 208-257.

[26] Coates, M. E., Ford, J. E., and Harrison, G. F., Activity for chicks and microorganisms of some analogues of cyanocobalamin, in 2. Europäisches Symposion, Hamburg 1961, Ed. Heinrich, H. C., Ferdinand Enke Verlag, Stuttgart, 1962, pp. 276-280.

[27] Chanarin, I., in The Megaloblastic Anaemias, Blackwell Scientific Publications, Oxford 1969, pp. 193-206.

[28] Hutner, S. H., Bach, M. K., Ross, G. I. M., A sugar-containing basal medium for vitamin B_{12}-assay with *Euglena:* application to body fluids, J. Protozool. **3**, 101, 1956.

[29] Anderson, B. B., Investigations into the *Euglena* method for the assay of the vitamin B_{12} in serum, J. Clin. Path. **17**, 14, 1964.

[30] Anderson, B. B., Investigations into the *Euglena* method of assay of vitamin B_{12}: the results obtained in human serum and liver using an improved method of assay, Ph. D. thesis, University of London, 1964.

[31] Skeggs, H. R., Huff, J. W., Wright, L. D., Bosshardt, D. K., The use of *Lactobacillus leichmannii* in the microbiological assay of the animal protein factor, J. Biol. Chem. **176**, 1459, 1948.

[32] Skeggs, H. R., Driscoll, C. A., Charney, J., Wright, L. D., The use of cyanide in the microbiological determination of vitamin B_{12}, J. Am. Chem. Soc. **19 A**, 1951.

[33] Skeggs, H. R., Microbiological assay of vitamin B_{12}, Meth. Biochem. Analysis **14**, 53, 1966.

[34] Skeggs, H. R., *Lactobacillus leichmannii* assay for vitamin B_{12}, Anal. Microbiol., Academic Press, New York 1963, p 551.

[35] Skeggs, H. R., Nepple, H. M., Valentik, K. A., Huff, J. W., Wright, L. D., Observation on the

use of *Lactobacillus leichmannii* 4797 in the microbiological assay of vitamin B_{12}, J. Biol. Chem. **184,** 211, 1950.
[36] Shorb, M.S., Activity of vitamin B_{12} for the growth of *Lactobacillus lactis,* Science N.Y. **107,** 397, 1948.
[37] Shorb, M.S., Unidentified growth factors for *Lactobacillus lactis* in refined liver extracts, J. Biol. Chem. **169,** 455, 1947.
[38] Rosenthal, H.L., Sarett, H.P., The determination of vitamin B_{12} activity in human serum, J. Biol. Chem. **199,** 433, 1952.
[39] Thompson, H.T., Dietrich, L.S., Elvehjem, C.A., The use of *Lactobacillus leichmannii,* in the estimation of vitamin B_{12} activity, J. Biol. Chem. **184,** 175, 1950.
[40] Matthews, D.M., Observations on the estimation of serum vitamin B_{12} using *Lactobacillus leichmannii,* Clin. Sci. **22,** 101, 1962.
[41] Brandt, V., Metz, J., Serum vitamin B_{12} levels in South African White and Bantu subjects, S. Afr. J. Med. Sci. **26,** 1, 1961.
[42] Linnell, J.C., Mackenzie, H.M., and Matthews, D.M., J. Clin. Pathol. **22,** 506, 1969.
[43] Linnell, J.C., Mackenzie, H.M., Wilson, J., Matthews, D.M., J. Clin. Pathol. **22,** 545, 1969.
[44] Linnell, J.C., Hussein, H.A., Matthews, D.M., J. Clin. Pathol. **23,** 820, 1970.
[45] Linnell, J.C., Hoffbrand, A.V., Hussein, H.A., Wise, I.J., Matthews, D.M., Clin. Sci. Mol. Med. **46,** 163, 1974.
[46] Green, R., Newmark, P.A., Musso, A.M., Mollin, D.L., J. Haemt. **27,** 507-526. 1974.
[47] Sourial, N.A., Mollin, D.L., Differential assay of cobamides in serum using R-protein radioisotopic dilution assay, *E. Coli* and *E. gracilis* assays, in Vitamin B_{12}, Eds. B. Zagalak, W. Friedrich, Walter de Gruyter, Berlin 1979, pp. 657-662.
[48] Grossowicz, N., Aronovitch, J., Rachmilewitz, M., Proc. Soc. Exp. Biol. Med. **87,** 513-514, 1954.
[49] Marqvering, B., Menke, K.H., Steffens, W., Breloh, B., Kobalt- und Vitamin-B_{12}-Stoffwechsel VIII. der Einfluß einer 5,6-Dimethylbenzimidazol-Zulage zu einer B_{12}-Mangelration auf das Wachstum von Masthähnchen, Zeitschrift für Tierphysiologie. Tierernährung und Futtermittelkunde **24,** 193-199, 1968.
[50] Wagle, S.R., and Johnson, B.C., Arch. Biochem. Biophys. **70,** 619, 1957.
[51] Wagle, S.R., Metha, R., and Johnson, B.C., Arch. Biochem. Biophys. **72,** 241, 1957.
[52] Wagle, S.R., and Johnson, B.C., J. Am. Chem. Soc. **79,** 4249, 1957.
[53] Johnson, B.C., Studies on the mechanism of action of vitamin B_{12} in animal nutrition, Am. J. Clin. Nutr. **6,** 34, 1958.
[54] Ott, W.H., Rickes, E.L., Wood, T.R., Activity of crystalline vitamin B_{12} for chick growth, J. Biol. Chem. **174,** 88, 1964.
[55] Hallick, J.V., Reid, B.L., Brown, C.L., and Couch, J.R., The vitamin B_{12} content of egg yolk as influenced by oral and parenteral administration of the vitamin, J. Nutr. **50,** 331, 1953.
[56] Fischer, N.A., Benson, E.M., and Swendseid, M.E., The distribution of vitamin B_{12} in the developing chick egg, Arch. Biochem. Biophys. **74,** 458, 1958.
[57] Mapson, L.W., Evidence for the existence of a dietary principle stimulating general growth and lactation, Biochem. J. **26,** 970, 1932.
[58] Ott, W.H., Rickes, E.L., and Wood, T.R., Activity of crystalline vitamin B_{12} for chick growth, J. Biol. Chem. **174,** 1047, 1948.
[59] Voigt, M.N., Eitenmiller, R.R., and Wate, G.O., Vitamin assay by microbial and protozoan organisms: Response to vitamin concentration, incubation time and assay vessel size. J.Food Sci. **43,** 1418-1428, 1978.
[60] Swift, D.G., Vitamins and phytoplankton growth. In: The physiological ecology of phytoplankton, ed. I. Morris, Blackwell Scientific Publications, Oxford 1980, pp. 329-368.
[61] Guillard, R.R.L., and Cassie, V., Minimum cyanocobalamin requirements of some marine centric diatoms, Limnol. Oceanog. **8,** 161-165, 1963.
[62] Ryther, J.M. and Guillard, R.R.L., Studies of marine planktonic diatoms, 2. Use of *Cyclotella nana* Husted for assay of vitamin B_{12} in seawater. Can. J. Microbiol. **8,** 437-445, 1962.
[63] Gold, K., A microbiological assay for vitamin B_{12} in sewater using radiocarbon. Limnol. Oceanog. **9,** 343-347, 1964.
[64] Carlucci, A.F. and Silbernagel, S.B. Bioassay of seawater, 1. A ^{14}C uptake method for the determination of concentrations of vitamin B_{12} in seawater, Can. J. Microbiol. **12,** 175-183, 1966.
[65] Swift, D.G., and Guillard, R.R.L., Diatoms as tools for assay of total B_{12} activity and cyanocobalamin activity in sea water. J. Mar. Res. **35,** 309-320, 1977.
[66] Haines, K.C., and Guillard, R.R.L., Growth of vitamin B_{12}-requiring marine diatoms in mixed laboratory cultures with vita-

min B_{12}-producing marine bacteria. J. Phycol. **10**, 245-252, 1974.

[67] Guillard, R. R. L., B_{12} specificity of marine centric diatoms, J. Phycol. **4**, 59-64, 1968.

[68] Droop, M. R., A suggested method for the assay of vitamin B_{12} in sea water. J. Mar. Biol. Ass. U. K. **34**, 435-440, 1955.

[69] Martin, J. L. Y., Mise au point d'une methode de dosade de la vitamine B_{12} dans l'eau de mer. Tèthys **5**, 95-105, 1973.

[70] Provasoli, L. and Carlucci, A. F., Vitamins and growth regulators, in Algal Physiology and Biochemistry, ed. W. D. P. Stewart, Blackwell Scientific Publications, Oxford 1974, pp. 741-781.

[71] Cattell, S. A. , Dinoflagellates and vitamin B_{12} in the Strait of Georgia, British Columbia, Ph. D. thesis Univ. British Columbia, Vancouver 1969.

[72] Yamamoto, M., Okamoto, R. and Inui, T., Application of a marine methanol-utilizing bacterium for bioassay of vitamin B_{12} in seawater, J. Ferment. Technol. **57**, 400-407, 1979.

5.3.3 Radio-isotope dilution assay (RIDA)

This method introduced in the early 1960's has been since continuously improved. It permtis the assay of femtomolar (picogram) quantities of vitamin B_{12} in a volume of about 200 µl. It is particularly suitable for a direct assay of the vitamin in serum, tissues and body fluids. Its sensitivity may be compared with the microbiological assays considered to be the most sensitive so far. The specificity depends on the B_{12}-binding proteins applied. The assay is very convenient due to its rapidity and simplicity. A set of ten samples and standards may be completed within 2 hours. The method employs suitable B_{12}-binding proteins and labelled cyanocobalamin. The animal B_{12}-binding proteins *cobalophilin* (CP) and *intrinsic factor* (IF) are applied as reagents (for characteristics of these proteins see Chapter 8). [^{57}Co] labelled cyanocobalamin (specific activity 180-300 mCi/µg, Amersham) is a convenient radiochemical for this purpose since it permits a direct assay of the radioactivity in the aqueous solution of the protein bound vitamin. Besides ^{57}Co two other radioactive isotopes of cobalt ^{58}Co and ^{60}Co are being introduced to Cbl for analytical and diagnostic purposes. All cobald labelled Cbl's are γ radioactive and thus require strict safety precautions. The ^{57}Co-la-

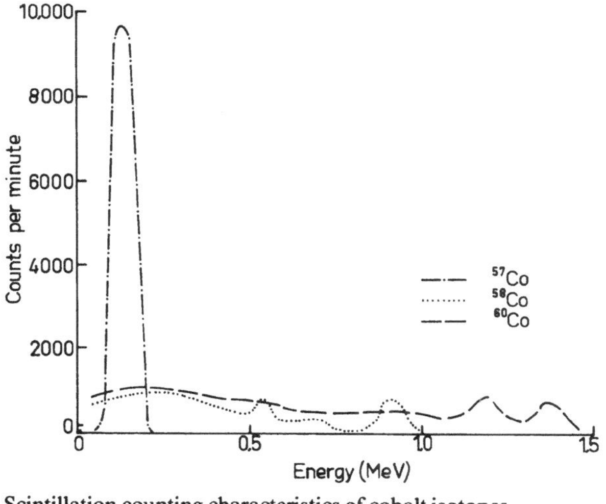

Fig. 5-4 Scintillation counting characteristics of cobalt isotopes.

belled Cbl with its fairly long half-life (270 days) and the lowest radiation energy is advantageous. As shown in Fig. 5-4 ^{57}Co has a scintillation spectrum with a single energy peak at 0.123 MeV (see also Table 3-4). ^{125}I-labelled cobalamin has been also obtained for analytical purposes [6b].

Method of assay. The method is based on the competition of the assayed cobalamin with the radioactive cobalamin for binding to a limited amount of B_{12}-binding protein. The excess of unbound B_{12} is separated from the protein bound B_{12} by dialysis or adorption on hemoglobin-coated charcoal. The radioactivity of the protein bound cobalamin remaining in solution is measured and the amount of the vitamin B_{12} is read from a standard curve. The standard curve is obtained by running, simultaneously with the analysis, samples with standard vitamin B_{12} (from 50 to 2000 pg) (see Fig. 5-5). Prior to the assay, the analysed vitamin B_{12} must be liberated from complexes with proteins and converted into the cyanide form. This may be achieved by acidification of the analysed sample and heat denaturation in presence of cyanide, light and air. The latter two factors are particularly obligatory in cyanide treatment of MeCbl (see section 5.1 and Table 5-5). The assay procedure was described in detail by Lau et al. [1] as well as in a number of other papers [2-13, 21-18, 32-49]. A vitamin B_{12} radioassay kit for estimation of cobalamin in serum is available from the Radiochemical Centre in Amersham, England. The procedure of assay with the use of this kit is summarized in Table 5-6. The concentration of CN-Cbl in serum is expressed in pg/ml or pmol/l (Fig. 5-6).

Modifications of the RIDA method. There is a potential to utilize transcobalamin (TC), the third animal B_{12}-binding protein, for the radiodilution assay technique. The TC-Cbl complex adsorbs strongly to cellulose nitrate filters. This feature has been applied for TC estimation in serum [29, 30]. The [^{57}Co]Cbl-TC complex may also be rapidly separated from the free [^{57}Co]Cbl by chromatography on cellulose nitrate membrane strips [31]. This method permits the measurement of both bound and free Cbl,

Table 5-6 Vitamin B_{12} radioassay kit protocol (Amersham).

	Standards pg B_{12}/ml						Unknowns		
	0	50	200	500	1000	2000	1	2	3 etc
tube nos.	1, 2	3, 4	5, 6	7, 8	9, 10	11, 12	13, 14	15, 16	17, 18
	µl	µl	µl	µl	µl	µl	µl	µl	µl
Standard solutions	200	200	200	200	200	200	–	–	–
Unknown	–	–	–	–	–	–	200	200	200
Denaturing reagent (contains KCN, HCl, [^{57}Co]CN-Cbl)	1000	1000	1000	1000	1000	1000	1000	1000	1000
	Vortex mix and boil for 5 minutes. Cool exactly 5 minutes								
Intrinsic factor	200	200	200	200	200	200	200	200	200
	Vortex mix and incubate for 30 minutes at room temperature								
Charcoal	200	200	200	200	200	200	200	200	200
	Vortex mix and stand for 5 minutes. Centrifuge for 10-15 minutes at 1500 g. Decant into counting tubes.								

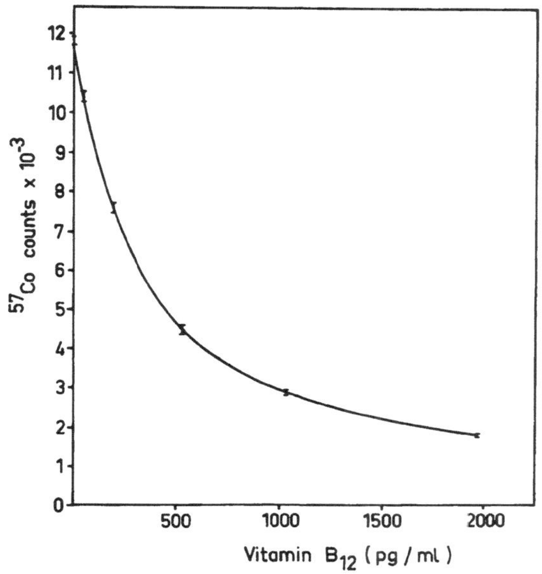

Fig. 5-5 Typical standard curve obtained in the RID-assay (see Table 5-6). The curve in this Figure serves as an example and must not be used to read the Cbl concentration in a particular assay. A set of assayed samples should accompany standards (see section 5.3.3) (with permission of Amersham International plc, England)

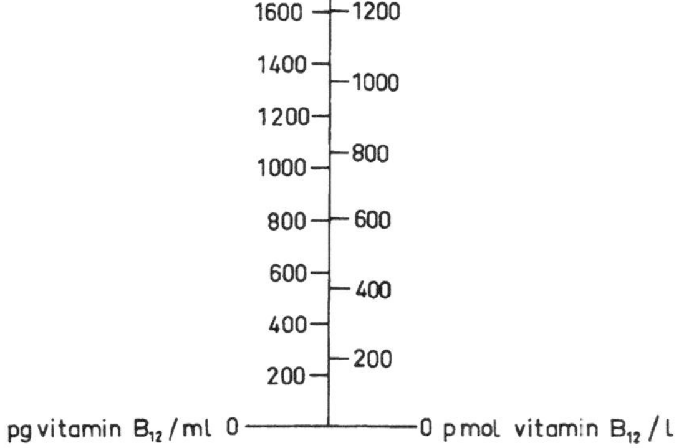

Fig. 5-6 Conversion rate of CN-Cbl concentration in pg/ml into pmol/l. This diagram cannot be applied to other cobalamins which differ from CN-Cbl by molecular mass (for molecular weights of corrinoids see Table 3-1).

thus increasing the accuracy of the assay. A less hazardous and cheaper tritium labelled Cbl may be applied in this technique as well.

The separation of free Cbl from the protein-bound can also be achieved by two other methods:

1. Adsorption of the holoprotein on zirconyl phosphate gel [16].
2. Immobilization of the Cbl-protein binder on Sepharose [14, 15], polyacrylamide [35 d, 44] or anion exchange resin [40].

From Bio-Rad and Becton Dickinson [42] radioassay kits are available which utilize Sepharose immobilized intrinsic factor. The application of immobilized binding protein allows to omit the adsorption on charcoal, a procedure not very reliable in separation of the protein bound ^{57}Co-vitamin B_{12} from the free one [45]. Solid phase radioassays using polyacrylamide-bound binding proteins have been developed as well [35 c, d]. Immobilized specific binding proteins are applied for concentration of vitamin B_{12} and its separation from B_{12} analogues [35 d].

Application and evaluation of the method. The values obtained depend on the type of binding protein used (see Table 5-7). The two binding proteins applied in this method, CP of saliva and hog IF, differ in specificity toward corrinoids [7, 11-14, 24-27, 32-34, 36, 37, 39-48]. CP is one of the least specific binders. It also binds effectively corrinoids inactive for man and animals such as cobinamide and pseudovitamin B_{12}. Thus the application of CP is limited to the assay of material which is not contaminated with other corrinoids. Until recently it was commonly accepted that the human and animal tissues and serum contain only cobalamin. The validity of this view has been questioned [35 a-d] since the B_{12} serum values obtained by RIDA with the use of saliva CP are higher than those obtained by the more specific microbiological assay [24, 25, 32-37, 41, 42]. The discrepancies of these two methods are particularly evident

Table 5-7 Serum B_{12} (pg/ml) in 10 normal humans assayed with the RIDA and the microbiological method (from Begley and Hall [11]).

Subject	Binder Diluted in Buffer[a]		Binder Diluted in Buffer[a] Containing 100 μg PVP[c]/0.5 ml		Bioassay *E. gracilis*
	R[b]	IF	R[b]	IF	
1	429	192	756	484	360
2	456	160	812	536	517
3	420	156	780	504	443
4	368	4	640	252	190
5	360	156	696	408	273
6	504	156	840	520	403
7	300	48	620	324	257
8	276	88	600	408	403
9	464	136	860	480	456
10	360	80	620	380	367

[a] 0.3 M Sodium acetate, pH 4.8
[b] Cobalophilin of saliva (see also section 8.3)
[c] PVP = Polyvinyl-pyrolidone (MW 40000)

Table 5-8 Mean serum B_{12} concentration (pg/ml) using *E. gracilis* and RIDA assays (from Sourial and Mollin [47]).

	E. gracilis mean ± S. D.	RIDA mean ± S. D.	Difference mean ± S. D.
Normals [73]	435 ± 149	715 ± 186	280 ± 119
Partial Gastrectomy [7]	134 ± 100	472 ± 208	343 ± 127
Folate Deficiency [19]	200 ± 91	532 ± 139	317 ± 126
Pregnancy [6]	104 ± 23	394 ± 110	*290 ± 69
Pernicious Anaemia [36]	38 ± 18	174 ± 66	136 ± 48

in pernicious anaemia cases, when there is a low true cobalamin level in serum [24-25, 41, 42] (see Table 5-8). The addition of polyvinyl pyrrolidone (PVP) to the assayed sample and the application of IF, instead of the less specific CP, produces values very close to the microbiological assays (Tables 5-7 and 5-8).

The three forms of cobalamin OH-Cbl, MeCbl and AdoCbl occurring in serum, tissues and gastric juice are tightly bound to proteins. Their liberation from the complexes and conversion to CN-Cbl is essential for this method. Drastic treatments often necessary to detach cobalamin from the binding proteins, such as very low pH and elevated temperature at acidic pH, may cause partial inactivation of cobalamin due to deamidation at the corrin ring. Reducing agents and strong oxidizers inactivate cobalamin as well. The susceptibility to chemical changes of various forms of cobalamin is different. Particularly labile is OH-Cbl. Concentrated ascorbic acid inactivates OH-Cbl within minutes [27, 28, 38]. CN-Cbl is the most inert form of cobalamin. Thus the liberation of cobalamins from the protein complexes should accompany their immediate conversion to the cyano-form. There are problems in keeping constant binding capacity of the binding proteins in the course of storage. High reproducibility in this regard was achieved when the proteins were stabilized by drying them with Sephadex G-50 at temperature > 0 °C [17-19].

References

[1] Lau, K. S., Gottlieb, C. W., Wasserman, L. R., Herbert, V., Measurement of serum vitamin B_{12} level using radioisotope dilution and coated charcoal assay, Blood **26**, 202, 1965.

[2] Hillman, R. S., Oakes, M., Finholt, C., Hemoglobin coated charcoal radioassay for serum vitamin B_{12}. A simple modification to improve intrinsic factor reliability, Blood **34**, 385, 1969.

[3] Herbert, V., Gottlieb, C. W., Lau, K. S., Hemoglobin-coated charcoal assay for serum vitamin B_{12}, Blood **28**, 130-132, 1966.

[4a] Frenkel, E. P., White, J. D., Reisch, J. S., Sheehan, R. G., Comparison of two methods for radioassay of vitamin B_{12} in serum, Clin. Chem. **19**, 1357-1360, 1973.

[4b] Green, R., Newmark, P. A., Musso, A. M., Mollin, D. L., The use of chicken serum for measurement of serum vitamin B_{12} concentration by radioisotopic dilution: description of method and comparison with microbiological assay results, Br. J. Haematol. **27**, 507-533, 1974.

[5] Lin, Y. K., Sullivan, L. W., An improved radioisotope dilution assay for serum vitamin B_{12} using hemoglobin-coated charcoal, Blood **39**, 426-432, 1972.

[6a] Rothenberg, S. P., in methods in radioimmunoassay, toxicology and related areas, Eds. I. L. Simmons and G. W. Ewing, New York, Plenum Press 1974, pp. 76-96.

[6b] Houts, T. M., ^{125}J-labelled cobalamin deriva-

tive compatible with intrinsic factor, Clinica Chemica Acta **126,** 315-322, 1982.

[7] Kolhouse, J. F., Kondo, H., Allen, N. C., Podell, E., and Allen, R. H., Cobalamin analogues are present in human plasma, and can mask cobalamin deficiency because current radioisotope dilution assays are not specific for true cobalamin, New Engl. J. Med. **299,** 785-792, 1978.

[8] Wierusz-Wysocka, B., Binding of vitamin B_{12} by serum proteins in chronic liver disease, Pol. Arch. Med. Wewn. **58,** 139-147, 1977.

[9] Wierusz-Wysocka, B., Vitamin B_{12} binding by serum proteins in virus hepatitis, Pol. Arch. Med. Wewn. **56,** 445-455, 1976.

[10] Beck, R. A., Essential prerequisites for the analysis of cyanocobalamin in biochemically complex samples using radiometric competitive binding assays, in Vitamin B_{12}, Eds. B. Zagalak and W. Friedrich, Walter de Gruyter, Berlin 1979, pp. 675-679.

[11] Begley, J. A., Hall, C. A., Effect of residual extract products and the type of binders (R or IF) on serum vitamin B_{12} levels by radioisotope dilution assay, in Vitamin B_{12}, Eds. B. Zagalak and W. Friedrich, Walter de Gruyter, Berlin 1979, pp. 655-656.

[12] Cooper, B. A., Peyman, J., Jonas, E., and Whitehead, V. M., Vitamin B_{12} assay: An evaluation of radiodilution assay using cobinamide to increase specificty, in Vitamin B_{12}, Eds. B. Zagalak and W. Friedrich, Walter de Gruyter, Berlin 1979, pp. 649-654.

[13] Sourial, N. A., Mollin, D. L., Differential assay of cobamides in serum using R-protein radioisotopic dilution assay, *E. coli* and *E. gracilis* assays, in Vitamin B_{12}, Eds. B. Zagalak and W. Friedrich, Walter de Gruyter, Berlin 1979, pp. 657-662.

[14a] Kolhouse, J. F., Allen, R. H., Isolation of cobalamin and cobalamin analogs by reverse affinity chromatography. Anal. Biochem. **84,** 486-490, 1978.

[14b] Dekker, B. G., Van Dongen, C. J., Schipper, N. P. M. and De Ligny, C. L., The influence of experimental conditions on the association constant for binding of vitamin B_{12} by human intrinsic factor coupled to Sepharose 4B, Int. J. Appl. Radiat. Isot. **34,** 1351-1356, 1983.

[15] Cotter, R., Rothenberg, S. P., Weiss, J. P., Purification of the intestinal receptor for intrinsic factor by affinity chromatography, Biochim. Biophys. Acta **490,** 19-26, 1977.

[16] Hansen, H. J., Miller, O. N., Gallo-Torres, H., and Goldsmith, G. A., Assay of intrinsic factor activity of human gastric juice with zirconyl phosphate gel, Anal. Biochem. **16,** 267, 1966.

[17] Schneider, Z., Friedmann, H. C., Stabilization of enzymes in Sephadex — a long-term experiment. J. Appl. Biochem. **3,** 135-146, 1982.

[18] Schneider, Z., Stroiński, A., Pawełkiewicz, J., Thermostability of enzyme in the three-dimensional network of polysaccharide chains, Bull. Acad. Polon. Sci. **16,** 203-204, 1968.

[19] Fenrych, W., Białka wiążące witaminę B_{12} w osoczu. Wpływ czynników fizyko-chemicznych oraz komórkowych na transkobalaminy i ich wzajemną zależność, Polski Tygodnik Lekarski **25,** 973-975, 1977.

[20] Carmel, R., A rapid ammonium sulfate precipitation technic for separating serum vitamin B_{12}-binding proteins, Am. J. Clin. Pathol. **62,** 367, 1974.

[21] Gottlieb, C., Lau, K. S., Wasserman, L., Herbert, V., Rapid charcoal assay for intrinsic factor (IF), gastric juice unsaturated B_{12}-binding capacity, antibody to IF, and serum unsaturated B_{12}-binding capacity, Blood **25,** 875, 1965.

[22] Scott, J. M., Bloomfield, F. J., Stebbins, R., Herbert, V., Studies on derivation of transcobalamin III from granulocytes, J. Clin. Invest. **53,** 228, 1974.

[23] Mollin, D. L., Anderson, B. B., Burman, J. F., The serum vitamin B_{12} level: its assay and significance, Clinics in Haematology **5,** 521-546, 1976.

[24] Cooper, B. A., Whitehead, V. M., Evidence that some patients with pernicious anaemia are not recognized by radiodilution assay for cobalamin in serum, New Eng. J. of Med. **299,** 816-818, 1978.

[25] Cooper, B., Frenkel, E. P., Colman, N., Herbert, V., Akerkar, A., Block, E., Weaver, D., Multi-laboratory evaluation of "serum vitamin B_{12} level" measured by radioassay for "total B_{12}" vs. "true cobalamin" vs. microbiologic assay with *Euglena gracilis*. Amer. Assoc. Clin. Chem. Abstracts, New Orleans 1979.

[26] Gottlieb, C. W., Retief, F. P., Herbert, V., Blockade of vitamin B_{12} binding sites in gastric juice, serum and saliva by analogues and derivatives of vitamin B_{12} and by antibody to intrinsic factor. Biochim. Biophys. Acta **141,** 560-572, 1967.

[27] Herbert, V., Jacob, E., Wong, K. T., Scott, J., Pfeffer, R. D., Low serum vitamin B_{12} levels in patients receiving ascorbic acid in megadoses: Studies concerning the effect of ascorbate on

radioisotope vitamin B_{12} assay. Am. J. Clin. Nutr. **31,** 253-258, 1978.

[28] Herbert, V., Landau, L., Bash, R., Grosberg, S., Colman, N., Ability of megadoses of vitamin C to destroy vitamin B_{12} and to reduce absorption of vitamin B_{12} with a note on B_{12} radioassays, in, Vitamin B_{12}, Eds. B. Zagalak and W. Friedrich, Walter de Gruyter, Berlin 1979, pp. 1069-1077.

[29] Selhub, J., Toporek, M., Rachmilewicz, B., Grossowicz, N., FEBS Lett. **44,** 71-74, 1974.

[30] Selhub, J., Rachmilewicz, B., Grossowicz, N., Fractionation of serum transcobalamins on charged cellulose filters, Proc. Soc. Exp. Biol. Med. **152,** 204-209, 1976.

[31] Schneider, Z., Aliphatic alcohols improve the adsorptive performance of cellulose nitrate membranes-application in chromatography and enzyme assays, Anal. Biochem. **108,** 96-103, 1980.

[32] Rahalkar, A. D., Lal. R., Shah, K. B., and Manc, R. S., Radioassay of serum vitamin B_{12} using a competitive protein-binding technique, J. Radioanal. Chem. **52,** 21-30, 1979.

[33] Molin, D. L., Hoffbrand, A. V., Ward, P. G., and Lewis, S. M., Interlaboratory comparison of serum vitamin B_{12} assay, J. Clin. Pathol. (Lond.) **33,** 243-248, 1980.

[34] Dawson, D., Delamore, I. W., Fish, D. I., Flaherty, T. A., Gowenlock, A. H., Hunt, L. P., Maciver, J. E., Thornton, J. A., and Waters, H. M., An evaluation of commercial radioisotope methods for the determination of folate and vitamin B_{12}, J. Clin. Pathol. (Lond.) **33,** 234-242, 1980.

[35 a] Kondo, H., Kolhouse, J. F., and Allen, R. H., Presence of cobalamin analogs in animal tissues, Proc. Natl. Acad. Sci. USA **77,** 817-821, 1980.

[35 b] Begley, J. A., Hall, C. A., Forms of vitamin B_{12} in radioisotope dilution assays, J. Clin. Pathol. **34,** 630-636, 1981.

[35 c] Zacharakis, R. E. A., Muir, M., Chanarin, I., Comparison of serum vitamin B_{12} estimation by saturation analysis with intrinsic factor and with R-protein as binding agents, J. Clin. Pathol. **34,** 357-417, 1981.

[35 d] Muir, M., Chanarin, J., Separation of cobalamin analogs in human sera by binding to intrinsic factor and to R-type vitamin B_{12} binders, Br. J. Haematol. **54,** 613-622, 1983.

[36] Kubasik, N. P., Ricota, M., Sine, M. R., Commercially supplied binders for plasma cobalamin (vitamin B_{12}) analysis purified intrinsic factor cobinamide-blacked R-protein binder and nonpurified intrinsic factor-R-protein binder compared to microbiol. assay. Clin. Chem. **26,** (5), 598-600, 1980.

[37] Le Febvre, R. J., Virji, A. S., Mertens, B. F., Erroneously low results due to high nonspecific binding encountered with a radioassay kit that measures true serum vitamin B_{12}, Am. J. Clin. Pathol. **74,** 209-222, 1980.

[38] Marcus, M., Pradhudesai, M., and Wassef, S., Stability of vitamin B_{12} in the presence of ascorbic acid in food and serum: Restoration by cyanide of apparent loss. Am. J. Clin. Nutr. **33,** 137-143, 1980.

[39] Lindemaus, J., Van Kapel, J., and Abels, J., Preparation of vitamin B_{12}-free serum for the use in vitamin B_{12} radioassays. Clin. Chim. Acta. **95,** 29-34, 1979.

[40] Castro, A., Cid, A., Buschbaum, P. A., and Clark, L., A new radioassay for serum vitamin B_{12} determination using chicken serum as a binder and an anion exchange resin, Res. Commun. Chem. Pathol. Pharmacol. **24,** 583-594, 1979.

[41] Sturgeon, M. F., Erroneously low results due to high non-specific binding encountered with a radioassay kit that measures true serum vitamin B_{12}, Am. J. Clin. Pathol. **75,** 767-775, 1981.

[42] Bain, B., Broom, G. N., Woodside, J., Litwinczuk, R. A., and Wickramasinghe, S. N., Assessment of a radioisotopic assay for vitamin B_{12} using an intrinsic factor preparation with R-proteins blocked by vitamin B_{12} analogs, J. Clin. Pathol. Clond. **35,** 1110-1113, 1982.

[43] Gijzen, A. H. J., De Kock, H. W., Meulenchjk, P. N., Schmidt, N. A., Schopman, W., Tertoden, J. F. W., Need for a sufficient number of low level sera in comparison of different serum vitamin B_{12} assays, Clin. Chim. Acta **127,** 185-196, 1983.

[44] Muir, M., Chanarin, I., Solid phase vitamin B_{12} assays using polyacrylamide-bound intrinsic factor and polyacrylamide bound R-binder, Br. J. Haematol. **53,** 423-436, 1983.

[45] Jacob, E., Wong, K-T. J., Limitations of protein-coated charcoal in the separation of free from bound vit. B_{12} in serum, J. Clin. Pathol. **36,** 1022-1027, 1983.

[46] Chen, I-W., Silberstein, E. B., Maxon, H. R., Volle, C. P., Sohnlein, B. H., Semiautomated system for simultaneous assays of serum vit. B_{12} and folic acid in serum evaluated, Clin. Chem. **28,** 2161-2165, 1982.

[47] Cooper, B. A., Whitehead, V. M., Evidence that some patients with pernicious anaemia

are not recognised by radio dilution assay for cobalamin in serum, N. Engl. J. Med. **299**, 819-824, 1978.

[48] Chen, I.W., Silberstein, E.B., Maxon, H.R., Sperling, M., Barnes, E., Clinical significance of serum vitamin B_{12} measured by radioassay using pure intrinsic factor, J. Nucl. Med. **22**, 447-498, 1981.

[49] Fogel, Y., Prospects for using radiopharmaceutical preparations in West Germany, Med. Radiol. **27**, 69-71, 1982.

5.3.4 Enzymatic assay of corrinoids

Assay of coenzymatic forms of cobamides. The method employs suitable apoproteins of the vitamin B_{12}-dependent enzymes such as glycerol dehydratase, (EC 4.2.1.30) dioldehydratase (EC 4.2.1.28) or ethanolamine deaminase (EC 4.3.1.7) (see Chapter 7). The apoproteins combine with the assayed cobamide coenzyme into an active holoenzyme. The activity of the enzyme thus obtained is proportional to the concentration of the cobamide coenzyme providing the presence of the excess of the apoenzyme in the assayed sample. Depending on the sensitivity of enzyme assay nanomolar-to picomolar quantities of coenzymatic forms of cobamides may be estimated.

Assay of non-coenzymatic forms of cobamides. The method is based on the observation that cyanocobalamin, hydroxocobalamin and many other non-coenzymatic forms of cobamides form a stable and inactive complex with the glycerol dehydratase apoenzyme in a definite stoichiometric relationship [2, 4, 5]. Since the complex is enzymatically inactive, the amount of vitamin B_{12} bound is proportional to the amount of inactivated apoenzyme. The method permits the assay of picomolar quantities (nanograms) of vitamin B_{12} and some of its derivatives within 1-2 hours [7, 8]. The incomplete forms of corrinoids such as cobinamides and deamidated cobalamin do not bind effectively to the apoenzyme and thus do not interfere with the assay if present at equal or several-fold higher concentrations.

Principle of assay. The assayed corrinoid is preincubated with an excess of a fixed amount of glycerol dehydratase apoenzyme. Part of the apoenzyme forms an inactive complex with the corrinoid. The remaining active apoenzyme is assayed enzymatically by addition of substrate and AdoCbl. The product β-hydroxypropionaldehyde formed from glycerol is measured in a color reaction which is produced after heating it with tryptophan and conc. HCL [2, 4, 7]. Its amount is inversely proportional to the amount of the determined corrinoid (Fig. 5-7). Standard samples with cyanocobalamin are run simultaneously with the assayed sample. Under conditions used the glycerol dehydratase holoenzyme undergoes selfinactivation with a half inactivation time of about 7 minutes. Thus after 30 minutes of incubation the reaction practically stops itself. The details of assay are described in Table 5-9. The procedure is a modification of the method described in [6-8].

Source of enzyme. Glycerol dehydratase apoenzyme is obtained from the bacterial strain *Klebsiella pneumoniae* ATCC 25955, formerly known as *Aerobacter aerogenes* 572 PZH Warsaw, by a method described in [2, 4, 7]. Crude preparation of the apoenzyme is sufficient for the assay. Bacteria obtained from a 10 liter culture yield enzyme sufficient for about 10000 assays. The apoenzyme in 10% of glycerol and 0.01 M K_2HPO_4 (pH 8.6) can be stored for several years at $-20\,°C$. Preparations of the apo-

Table 5-9 Protocol of enzymatic assay of non coenzymatic cobamides (for principles of assay see text in section 5.3.3).

Tube no.	1, 2	3, 4	5, 6	7, 8	9, 10	11, 12	13, 14	15, 16	17	comments
standard B_{12}:	µl of component added:									use glass tubes only
10 pmoles/ml	–	50	–	–	–	–	–	–	–	
40 pmoles/ml	–	–	50	–	–	–	–	–	–	
80 pmoles/ml	–	–	–	50	–	–	–	–	–	
100 pmoles/ml	–	–	–	–	50	–	–	–	–	
analyzed sample (10–1000 pmoles of B_{12}/ml):	–	–	–	–	–	5	20	50		avoid excess of KCN in the sample
H_2O	–	–	–	–	–	45	30	–	80	
2 M K_2HPO_4 (pH 8.6)	20	20	20	20	20	20	20	20	20	
Enzyme ($\approx$40 units/ml)[a]	30	30	30	30	30	30	30	30	–	The enzyme stabilized in 10% of glycerol should be dialyzed before use for 1 hour against a 100-fold greater volume of 0.05 M K_2HPO_4 pH 8.6.
	Incubate for 30 minutes at 5–8 °C followed by 10 minutes of incubation at 30 °C									Protect from strong light.
mixture: 5% glycerol, 20 mM Na_2SO_3, and 20 µM AdoCbl	100	100	100	100	100	100	100	100	100	The components should be mixed before use. The mixture prewarmed to 30 °C is added to tubes in dimmed red light followed by immediate mixing.
	Incubate for 30 minutes at 30 °C in dark or dimmed red light followed by cooling on ice bath.									
tryptophan solution (300 mg/100 ml of 0.1 N HCl)	100	100	100	100	100	100	100	100	100	Prepare the tryptophan solution every 2 weeks. Keep it protected from light cool or frozen
Conc. HCl	600	600	600	600	600	600	600	600	600	Mix each tube immediately.
	Heat for 5 minutes at 60 °C followed by cooling on ice. Measure the optical density at 555 nm within 1 hour against tube No 17. Draw the standard curve as shown in Fig. 5-7 and use it to read up the quantity of CN-Cbl in the sample.									

[a] one unit is defined here as the amount of apoenzyme which produces an optical density = 1.0 in the color reaction under conditions given in the table

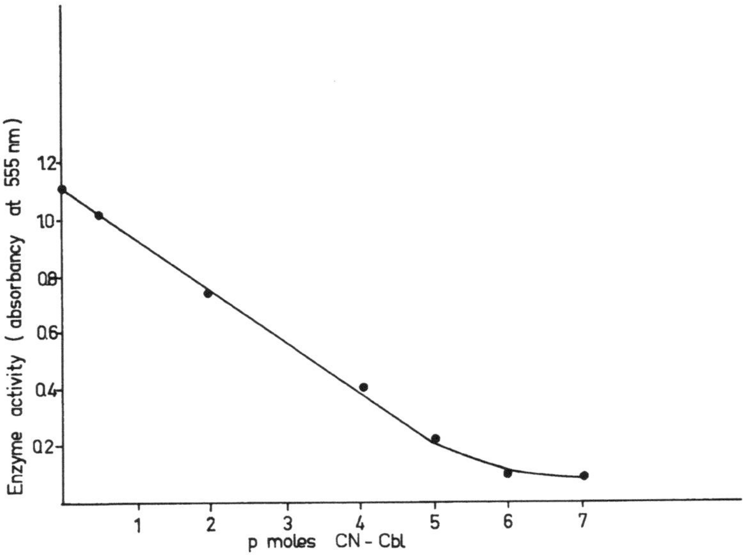

Fig. 5-7 Typical standard curve obtained in the assay of cobamides by the enzyme inhibition method (see section 5.3.4 and Table 5-9).

enzyme dried in Sephadex G-50 as described in [18, 19] are extremely stable. They may be kept at room temperature for more than 10 years. On the other hand glycerol dehydratase solution is labile at room temperature. It also loses its activity if liophylized.

References

[1] Schneider, Z., Pech, K., Pawełkiewicz, J., Enzymatic transformation of glycerol to β-hydroxypropionic aldehyde, II. Dissociation of the enzyme into two protein fragments, Bull. Acad. Pol. Sci. **13**, 99, 1965.

[2] Schneider, Z., Pawełkiewicz, J., The properties of glycerol dehydratase isolated from *Aerobacter aerogenes*, and the properties of the apoenzyme subunits, Acta Biochim. Polon. **13**, 311, 1966.

[3] Stroiński, A., Schneider, Z., Pawełkiewicz, J., Coenzyme B_{12}-dependent glycerol dehydratase from *Aerobacter aerogenes* properties of self-inactivated holoenzyme, Bull. Acad. Pol. Sci. **15**, 12, 1967.

[4] Schneider, Z., Larsen, E.C., Jacobson, G.J., Johnson, B.C., Purification and properties of glycerol dehydrase, J. Biol. Chem. **245**, 3388, 1970.

[5] Johnson, B.C., Stroiński, A., Schneider, Z., Glycerol dehydratase, properties and subunits structure, Methods in Enzymology, Carbohydrates, **42 C**, 1975.

[6] Pawełkiewicz, J., Schneider, Z., Enzymatic determination of vitamin B_{12}, coenzyme B_{12}, and other cobamide derivatives in picomole quantities by means of glycerol dehydratase from *Aerobacter aerogenes*, Bull. Acad. Pol. Sci. **15**, 2, 1967.

[7] Schneider, Z., Friedmann, H.C., Studies on the enzymatic dephosphorylation of vitamin B_{12} 5′-phosphate, Arch. Biochem. Biophys. **152**, 488, 1972.

[8] Schneider, Z., Enzymatic estimation of vitamin B_{12}, in Vitamin B_{12}, Eds. B. Zagalak and W. Friedrich, Walter de Gruyter, Berlin 1979, pp. 673-675.

[9] Schneider, Z., Johnson, B.C., Studies on the thermal dissociation of glycerol dehydratase, Arch. Biochem. Biophys. **144**, 491, 1971.

[10] Smiley, K.L., Sobolov, M., A cobamide-requiring glycerol dehydrase from an acrolein-form-

ing *Lactobacillus*, Arch. Biochem. Biophys. **97**, 538-543, 1962.

[11] Toraya, T., Honda, S., Kuno, S., Fukui, S., Coenzyme B_{12}-dependent diol dehydratase: regulation of apoenzyme synthesis in *Klebsiella pneumoniae (Aerobacter aerogenes)* ATCC 8724, J. Bacteriol. **135**, 726-729, 1978.

[12] Wagner, O.W., Lee, H.A., Frey, J., Abeles, R.H., Studies on the mechanism of action of cobamide coenzymes. Chemical properties of the enzyme-coenzyme complex, J. Biol. Chem. **241**, 1751-1762, 1966.

[13] Toraya, T., Kuno, S., Fukui, S., Distribution of coenzyme B_{12}-dependent diol dehydratase and glycerol dehydratase in selected genera of *Enterobacteriaceae* and *Propionibacteriaceae*, J. Bacteriology **141**, 1439-1442, 1980.

[14] Toraya, T., Honda, S., Fukui, S., Fermentation of 1,2-propanediol and 1,2-ethanediol by some genera of *Enterobacteriaceae*, involving coenzyme B_{12}-dependent diol dehydratase, J. Bacteriol. **139**, 39-47, 1979.

[15] Honda, S., Toraya, T., Fukui, S., In situ reactivation of glycerol-inactivated coenzyme B_{12}-dependent enzymes, glycerol dehydratase and diol dehydratase, J. Bacteriol. **143**, 1458-1465, 1980.

[16] Poznanskaya, A.A., Yakusheva, M.I., Yakovlev, V.A., Study on the mechanism of action of adenosylcobalamin-dependent glycerol dehydratase from *Aerobacter aerogenes*. II. The inactivation kinetics of glycerol dehydratase complexes with adenosylcobalamin and its analogs, Biochim. Biophys. Acta **484**, 236-243, 1977.

[17] Bachovchin, W.W., Moore, K.W., Richards, J.H., Mechanism of action of adenosylcobalamin: hydrogen transfer in the inactivation of diol dehydratase by glycerol, Biochemistry **17**, 2218-2224, 1978.

[18] Schneider, Z., Stroiński, A., Pawełkiewicz, J., Thermostability of enzyme in the three-dimensional network of polysacharide chains, Bull. Acad. Pol. Sci. **16**, 4, 1968.

[19] Schneider, Z., and Friedmann, H.C., Stabilization of enzymes in Sephadex – a long-term experiment, J. of App. Biochem. **3**, 135-145, 1982.

5.3.5 Spectroscopic assay

Relatively pure and concentrated samples of cobalamins are convenient to assay spectroscopically against a standard sample. Instead of using standards, molar extinction coefficients of particular absorption bands indicated in Table 3-1 may be used for calculation of vitamin B_{12} concentration. To make sure that there are no interfering colored impurities a spectrum of the sample may be run and compared with the standard one (see Figs. 3-8 to 3-10). It should be stressed, however, that many of the vitamin B_{12} analogues and hydrolysis- or oxydation products biologically inactive are nearly indistinguishable spectroscopically from vitamin B_{12} itself. Thus it is advisable to use this technique for total corrinoid estimation only, unless the purity and identity of the sample is proved. For total corrinoid estimation it is very useful to convert the corrinoids into the dicyano form in presence of 0.1 M KCN and light. The molar extinction coefficients in the range of 365-368 nm of most corrinoids in their dicyano form approach the value of 30400. The concentration (C) of the corrinoid may be calculated from the measured absorbancy (A) in a 10 mm cuvette and the molar extinction *(ε)* by the following equation:

$$C = \frac{A}{\varepsilon}\ [\text{mmol/ml}] \text{ or } \frac{A \times 1000}{\varepsilon}\ [\mu\text{mol/ml}]$$

A spectroscopic method for estimation of AdoCbl in presence of CN-Cbl and OH-Cbl based on its conversion rate into the dicyano form was elaborated as well [1].

References

[1] Cavallini, D., and Scandurra, R., Spectroscopic determination of CoB_{12} in the presence of B_{12} compounds, Methods in Enzymology **18**, 3-5, 1971.

5.3.6 Neutron activation assay

The technique is based on nuclear conversion of isotope ^{59}Co naturally occurring in vitamin B_{12} into radioactive ^{60}Co by neutron irradiation. The radioactivity of ^{60}Co is measured and the amount of vitamin B_{12} is calculated from a standard sample irradiated simultaneously (see Table 5-10). This very sensitive method has been used to confirm other methods applied to estimate the vitamin B_{12} complexed with enzyme protein [1]. It has limited application because of high costs. Contamination of the sample with Co ions interferes with the assay.

Table 5-10 Vitamin B_{12} assay in the glycerol dehydratase-B_{12} complex by means of neutron activation and by the enzymatic method (data from Schneider et al. [1]).

Sample	Neutron activation assay nanomoles Co per sample	Enzymatic assay nanomoles B_{12} per sample
A	2.28 ± 0.08	1.85 ± 0.15
B	2.32 ± 0.07	1.95 ± 0.12
C	1.14 ± 0.05	0.96 ± 0.05
D	1.12 ± 0.05	0.91 ± 0.07
E (albumin)	$\leqslant 0.03$	none

The samples in glass vials and a cobalt standard were placed in the rotary specimen rack of the TRIGA Mark I nuclear reactor (acitvation analysis service by Gulf General Atomic Inc., San Diego, USA). They sustained a thermal-neutron dose of about 7×10^{16} thermal neutrons/cm^2. The gamma-ray activity at 1.17 MeV arising from the decay of ^{60}Co was used as the analytical indicator for the element. The enzymatic assay of vitamin B_{12} was done as described in section 5.3.3 after acidification of the sample with HCl to pH about 3 followed by immediate neutralization with KOH and dilution with water to a concentration appropriate to the sensitivity of the method.

References

[1] Schneider, Z., Larsen, E.C., Jacobson, G.J., Johnson, B.C., Purification and properties of glycerol dehydrase, J. Biol. Chem. **245**, 3388-3396, 1970.

[2] Stock, G.J.Jr. and Schweikert, E.A., Microanalysis of biometallic compounds using nuclear and atomic activation, J. Radioanal. Chem. **52**, 117-126, 1979.

5.3.7 Chemical quantitation of corrinoids

The chemical methods are based mainly on determination of fragments of vitamin B_{12} with the use of modern analytical techniques.

Mass fragmentography is used for the determination of 5′-adenosyl corrinoids. The assay is based on anaerobic photocyclization of the 5′-deoxyadenosyl part of the corrinoid and quantitative estimation of the product in a coupled gas chromatography-mass spectrometry system (see also photodecomposition of corrinoids in section 3.1.4.1).

Methylcobalamin (MeCbl) and other methyl-corrinoids are assayed quantitatively by an indirect method based on the ability of methylcorrinoids to methylate $HgCl_2$ stoichiometrically. The methylmercury chloride produced in this reaction can be determined at picogram level using gas chromatography. For methylation of heavy metals by MeCbl see section 3.1.4.5.

The cobalt-containing corrinoids may be assayed at µg level by determination of the inorganic cobalt released from the corrinoids by means of chemical decomposition [4].

References

[1] Zagalak, B., Chemical quantitation of corrinoids, in Vitamin B_{12}, Eds. B. Zagalak and W. Friedrich, Walter de Gruyter, Berlin 1979, pp. 671-672.

[2] Zagalak, B., Redweik, U., Kempken, B., Curtius, H. Ch., in Quantitative mass spectrometry in life sciences, Elsevier, Amsterdam 1977, p. 157.

[3] Zagalak, B., Curtius, H. Ch., FEBS Lett. **72**, 29, 1976.

[4] Bruno, P., A new spectrophotometric method for determination of vitamin B_{12} as cobalt, Anal. Lett. **14**, 1493-1500, 1981.

5.4 Procedures for microbial production and purification of vitamin B_{12} and other cobamides and cobinamides

5.4.1 Isolation of cobamides from Propionibacterium streptomyces and Pseudomonas

(according to D. Perlman, Methods in Enzymology **18 C**, 75, 1971)

The nutritional value and elucidation of the structure of cobamides led to interest in the mechanisms of biosynthesis of these important vitamins. Practically all the cobamides found in nature can be traced to microbial biosynthesis, and the commercial production is based on processes using species of *Propionibacterium, Streptomyces, or Pseudomonas* [1-4]. The purpose of this section is to provide information on the laboratory-scale preparation of cobamides so that investigations of the biosynthetic pathways and use of these pathways to prepare "unnatural" cobamides will be expedited (see Chapter 4).

Three types of processes have been described in the literature: In the first, cobamide-producing cultures are grown in cobalt salt-containing media, the cells are harvested, and the cobamide is recovered by extraction from the cells. In the second, the cobamide-producing cultures are grown in precursor-containing media, the cells are harvested, and the desired cobamide is recovered by extraction and eventual fractionation of the cellular material. In the third, cells of various microorganisms are used to join together portions of the cobamide structure to complete the biosynthesis of the cobamide.

In the processes using the growing cultures, the organisms are frequently grown in very complex media containing such heterogeneous substances as cornsteep liquor, soybean meal, and/or hydrolyzed casein. Under these circumstances, addition of amino acid precursors or similar materials often makes little difference in the biosynthesis of the cobamides, and mechanistically the "dilution factor" is so high that it is difficult to obtain useful information on the biosynthetic pathway. However, these cells do have all the enzymes involved in cobamide biosynthesis and can be used as sources of these systems. Many of the cobamide-producing cultures will grow well in chemically defined media, and although the cobamide production is significantly lower (often a factor of 5-fold), the data obtained can be useful, too. The cobamide yields in fermentations using complex media range from 3 to 30 mg/liter, as determined by bioassay [5, 6].

In the second type of process certain propionibacteria strains or other organisms are grown in media containing compounds thought to be (or desired to be) incorporated into the cobamide (nucleotide portion). The cells are harvested when ionophoretic or paper chromatographic analyses show the biosynthesis of the "new" substance to be maximal, and the cobamide is recovered from the extract [2]. Usually the culture produces a number of related cobamides, and these must be separated by column chromatography or continuous ionophoresis in order to obtain the desired compound. Not all cultures will "accept" the added precursors [2, 7] and some judgment is necessary when embarking on such a program.

In the third type of process, certain strains of *Escherichia coli* (or other organisms) are grown in selected media, and once maximal cell population is reached, precursor compounds are added to the cell suspensions. The enzymes in the cells join these together to form the cobamides.

Analytical methods

The samples from the fermented media are prepared as follows before examination by filter paper chromatography and paper ionophoresis (followed by bioautography): Approximately 25 ml of the sample (pH about 6.5) is heated for 15 minutes in a boiling hot water bath to release the cobamides from the microbial cells. The suspension is then centrifuged, and the collected solids are discarded. Approximately 0.5 ml of an aqueous solution of KCN (5 g/100 ml) is mixed with a 10-ml aliquot of the clear supernatant (this converts all the cobamides present to the cyano form). An aliquot of this solution is assayed for cobamide content by the bioassay method described in the *U.S. Pharmacopeia* (15th edition), which is based on the growth response of *Lactobacillus leichmannii* (ATCC 7830) with pure 5,6-dimethylbenzimidazolylcobamide cyanide as standard (see section 5.3.1). A second aliquot is analyzed by a modification of the bioassay method described by Ford and Porter [8] which uses the growth response of *Ochromonas malhamensis* and the same reference standard. (Prepared bioassay media distributed by Difco Laboratories, Detroit, Michigan 48201, may be used for these bioassay tests; the assay organisms can be obtained from the American Type Culture Collection, Rockville, Maryland 20852). A third aliquot (approximately 10 ml in volume) is thoroughly mixed with 10 ml of a phenol-benzene solution (30 parts 88% liquefied phenol, 70 parts benzene). The emulsion is separated by centrifugation, and the nonaqueous layer (containing the cobamides) is transferred to a second tube; 10 ml of *n*-butanol and 5 ml of distilled water are added, and the two phases are well mixed by shaking. The emulsion is broken by centrifugation, and the bottom aqueous layer containing the cobalamins present in the fermentation sample is removed for analysis (extraction of cobamides see section 5.1).

Aliquots of the aqueous concentrate are analyzed by a paper ionophoretic method described by Holdsworth [9] using 0.5 N acetic acid solution containing 0.02% KCN (w/v) as electrolyte and a voltage gradient of 8 V/cm for 18 hours across the strip of Whatman 3 MM paper. Aliquots of solutions of 5,6-dimethylbenzimidazolylcobamide cyanide, 2-methyladenylcobamide cyanide, cobinamide cyanide, and adeninylcobamide cyanide are spotted on the paper strip (approximately 0.1 μg of cobamide per spot) parallel to the areas where the fermentation extracts are applied [see section 5.3.1). The paper strips are dried after the 18-hour electrophoresis period and then placed for 15 minutes on the surface of large agar plates which have been seeded with a cobamide-requiring culture of *Escherichia coli* (ATCC 11105). [The agar medium contains 20 g glucose, 1.2 g citric acid, 0.4 g $(NH_4)_2HPO_4$, 0.08 g KCl, 0.42 g $MgCl_2 \cdot 4H_2O$, 0.04 g $MnCl_2 \cdot 4H_2O$, 0.23 g $FeCl_3 \cdot 6H_2O$, 0.02 g $ZnCl_2$, 0.04 g $CoCl_2 \cdot 6H_2O$, 15 g agar, 0.1 g triphenyltetrazolium chloride, and sufficient distilled water to dilute the mixture to 1 liter. The triphenyltetrazolium chloride as well as the glucose are sterilized separately from the other ingredients of the media and added to the agar just prior to pouring the agar solution into a 3-quart Pyrex baking dish.] After the 15-minute contact period the paper strips are removed and the agar plates, covered with aluminium foil, are incubated at 37 °C for 18 hours. Zones of growth of the *E. coli* culture (visible as red zones due to the reduction of the tetrazolium dye to the colored formazan) are found on the agar plate in the areas corresponding to the areas of the paper strip containing the cobamides. In this ionophoretic system, 5,6-dimethylbenzimidazolylcobamide cyanide is essentially neutral, while adeninylcobamide cyanide, 2-methyladeninylcobamide cyanide, and cobinamide cyanide move approximately 4, 8, and 10 cm from the point of application to the paper. This paper ionophoretic method may be used preparatively to recover and purify the cobamides present in the fermentation extracts. If this is the objective, larger fermentation samples, e.g., 100 ml, are extracted, and the final aqueous concentrate is evaporated to a small volume before it is applied on the paper strip. The cobinamide cyanide is yellow orange, while the other "complete" cobamides have a characteristic red color.

Table 5-11 Chromatographic mobilities of cobamides.

Cobamide	R_s (relative)	
	sec-Butanol-acetic acid-water-KCN	*sec*-Butanol-water-KCN-$KClO_4$
5,6-Dimethylbenzimidazole	1.0	1.0
Benzimidazole	0.95	0.90
2,6-Diaminopurine	0.3	1.05
Adenine	0.3	0.25
2-Methyladenine	0.4	0.45

Paper chromatography using *sec*-butanol-acetic acid-water-5% KCN (100:1:50:0.25) or *sec*-butanol-water-5% KCN (77:23:0.25) saturated with $KClO_4$ as developing solvent can be used as an alternative to the paper ionophoresis. The mobilities of the common cobamides in these systems are as given in the Table 5-11.

Preparation of cobamides by *Propionibacterium* fermentations

Stock cultures of propionibacteria are maintained in stabs of a tomato juice-peptone-yeast extract agar and stored at 5 °C until needed. Inoculum is prepared by transferring viable cells to flasks of a medium containing yeast extract (10 g/liter), glucose (5 g/liter), and calcium carbonate (5 g/liter), followed by incubation at 30 °C for 3-5 days. After two transfers in this medium, a few milliliters of the cell suspension are used as inoculum for the cobamide fermentations. The fermentation medium (patterned after that of Speedie and Hull [10]) contains 80 ml of cornsteep liquor (from A. E. Staley Co. or Penick and Ford), (neutralized to pH 7.0 with NaOH), 100 g of glucose (sterilized separately as a 50% solution), and 0.01 g of $CoCl_2 \cdot 6H_2O$, with water to 1 liter. This medium is sterilized by autoclaving at 121 °C for 20 minutes. During the fermentation period, the pH of the growing culture is maintained between pH 6 and 7 by addition of 7 N NH_4OH (prepared by diluting concentrated NH_4OH with sterile water in a sterile flask). Sometimes addition of $CaCO_3$ (sterilized as a dry powder) at 20 g/liter is helpful in pH control.

For the preparation of 5,6-dimethylbenzimidazolylcobamide, the following cultures may be used: *P. freudenreichii* (ATCC 6207), *P. shermanii* (ATCC 8262, ATCC 9614, ATCC 9615, ATCC 9616, ATCC 9617), or *P. technicum* (ATCC 14073, ATCC 14074) (The various forms of corrinoids produced by propionic bacteria are given in Table 6-1). In these fermentations, the laboratory stages may include an initial anaerobic fermentation, at 30 °C for 3 days, of 250 ml of medium/500 ml Erlenmeyer flask, or 3 liters of medium/4 liter Erlenmeyer flask followed by an aerobic fermentation of 100 ml of medium/250 ml Erlenmeyer flask on a rotary shaker operating at 280 rpm (1 inch throw) for 2-3 days. The pH should be adjusted at least every 12 hours (using the sterile NH_4OH) to pH 6-7. Contamination by streptococci, micrococci, and other Gram-positive bacteria is a common problem and can be detected by making smears of the growing culture and staining with Gram stain. The propionibacteria grow as rods in the anaerobic phase and as very short rods (practically cocci) during the aerobic period.

Analyses of samples taken from the anaerobic phase of the process show the presence of only cobinamide in high-producing fermentations, while 5,6-dimethylbenzimidazolylcobamide predominates in the aerated cultures (see Chapter 4). All the cobamides are found associated with the cellular material, and centrifugation of the cells yields a sludge which when dispersed in a minimum of water and heated for 15 minutes in a boiling water bath releases the cobamides into the solution. Maximal yields of 5,6-dimethylbenzimidazolylcobamide by this process range from 10 to 25 mg/liter, and depend in part on the effectiveness of the anaerobiasis during "phase one", the careful adjustment of the pH (growth stops when the pH drops below pH 5.8), and addition of glucose to maintain at least a level of 15 g/liter during the aerobic phase. At pH of the medium below 5 a partial release of cobamides into the medium occurs.

When *P. arabinosum* (ATCC 4965), *P. pentosaceum* (ATCC 4785), *P. petersonii* (ATCC 4870), or *P. thoenii* (ATCC 4874) are substituted for the *P. freudenreichii, P. shermanii,* or *P. technicum,* and the fermentations are grown anaerobically, the cultures produce mixtures of adeninylcobamide, 2-methyladeninylcobamide, cobinamide, and traces of other factors related to cobamides. No 5,6-dimethylbenzimidazolylcobamide is found. Addition of various purines of benzimidazoles (or other imidazoles, see references 2 and 11) at levels of 0.5–2 g/liter (usually added in increments during the 5-day incubation period) results in biosynthesis of cobamides containing these bases (or bases derived from the metabolism of these supplements) in the nucleotide portion of the molecule (see Table 6–1). The maximal yields of these cobamides may range from 0.5 to 2 mg per liter of fermented medium after a 5-day incubation period. (This may be difficult to determine with accuracy since many of these cobamides (see reference 2) are less effective promoters of the growth of *L. leichmanii* or *O. malhamensis* than the 5,6-dimethylbenzimidazolylcobamide cyanide used as reference standard in bioassays). Aeration of the cultures results in somewhat lower cobamide yields, as, in general, the cobamide synthesis is a direct function of the amount of cellular material produced.

An alternative to the process for producing "new" cobamides by adding purines, etc. to the growing *P. arabinosum* cultures or to *P. freudenreichii* (both growing anaerobically) is to grow the cells anaerobically, collect them by centrifugation, and resuspend them in a buffer containing the purine (or other base). These cells will contain cobinamide and will insert the base into the nucleotide portion of the molecule. This method is useful in producing cobamides containing radioactive bases in the nucleotide portion of the molecule, and also when the amount of base available is quite limited [12] (see Chapter 4).

Preparation of cobamides by streptomycete fermentations. Stock cultures of streptomycetes are maintained on slants of yeast extract, meat extract, peptone, glucose agar or Berger's agar (tomato paste and oatmeal agar) and stored at 5 °C until needed. Inoculum is prepared by transferring viable cells and spores to 250-ml flasks containing soybean meal (3 g/100 ml), glucose (3 g/100 ml), and $CaCO_3$ (1 g/100 ml) with water to 100 ml. The inoculated flasks are placed on a rotary shaker (280 rpm, 1 inch throw) located in a 28 °C incubator. After 3 days of incubation, when considerable, cellular growth has occurred, 5-ml aliquots are aseptically transferred to a second series of flasks of this medium containing in addition to the soybean meal (Archer Daniel Midland Co. or A.E. Staley Co.), glucose, $CaCO_3$, and 0.025 g/liter $CoCl_2 \cdot 6H_2O$. The second series of flasks are incubated for 5–7 days on the shaker, and the cells are harvested by filtration or centrifugation. All the cobamides produced are found associated with the cellular material. They may be "released" from the cells by adjustment of the pH to 2, by heating to 80–90 °C, by sonication, or by adding NaCl to give a concentration of 20 g/liter [13].

All cultures of streptomycetes tested will form 5,6-dimethylbenzimidazolylcobamide when grown in the above medium [14]. The highest yields noted, ca. 5 mg/liter, were obtained with the streptomycete ATCC 11072; yields with other species ranged from 1 to 3 mg/liter.

A chemically defined medium containing, per liter, 20 g glucose, 2.6 g glycine, 2.2 g monosodium glutamate, 0.5 g $K_2HPO_4 \cdot 3H_2O$, 0.5 g $MgSO_4 \cdot 7H_2O$, 0.015 g $CuSO_4 \cdot 5H_2O$, 0.016 g $MnSO_4 \cdot 4H_2O$, 0.03 g $ZnSO_4 \cdot 7H_2O$ 0.025 g $FeSO_4 \cdot 7H_2O$, 0.05 g $CaCl_2.2H_2O$, and 0.025 g $CoCl_2.6H_2O$ is useful in studies of the biosynthetic mechanisms [14] operative in cobamide synthesis by streptomycetes, even though the cobamide yields rarely exceed 1.5 mg per liter of fermented medium.

Preparation of cobamides by Pseudomonad fermentations. High yields of cobamides are obtained with selected pseudomonads when grown in chemically defined media [15, 16]. Using *P. denitrificans* growing in a medium containing, per liter, 5 g glutamic acid, 30 g sucrose, 2 g $(NH_4)_2HPO_4$, 0.8 g KCl, 0.5 g $MgSO_4.7H_2O$, 0.2 g $MnSO_4.4H_2O$, 0.02 g $FeSO_4.7H_2O$, 0.02 g $ZnSO_4.7H_2O$, 0.04 g $(Co(NO_3)_2.6H_2O$, 0.025 g 5,6-dimethylbenzimidazole, and 5 g betaine at 30 °C under aeration (40 ml of medium per 250-ml Erlenmeyer flask on a rotary shaker), yields of the order of 15 mg/liter are obtained [15]. Substitution of 5-hydroxybenzimidazole for the 5,6-dimethylbenzimidazole resulted in biosynthesis of 5-hydroxybenzimidazolylcobamide [16], and presumably this organism will accept other purines and benzimidazoles.

Preparation of cobamides by Escherichia coli fermentation. Washed cells and growing cultures of *E. coli* will produce cobamides when supplemented with cobinamide and 5,6-dimethylbenzimidazole [17, 18]. The efficiency of the process may be quite high when the amount of cobinamide limits the amount of cobamide formed. Wild-type [17] or *E. coli* (Davis 113-3) [18] can be used in this process.

References

[1] H. M. Wuest and D. Perlman, in The Vitamins, eds. W. H. Sebrell, Jr. and R. S. Harris, Vol. 2, Academic Press, New York, 1968, p. 139.
[2] L. Mervyn and E. L. Smith, Progr. Ind. Microbiol. **5,** 161, 1964.
[3] P. B. Riley, P. W. Jackson, D. Ross, and P. A. Savage, Soc. Chem. Ind. (London) Monogr. **12,** 127, 1961.
[4] D. Perlman, Chem. Weck, Dec. 16, 1967.
[5] H. L. Rosenthal, in The Vitamins, eds. W. H. Sebrell, Jr. and R. S. Harris, Vol. 2, Academic Press, New York, 1968, p. 145.
[6] H. R. Skeggs, in Analytical Microbiology, ed. F. Kavanagh, Academic Press, New York, 1963, p. 551.
[7] D. Perlman, J. M. Barrett, and P. W. Jackson, in Vitamin B_{12} und Intrinsic Factor, 2. Europäisches Symposion über Vitamin B_{12} und Intrinsic Factor, ed. H. C. Heinrich, Ferdinand Enke, Stuttgart, 1962, p. 58.
[8] J. E. Ford and J. W. G. Porter, Brit. J. Nutr. 7, 299, 1953.
[9] E. S. Holdsworth, Nature **171,** 148, 1953.
[10] J. D. Speedie and G. W. Hull, U.S. Patent, 2,951,017, 1960.
[11] D. Perlman and J. M. Barrett, Can. J. Microbiol. 4, 9, 1958.
[12] D. Perlman and J. M. Barrett, J. Bacteriol. **78,** 171, 1959.
[13] R. B. McCormack, A. F. Langlykke, and D. Perlman, U.S. Patent 2,656,300, 1954.
[14] R. Bray and D. Shemin, Biochim. Biophys. Acta **30,** 647, 1958.
[15] A. L. Demain, H. J. Daniels, L. Schnable, and R. F. White, Nature **220,** 1324, 1968.
[16] I. M. Miller, U.S. Patent 2,939,822, 1960.
[17] J. E. Ford, E. Holdsworth, and S. K. Kon, Biochem. J. **59,** 86, 1955.
[18] H. Dellweg, E. Becher, and K. Bernbauer, Biochem. Z. **328,** 88, 1956.

5.4.2 Preparation of cobinamide-guanosine-diphosphate coenzyme (Ado-Cbi-GDP)

W. Walerych, A. Gawrysiak and Z. Schneider

General remarks. Cobinamide-guanosine-diphosphate was already isolated and identified by Di Marco and coworkers [1] in 1957 (for formula of this corrinoid see Table 3-1). This compound - accompanying, in small amounts, crude cobamide preparations obtained from various fermentations - has been independently isolated in the cyanoform (CN-Cbi-GDP) by other investigators as well [2-5]. In our laboratory it has been isolated in the coenzymatic form AdoCbi-GDP from *Propionibacterium shermanii* for the purpose of studying the biosynthesis of corrinoids [6, 7]. Studies in vitro proved that this corrinoid is the precursor of cobalamin (see Fig. 4-2). More recent investigations show that it can be converted enzymatically into AdoCbl-5-phosphate by an enzyme system in which ribosomal proteins play a significant role [7]. Only the coenzymatic forms AdoCbi-GDP or MeCbi-GDP (but not the cyano form) may serve as substrates in this enzyme system. The amount of this precursor corrinoid in bacteria is usually low (see Table 6-1). Its content in *P. shermanii* grown for production of cobalamin amounts to 1-2% of the whole corrinoid pool. The procedure described below allows for the production of AdoCbi-GDP from *P. shermanii* at a yield exceeding 30% of all accompanying corrinoids. This is achieved by controlled addition of cobalt to the growth medium and appropriate growth procedure.

Growth of *P. shermanii*. The bacteria were maintained and grown for AdoCbi-GDP in the same medium. The medium contained: 50 g of caseine hydrolysate obtained by acid hydrolysis, 20 g of casein hydrolysate obtained by enzymatic hydrolysis, 10 g of yeast extract (Difco), 30 g KH_2PO_4, 4 g $MgCl_2$, 4 mg calcium pantothenate and 3 mg biotin in 8 liter of tap water. One ml of the stock culture was transferred to 400 ml of the medium in a 0.5 liter. Erlenmeyer flask and grown for 5 days at 28 °C. The culture was daily supplemented with 5 ml of 50% glucose and the pH was adjusted to 6.8 with 25% ammonia solution. Subsequently the whole culture was transferred to a 1 liter flask containing 400 ml of medium and the culture was kept for another 5 days as described above.

The whole culture was finally transferred to a 10 liter flask containing 8 liter of medium. The bacteria were grown for 9 days as described above except that daily a larger portion of glucose was added (10-20 ml). The amount of glucose was adjusted to the rate of metabolism which was judged by the amount of ammonia used for neutralization of the fermentation medium. The metabolism usually slowed down significantly in the last two days. On the ninth day cobalt (100 mg $CoSO_4 \cdot 7H_2O$) was added and after 18 hours the cells were harvested by centrifugation. The cells were immediately frozen if not used at the same day for extraction of the corrinoid.

Isolation of AdoCbi-GDP. At this stage the procedure was carried out in red light due to light sensitivity of the corrinoids. The bacteria, about 50-100 g of the thick paste, were suspended in 800 ml of distilled water and adjusted to pH 4 with HCl. The suspension was heated at 70 °C for 20 minutes, cooled and centrifuged. Under these conditions the corrinoids are released from the bacteria into the medium; thus, they are recovered from the supernatant solution. The extraction was repeated twice and the combined supernatant solutions were used for phenol extraction. AdoCbi-GDP and the other accompanying corrinoids were extracted with a small amount of phenol-chloroform mixture 1:1 (see also the procedures in sections 5.1 and 5.4.1). The heavier phenol-chloroform layer containing the corrinoids was separated from the aqueous layer by centrifugation. The extraction with phenol was repeated 4–5 times. The phenolic solutions were combined and the corrinoids were then reextracted with water by addition of 4 parts of a chloroform-butanol mixture 1:2 and a small amount of water. After vigorous shaking the corrinoids were driven to the aqueous layer at the top. The aqueous layer was transferred to another tube; a fresh portion of water was added and the extraction was repeated. The aqueous solutions of corrinoids were combined and washed several times with an equal volume of diethylether in order to remove phenol. The solution was then concentrated by evaporation to a volume appropriate to applying on paper for ionophoresis. The corrinoid extract was subjected to paper ionophoresis (3 MM Whatman) in 1.0 M acetic acid using a voltage gradient of 8 V/cm. The orange-yellow AdoCbi-GDP spot migrated rather slowly towards the cathode and after 16 hours reached a distance of 3.5 cm while the accompanying yellow AdoCbl spot was found 14.5 cm from the start.

Other corrinoids which accompany the crude extract in minor percentages, such as AdoCbi and AdoCbi-P, show higher mobility than AdoCbi-GDP as well. CN-Cbl added to the extract as a marker and assumed to be electrically neutral moved under these conditions 0.5 cm toward the cathode due to electroendoosmosis. The band containing AdoCbi-GDP was cut off the paper, eluted with water, concentrated and subjected to descending paper chromatography (3 MM Whatman) using n-butanol, isopropanol, water and acetic acid (100:65:100:1). The main spot ($R_f = 0.18$) was cut off and the corrinoid was eluted with water. The concentration of the corrinoid was determined spectrophotometrically. Known portions of the corrinoid solution were transferred to vials and dried under vacuum. The vials protected from light and moisture were stored at +5 °C for several years with no noticable change. Similarly radioactive AdoCbi-GDP has been obtained by addition of radioactive cobalt salts to the growth medium.

Properties of AdoCbi-GDP. The corrinoid contains GDP attached to the hydroxyl of the isopropanol-amine residue of cobinamide instead of the dimethylbenzimidazole nucleotide present in cobalamin (see Fig. 4-2). The GMP moiety may be split off by catalytic controlled hydrolytic procedure. The guanosine moiety can also be removed by enzymatic means using snake venom or yeast pyrophosphatases. Unlike cobalamin coenzyme, which is yellow only at acidic pH, the aqueous solutions of cobinamide-GDP coenzyme remain yellow-orange at acidic – as well as at alcaline pH. The electronic visible spectrum of this corrinoid remains almost unchanged at both acidic-(1.5) and alcaline pH (12.5). It is similar to other spectra of coenzymatic forms at acidic pH. The visible spectrum at pH 4 is characterized by a distinguished maximum at 462 nm, a minimum at 420 nm and two less distinguished peaks at 365 nm and 307 nm. The latter two peaks may appear as shoulders at extreme pH's. The millimolar extinction coefficient for this corrinoid at 462 nm is 9.8. In the UV range only one wide peak appears in the range of 260–270 nm. The spectra of the two forms: AdoCbi-GDP and CN-Cbi-GDP differ significantly from each other. The cyanoform is distinguished by several peaks at: 272, 321, 355, 405, 498, millimolar extinction coefficient 8.88 at 530 nm. The spectrum cannot be applied for identification of this corrinoid since other corrinoids in the coenzymatic forms produce similar spectra. Particularly indistinguishable spectrum, in the visible range, is produced by AdoCbi-P, the product obtained by hydrolytic cleavage of GMP from AdoCbi-GDP (for formula of AdoCbi-P see Table 3-1). The two corrinoids may be distinguished quickly by their different ionophoretic mobility (see above). The concentration of the corrinoid can be determined spectrophotometrically by direct measurement of optical density at 462 nm applying the above mentioned millimolar extinction coefficient 9.8 or after its conversion into the dicyano-form (for details see section 5.3.5). The biological activity of cobinamide-GDP depends on the specificity of the microbial species used for testing. The less specific *E. coli* shows full biological activity, while *Euglena gracilis*, a more specific organism for cobalamin, responds with 10% growth only relative to vitamin B_{12} standard. The corrinoid is not absorbed by the animal digestion system, nor does it cure pernicious anaemia. It is not active as a coenzyme in the known enzyme systems. In contrast to the cobamides it does not form complexes with the apoenzyme of glycerol dehydratase. This property has been applied for monitoring enzymatic conversion of the corrinoid into cobamides (see section 5.3.4).

References

[1] DiMarco, A., Boretti, G., Ghione, M., Migliacci, A., Sanfilippo, A., Ital. J. Biochem. **6,** 259-269, 1957.

[2] Pawełkiewicz, J., Walerych, W., Bartosinski, B., Badania nad biosynteza związków korfirynowych. 3. Identyfikacja kobalaminy Y_1 jako guanozynodwu fosforan u Witamin B_{12} (czynnika B), Acta Biochim. Pol. **6,** 431-440, 1959.

[3] Zodrow, K., Stefaniak, O., The Biosynthesis of factors Y_1 and Y_2 by *Propionibacterium shermanii* and *P. freudenreichii*. Acta Microbiol. Pol. **12,** 267-270, 1963.

[4] Irion, E., Ljungdahl, L., Isolation of Co-5-deoxyadenosylcobinamide guanosine diphosphate, α-(2-methyladenyl)-Co-5′-deoxyadenosylcobamide and Co-methylcorrinoids from *Clostridium thermoaceticum*. Biochemistry **7,** 2350-2355, 1968.

[5] Pawełkiewicz, J., and Bartosiński, B., The enzymatic synthesis of vitamin B_{12}. Bull. Acad. Polon. Sci. **8,** 5-7, 1960.

[6] Walerych, W., Kato, T., and Pawełkiewicz, J., Enzymic synthesis of vitamin B_{12}. Biochem. Biophys. Res. Commun. **31,** 328-332, 1968.

[7] Pezacka, E. and Walerych, W., Biosynthesis of vitamin B_{12} Part I. Role of the ribosomal proteins in vitamin B_{12} biosynthesis, Biochim. Biophys. Acta **678,** 300-310, 1981.

5.4.3 Preparation of the vitamin B_{12} coenzyme analogues containing a cobalt-carbon bond

(according to D. Dolphin [1])

General considerations. Reduction of the cobalamins, which contain trivalent cobalt, proceeds in two distinct steps. The first one-electron reduction product is cob(II)alamin (B_{12r}) which undergoes a further one-electron reduction to cob(I)alamin (B_{12s}) which may be considered as either containing monovalent cobalt or being a trivalent cobalt hydride:

$$H^+ \rightleftharpoons H^-$$
$$Co(I) \rightleftharpoons Co(III)$$

B_{12s} is the most nucleophilic species known to exist in aqueous solution [2], and this high nucleophilicity results in the ease of preparation of the B_{12} coenzyme and its analogs, since B_{12s} undergoes both rapid substitution and addition reactions with a variety of electrophiles (see section 3.1.4.2.) [3].

Preparation of B_{12s}

The most convenient reduction is that brought about by sodium borohydride. Although such reductions are usually complete within a few minutes, they occasionally stop at the brown B_{12r} stage, and the further reduction to B_{12s} then requires both a large excess of sodium borohydride and an extended reaction time which results in some irreversible reduction of the corrin chromophore. It has been reported that addition of copper catalyzes the reduction, and we have found this to be so, but this reducing system also reduces the alkylcobalamins which are normally stable toward sodium borohydride by itself. Thus, the copper-catalyzed reduction is unsuitable for the preparation of B_{12s} as a precursor of the alkylcobalamins. If, however, cobaltous ions are added to a cobalamin borohydride solution, reduction is catalyzed without further reduction of alkylcobalamins. Thus, this reduction is suitable for the syntheses of both the vitamin B_{12} coenzyme and its analogues. B_{12s} is a highly unstable compound even if in oxygen free solution. Aqueous solutions decompose spontaneously to B_{12r} and hydrogen. The decomposition rate may be slowed down significantly at pH 9 (see Table 5-12).

Table 5-12 Half-life of B_{12s} as a function of pH

pH	Half-life (min)
6.98	22
8.01	67
9.01	140
9.93	>240

Procedure. A solution of cyano- or hydroxocobalamin (50 mg) and cobalt nitrate (1 mg) in water (10 ml) is deoxygenated with a stream of argon. After 10 min, a deoxygenated solution of sodium borohydride (15 mg) in water (1 ml) is added to the red cobalamin solution, which immediately turns brown and then blue-green, signifying that the reduction is complete. A wide variety of reducing agents are available for the reduction of vitamin B_{12} to B_{12s}, and further examples are given in the following section on the preparation of the alkylcobalamins.

Preparation of alkyl- and acyl analogues of the vitamin B_{12} coenzyme

The preparation of such analogs is based upon the reaction of B_{12s} with suitable electrophiles, and the principal reactions are summarized in section 3.1.4.2. The limiting factors in the synthesis of B_{12} coenzyme analogues by nucleophilic displacement reactions are primarily steric, and they may either limit the rate of the reaction or effect the stability of the coenzyme analog once it has been formed. Thus, no reaction is observed between B_{12s} and neopentyl chloride, while the reaction between secondary halides such as α-chloropropionic acid and 2-bromobutane is fast, but the resulting coenzyme analogues are not sufficiently stable to survive the purification procedures.

Although steric considerations are of primary importance in determining stability, certain coenzyme analogues that are sterically feasible require considerable care during their isolation. Thus, benzylcobalamin is rapidly decomposed by oxygen, due to the ready homolytic cleavage of the cobalt carbon bond, and can be isolated only when oxygen is vigorously excluded during all stages of its preparation. Similarly, the analog derived from methyl acrylate is base labile, due to the acidity of the β-hydrogen atoms and care must be taken to maintain neutrality during its preparation.

The addition of B_{12s} to alkenes and alkynes also results in the formation of coenzyme analogues, and as well as steric considerations concerning the stability of the products, the alkenes which are less susceptible to nucleophilic attack than the corresponding alkynes must be activated by being in conjugation with an electron withdrawing group. Thus, whereas acetylene reacts with B_{12s} to give vinylcobalamin, no reaction is observed with ethylene, although acrylonitrile and methyl acrylate react rapidly to give the corresponding alkylcobalamin.

The low solubility of the cobalamins in most solvents, coupled with the need for solvents that are stable toward reducing agents, restricts the choice for the preparation of coenzyme analogues to water, methanol, ethanol, and aqueous acetic acid.

Both the vitamin B_{12} coenzyme and its analogues are light sensitive, especially in solution, and the manipulations given below should be carried out using the minimum intensity of light that one can conveniently work in as well as shielding reaction flasks, chromatographic columns, etc., with aluminium foil.

Purification by extraction through phenol (see section 5.1.). The procedure of extraction through phenol removes water-soluble salts from cobalamins. A number of modifications of the original procedure [4] have been made, but in our hands many such procedures often result in emulsions. We have found, however, that the following technique is both reproducible and free from the problem of emulsions.

The aqueous cobalamin solution is extracted with one-fifth of its volume of a solution of phenol in methylene chloride. A stock solution may be prepared by dissolving phenol (100 g) in methylene chloride (100 ml). The organic layer is separated, and the aqueous layer is reextracted with successive aliquots of phenol-methylene chloride until no further color is extracted. The combined organic extracts are washed with distilled water 2 × ⅕ the volume of the organic layer, and the organic layer is diluted with methylene chloride to 10 times its original volume. The cobalamin is reextracted from the organic layer with aliquots of distilled water 1/20 the volume of the organic layer until no color remains in the organic layer. The combined aqueous extracts are then washed with methylene chloride (3 times the volume of the combined aqueous layer) to remove traces of phenol. The methylene chloride remaining in the aqueous phase is removed when the solution is reduced in volume, or may be removed by passing a stream of nitrogen through the solution.

Chromatography. Thin-layer chromatography on cellulose [5] or silica gel [5] is useful for the rapid analysis of cobalamins, but the close similarity in R_f values of many cobalamins makes paper chromatography more reliable, and it is advisable to use three solvent systems when determining the composition or purity of cobalamins (see also section 5.2).

Ascending chromatography was carried out on Whatman No. 1 paper using the following three solvent systems (ratios given by volume, the top layer being used for systems I and II):

Solvent I: Butan-2-ol-water-25% ammonium hydroxide (50:36:14)
Solvent II: *n*-Butanol-ethanol-water (50:15:35)
Solvent III: *n*-Butanol-propan-2-ol-water (37:26:37)

Mobilities are quoted relative to cyanocobalamin (R_{CN}).

Procedures. The following specific procedures are presented as representing typical syntheses of coenzyme analogs, and although a number of different reducing agents are employed to show the scope of the reactions, we recommend using the cobalt-catalyzed sodium borohydride reaction see (3.1.3).

Preparation of methylcobalamin [6, 7]. A solution of cyano- or hydroxocobalamin (100 mg) and cobalt nitrate (1 mg) in water (10 ml) is placed in a 25-ml Erlenmeyer flask, stoppered with a serum cap, and deoxygenated with a gentle stream of argon. After 10 minutes, a deoxygenated solution of sodium borohydride (20 mg) in water (0.5 ml) is added (foaming may occur at this stage and the rate of flow of argon may need to be adjusted). The solution immediately turns brown and then blue-green. After a further 5 minutes, methyl iodide (200 mg) is added whereupon the solution turns yellow-orange. The cobalamin in solution is purified by extraction through phenol (see above), and the resulting aqueous solution is reduced to 5 ml on a rotary evaporator, the heating bath being kept below 50 °C. This solution is placed on a column of carboxymethyl cellulose (30 × 2 cm) which has been washed with 0.1 N hydrochloric acid (20 ml) and then water until the washings are neutral, and eluted from the column with water. The eluate is reduced in volume to 1 ml and treated with acetone until the solution shows a faint turbidity. On standing overnight, bright red crystals of methylcobalamin are deposited. These are collected by filtration, washed with acetone, and air dried. Yield: 91 mg.

Light absorption: Fig. 3-10
Principal absorption bands ($\varepsilon \times 10^4$ in parentheses)
- at neutral pH: 340 (1.31), 375 (1.21), 528 (0.30)
- at acidic pH: 264 (2.48), 278 (1.91), 286 (2.04), 304 (2,24), 317 (1.85), 333 (1.27), 378 (0,87), 384 (0.80), 461 (0.89).

R_{CN} values: solvent I, 1.9; solvent II, 2.7; solvent III, 1.3.

Preparation of acetylcobalamin [6, 7]. Cyano- or hydroxocobalamin (50 mg) in 10% acetic acid (10 ml) is magnetically stirred and deoxygenated with a stream of argon. After 10 minutes, zinc dust (1 g) is added. The solution rapidly turns brown and then after a further 10 minutes, blue-green. Acetyl chloride, or acetic anhydride (150 mg), is added and the resulting yellow-brown solution is filtered immediately to remove unreacted zinc. The product in the filtrate is purified by extraction through phenol, and the resulting aqueous layer is treated in the manner described above for methylcobalamin. The product crystallizes from aqueous acetone as large red needles. Yield: 37 mg.

R_{CN} values: solvent I, 0.444 (decomposes to B_{12b}); solvent II, 1.8; solvent III, 1.3.

Preparation of carboxymethylcobalamin [6, 7]. A magnetically stirred solution of cyano or hydroxocobalamin (100 mg) in 15% ammonium chloride (15 ml) is deoxygenated with a stream of argon: after 10 minutes, zinc dust (1 g) is added. The solution turns brown and then blue green. Monochloroacetic acid 200 mg is added to the stirred solution which immediately turns orange-brown. This solution is filtered, and the filtrate is purified by extraction through phenol. The resulting aqueous solution is treated in the manner described for methylcobalamin, and the product is crystallized from aqueous acetone as red needles. Yield: 71 mg.

R_{CN} values: solvent I, 0.81; solvent II, 1.6; solvent III, 1.0.

Preparation of vinylcobalamin [7]. To a solution of cyano- or hydroxocobalamin (50 mg) in methanol (20 ml) is added cobalt nitrate (1 mg) in water (0.2 ml). The resulting solution is deoxygenated with a stream of argon, and after 10 minutes a deoxygenated solution of sodium borohydride (10 mg) in water (0.5 ml) is added. The solution rapidly turns blue-green after passing through a brown stage. Acetylene is passed through this solution for 2 minutes: by this time the solution has turned orange red. The flow of acetylene is then stopped, but the argon is continued for 10 minutes longer. The resulting solution is taken to dryness, and the residue is dissolved in water (5 ml). This product is purified from aqueous solution by extraction with phenol, and treated in the manner described for methylcobalamin. The product crystallizes from aqueous acetone as very small red needles. Yield: 41 mg.

Principal absorption bands ($\varepsilon \times 10^4$ in parentheses)
- at neutral pH: 340 (1.31), 375 (1.21), 528 (0.30)
- at acidic pH: 260 (2.18), 278 (2,10), 286 (2.08), 325 (1.45), 350 (1.07), 460 (0.79)

R_{CN} values: solvent I, 2.0; solvent II, 2.4; solvent III, 1.5.

Preparation of ethynylcobalamin [7]. A solution of cyano- or hydroxocobalamin (50 mg) in 0.2 M EDTA (10 ml) is deoxygenated with argon for 10 minutes. Chromous acetate [8] (20 mg) is added, and when the solution had turned purple green, bromoacetylene (100 mg) is swept into the reaction mixture with a stream of argon. This is continued for a further 10 minutes after the bromoacetylene has been passed into the reaction vessel. The red solution is purified by extraction through phenol, and the resulting aqueous solution is treated in the manner described for methylcobalamin. The product crystallizes from aqueous acetone as deep red needles. Yield: 38 mg.

Principal absorption bands ($\varepsilon \times 10^4$ in parentheses)
- at neutral pH: 340 (1.01), 369 (1.52), 538 (0.68), 552 (0.73),
- at acidic pH: 266 (2.02), 277 (1.78), 287 (1.73), 320 (1.40), 350 (1.00), 457 (0.75)

R_{CN} values: solvent I, 1.45; solvent II, 2.2; solvent III, 1.4.

References

[1] Dolphin D., Methods in Enzymology **18**, 34-65, 1971.
[2] Schrauzer, G. N., Deutsch, E., and Windgassen, R. J., J. Am. Chem. Soc. **90**, 2441, 1968.
[3] Johnson, A.W., Mervyn, L., Shaw, N., and Smith, E. L., J. Chem. Soc. 4146, 1963.
[4] Laland, P., and Klem, A., Acta Med. Scand. **88**, 620, 1936.
[5] Brenner, M. and Niederwieser, A., Methods in Enzymology **11**, 39, 1965.
[6] Müller, O. and Müller, G., Biochem. Z. **336**, 299, 1962.
[7] Smith, E. L., Mervyn, L., Muggleton, P. W., Johnson, A. W., and Shaw, N., Ann. N. Y. Acad. Sci. **112**, 565, 1964.
[8] Hatfield, M. R., in Organic Syntheses McGraw-Hill, New York, 1950, vol 3, p. 148.

5.4.4 Synthesis of Vitamin B_{12} 5′-Phosphate coenzyme and its alkyl analogues*

by H. C. Friedmann and Z. Schneider

Principle. Cyanocobalamin reacts with β-cyanoethyl phosphate at room temperature under anhydrous conditions in the presence of an excess of dicyclohexylcarbodiimide to form β-cyanoethylcyanocobalamin 5′-phosphate [1]. The cyanoethyl group is removed by slow hydrolysis with ammonia at 0°. The cyanocobalamin 5′-phosphate is purified from the resulting mixture of corrinoids on Dowex 1-acetate columns, and crystallized from acetone. The crystalline preparation is used subsequently for synthesis of its coenzymatic form and alkyl analogues.

Reagents

Cyanocobalamin (vitamin B_{12} cryst.)
β-Cyanoethyl phosphate. May be synthesized according to Tener [2]. Available commercially as Ba salt, dihydrate, M_r 322.5 or trihydrate, M_r 340.5.
Dowex 50W X12, 100-200 mesh, cycled, H^+ form
Dowex 1 X2, 200-400 mesh, cycled, acetate form
Dimethyl formamide and pyridine, both dried by shaking with Linde molecular sieve type 5 A, 1/16 inch pellets. Storage over the pellets settles "fines".
Dicyclohexylcarbodiimide, M_r 206 (Aldrich Chemical Company, Inc.)
5′-iodo-5′-deoxyadenosine
CH_3I and C_2H_5I
$Mg(ClO_4)_2$
Liquefied phenol, USP
Diethyl ether
Acetone
Conc. ammonia
Acetic acid
Sodium acetate buffer, 2 N, pH 3.9, diluted to 0.1 N and to 0.25 N
Sodium chloride

Preparation of free β-cyanoethyl phosphate. Ion exchange with excess Dowex 50 (H^+) is used [2]. For example, 5 millimoles of barium β-cyanoethyl phosphate (1.7 g

* The text is partly reprinted from: M. C. Friedmann, synthesis of vitamin B_{12} 5′-phosphate, Methods in Enzymology **18 C**, 54-59, 1971.

trihydrate) in 5 ml of water are mixed with 5 g of Dowex 50W (H^+) (exchange capacity, 5 meq/g). The mixture is applied to a column, 1.5 × 4.5 cm, of Dowex 50W(H^+) and washed out with about 2½ column volumes of water. The effluent after addition of 4 ml of pyridine is concentrated *in vacuo* at about 35 °C to about 4 ml. The volume is made up to 10 ml with pyridine. The solution, stored at 4 °C was found to be still usable after 15 months.

Preparation of vitamin B_{12} 5′-Phosphate. To 230 mg of vitamin B_{12} (0.17 millimole) add 2.04 ml of the above solution of β-cyanoethyl phosphate (6 × 0.17 = 1.02 millimoles). Evaporate the mixture to dryness *in vacuo* at 40–44 °C in a 50-ml round-bottom boiling flask with ground-glass neck. Repeat the evaporation twice with about 5 ml of dry pyridine. Complete the drying overnight in a vacuum desiccator over $Mg(ClO_4)_2$ at room temperature.

Place a small magnetic stirring bar in the bottle. Dissolve the red material in 12.7 ml of dry dimethyl formamide and 8.5 ml of dry pyridine. Add a solution of 272 mg dicyclohexylcarbodiimide (7.8 × 0.17 = 13.3 millimoles) in 4.2 ml of dry pyridine. (Use rubber gloves to handle the dicyclohexylcarbodiimide to prevent contact with skin and possible sensitization or allergic reaction.) Close the bottle with a ground-glass stopper and stir the mixture at room temperature. A white precipitate of dicyclohexyl urea begins to form within 1 or 2 hours. The reaction is interrupted after about 120 hours by placing the bottle in ice water and adding 25 ml of cold water. The precipitated dicyclohexyl urea is removed by centrifugation, and the supernatant solution is evaporated to a small volume *in vacuo* at room temperature. The precipitate is washed with water until almost colorless, and the washings are added to the concentrate. The evaporation *in vacuo* is repeated. Further evaporation with a small volume of water may be necessary to remove all odor of pyridine. A scant white precipitate is removed from the concentrate by centrifugation and may be washed to obtain some slight additional corrinoid material.

Purification on Dowex-1 Columns. The corrinoids in the combined supernatant solutions (about 30 ml) are desalted [3] by extraction into a small volume of liquefied phenol (2 ml, followed by 1 ml). The phases may be separated by centrifugation. The colorless supernatant solution after the second extraction is discarded. The combined phenolic phases are washed 2 or 3 times with about 4 ml of water. The faintly pink aqueous wash is discarded, and the corrinoids are returned to water by the addition of 2–3 volumes of diethyl ether [4]. Phenol is removed from the aqueous phase by four washes with about 5 ml of diethyl ether. The ether is removed *in vacuo* at room temperature, and the red solution is concentrated to about 2 ml. The solution is transferred to a 25-ml measuring cylinder and cooled in ice; 9 ml of concentrated ammonia solution (15 N) is added. The solution is diluted to 15 ml with cold water and mixed. After hydrolysis at 0 °C for 65 hours the mixture is diluted with about 50 ml of water, and evaporated *in vacuo* at room temperature to about 10 ml. The procedure is repeated until there is no more perceptible odor of ammonia, and the evaporation is then continued to near dryness.

The residue is dissolved in about 1.5 ml of water and applied to a Dowex-1 acetate column, 13 × 205 mm. Elution is performed at the rate of about 0.9 ml per minute. After removal of some orange and red material by water (about 50 ml), traces of orange and red substances are eluted by the 0.1 N sodium acetate buffer solution (about 50 ml). Continuation of the elution with the 0.25 N sodium acetate buffer (about 90 ml) removes the main red fraction from the column. This fraction consists mainly of vitamin B_{12} 5′-phosphate and some material exhibiting the orange color and ionophoretic neutrality at acid pH of cobinamide phosphate [5, 6]. The mixture is desalted via phenol extraction, passed back into water with the help of diethyl ether, freed of phenol and ether as before, and applied to a fresh Dowex-l acetate column, 13 × 190 mm. Elution is performed at the rate of 1.5 ml per minute. Some red material is removed with water (about 100 ml). Appreciable amounts of orange and red material are now eluted with 0.5 N acetic acid (about 360 ml). After this eluate has become essentially colorless, the main red band containing principally vitamin B_{12} 5′-phosphate is readily eluted by a solution of 0.02 M NaCl in 0.5 N acetic acid [7] (about 130 ml). The material in the pooled eluate is again desalted via phenol treatment. It still contains traces of presumed cobinamide phosphate. These are removed to barely detectable traces when the vitamin B_{12} 5′-phosphate is crystallized from acetone.

Crystallization. For successful crystallization it is necessary to use an aqueous solution with a theoretical absorbance at 361 nm between 77 and 82. Stronger solutions have to be diluted to this value. The theoretical absorbance is calculated from the absorbance at 361 nm of a small aliquot diluted with water by a known factor to an absorbance of less than 0.5. In order to induce crystallization, 8 ml of acetone are added per milliliter with gentle mixing in unscratched Corex glass centrifuge tubes. The tubes are closed with foil and allowed to stand away from direct light at room temperature (about 25 °C). Crystallization in the form of needles about 2 mm long is usually complete overnight. The temperature is important. At 15 °C only a scant and fine precipitate is formed. The crystals are centrifuged down and washed a few times with acetone. The acetone is removed *in vacuo* at room temperature, and the crystals are allowed to stand about 1 hour over $Mg(ClO_4)_2$. The yield of the first crop of crystals is 52 mg. More material can be obtained from the mother liquor.

Preparation of vitamin B_{12}-5′-phosphate coenzyme (AdoCbl-5′-P). The coenzymatic form is conveniently synthetized from the reduced form B_{12s}-P and 5′-iodo-5′-deoxyadenosine following the method described in [8]. The 5′-iodo-5′-deoxyadenosine is a commercial product. It may be obtained from 2′,3′-isopropylidene adenosine according to [9]. The following procedure is used: 25 mg of CN-B_{12}-P is dissolved in 5 ml of water and the solution is neutralized to pH 7–8 with KOH. Argon is bubbled through for 15 minutes, followed by addition of 10 mg of $NaBH_4$ dissolved in 2 ml. The solution is kept in an argon atmosphere for about 10 minutes followed by injection of 10 mg 5′-iodo-5′-deoxyadenosine dissolved in 2 ml of water and deoxygenated with argon. The solution changes color at this point from blue-green to red. The procedure is done at this stage in dim red light. After approx. two minutes the excess of $NaBH_4$ is decomposed by gradual addition of 500 µl of 1.0 N acetic acid. The material is dried in a vacuum desiccator over P_2O_5 and after being dissolved in a small volume of water is applied in a streak on Whatman No 3 MM paper, at a distance of about one third of the length from one end. Ionophoresis in 0.5 N acetic acid with the corrinoids initially near the anode follows. A potential of 300 V is applied over a distance of 18 cm for 3 h at room temperature in the dark. Three bands separate: a major fast yellow band and a slower red band move toward the cathode, while unconverted OH-B_{12}P remains near the origin. The yellow substance is eluted from the dried paper with water. The yield is about 60%. The vitamin B_{12}-5′-phosphate coenzyme shows coenzyme activity with glycerol dehydratase apoenzyme [8] (see Chapter 7). The MeCbl-5′-P or EtCbl-5′-P are synthetized and purified similarly to AdoCbl-5′-P by using instead of 5′-iodo-5′-deoxyadenosine 500 µl of CH_3I or C_2H_5I respectively. Under a voltage gradient of 16.5 V/cm applied for 3 h, the following mobilities relative to CN-Cbl were measured in 0.5 N acetic acid ("−" denotes net movement toward anode) [8]: CN-Cbl-5′-P,-41 mm; AdoCbl-5′-P, 34 mm; AdoCbl, 74 mm; MeCbl-5′-P,-10 mm; EtCbl-5′-P,-1 mm.

References

[1] Wagner, F., Biochem Z. **336**, 99, 1962
[2] Tener G. M., J. Am. Chem. Soc. **83**, 159, 1961
[3] H. A. Barker, R. D. Smyth, H. Weissbach, A. Munch-Petersen, J. I. Toohey, J. N. Ladd, B. E. Volcani, and R. M. Wilson, J. Biol. Chem. **235**, 181 1960.
[4] J. T. Toohey, Federation Proc. **25**, 1628, 1966.
[5] K. Bernhauer, F. Wagner, H. Dellweg, and P. Zeller, Helv. Chim. Acta **43**, 700, 1960.
[6] H. C. Friedmann, J. Biol. Chem. **243**, 2065, 1968.
[7] R. L. Sinsheimer, J. Biol. Chem. **208**, 445, 1954.
[8] Schneider, Z., and Friedmann, M. C., Arch. Biochem. Biophys. **152**, 488-495, 1972.
[9] Jahn, W., Chem. Ber. **98**, 1705, 1965.

6 The Occurrence and Distribution or Corrinoids

Z. Schneider

Vitamin B_{12} forms present in living systems and their environments orginate exclusively in microorganisms. The corrin ring and the nucleotide moiety are synthetized only by bacteria and some algae while the beta ligands may also be attached to cobalt by enzyme systems present in animals and man. Concerning the nucleotide moiety, microorganisms contain various forms of corrinoids while man and animals utilize under natural conditions almost entirely a cobamide which contains 5,6-dimethylbenzimidazole in the nucleotide part.

6.1 Microorganisms

Various microorganisms, but not all, are able to synthetize corrinoids. Many microorganisms (may be all), which do not produce these cofactors, require them for growth. The biosynthesis of the rather large and complex compound is a multistep process requiring many enzymes. A number of environmental factors are known to affect the synthesis of the precursors. Their synthesis in adequate proportion seems to be often disturbed, since, besides biologically active corrinoids, the microorganisms contain incomplete forms of corrinoids, in particular such which are deprived of the nucleotide moiety, like: cobyric acid (Ado-Cby), cobinamide (Ado-Cbi), cobinamide phosphate (Ado-Cbi-P) and cobinamide guanosine diphosphate (Ado-Cbi-GDP). Under unfavourable conditions for the synthesis of the nucleotide moiety precursor (base) the incomplete forms, mainly cobinamide, dominate in the corrinoid pool of the cell. (see Table 6-1). The complete forms increase in older (or aerated) cultures on the expense of the incomplete forms. The aerobes e.g. *Bacillus megaterium*, *Streptomyces sp.* and *Nocardia rugosa* and the aerotolerant anaerobes e.g. *Propionibacterium sp.* utilize riboflavin as precursor of the base, 5,6-dimethylbenzimidazole [15a]. The conversion of riboflavin to dimethylbenzimidazole is oxygen dependent, thus aeration of the culture especially before the harvest of the cells, results in high yields of cobalamin (see Table 4-1). The obligate anaerobes like *Butiribacterium rettgeri* and *Clostridium barkeri* synthetize 5,6-dimethylbenzimidazole cobamide as well. They do not utilize, however, riboflavin as precursor of the nucleotide moiety [15b] Besides cobalamin, the cobamide which is utilized by man and animals, the microorganisms synthetize a number of other cobamides active as cofactors in bacterial enzyme systems [16-29]. The naturally occurring base-containing corrinoids can be divided into three groups: those containing a benzimidazole base, those with a purine base and those which contain a nitrogen-free compound like phenol or p-cresol bound O-glycosidically to ribose [55]. With regard to the latter group of corrinoids found in trace amounts there is so far no evidence for its biological function. Concerning the cobamides which occur naturally and contain the benzimidazole base, the following are found: 5,6-dimethylbenzimidazolylcobamide (5,6-Me_2Bza-Cba or Cbl), 5-hydroxybenzimidazolylcobamide (5-OHBza-Cba) and 5-methoxybenzimidazolylcobamide (5-MeOBza-Cba) [26, 27]. The latter two are synthetized by methane bacteria in sewage [6, 7] and acetic acid forming bacteria like *Clostridium thermoaceticum* (see Table 3-4 for characteristics of those cobamides). The 5-hydroxybenizimidazole base seems to be somewhat slower synthetized than the corrin ring, since the greater part of corrinoids in *C. thermoaceticum* is composed of cobyric acid, while only about 17 per cent of corrinoids is in the form of 5-MeOBza-Cba (named also Factor IIIm) [27]. The precursor of 5-hydroxybenzimidazole is not known yet [15]. From activated sludge two forms have been isolated: benzimidazolyl cobamide (Bza-Cba) and naphtimidazolyl cobamide (Nfta-Cba), which do not occur, however, in true natural sources. Sewage and activated sludge may contain traces of a number of vitamin B_{12} analogues modified in the nucleotide part rarely obtained from other sources (Table 4-2) [6, 7, 24, 47-50]. It seems that bacteria utilize, for synthesis of analogues, the heterocycles present already there (see guided biosynthesis of corrinoids in Chapters 3 and 4).

Concerning the purine-containing corrinoids two main cobamides are found in microorganisms: adenyl cobamide (Ade-Cba) and 2-methyladenylcobamide (2-MeAde-Cba) named also pseudovitamin B_{12} and Factor A respectively. Other cobamides containing such bases like: guanine, xanthine, hypoxanthine (Factor G), 2-methylhypoxanthine (Factor H), 2-methylmercaptoadenine (Factor S) and 2-methylmercaptohypoxanthine (Factor S′) are isolated in trace amounts from various sources (sewage) [6, 7, 24, 47-51]. Ade-Cba and 2-MeAde-Cba are synthetized concomitantly with the benzimidazole base containing cobamides. The purine-containing corrinoids are usually synthetized predominantly when there is no 5,6-dimethylbenzimidazole available. There are however strains e.g. *P. shermanii, P. freudenrcichii* or *Rhizobium sp.* which synthetize little of the purine containing cobamides, and under conditions uncongenial for the benzimidazole base synthesis at sufficient rate, part of the corrinoid remains in the cell rather in the nucleotide free form [30, 31]. On the other hand such species as *P. arabinosum, E. coli* and most species living in digestive systems of man and animals synthetize predominantly purine containing cobamides (Tables: 6-1, 6-13 and 6-14). It seems that these cobamides and the incomplete forms (cobinamides) dominate in the whole pool of naturally occurring corrinoids [16, 20-24, 32, 33]. The strains utilizing purines are flexible, however, since they can produce cobalamin at the expense of other forms when there is adequate synthesis of dimethylbenzimidazole (aeration) or the base is added to the growth medium. Indeed much higher level of cobalamin is observed in cultures grown in presence of the benzimidazole base (see Table 6-1). the factories producing cobalamin take advantage of this feature. Very efficient strains of *Propionibacterium sp., Streptomyces sp., Rhizobium sp.* and methane forming bacteria have been selected for preparative and industrial production of the vitamin [2, 3, 9, 10, 34, 35, 42, 47-50]. They yield in appropriate medium up to 25 mg of cobalamin per liter [51-54a-g] of growth medium. The yield of cobalamin production in rumen is significantly higher when the feed is supplemented with 5,6-dimethylbenzimidazole [36]. Under natural conditions the cobalt is often the limiting factor of vitamin B_{12} synthesis in rumen [37-39]. In addition the greater part of

Table 6-1 Corrinoids produced by various microorganisms Data selected by K. Trojanowska and Z. Schneider

Microorganism	Additive		Corrinoids µmoles per 1 liter of growth medium (in parentheses mg/l)						References
	5,6-Me_2Bza	Co^{2+}	Cobalamins Cbl	Adenine cobamides		Cbi	Cbi-GDP	Total	
				Ade-Cba	2-MeAde-Cba (factor A)				
1	2		3	4	5	6	7	8	9
Propionibacterium shermanii	+	+	6.85 (9.29)	–	–	0.57 (0.59)	–	7.4 (9.88)	Pędziwilk [1]
	–	+	0.35 (0.47)	0.098 (0.13)	–	4.38 (4.52)	–	4.83 (5.12)	Pędziwilk [1]
P. shermanii ATCC 8262, ATCC 9616, ATCC 9617	+	+	7.4–18.4 (10–25)	–	–	–	–	–	Perlman [2]
P. freudenreichii	+	+	6.9 (9.36)	–	–	0.55 (0.57)	–	7.45 (9.93)	Pędziwilk [1]
P. freudenreichii	–	+	0.26 (0.35)	0.074 (0.10)	–	3.58 (3.70)	–	3.91 (4.15)	Pędziwilk [1]
P. freudenreichii ATCC 6207	–	+	7.4–18.4 (10–25)	–	–	–	–	–	Perlman [2]
P. petersonii	+	+	0.58 (0.79)	–	–	5.21 (5.38)	–	5.79 (6.17)	Pędziwilk [1]
	–	+	0.51 (0.69)	0.11 (0.15)	–	5.76 (5.95)	–	6.38 (6.79)	Pędziwilk [1]
P. thoemii	+	+	1.13 (1.53)	0.074 (0.19)	0.048 (0.07)	0.068 (0.07)	0.085 (0.12)	1.41 (1.89)	Pędziwilk [1]
	–	+	0.06 (0.089)	0.316 (0.43)	0.076 (0.10)	1.22 (1.26)	–	1.68 (1.88)	Pędziwilk [1]
P. rubrum	+	+	1.17 (1.59)	–	–	0.22 (0.23)	–	1.39 (1.82)	Pędziwilk [1]
	–	+	0.21 (0.28)	–	0.14 (0.19)	1.26 (1.30)	–	1.61 (1.77)	Pędziwilk [1]
P. Jensenii	+	+	1.45 (1.97)	0.10) (0.13)	–	0.11 (0.11)	0.11 (0.16)	1.77 (2.37)	Pędziwilk [1]
	–	+	0.038 (0.05)	0.19 (0.26)	0.086 (0.17)	0.57 (0.59)	–	0.88 (1.07)	Pędziwilk [1]
P. sanguineum	+	+	1.78 (2.41)	1.21 (1.63)	0.17 (0.23)	0.41 (0.42)	–	3.57 4.69)	Pędziwilk [1]
	–	+	–	2.49 (3.35)	0.73 (0.99)	0.84 (0.87)	–	4.06 (5.21)	Pędziwilk [1]
P. raffinosaceum	+	+	0.39 (0.53)	0.21 (0.28)	–	0.11 (0.11)	–	0.71 (0.92)	Pędziwilk [1]
	–	+	0.02 (0.03)	0.67 (0.90)	–	0.16 (0.17)	–	0.85 (1.1)	Pędziwilk [1]

Table 6-1 continued

Microorganism	Additive		Corrinoids μmoles per 1 liter of growth medium (in parentheses mg/l)						References
	5,6-Me_2Bza	Co^{2+}	Cobalamins Cbl	Adenine cobamides		Cbi	Cbi-GDP	Total	
				Ade-Cba	2-MeAde-Cba (factor A)				
1	2		3	4	5	6	7	8	9
P. arabinosum	+	+	0.93 (1.26)	1.30 (1.75)	–	0.30 (0.30)	–	2.53 (3.31)	Pędziwilk [1]
	–	+	0.12 (0.16)	2.92 (3.93)	0.14 (0.19)	0.24 (0.25)	–	3.42 4.53)	Pędziwilk [1]
P. pituitosum	+	+	0.81 (1.10)	0.50 (0.67)	–	0.24 (0.25)	–	1.55 (2.02)	Pędziwilk [1]
	–	+	0.12 (0.16)	1.12 (1.51)	–	0.56 (0.58)	–	1.80 (2.25)	Pędziwilk [1]
P. wentii	+	+	0.07 (0.09)	0.51 (0.69)	–	0.33 (0.34)	–	0.91 (1.12)	Pędziwilk [1]
	–	+	0.13 (0.18)	0.35 (0.47)	0.10 (0.14)	0.43 (0.44)	–	1.01 (1.23)	Pędziwilk [1]
P. pentosaceum	+	+	1.24 (1.68)	–	0.22 (0.30)	–	–	1.46 (1.98)	Pędziwilk [1]
	–	+	0.77 (1.04)	0.10 (0.13)	0.05 (0.07)	0.06 (0.06)	–	0.98 (1.3)	Pędziwilk [1]
P. technicum	+	+	1.15 (1.56)	0.08 (0.11)	0.01 (0.01)	0.05 (0.05)	0.11 (0.16)	1.40 (1.89)	Pędziwilk [1]
	–	+	0.08 (0.11)	0.35 (0.47)	0.06 (0.08)	0.72 (0.74)	–	1.21 (1.4)	Pędziwilk [1]
P. technicum ATCC 14073	–	+	7.4–18.4 (10–25)	–	–	–	–	–	Perlman [2]
P. zeae	+	+	0.99 (1.34)	0.76 (1.02)	–	0.55 (0.57)	–	2.30 2.93)	Pędziwilk [1]
	–	+	0.34 (0.046)	1.19 (1.60)	0.37 (0.50)	0.56 (0.58)	–	2.15 (2.73)	Pędziwilk [1]
Propionibacterium sp.	+	+	dominate	none	none	present	present	22.1 (30)	Bielecka et al. [48]
Bacillus megatherium	–	+	0.33 (0.44)	–	–	–	–	–	Perlman [3]
Corynebacterium diphtheriae	–	+	traces	0.19 (0.26)	0.18 (0.24)	other 1.2	1.6	1.55 (2.1)	Pawełkiewicz, Zodrow [4, 5]
No 44 PZH Warsaw	+	+	1.3 (1.8)	traces	traces	–	–	1.3 (1.8)	
Flavobacterium solare			0.44 (0.59)	–	–	–	–	–	Perlman [3]

Flavobacterium devorans	–	+	0.44 (0.59)	–	–	–	–	–	Perlman [3]
Streptomyces griseus	–	+	0.22 (0.30)	–	–	–	–	–	Perlman [3]
Streptomyces fradiae	–	+	0.52 (0.71)	–	–	–	–	–	Perlman [3]
Streptomyces olivaceus	–	+	2.4 (3.3)	–	–	–	–	–	Perlman [3]
Streptomyces species ATCC 11072	–	+	0.75–2.2	–	–	–	–	–	Perlman [3]
			(1–3)			–	–	–	Perlman [3]
Clostridium acetobutylicum Weizmann	–	+	none	0.044 (0.06)		–	–	–	Neujahr [6]
	+	+	0.025 (0.035)	none	none	traces	–	–	Neujahr [6]
Clostridium butylicum	–	+		0.036 (0.05)		–	–	–	Neujahr [6]
	+	+	0.018 (0.024)	none	none	traces	–	–	Neujahr [6]
Methanobacterium Omelianskii	+	+	4.4 (5.9)		–	–	–	–	Neujahr [6]
Desulphovibrio desulphuricans			0.22 (0.30)	–	–	–	–	–	Neujahr [7]
Pseudomonas denitrificans	+	+	11.1 (15)	–	–	–	–	–	Perlman [2]
Rhizobium meliloti	–	+	0.017–0.165	–	– –	–	–		Kaszubiak et al [8]
Rhizobium trifolii	–	+	0.05–0.02	–	– –	–	–		Malińska and Pędziwilk [13]
various straims from *root nodules*	–	+	0.002–0.14	–	– –	–			Burton and Lochhead [14]
Rhizobium lupini	–	+	0.02–0.05	traces	traces –	–	–		Malińska and Pędziwilk [13]

Table 6-1 continued

Microorganism	Additive		Corrinoids µmoles per 1 liter of growth medium (in parentheses mg/l)						References
	$5,6\text{-}Me_2Bza$	Co^{2+}	Cobalamins Cbl	Adenine cobamides		Cbi	Cbi-GDP	Total	
				Ade-Cba	2-MeAde-Cba (factor A)				
1	2		3	4	5	6	7	8	9
various strains from *root nodules*	–	+	0.001–0.01	–	– –	–	–		Burton and Lochhead [14]
Rhizobium meliloti	+	+							
mutant Fxx			6.4 (8.7)	–	– –	–	–		Szucs [9]
mutant F312			0.4 (0.55)	–	– –	–	–		
Mixed bacterium population isolated from sewage methane producing species	+ methanol	+ 1.4%	5.9–7.4 (8–10)	–	– –	–	–		Szontagh et al [10]
Escherichia coli sp.	+	cobinamide	depends on amount of cobinamide added	–	– –	–	–		Dellweg et al [11] and Ford et al [12]
Rhodopseudomonas gelatinosa	+	+	33 µg/g cell						Sasaki et al [51, 52]
Methanol-utilizing bacterium strain XF			0.17 (0.24)	–	– –	–	–	–	Dumenil et al [53]

cobalt present in feed is wasted in the rumen for synthesis of pseudovitamin B_{12}, the corrinoid inactive for animals (Tables 6-13, 6-14 and text in section 6.5). Due to low concentrations of cobalt in such sources as: soil, water and plants the microorganisms developed a very effective mechanism for its uptake: e.g. the *Rhizobium sp. and Azotobacter vinelandii* which require cobalt (or B_{12}) for dinitrogen fixation respond with growth to concentrations of cobalt in the range of 10^{-12} to $1o^{-13}$ M [40, 44]. Concerning various strains of the Rhizobium species there is a positive correlation between the vitamin B_{12} synthesis and its ability to fix dinitrogen during symbiosis with plants (Fig. 6-1 and Table 6-6). For information on B_{12} production by microorganisms in soil see text below.

There is a large group of microorganisms named blue-green bacteria traditionally still classified with algae under the class *Cyanophyceae* (see text in section 6.3.1 and Tables 6-3 A and 6-3 B) which, considering their wide distribution in natural waters, seem to be one of the major producers of corrinoids in nature.

Besides corrinoids containing cobalt, some photosynthetic bacteria like: *Rhodospirillum rubrum, Rhodospseudomonas palustris, Rhodomicrobium vannielii* and *Chromatium* produce cobalt free (or descobalt) corrinoids [45, 46, 55]. The latter strain yields approximately 1 mg of these compounds per 100 g of cell paste [45]. The biological function of these corrinoids is unknown. Since they incorporate readily also other than cobalt ions into the corrin ring they are utilized for synthesis of cobalt substituted analogues (for details see Table 3-1 and text in 3.1.1.7).

References

[1] Pędziwilk, F., Biosynthesis of cobalamin by propionic acid bacteria, The Poznań Society of Friends of Science, Department of Agricultural and Sylvicultural Sciences Publications **11,** 141-191, 1962.

[2] Perlman, D., Processes for the biosynthesis of cobamides, in Methods in Enzymology **18 C,** 75-81, 1971.

[3] Perlman, D., Microbial Synthesis of cobamides, Advances in Applied Microbiology **1,** 87-112, 1959.

[4] Pawełkiewicz, J. and Zodrow, K., Studies on the biosynthesis of cobalamin compounds by *Corynebacterium,* Acta Biochim. Polon. **4,** 203-210, 1957.

[5] Pawełkiewicz, J. and Zodrow, K., Precursors of the biosynthesis of the nucleotidocyanocobalamins, Acta Microbiol. Polon. **6,** 9-15, 1957.

[6] Neujahr, H. Y., On vitamin B_{12} factors in sewage sludge, Almqvist and Wiksells Boktryckeri AB, Uppsala 1960.

[7a] Neujahr, H. Y., On vitamins in sewage sludge. Production of vitamin B_{12} by sulphatereducing bacteria, Acta Chemica Scand. **13,** 1960-1970, 1959.

[7b] Sjöström, A. G. M., Neujahr, H. Y., Lundin, H., Vitamin B_{12}, folic acid, and folinic acid factors in digested municipal sludge, Acta Chem. Scand. **7,** 1036-1040, 1953.

[7c] Neujahr, H. Y., On vitamins in sewage sludge. II. Formation of vitamin B_{12}, folic acid-, and folinic acid factors in municipal sludge, Acta Chem. Scand. **9,** 622-630, 1955.

[7d] Neujahr, H. Y., On vitamin in sewage sludge. III. Vitamin B_{12} activity of different types of sewage sludge, Acta Chem. Scand. **9,** 803-806, 1955.

[7e] Neujahr, H. Y., On vitamins in sewage sludge. IV. Isolation of new vitamin B_{12}-like factors, Acta Chem. Scand. **10,** 917-927, 1956.

[7f] Neujahr, H. Y., On vitamins in sewage sludge. V. The effect of antibiotics on vitamin B_{12} contents of microbially decomposing sewage sludge, Acta Chem. Scand. **11,** 1191-1200, 1957.

[7g] Neujahr, H. Y., On vitamins in sewage sludge. VI. The effect of cobalt chloride and copper sulphate on the vitamin B_{12} activity of microbially decomposing sewage sludge, Acta Chem. Scand. **12,** 408-414, 1958.

[7h] Neujahr, H. Y. and Callieri, A. D., On vita-

mins in sewage sludge. VII. Production of vitamin B_{12} by an enrichment culture of *Methanobacterium omelianskii*. Acta Chem. Scand. **12**, 1153-1166, 1958.

[7 i] Neujahr, H. Y., and Callieri, A., On vitamins in sewage sludge. VIII. Utilization of some vitamin B_{12} factors by an enrichment culture of *Methanobacterium omelianskii, Acta Chem. Scand.* **12**, 1167-1175, 1958.

[7 j] Neujahr, H. Y. and Callieri, A., On vitamins in sewage sludge. IX. Production of vitamin B_{12} by methane bacteria. Acta Chem. Scand. **13**, 1453-1459, 1959.

[7 k] Neujahr, H. Y., On vitamins in sewage sludge. XI. Further studies on the production of vitamin B_{12} by some enrichment cultures of methane bacteria, Acta Chem. Scand. **14**, 28-42, 1960.

[7 l] Neujahr, H. Y. and Rossi-Ricci, G., On vitamins in sewage sludge. XII. Production of vitamin B_{12} by certain *Clostridia*, Acta Chem. Scand. **14**, 43-47, 1960.

[7 m] Neujahr, H. Y., Sewage as a substitute for casein hydrolyzates in fermentations with *Propionibacterium shermannii*, Acta Chem. Scand. **13**, 2158-2160, 1959.

[7 n] Neujahr, H. Y., Kurz, W. G. and Rossi-Ricci, G., Studies on vitamin B_{12} production by *Propionibacterium shermannii* grown in sewage, Arkiv. Kemi **15**, 363, 1960.

[8] Kaszubiak, H., Malinska, E., Pędziwilk, Z., Studies on production and structure of corrinoids from *Rhizobium meliloti*, Acta Microbiol. Polon. **6**, 239-251, 1957.

[9] Szücs, J. S., Phage infection of vitamin B_{12} producing *Rhizobium* cultures, Acta Microbiol. Acad. Sci. Hung. **18**, 81-86, 1971.

[10] Szontagh, T., Simonovits, E. and Johan, B., Vitamin B_{12} producing fermentations with a mixed bacterium population of sewage sludge origin, Acta Microbiol. Acad. Sci. Hung. **19**, 87-96, 1972.

[11] Dellweg H. Becher, E. and Bernhauer, K., Biochem Z. **328**, 88, 1956.

[12] Ford, J. E., Holdsworth, E. and Kon, S. K., The biosynthesis of vitamin B_{12}-like compounds, Biochem J. **59**, 86, 1955.

[13] Malinska, E. and Pędziwilk Z., Studies on synthesis and composition of corrinoids in *Rhizobium*, Acta Microbiol. Polon. **7**, 125-130, 1958.

[14] Burton, M. O. and Lochhead A. G., Production of vitamin B_{12} by *Rhizobium sp.*, Canad. J. Bot. **30**, 521, 1952.

[15 a] Renz, P., Hörig J. and Wurm R., On the biosynthesis of the 5,6-dimethylbenzimidazole moiety, in, Vitamin B_{12}, Eds. B. Zagalak and W. Friedrich, Walter de Gruyter, Berlin 1979, pp 317-322.

[15 b] Hoellrigl, V., Lamm, L., Rowold, J., Hoerig, J. and Renz, P., Biosynthesis of vitamin B_{12}: Different pathways in some aerobic and anaerobic microorganisms. Arch. Microbiol. **132**, 155-158, 1982.

[16] Perlman, D., and Barrett, J. M., Biosynthesis of cobalamin analogues by *Propionibacterium arabinosum*, Can. J. Microbiol., **4**, 9-15, 1958.

[17] Kurnatowski, R., Szymkowiak, A., Wisniewski, W., i Pawełkiewicz, J., Prekursory biosyntezy nukleotydocyjanokobalamin. Biosynteza vitaminy B_{12} i innych nukleotydocyjanokobalamin., Acta Biochim. Polon. **5**, 19-25, 1958.

[18] Pawełkiewicz, J., Zagadnienia witaminy B_{12} i innych kobalamin na IV Międzynarodowym Kongresie Biochemii w Wiedniu, Postępy Biochemii **5**, 385-396, 1959.

[19] Perlman, D., and Semar, J. B., Production of cobamides by *butyribacterium rettgeri*, Biotechnology and Bioengineering, **5**, 21-25, 1963.

[20] Dryden, L. P., Hartman, A. M., Variations in the amount and relative distribution of vitamin B_{12} and its analogs in the bovine rumen, J. Dairy Sci. **54**, 235-246, 1971.

[21] Dryden, L. P., Hartman, A. M., Bryant, M. P., Robinson, J. M., Moore, L. A., Production of vitamin B_{12} and vitamin B_{12} analogues by pure cultures of ruminal bacteria, Nature **195**, 201-202, 1962.

[22] Ford, J. E., Porter, J. W. G., Brit, J., Vitamin B_{12}-like compounds. Some properties of compounds isolated from bovine gut contents and faeces, Nutr. **7**, 326, 1953.

[23] Pawełkiewicz, J., Vitaminy grupy B_{12} i ich biosynteza., Postępy Biochemii. **3**, 3-38, 1957.

[24] Friedrich, W., Vitamin B_{12} und verwandte Corrinoide, Fermente Hormone Vitamine, Band III/2, Eds. Ammon, R. and Dirscherl, W., Georg Thieme Verlag, Stuttgart 1975, pp. 169-186.

[25] Janicki J., Pawełkiewicz J. Nowakowska H., Production of vitamin B_{12} concentrates and crystalline B_{12} from comunal sewage exposed to methane fermentation, Acta Biochim. Polon. **3**, 161, 1956.

[26] Friedrich W., Bernhauer, K., Vitamin B_{12}-Analoga des Belebtschlammes, Angew. Chemie. **71**, 284, 1959.

[27 a] Ljungdahl, L., Irion, E. and Wood, H. G., To-

tal synthesis of acetate from CO_2. I. Co-methylcobyric acid and Co-methyl-5-metoxy-benzimidazolyl cobamide as intermediates with *Clostridium thermoaceticum*, Biochemistry **4**, 2771-2779, 1965.

[27b] Irion, E., Ljungdahl, L., Isolation of Factor III m coenzyme and cobyric acid coenzyme plus other B_{12}-factors from *Clostridium thermoaceticum*, Biochemistry **4**, 2780-2790, 1965.

[28] Blaylock, B.A., Stadtman H., T.C., Biosynthesis of methane from the methyl moiety of methylcobalamin, Biochem. Biophys. Res. Commun. **11**, 34-38, 1963.

[29] Salem, A.R., Foster, M.A., The microbial biosynthesis of methionine, Biochem. J. **127**, 845-853, 1972.

[30] Barchielli, R., Boretti, G., Di Marco, A., Julita, P., Migliacci, A., Minghetti, A., Spalla, C., Isolation and structure of a new factor of the vitamin B_{12} group: guanosine diphosphate factor B., Biochem. J. **74**, 382-387, 1960.

[31] Boretti, G., Di Marco, A., Fuoco, L., Marnati, M.P., Migliacci, A., Spalla, C., Intermediates in the biosynthesis of vitamin B_{12}, Biochem. Biophys. Acta **34**, 379-380, 1960.

[32] Zodrow, K., Stefaniak, O., Chelkowski, J., Malinska, E., Isolation of corrinoids from soil, Acta Microbiol. Polon. **11**, 341-348, 1962. (see also references in sections 6.2 and 6.3)

[33] Kliewer, M., Evans, H.J., Effect of cobalt deficiency on the B_{12} coenzyme content of *Rhizobium meliloti*, Arch. Bioch. Biophys. *97*, 428-429, 1962.

[34] Maitra, P.K., Ganguly, S., and Roy, S.C., Studies on the production of vitamin B_{12} by *Streptomyces olivaceus*, Ann. Bioch. Exp. Med. **15**, 212-214, 1955.

[35] Schindler, O., und Reichstein, T., Isolierung von Vitamin B_{12} aus Leberextrakt sowie aus Rückständen der Streptomycinfabrikation, Helvetica Chimica Acta **XXXV**, 308-316, 1952.

[36] Rickard, T.R., Bigger, G.W. and Elliot, J.M., Effect of 5,6-dimethylbenzimidazole, adenine and riboflavin on ruminal vitamin B_{12} synthesis, J. Anim. Sci. **40**, 1199-1204, 1975.

[37] Gawthorne, J.M., The effect of cobalt intake on the cobamide and cobinamide composition of the rumen contents and blood plasma of sheep, Austr. J Exp. Biol. Med. Sci. **48**, 285, 1970.

[38] Gawthorne, J.M., *In vitro* studies of the factors affecting the metabolism of cobamides and cobinamides by sheep rumen microorganisms, Austr. J. Exp. Biol. Med. Sci. **48**, 293, 1970.

[39] On effects of cobalt on B_{12} production in rumen see the references in section 6.5.

[40] Kliewer, M. and Evans, H.J., Cobamide coenzyme contents of soybean nodules and nitrogen fixing bacteria in relation to physiological conditions, Plant Physiol. **38**, 99-104, 1963. (see also related references of these authors in sections 6.2 and 6.3)

[41] Nicholas, D.J.D., Kobayashi M., Wilson P.W., Cobalt requirement for inorganic nitrogen metabolism in microorganisms, Bacteriol. Proc. **101**, 1962.

[42] Buksowicz, K., Grzybowski, E., Przemysł Spożywczy **25**, 308, 1971 and Nutr. Abstr. Rev. **42**, 503, 1972.

[43] Iswaran, V., Sundara, Rao, W.V., Mathur, S.P., Role of vitamin B_{12} in nitrogen fixation by *Azotobacter chroococcum*, Current Sci. **29**, 2-63, 1960.

[44] Wilson, D.O., Reisenauer, H.M., Cobalt requirement of symbiotically grown alfalfa, Plant and Soil **19**, 364-373, 1967.

[45] Toohey, J.I., Purification of descobalt corrins from photosynthetic bacteria, Methods in Enzymology **18 C**, 71-75, 1971.

[46] Koppenhagen, V.B., Wagner, F., Pfiffner, J.J., J. Biol. Chem. **249**, 6532-6540, 1974.

[47] Stoyanova, L.G., Vorobieva, L.I., and Lobzow, K.I., Physiological and biochemical characteristics of *Propionibacterium shermanii* growth in egg albumin. Mikrobiologiya **48**, 1011-1016, 1979.

[48] Bielecka, M., Czarnocka-Roczniakowa, B., and Gradzinska, A., Optimalization of whey medium composition for biosynthesis of vitamin B_{12} by propionibacteria, Acta. Aliment. Pol. **4**, 133-140, 1978.

[49] Datsyuk, N.M., Kucheras, R.V., Rurik, V.D., Gudz, S.P., Kostruba, M.F., and Kremenetsskaya, S.Ya., Pool of flavins in *Propionibacterium shermanii* and the production of corrinoids by bacterial cultures. Mikrobiologiya **47**, 451-455, 1978.

[50] Mantere-Alhonen, S., New investigations on propionic acid bacteria: A review of the literature, Meijeritiet Aikak **35**, 88-99, 1977.

[51] Sasaki, K. and Nagai, S., The optimum pH and temperature for the aerobic growth of *Rhodopseudomonas gelatinosa* and vitamin B_{12} and ubiquinone formation on a starch medium, J. Ferment. Technol. **57**, 383-386, 1979.

[52] Sasaki, K., Noparatnaraporn, N., Hayashi, S., Nishizawa, Y. and Nagai, S. Single cell pro-

tein production by treatment of soybean wastes with *Rhodopseudomonas gelatinosa*. J. Ferment. Technol. **59**, 471-478, 1981.

[54] Dumenil, G., Cremieux, A., Chevalier J. and Guiraud, H., Vitamin B_{12} formation by a gram-variable methanol-utilizing bacterium, Biotechnol. Lett. **3**, 285-290, 1981.

[55] Dinglinger, F. and Braun, J., O-(α-D-Ribofuranosyl)-p-kresol, das Nukleosid des Corrinoidfaktors Ib aus Faulschlamm. Hoppe Seyler's Z. Physiol. Chem. **351**, 1157-1160, 1970.

[56] Pędziwilk, F., Skupin, J., Trojanowska, K., Biosynthesis of vitamin B_{12} by the mutants of *Propionibacterium shermanii* 1. Induced with ethylmethanesulfonole and dimethyl sulfate. Acta. Aliment. Pol. **9** 113-119, 1984.

[57] Yongsmith, B., Chutima, K., Production of vitamin B_{12} by living bacterial cells immobilized in calcium alginate gels, J. Ferment. Technol. **61**, 593-598, 1983.

[58] Abdel-Hafez, A.M., Mashhoor, W.A., Gamal, R.F., Refaat, A.A. and El-Sayed, A.A., Production of vitamin B_{12} by propionic acid bacteria: 2. Effect of nutritional and environmental factors on vitamin B_{12} production by *Propionibacterium shermanii P-16*. Shams Univ. Fac. Agric. Res. Bull. **1686**, 1-12, 1981.

[59] Marwaha, S.S., Sethi, R.P. and Kennedy, J.F., Role of amino acids, betaine and choline in vitamin B_{12} biosynthesis by strains of *Propionibacterium*. Enzyme Microb. Technol. **5**, 454-456, 1983.

[60] Marwaha, S.S., Sethi, R.P., Kennedy, J.F., Influence of 5,6-dimethylbenzimidazole on vitamin B_{12} biosynthesis by strains of *Propionibacterium*. Enzyme. Microb. Technol. **5**, 361-364, 1983.

[61] Shteinberg, B.I., Gebgardt, A.G., Datsyuk, N.M., The effect of organic compounds on growth and cobalaminogenesis of *Achromobacter cobalamini*, Mikrobiologiya (Russ) **51**, 910-914, 1982.

[62] Kuznetsova, R.A., Datsyuk, N.M., Matviiko, S.G. and Franchuk, E.K., Iron effect on fermentation intensity and catalase formation by propionic acid bacteria. Mikrobiol. Zh. Kier. **45**, 3-7, 1983.

[63] Fa, Y-H., Kusel, J.P., Demain, A.L., Dependence of betaine stimulation of vitamin B_{12} overproduction on protein synthesis. Appl. Environ. Microbiol. **47**, 1067-1069, 1984.

[64] Kusel, J.P., Fa, Y.H. and Demain, A.L., Betaine stimulation of vitamin B_{12} biosynthesis in *Pseudomonas denitrificans* may be mediated by an increase in activity of δ-aminolevulinic acid synthase (EC 2.3.1.37). J. Gen. Microbiol. **130**, 835-842, 1984.

[65] Dumenil, G., Andriantsoa, M., Laget, M. and Cremieux, A., Study of some factors influencing growth and vitamin B_{12} production of a facultative methylotrophic *Corynebacterium*. Biotechnol. Lett. **5**, 813-818, 1983.

[66] Siefert, E. and Koppenhagen, V.B., Studies on the vitamin B_{12} auxotrophy of *Rhodocyclus purpureus* and two other vitamin B_{12}-requiring purple nonsulfur bacteria. Arch. Microbiol **132**, 173-178, 1982.

6.2 Occurrence of corrinoids in soil, water, sediments and sewage

6.2.1 Soil

Following the crystallization of vitamin B_{12} in 1948 and developement of sensitive microbiological methods of B_{12}-assay a great deal of work has been done to assay corrinoids in natural sources such as soil. The interest of vitamin B_{12} in soil reached its apogeum in course of fifties and in that period most of the publications on this subject originate [1-19]. It was early established that the vitamin B_{12} content in soil depends on the presence of organic matter [12, 40] and the number of microorganisms, however, other environmental factors affect its concentration as well (Table 6-2) [39]. Not all microorganisms occurring in soil produce corrinoids but many (if not all) require them for growth [1-3, 6, 8, 13, 18]. Burton and Lochhead [3] found that 64 per cent of bacterial strains and 66 per cent of *Actinomycetes* strains isolated from soils

Table 6-2 Vitamin B_{12} content in soil (data from Duda et al. [12])

Characteristic of soil					Number of microorganisms in thousands per 1 g of dry matter				Vitamin B_{12} content assayed with *E. gracilis* in nanogram (ng) per:		
Source	Depth cm	Dry weight %	Organic matter %	pH	Total	Actino-myces	Spores	Fungi	10^6 of bacteria	1 g of organic matter	1 g of dry soil
1. Meadow peaty soil	5–10	46.6	45.6	6.70	23.200	22.6	1.2	3	2.3	117.5	53.6
	25–30	40.1	27.6	6.70	14.900	1.8	0.8	2			11.9
	45–50	65.5	7.4	6.50	3.200	1.4	1.0	1			5.4
2. Podsol soil forest: oak, pine	5–10	93.8	3.1	4.55	1.600	17.6	2.7	49	1.0	52.9	1.7
	10–15	93.9	1.7	4.70	2.300	34.0	1.1	17			0.8
	30–35	93.1	2.3	4.80	1.200	2.5	1.8	22			0.6
	45–50	95.5	1.3	4.70	2.200	42.2	0.04	5			0.7
3. Podsol soil forest: beech trees	5–10	97.4	1.8	4.90	2.700	6.4	1.3	80	0.3	47.1	0.9
	20–25	95.5	1.2	5.35	1.00	20.4	2.7	18			0.5
	38–42	97.8	0.9	5.55	6.100	2.7	0.2	6			0.4
4. Podsol soil field: alfalfa	10–15	89.2	2.2	7.18	8.700	50.3	4.5	9	1.0	396.9	8.7
	34–37	92.5	2.1	7.00	7.900	18.9	0.8	0			0.5
5. Podsol soil field: oat	10–15	90.8	2.1	6.60	7.400	50.0	14.8	32	0.5	165.7	3.6
	34–37	95.6	1.6	6.25	4.100	24.4	5.5	8			0.9
6. Muck	10–15	43.4	28.2	8.10	10.200	12.2	16.8	0.7	5.2	188.7	53.2
	40–45	74.6	5.5	8.10	900	28.2	26.7	0.0			3.7
7. Muck park: maple and linden trees and	10–15	85.7	3.7	7.98	5.600	17.0	22.6	3	4.2	622.9	23.3
undergrowth	40–45	81.0	4.5	8.20	5.100	18.8	32.0	7			6.2
8. Podsol soil field: oat barley	10–15	95.5	2.7	5.93	12.600	31.5	18.0	53	0.24	112.2	3.0
	43–47	94.1	0.5	7.10	2.400	12.6	7.1	2			0.05
9. Podsol soil field: rye	10–15	96.1	1.9	7.15	16.500	31	9.9	22	0.29	258.8	4.8
	50–55	98.0	0.1	7.45	1.900	28.0	16.5	1			0.08
10. Podsol soil young suburban forest with	5–10	98.4	1.0	5.50	700	51.9	7.9	203	0.08	5.0	0.05
various tree species	30–35	97.1	1.1	4.91	100	11.6	18.1	24			0.03

and manures had the ability for biosynthesis of corrinoids. Similar results were obtained by other authors [8]. The acid forest soils, in which fungi dominate, contain little vitamin B_{12} despite high content of organic matter (see Table 6-2) [12]. Table 6-2 indicates that B_{12} content in soil varies with depth. The bacteria of the earth worms alimentary tracts produce appreciable quantities of corrinoids. The corrinoids are deposited in the animal remnants at various depth of the soil [12b]. The layers from 5-15 cm are the richest. As mentioned in section 6.1 microorganisms produce a variety of corrinoids of which only one, the cobalamin, is active in man and animals [20, 21]. The quantities of corrinoids estimated in soil depend very much on the microorganism used for assay. The least specific strain *E. coli 113-3* responds by growth to many corrinoids not active in man giving estimates up to 10 fold higher than the more specific for cobalamin strain *Euglena gracilis* [22, 23]. Since among corrinoids cobalamin was the main interest in the fifties the assays were done mainly with the latter strain giving less attention to assay of other corrinoids which are, in contrast to man and animals, growth factors to a great number of micoroorganisms in soil. The data in Table 6-2 are obtained with the *Euglena* test thus do not indicate the total content of corrinoids in soil. Other authors, also on the basis of *Euglena* test, found B_{12} content (per g of fresh soil) for garden-, greenhouse- and meadow soils in quantities 2-15 ng [24] and for marsh type soil 24 ng [21]. Corrinoids isolated from 700 kg of marsh type soil and separated by chromatographic means yielded 60.5 per cent of cabalamin (4.85 mg) and 39.5 per cent of cobinamide [21]. Another factor which greatly affects the amount of corrinoids in soil is the concentration of cobalt. This element is not evenly distributed in soil [32-38] and occurs there in the concentration range 1-40 ppm [32]. Since plants require cobalt as micronutrient they compete with bacteria for cobalt ions particularly in soil poor in this element [25-32, 35]. Due to very high affinity for cobalt ions, bacteria give little chance to roots in this competition [33]. The cobalt ions once trapped by the strong chelating corrin ring are not accessible to plants. The corrinoids are relatively resistant to chemical and microbial degradation and cobalt does not dissociate from the corrin ring. On the other hand, even the microorganisms which produce corrinoids take them up from the environment. It seems thus, that in cobalt poor soil there is a dynamic turnover of corrinoids (interfeeding) and there is a very small concentration of them outside the cells of microorganisms. The ecological aspects of corrinoids in soil have not been studied extensively.

References

[1] Lochhead, A. G., and Thexton, R. H., Vitamin B_{12} as a growth factor for soil bacteria, Nature **167,** 1034, 1951.

[2] Lochhead A. G., and Thexton, R. H., Qualitative studies of soil microorganisms X. Bacteria requiring vitamin B_{12} as growth factor, J. Bact. **63,** 219, 1952.

[3] Burton, M. O., Lochhead, A. G., Studies on the production of B_{12}-active substances by microorganisms, Canad, J. Bot. **29,** 352, 1951.

[4] Burton, M. O., Lochhead, A. G., Production of vitamin B_{12} by *Rhizobium sp.,* Canad. J. Bot. **30,** 521, 1952.

[5] Lochhead, A. G., and Burton, M. O., Incidence in soil of bacteria requiring vitamin B_{12} and the terregens factor, Soil. Sci. **82,** 237, 1956.

[6] Lochhead, A. G., Burton, M. O., Soil as a habitat of vitamin-requiring bacteria, Nature **178,** 144, 1956.

[7] Vyas, S. R., Prasad, N., Production of vitamin B_{12} by *Rhizobia* in Gujarat soil, J. Sci. Ind. Res. (B-C) **15,** 12, 1956.

[8] Janicki, J., Pawełkiewicz, J., Stawicki, S., Zodrow, K., Production of vitamin B_{12} by *Streptomyces* strains, Acta Microbiol. Polon. **3,** 1954.

[9] Kaszubiak, H., Malinska, E., Pędziwilk, Z., Studies on the production and structure of cobalamins in *Rhizobium meliloti,* Acta Microbiol. Polon. **6,** 239-251, 1957.

[10] Lochhead, A.G., and Burton, M., Qualitative studies of soil microorganisms. XII Characteristics of vitamin B_{12}-requiring bacteria, Canad. J. Microbiol. **1,** 319, 1955.

[11] Krasilnikow, H.A., Mikroorganizmy i plodorodije pochvy (Russian), Izvestija, Akad. Sci. USSR **2,** 14, 1954.

[12a] Duda, J., Malinska, E., Pędziwilk, Z., Relationship between the content of vitamin B_{12} and the number of microorganisms in soil, Acta Microbiol. Polon. **6,** 355-365, 1957.

[12b] Atlavinyte, O., Galvelis, A., Dacinlyte, J. and Luganskas, A., Effects of entobacterin *Bacillus thuringiensis* var. galleriae on earthworm activity, Pedobiologia **23,** 372-379, 1982.

[13] Goldberg, M.K., Hutner, S.H., Ford, J.E., Nutrition of a cobalamin-requiring soil bacterium, Can. J. Microbiol. **3,** 329, 1957.

[14] Lochhead, G.A., and Burton M.O., Qualitative studies of soil microorganisms. XIV. specific vitamin requirements of the predominant bacterial flora, Can. J. Microbiol. **3,** 35, 1957.

[15] Malinska, E., Pędziwilk, Z., Studies on synthesis and composition of cobalamins in *Rhizobium,* Acta Microbiol. Polon. **7,** 125-130, 1958.

[16] Sanders, F., Seaman, G.R., Metabolism of cobalamins by soil microorganisms, J. Bact. **79,** 619, 1960.

[17] Lochhead, A.G., Soil bacteria and growth-promoting substances, Bact. Rev. **22,** 145, 1958.

[18] Ford, J.E., Hutner, S.H., On the nature of the vitamin B_{12} requirement in soil bacteria isolated by Lochhead and his co-workers, Can. J. Microbiol. **3,** 319, 1957.

[19] Levin, A.P., Funk, M.B., Tendler, M.D., Vitamin B_{12}, *Rhizobia* and leguminous plant, Science **120,** 784, 1954.

[20] Holdsworth, E.S., Differentiation of vitamin B_{12} active compounds by ionophoresis and microbiological assay, Nature **171,** 148, 1953.

[21] Zodrow, K., Stefaniak, O., Chełkowski, J., Isolation of corrinoids from soil, Acta Microbiol. Polon. **11,** 341-348, 1962.

[22] Heinrich, H.C., Lahann, H., Die mikrobiologische Bestimmung der B_{12}-Vitamine mit der Grünalge *Euglena gracilis var. bacillaris,* Z. Naturforschung **8b,** 549, 1953.

[23] Ford, J.E., The microbiological assay of vitamin B_{12}. The specificity of the requirement of *Ochromonas malhamensis* for cyanocobalamin, Brit. J. Nutrit. **7,** 299, 1953.

[24] Robbins, W.J., Hervey, A., and Stebbins, M.E., Studies on *Euglena* and vitamin B_{12}, Bull. Torrey Bot. Club **77,** 423, 1950.

[25] Vogt, P., Jaakkola, A., The effect of mineral elements added to finnish soils on the mineral contents of cereal, potato and hay crops. Aluminium, boron, molybdenum, strontium, chromium, cobalt, lead, nickel. Acta Agr. Scandin. supp. **20,** 69-79, 1978.

[26] Horak, O., Determination of molybdenum and cobalt in plants, soils and forages with flameless atomic absorption-spectrometry, Bodenkultur **28,** 405-414, 1977.

[27] Adriano, D.C., Delaney, M., Paine, D., Availability of cobalt-60 to corn and bean seedlings as influenced by soil type lime and DTPA, Comm. Soil Sci. and Plant Anal. **8,** 615-628, 1977.

[28] Potatueva, Yu.A., Yanchuk, I.A., Agrochemical significance of cobalt, Khimya Selskom Khozyaistve **18,** 15-20, 1980.

[29] Krunic, D., Pantoric, M., Popovic, Z., The effect of trace elements on the yield and quality of sugar beet on chrnozem, brown forest, black swamp and allurium soils, Agrohemija **9/10,** 345-383, 1980.

[30] Jaakkola, A., Vogt, P., The effect of mineral elements added to finnish soils on the mineral contents of cereal, potato, and hay crops. Calcium, magnesium phosphorus, potassium, copper, iron, manganese, sodium and zinc. Acta Agriculturae Scan., supp. **20,** 53-68, 1978.

[31] Sorteberg, A., Effects of some heavy metals on oats in pot experiments with three different soil types, J. Sci. Agr. **50,** 317-334, 1978.

[32] Mengel, K., and Kirkby, E.A., Principles of plant nutrition, International Potash Institute, Worblaufen-Bern, Switzerland, 1980 (see also references 50, 52 in section 6.3).

[33] Kliewer, M., Evans, H.J., Cobamide coenzyme contents of soy bean nodules and nitrogen fixing bacteria in relation to physiological conditions, Plant. Physiol. **38,** 99-104, 1963.

[34] Iswaran, V., Sundara-Rao, W.V., Mathur, S.P., Role of vitamin B_{12} in nitrogen fixation by *Azotobacter chroococum,* Current Sci. **29,** 2-63, 1960.

[35] Kubota, J. and Allaway, W.H., Geographic distribution of trace element problems, in Micronutrients in agriculture, eds. J.J. Mortvedt, P.M. Giordano and W.L. Lindsay, Soil Sci. Soc. America, Madison/USA 1972, pp. 525-554.
[36] McKenzie, R.M., The mineralogy and chemistry of soil cobalt. In: Trace Elements in Soil-Plant-Animal Systems. Eds. D.J.D. Nicholas and A.R. Egan, Academic Press, New York 1975, pp. 83-93.
[37] Mitchell, R.L., Cobalt in soil and its uptake by plants, Agrochicima, **16**, 521-532, 1972.
[38] See also references: 50, 52-54 in section 6.3.2.
[39] Baya, A.M., Boethling R.S. and Cormenzana, R., Vitamin production in relation to phosphate solubilization by soil bacteria, Soil Biol. Biochem. **13**, 527-532, 1981.
[40] Avakyan Z.G. and Afrikyan E.K., Vitamin B_{12} in soils (in Russian), Biol. Zh. Arm. **34**, 253-258, 1981.

6.2.2 Water

Vitamin B_{12} occurs in sea- [1-16, 38, 39] and freshwater [17-21] in concentrations up to 30 ngl^{-1} as measured by a number of investigators with various microbiological methods (Table 6-3 A). This level is about a thousand times lower than that in soil. Because of the large volume ocean water seems to be the greatest B_{12}-pool in nature. The soluble form of the vitamin is secreted into water by B_{12}-producing algae [22, 23] and bacteria as well as being released from dying cells and organic particles of the sediments. Besides processes leading to input of the vitamin, dynamic opposite processes occur such as uptake by auxotrophic organisms [7-28], photo-chemical decomposition by solar radiation [37] and biological degradation. Thus the concentration of vitamin B_{12} measured in water samples represents the balance between input and output rates. The intensity of these processes are affected by various environmental factors - mainly temperature (season), nutrients and the species dominating in the plankton. Vitamin B_{12} is one of the growth factors for a number of algal and bacterial species. Its high concentration in water under favourable conditions causes rapid growth (bloom) of auxotrophic algae (auxotrophic with respect to vitamin B_{12}). The dynamic growth of auxotrophs is accompanied by a drop in vitamin B_{12} in water to levels often undetectable by the sensitive microbiological tests (below 0.1 $ng \cdot l^{-1}$). Usually the bloom of auxotrophic species is followed by the development of algae which produce and secrete vitamin B_{12}. On the basis of data so far available one may observe trends for periodic variations characterized by high concentration of the vitamin in winter, very low in spring, followed by a moderate level in summer (see Table 6-3). The level of corrinoids in water can be determined by a number of bacterial and algal [8, 14, 15, 24, 29] test organisms and also by a B_{12}-dependent fungus [11]. The values obtained with the use of various organisms can differ because of different growth response to vitamin B_{12} derivatives [13] which always accompany the true vitamin B_{12}. For proper evalutation of the analytic data presented in the literature the specificity of the test organism has to be considered (see section 5.3.1). Prior to assay the water samples are deprived of particulate matter by centrifugation or filtration. On the basis of this treatment it is assumed that free vitamin B_{12} is present in the sample only. This may not be true, since water may contain appreciable amounts of soluble proteins, not removable by this treatment, which firmly bind the vitamin and make it inaccessible to the test organisms [27, 30-34]. These proteins, secreted by algae, resemble the animal B_{12}-binding proteins: intrinsic factor and cobalophylin (see

Table 6-3 A Occurrence of vitamin B_{12} in natural waters

Source	Depth	Season	Vitamin B_{12} Content $ng \cdot l^{-1}$		Dominating Organisms in Water	Test Organism	Authors
			water	particulate matter			
1. Lake Washington		winter 1976	8–10			*Ochromonas malhamensis*	Parker [17]
		summer 1976	4		*blue-green algae Oscillatoria spp.*		
2. Lake Sagami, Japan		summer 1970	12		*Anacystis cyane ea*	*L. Leichmannii*	Ohwada Taga [21 a]
3. Lake Tsukui, Japan	upper layer	winter	4			*L. Leichmannii*	Ohwada et al. [21 b]
		March–May	0		*Cyclotella sp.*		
		August	7		*Anacystis cyanea*		
		summer 1971	0	10–20	*Fragilaria crotonensis Synedra acus*		
4. English Lake District			4–16			*Euglena gracilis*	Daisley [19]
5. Lake Nantua, France	upper 5 m	February 1971	26			*E. gracilis*	Pommel [20]
		May 1971	3		*Oscillatoria rubescens*		
		September	28		*Aphanizomenon flos-aquae Oscillatoria redekei*		
6. Lake Leman, Geneve			1–6		diatoms and flagellates	*E. gracilis*	Pommel [20]
7. Pacific Ocean, north-east	upper 60 m		0.1				Carlucci [1]
Subarctic Pacific			3–4				Natarajan [3]
9. Sagami Bay, Japan coast	upper 20 m	spring	0.6				Ohwada and Taga [4]
		summer	4.8		Seasonal peak of *chlorophyla*		
		fall	1.7				
		winter	0.6				
	20–200 m	winter, spring	1				
		summer, fall	1.8				

Table 6-3 A continued

Source	Depth	Season	Vitamin B_{12} Content $ng \cdot l^{-1}$		Dominating Organisms in Water	Test Organism	Authors
			water	particulate matter			
10. Sagami Bay, Aburatsubo Inlet	upper layer	winter	3.1				Ohwada and Taga [4]
		spring	1.1				
		summer	2				
		fall	2.7				
11. Barents Sea, Dalniye Zelentsy Inlet		winter	2			*E. coli*	Propp [12]
		early spring	0		*Skeletonema costatum*		
		late spring	0		*Leptocylindricus danicus*		
12. South Polar Seas			0–3.3				Carlucci and Cuhel [5]
13. North Sea and northern part of Atlantic			0.5–1				Cowey [7]
14. Sargasso Sea	upper 100 m		0.0–0.1				Menzel and Spaeth [10]
	1000–2600 m		0.08–0.2				
15. Long Island Sound		January	16			*Thraustochytrium globosum* (fungus)	Vishniac and Riley [11]
		February-April	4–5		*Skeletonema costatum*		
		August	12.5				
16. Strait of Georgia, British Columbia		winter	8–12			*Amphidinium cartare*	Cattel [9]
		summer	0.0				
17. Gulf of Mainc	20 m	winter	1.6			*Thalassiosira pseudonana 3H*	Swift [13]
	surface	winter, spring	0.9–0.1		*Thalassiosira nordenskioeldii*		
	surface	summer	0.3–0.8				
	over 100 m	summer	1.5				
18. Napeaque Bay	6–16 m	winter, spring	1–2		*Diatom S. costatum*		Bruno and Staker [6]
	6–16 m	summer	2–7				

Chapter 8). Additional treatment seems to be required such as heating the sample in the presence of cyanide in order to liberate the protein bound vitamin. Isotopic methods (see section 5.3.2) were also applied for the assay of B_{12} in ocean waters [35, 36]. They give results 4–10 times higher [35] than specific bioassays, presumably due to assay by this technique a wider spectrum of B_{12} derivatives.

References

[1] Carlucci, A. F., The ecology of the plankton of La Jolla, California, in the period April through September 1967. II. Vitamin B_{12}, thiamin and biotin, Bull. Scripps. Inst. Oceanogr. **17,** 23–31, 1970.

[2] Carlucci A. F. and Silbernagel S. B., Bioassay of seawater. III. Distribution of vitamin B_{12} in the north-east Pacific Ocean, Limnol, Oceanogr. **11,** 642–646, 1966.

[3] Natarajan, K. V., Distribution and significance of vitamin B_{12} and thiamin in the subarctic Pacific Ocean, Limnol, Oceanogr. **15,** 655–659, 1970.

[4] Ohwada, K. and Taga N., Distribution and seasonal variation of vitamin B_{12}, thiamin and biotin in the sea, Mar. Chem. **1,** 61–73, 1972.

[5] Carlucci, A. F. and Cuhel R. L., Vitamins in the South Polar Seas: distribution and significance of dissolved and particulate vitamin B_{12}, thiamin and biotin in the Southern Indian Ocean, in: Adaptations within Antarctic Ecosystems, Proceedings of Third SCAR Symposium on Antarctic Biology 1974, Ed. G. A. Llano, Gulf Printing Co., Houston 1977, pp. 115–132.

[6] Bruno, S. F., and Staker, R. D., Seasonal vitamin B_{12} and phytoplankton distribution near Napeaque Bay, New York (Block Island Sound), Limnol. Oceanogr. **23,** 1045–1050, 1978.

[7] Cowey, C. B., A preliminary investigation of the variation of vitamin B_{12} in oceanic and coastal water, J. Mar. Biol. Ass. U. K. **35,** 609–620, 1956.

[8] Cattell, S. A., Dinoflagellates and vitamin B_{12} in the Strait of Georgia, British Columbia Ph. D. thesis Univ. of British Columbia Vancouver British Columbia, 1969.

[9] Cattell, S. A., The seasonal cycle of vitamin B_{12} in the Strait of Georgia, British Columbia, J. Fish. Res. Bd. Can. **30,** 215–222, 1973.

[10] Menzel, D. W. and Spaeth, J. P., Occurrence of vitamin B_{12} in the Sargasso Sea, Limnol. Oceanogr. **7,** 151–154, 1962.

[11] Vishniac, H. S. and Riley G. A., Cobalamin and thiamin in the Long Island Sound: patterns of distribution and ecological significance, Limnol. Oceanogr. **6,** 36–41, 1961.

[12] Propp, L. N., Seasonal dynamics of vitamin B_{12} and phytoplankton variability in the Dalnezelenetskaya Guba (Inlet) of the Barents Sea. Oceanology **10,** 676–681, 1970.

[13 a] Swirft, D. G., Vitamin B_{12} as an ecological factor for centric diatoms in the Gulf of Maine, Ph. D. thesis, Johns Hopkins Univ., Baltimore 1973.

[13 b] Swift, D. G., Vitamins and phytoplankton growth, in The Physiological Ecology of Phytoplankton, Ed. I. Morris, Blackwell Scientific Publications, Oxford 1980, pp. 329–368

[13 c] Swift, D. G. and Guillard, R. R. L., Diatoms as tools for assay of total B_{12}-activity and cyano cobalamin activity in sea water, J. Mar. Res. **35,** 309–320, 1977.

[14] Nishiijima, T. and Yoshiniko, M., Seasonal variation in the concentration ov dissolved vitamin B_{12}, thiamine and biotin in Susaki Harbor and Nomi Inlet Japan, Rep. Usa. Mar. Biol. Inst. Kachi Univ. **01,** 1–8, 1979.

[15] Kabanova, Y. G. and Domanov, M. M., Influence of nutrients on the growth and functioning of phytoplankton in Cuban waters, Okeanologiya **19,** 692–698, 1979.

[16] Antia N. J., McAllister, C. D., Parsons, T. R., Stephens, V. and Strickland J. D. H., Further measurements of primary production using a large-volume plastic sphere, Limnol, Oceanogr. **8,** 166–183, 1963.

[17] Parker, M., Vitamin B_{12} in Lake Washington, USA: concentration and rate of uptake, Limnol. Oceanogr. **22,** 527–538, 1977.

[18] Ohwada, K. and Taga, N., Vitamin B_{12}, thiamin and biotin in Lake Sagami, Limnol. Oceanogr. **17,** 315–320, 1972.

[19] Daisley, K. W., Monthly survey of vitamin B_{12} concentrations in some waters of the English Lake District, Limnol. Oceanogr. **14,** 224–228, 1969.

[20] Pommel, B., Distribution et signification ecologique de la vitamine B_{12} et de la thiamine dans trois lacs subalpine et jurassien, Ann. Hydrobiol. **6**, 103-121, 1975.

[21a] Ohwada, K. and Taga, N., Vitamin B_{12}, thiamin and biotin in Lake Sagami, Limnol. Oceanogr. **17**, 315-320, 1972.

[21b] Ohwada, K., Otsuhata, M. and Taga, N., Seasonal cycles of vitamin B_{12}, thiamin and biotin in the surface water of Lake Tsukui, Bull. Jap. Soc. Sci. Fish. **38**, 817-823, 1972.

[22a] Aronson, S., DeAngelis B., Frank, O. and Baker, M., Secretion of vitamins and aminoacids into the environment by *Ochromonas danica*, J. Phycol. **7**, 215-218, 1971.

[22b] Carlucci, A.F. and Bowes P.M., Production of vitamin B_{12}, thiamin and biotin by phytoplankton, J. Phycol. **6**, 351-357, 1970.

[22c] Carlucci, A.F. and Bowes, P.M., Vitamin production and utilization by phytoplankton in mixed culture, J. Phycol. **6**, 393-400, 1970.

[22d] Carlucci, A.F. and Bowes, P.M., Vitamin B_{12}, thiamin and biotin content of marine phytoplankton, J. Phycol. **8**, 133-137, 1972.

[23a] Berland, B.R., Bonin, D.J., Durbec, J.-P. and Maestrini, S.Y., Bactéries hétérotrophes aerobies prélevées devant le delta du Rhone. IV. Besoins en vitamines et liberation de ces substances, Hydrobiologia **50**, 167-172, 1976.

[23b] Vacelet, E., E rôle des vitamines dans les relations bacteries plancton en eau de mer littorale et dans les flaques supralittorales. II. Evolution annuelle synthese des vitamines, Cah. Biol. Mar. **16**, 383-394, 1975.

[23c] Martin, J.L.Y., and Vacelet, Le rôle des vitamines dans les raltions bacteries-plancton en eau de mer littorale et dans les flaques supralittorales. III. Relation entre la quantite de vitamine B_{12} et les peuplements qui font la synthese, Cah. Biol. Mar. **16**, 511-519, 1975.

[24] Guillard, R.R.L., B_{12}-specificity of marine centric diatoms, J. Phycol. **4**, 59-64, 168.

[25] Varma, T.N.S., Abraham, A., and Hansen I.A., Accumulation of ^{58}Co-vitamin B_{12} by *Euglena gracilis*, J. Protozool. **8**, 212-216, 1961.

[26] Reeves R.B. and Fay, F.S., Cyanocobalamin (vitamin B_{12}) uptake by *Ochromonas malhamensis*, Am. J. Physiol **210**, 1273, 1966.

[27] Bradbeer, C., Transport of vitamin B_{12} in *Ochromonas malhamensis*, Arch. Bioch. Biophys. **144**, 184-192, 1971.

[28] Sarhan, F., Houde, M. and Cheneval, J.P., The role of vitamin B_{12} binding in the uptake of the vitamin by *Euglena gracilis*, J. Protozool. **27**, 235-238, 1980.

[29] Carlucci, A.F. and Silbernagel, S.B., Bioassay of seawater. I. A ^{14}C-uptake method for the determination of concentrations of vitamin B_{12} in seawater, Can. J. Microbiol. **12**, 175-183, 1966.

[30a] Droop, M.R., Vitamin B_{12} and marine ecology. IV. The kinetics of uptake, growth and inhibition in *Monochrysis lutheri*, J. Mar. Biol. Ass. U.K. **48**, 689-733, 1968.

[30b] Messina, D.S. and Baker, A.L., Interspecific growth regulation in species succession through vitamin B_{12} competitive inhibition, J. Plankton Res. **4**, 41-46, 1982.

[31] Pitner, I.J. and Almeyer, V.L., Vitamin B_{12}-binder and other algal inhibitors, J. Phycol. **15**, 391-398, 1979.

[32] Daislye, K.W., The occurrence and nature of *Euglena gracilis* proteins that bind vitamin B_{12}, Int. J. Biochem. **1**, 561-574, 1970.

[33] Pintner, I.J. and Altmyer V.L., Production of vitamin B_{12} binder by marine phytoplankton, J. Phycol. **9**, (suppl.), 13, 1973.

[34] Provasoli, L. and Carlucci, A.F., Vitamins and growth regulators, in Algal physiology and biochemistry, ed. W.D.P. Stewart, Blackwell Scientific Publications, Oxford 1974, pp. 741-787.

[35] Sharma, G.M. DuBois, H.R., Pastore, A.T. and Bruno, S.F., Comparison of the determination of cobalamins in ocean waters by radioisotope dilution and bioassay techniques, Anal. Chem. **51**, 196, 1979.

[36] Beck, R.A., Competetive intrinsic factor binding assay technique for cobalamins in natural waters, Anal. Chem. **50**, 200-202, 1978.

[37] Carlucci A.F., Silbernagel, S.B. and McNally, P.M., Influence of temperature and solar radiation on persistence of vitamin B_{12}, thiamine and biotin in seawater, J. Phycol. 5, 302-305, 1969.

[38] Kim, J.M., Cho, I.H., Park, Ch.U., Microbiological study on the preservation of marine environments. 1. Distribution of vitamin B_{12}, thiamin and biotin in the sea water of Kunsan Korca, Bull. Korean. Fish. Soc. **16**, 25-30, 1983.

[39] Fiala, M. and Oriol, L., Vitamin B_{12} and phytoplankton in the antarctic Ocean: Distribution and experimental approach, Mar. Biol. Berl. **79**, 325-332, 1984.

6.2.3 Sediments and sewage

Mud-type sediments [1, 2] contain a relatively high level of bound cobalamin and other corrinoids such as (Ade) Ado-Cba, and (5-OH Bza) Me-Cba. The corrinoids are produced mainly by bacteria engaged in anaerobic fermentation and methane formation. The methly-cobalamin and other methyl-cobamides have the capacity to methylate spontaneously heavy metal ions, in particular mercury ions. The methylated mercury appearing frequently in mercury polluted sediments is a strong poison, which affects the nervous system (see section 3.1.4.5). It may reach man via mercury polluted fish. Municipal sewage is one of the richest sources of corrinoids [3-5]. Sewage activated by addition of nutrients and by aeration contains the highest B_{12}-level of any other source (up to 90 mg per kg of sewage sludge). It also contains a considerable number of various corrinoids not occuring in other sources (see section 6.1 and Table 3-1). Since vitamin B_{12} is already critical for the growth of aquatic organisms at concentrations 10^7 times smaller, the release of municipal sewage into natural waters may affect drastically the balance of the aquatic ecosystem. The livestock waste is a particular rich source of corrinoids [6].

References

[1] Nishijima, T., and Hata, Y., Distribution of thiamine, biotin, and vitamin B_{12} in Lake Kojima. II. Distribution in the bottom sediments, Bull. Jpn. Soc. Sci. Fish **44,** 815-818, 1978.

[2] Burkholder, P.R. and Burkholder, L.M., Vitamin B_{12} in suspended solids and marsh muds collected along the coast of Georgia, Limnol. Oceanogr. **1,** 202-208, 1956.

[3] Beck, R.A. and Brink, J.J., Sensitive chemical method for routine assay of cobalamins in activated sewage sludge, Environ. Sci. Technol. **10,** 173-175, 1976.

[4] Beck, R.A. and Brink, J.J., Production of cobalamins during activated sewage sludge treatment, Environ. Sci. Technol. **12,** 435-438, 1978.

[5] For occurrence of corrinoids in sewage see also references: 6, 7, 10, 24, 25 in section 6.1.

[6] Oshida, T., Honjyo, T., Tanaka, K., The utilization of livestock waste 1. The content of vit B_{12} in fermented swine waste. Bull. Azaba. Univ. Vet. Med. **3,** 187-192, 1982.

6.3 Occurrence of vitamin B_{12} in algae and higher plants

6.3.1 Algae

The introduction of artificial media for growing pure cultures of algae (axenic culture) in the early fifties allowed the testing of these organisms for specific requirements. Studies with the use of axenic cultures revealed that vitamin B_{12} is one of the three vitamins, the other are thiamine and biotin, which stimulate or are essential for the growth of a number of algal species. As shown in Table 6-3 B, approximately 44% of the algae of various classes so far tested in axenic culture require the vitamin for growth [1-135]. The procariotic blue-green algae of the class *Cyanophyceae*[a], named also blue-green bacteria, seem to be mainly autotrophic for vitamin B_{12}, while the de-

[a] The blue-green bacteria are related to eubacteria and not to algae as has been revealed from studies of the primary structure of the 16 S ribosomal RNAs of these microorganisms. They are listed in this section for convenience, since they are referred in the literature together with algae.

Table 6-3 B Vitamin requirements of some algal species (data and the references [1-123] taken mainly from Provasoli and Carlucci [139])

Species**	B_{12}	Thiamine	Biotin*
CYANOPHYCEAE			
Agmenellum quadruplicatum (C) [115], *Anabaena cylindrica* [3], *A. gelatinosa* [15], *A. naviculoides* [15], *A. variabilis* [59], *Anacystis marina* [115], *A. nidulans* [59], *Aphanizomenon flos-aquae* [32], *Calothrix brevissima* [116], *C. parietina* [3], *C. scopulorum* [105], *Chroococcus turgidus* [3], *Coccochloris peniocystis* [32], *Gloeocapsa dimidiata* [32], *G. membranina* [119], *Lyngbya aestuarii* [3, 115], *Microcoleus chthonoplastes* [115], *M. tenerrimus* [115], *M. vaginatus fuscus* [6], *Microcystis (Diplocystis) aeruginosa* [32], *Nostoc entophytum* [105], *N. muscorum* [3, 59], *Oscillatoria brevis* [105], *O. lutea* [6], *O. rubescens* [121], *Phormidium foveolarum* [3], *P. luridum* [3], *P. tenue* [32], *Plectonema nostocorum* [32], *P. notatum* [3], *P. terebrans* [115], *Schizothrix calcicola* [6], *S. calcicola glomerata* [6], *S. calcicola mucosa* [6], *S. calcicola spiralis* [6], *S. calcicola vermiformis* [6], *Synechococcus cedrorum* [3].	0	0	0
Agmenellum quadruplicatum (C) [115], *Coccochloris elabens* (C) [115], *Lyngbya lagerheimii* [115], *Oscillatoria amphibia* [115], *O. lutea auxotrophica* [6], *O. lutea contorta* [6], *O. subtilissima* [115], *Phormidium persicinum* (C) [82], *Trichodesmium thiebautii* [131].	R	0	0
RHODOPHYCEAE			
Porphyridium purpureum (= cruentum) [18] *P. marinus* [127], *Rhodymenia palmata* [124]	0	0	0
Antithamnion sp. [111], *A. glanduliferum* [111], *A. spirographidis, (sarniense)* [111], *Asterocytis ramosa* [31], *Bangia fuscopurpurea* [92], *Erythrotrichia carnea* [30], *Goniotrichum alsidii (elegans)* [28], *Nemalion helmithoides (multifidum)* [29], *Polysiphonia urceolata* [30], *Porphyra tenera* (*Conchocelis* phase) [50], *Rhodella maculata* [114]	R	0	0
Enteromorpha linza [127], *Trailliella intricata* [127], *Grinnelia amerikana* [132]	R	–	–
BACILLARIOPHYCEAE			
Amphora coffaeiformis [64], *Asterionella formosa* [91], *Detonula confervacea* [36], *Fragilaria capucina* [91], *F.* sp. *(pinnata?)* [40], *Melosira* sp. *(nummuloides)* [38, 36], *Navicula corymbosa* [63, 64], *N. incerta* [64], *N. menisculus* [64], *N. pelliculosa* [60], *Nitzschia angularis affinis* [64], *N. curvilineata* [64], *N. filiformis* [64], *N. frustulum* [64], *N. hybridaeformis* [64], *N. laevis* [64], *N. lanceolata* [63, 64], *N. marginata* [64], *N. obtusa scalpelliformis* [64], *Phaeodactylum tricornutum* [46], *Stauroneis amphoroides* [64], *Tabellaria flocculosa* [91], *Thalassiosira nordenskioldii* [36].	0	0	0
Chaetocenis spp [126], *Thalassiosira rotula* [126], *T. aestivalis* [126], *Porosira glacialis* [126]	0	–	–
Achnanthes brevipes (M) [64], *Amphora coffaeiformis* [64] = *A. perpusilla* [48], *A. lineolata* [64], *Bellerochea* sp. *(polymorpha?)* [40], *B.* sp. *(spinifera?)* [40], Chaetoceros ceratosporum (M) [37], *C. gracilis* [13, 113], *C. lorenzianus* (M) [36], *C. pelagicus* (L) [36], *C. pseudocrinitus* (M) [37], *Coscinodiscus asteromphalus* (L) [36], *Cyclotella* sp. *(cryptica)* [64], *C.* sp. [64] *Cylindrotheca closterium (= Nitzschia closterium)* [37], *C. pseudonana (= Cyclotella nana)* (C) [39], *Ditylum brightwellii* (C) [37], *Fragilaria* sp. (L) [36], *F.* sp. *(pinnata?)* (M) [40], *F.* sp. *(rotundissima?)* [40], *Licmophora hyalina* [68], *Melosira* sp. [36], *M. juergensii* [108], *Nitzschia frustulum* (C) [64], *N. ovalis* (L) [64], *N. punctata* (C)			

Table 6-3 B continued

Species**	B_{12}	Thiamine	Biotin*
[64], *N. seriata* [36], *Opephora* sp. [64], *Rhizosolenia setigera* (M) [36], *Stephanopyxis turris* (C) [91], *Synedra affinis* [64] *(= Fragilaria brevistriata)* [68], *Synedra* sp. *(rotundissima?)* [40], *Thalassiosira fluviatalis* [36], *T. nordenskioldii* [36], *Skeletonema costatum* (C) [17, 130], *Nitzschia subpacifica* [133]	R	0	0
N. cuspitata [133], *Chaetoceros simplex* (L) [139]	R	–	–
Amphora coffaeiformis (C) [64], *A. paludosa duplex* [64], *Bellerochea* sp. *(polymorpha?)* [40], *Cylindrotheca fusiformis (= Nitzschia closterium)* [62, 64], *Navicula pelliculosa* [40].	0	R	0
Amphipleura rutilans [64, 68], Amphora coffaeiformis [64], *Bellerochea* sp. *(polymorpha?)* [40], *B.* sp. *(spinifera?)* [40], *Cylindrotheca fusiformis (= Nitzschia closterium)* [64, 37], *Cyclotella caspia* (C) [36, 37], *Nitzschia alba* [65], *N. leucosigma* [65], *N. putrida* [65].	R	R	0
Skeletonema turris [135]	R	–	–
CRYPTOPHYCEAE			
Cryptomonas ovata (M) [96], *Cyanophora paradoxa* (M) [96]	R	0	0
Chilomonas paramecium [70], *Rhodomonas lens* (C) [91]	0	R	0
Chroomonas salina [5], *Hemiselmis virescens* (M) [19],	R	R	0
Rhodomonas ovalis [55]	R	R	R
Chroomonas marina (M) [139]	R	–	–
CHRYSOPHYCEAE			
Stichochrysis immobilis [91]	0	0	0
Microglena arenicola (M) [19], *Monochrysis lutheri* (L) [19], *Ochromonas minuta* [85], *Pavlova gyrans* [84], *Syncrypta (Synochromonas) korschikoffii* [87]			
Synura sphagnicola [85]	R	R	0
Ochromonas danica [43, 86]	0	R	R
Synura caroliniana (M) [85], *S. petersenii* (M) [85]	R	0	R
Ochromonas malhamensis (M) [49, 86], *Poteriochromonas stipitata* [85]	R	R	R
Poterio-ochromonas malhamensis [125]	0	–	–
HAPTOPHYCEAE			
Coccolithus huxleyi [36], *Hymenomonas carterae* [96]	R	0	0
Apistonema aestuari (= Pontosphaera roscoffensis) [85], *Coccolithus* sp. [85], *C. huxleyi* [36, 84], *Dicrateria inornata* [85], *Hymenomonas* sp. No. 156 [84], *Ochrosphaera neapolitana* [84], *Pleurochrysis scherffelii* [91], *Syracosphaera* sp. No. 181 [84]	0	R	0
Chrysochromulina kappa [85], *C. brevefilum* [85], *C. strobilus* [85], *Cricosphaera (Syracosphaera) elongata* (M) [17], Isochrysis galbana (M) [91] *Prymnesium parvum* (M) [19]	R	R	0
DINOPHYCEAE			
Peridinium cinctum ovoplanum [12], *Symbiodinum microadriaticum* [72, 73]	0	0	0
Entomosigma sp. *(Heterosigma akashivo)* (M) [54], *Exuviaella* sp. (L) [53], *E. cassubica* (L) [94], *Glenodinium foliaceum* [19], *G. halli* [34], *G. monotis* [81], *Gonyaulax polyedra* [42, 112], *Gymnodinium splendens* [107], *Gyrodinium californicum* [95], G. resplendens (M) [94, 95], *G. uncatenum* (M) [94, 95], *Heterosigma inlandica* (M) [56], *Peridinium balticum* (L) [94], *P. chattoni* [94], *P. hangoei* (C) [51], *P. trochoideum* [19], *Woloszynskia limnetica* (= *Peridinium* sp.) (M) [91].	R	0	0
Crypthecodinium (Gyrodinium) cohnii [93].	0	0	R

Table 6-3 B continued

Species**	B_{12}	Thiamine	Biotin*
Cachonina niei [69].	R	R	0
Prorocentrum micans [18, 57].	R	0	R
Amphidinium carteri (Klebsi) (L) [71], *A. rhynchocephalum?* (L) [71], *Gymnodinium breve* [2, 120], *Oxyrrhis marina* [19, 20], Polykrikos schwartzii (C) [52].	R	R	R
Gymnodinium nelsoni (C) [51-56], *Peridiniopsis polonicum* [134]	R	–	–
PHAEOPHYCEAE			
Desmarestia ligulata [76], *D. viridis* [76], *Petalonia zoostericola* [110], *Scytosiphon lomentaria* [110], *Waerniella lucifuga* [18, 22].	0	0	0
Ectocarpus confervoides [9], *E. fasciculatus* [80], *E. parasitica* [18], *Litosiphon pusillus* [80].	0 to stimulatory	0	0
Pylaiella (Pilayella) littoralis [80, 12, 129]	R	0	0
XANTHOPHYCEAE AND EUSTIGMATOPHYCEAE			
Botrydiopsis intercedens [8], *Bamilleriopsis brevis* [8], *Chlorellidium tetrabotrys* [8], *Monodus subterraneus* [74, 75], *Polyedriella helvetica* [8], *Tribonema aequale* [7], *T. minus* [8].	0	0	0
Vaucheria dichotoma [127]			
EUGLENOPHYCEAE			
Euglena anabaena minor [1].	0	0	0
Trachelomonas abrupta (L) [97], *T. pertii* (L) [91], *Euglena spirogyra* [122]	R	0	0
Euglena pisciformis [24].	0	R	0
Astasia klebsii (= A. longa, = A. chattoni) [1, 91], *Euglena gracilis bacillaris* (L) [1], *E. gracilis typica* [1, 91], *E. gracilis urophora* [1, 91], E. klebsii [1], *E. mutabilis* [68], *E. stellata* [91], *E. viridis* (L) [91], *Peranema trichophorum* [106], Phacus pyrum (M) [91].	R	R	0
Eutreptiella sp. (C) [53].	R	R	R
CHLOROPHYCEAE, CHAROPHYCEAE AND PRASINOPHYCEAE			
Astrephomene gubernacilifera [104], *Bracteacoccus cinnabarinus* [79], *B. engadiensis* [79], *B. minor* [79], *B. terrestris* [79], *Chara aspera* [27], *C. globularis* [27], *C. zelanica* [27], *Chlamydomonas agloeformis* [1, 91]. *C. eugametos* [117, 118], *C. moewusii* [1, 47, 91], *C. pichinchae* [44], *C. reinhardtii* [68, 91], *Chlorella autotrophica* [102], *C. candida* [102], *C. ellipsoidea* [102], *C. emersonii* [102], *C. emersonii globosa* [102], *C. fusca* [102], *C. fusca vacuolata* [102], *C. infusionum* [102], *C. infusionum auxenophila* [102], *C. miniata* [102], *C. mutabilis* [102], *C. nocturna* [102], *C. photophila* [102], *C. pringsheimii* [102], *C. pyrenoidosa* (Emerson) [100], *C. regularis* [102], *C. regularis aprica* [102], *C. regularis imbricata* [102], *C. saccharophila* [102], *C. simplex* [102], *C. sorokiniana* [102], *C. vannielii* [102], *C. variabilis* [102], *C. vulgaris* [91, 102], *C. vulgaris luteoviridis* [102], *Chlorococcum aplanosporum* [79], *C. diplobionticum* [79], *C. echinozygotum* [79], *C. ellipsoideum* [79], *C. hypnosporum* [79], *C. macrostigmatum* [79], *C. minitum* [79], *C. multinucleatum* [79], *C. oleofaciens* [79], *C. perforatum* [79], *C. pinguideum* [79], *C. punctatum* [79], *C. scabellum* [79], *C. tetrasporum* [79], *C. vacuolatum* [79], *C. wimmeri* [79], *Chlorosarcinopsis auxotrophica* [35], *C. eremi* [35], *Cosmarium botrytis* [109], *C. impressulum* [109], *C. laeve* [109], *C. lundelli* [109], *C. meneghini* [109], *Cyanidium caldarium* [3, 4], *Cylindrocystis brebissonnii* [44], *Dictyochloris fragrans* [79], *Dunaliella primolecta* [91], *D. salina* [33, 91], *D. viridis* [33], *Eudorina elegans* [25],			

Table 6-3 B continued

Species**	B_{12}	Thiamine	Biotin*
Haematococcus pluvialis [91, 16], *Lobomonas pyriformis* [77], *Micrasterias cruxmelitensis* [109], *Nannochloris atomus* [91], *N. oculata* [19], *Nautococcus pyriformis* [79], *Neochloris alveolaris* [79], *N. aquatica* [79], *N. gelatinosa* [79], *N. minuta* [79], *N. pseudoalveolaris* [79], *Pandorina morum* [78], *P. unicocca* [98], *Polytoma obtusum* [1, 91], *P. uvella* [1, 91], *Radiosphaera dissecta* [79], *Scenedesmus obliquus* [91], *Selenastrum minutum* [66], *Spongiochloris excentrica* [79], *S. lamellata* [79], *S. spongiosis* [79], *Spongiococcum alabamense* [79], *S. excentricum* [79], *S. multinucleatum* [79], *S. tetrasporum* [79], *Staurastrum gladiosum* [109], *S. paradoxum* [109], *S. sebaldii ornatum* [109], *Staurodesmus pachyrhynchus* [109], *Stephanoptera gracilis* [33], *Stichococcus cylindricus* [91], *Stigeoclonium aestivale* [14], *S. helveticum* [14], *S. subsecundum* [14], *S. tenue* [14], *S. pascheri* [14], *S. variabile* [14], *S. farctum* [14].	0	0	0
Ulothrix subtilissima [128]	0	–	–
Balticola buetschlii (M) [21], *B. droebakensis* (M) [21], *Chlamydomonas chlamydogama* [47], *C. mundana* [26], *C. mundana astigmata* [26], *C. pallens* [88], *C. pulsatilla* (M) [21], *Chlorosarcinopsis sempervirens* [35], *Chlorosphaera consociata* [68], *Closterium braunii* Z 951/2 a [109], *C. macilentum* [109], *C. peracerosum* [109], *C. pusillum* [109], *C. siliqua* [109], *C. strigosum* [109], *C. turgidum* [109], *Cylindrocystis brebissonnii* [109], *Cosmarium turpini* [58], *Desmidium swartzii* [109], *Docidium manubrium* [109], *Gonium pectorale* [104], *Haematococcus droebakensis* [16], *H. buetschlii* [16], *H. pluvialis* [89], *Hyalotheca dissiliens* [109], *Lobomonas rostrata* [66], *L. sphaerica* [123], *Nannochloris* sp. [13, 113], *Platydorina caudata* [41], *Platymonas tetrathele* (M) [21], *Pleurotaenium trabecula* [109], *Stichococcus* sp. *(cylindricus?)* [67], *Volvox aureus* [90], *V. globator* [83], *V. tertius* (M) [83, 90], *Volvulina steinii* [11], *V. pringsheimii* [11], *Xanthidium cristatum* [109], *Haematococcus capensis typ* [89], *H. capensis borealis* [89]	R	0	0
Chlorella protothecoides [102], *C. protothecoides communis* [102], *C. protothecoides galactophila* [102], *C. protothecoides mannophila* [102], *Coelastrum morus* [68, 66], *Gloeocystis gigas* [68], *Gonium multicoccum* [99], *Polytoma caudatum* [70], *P. ocellatum* [70], *Polytomella caeca* [70], *Prototheca zopfii* [1, 91], *Raphidonema nivale* [44], *Selenastrum minutum* [68].	0	R	0
Acetabularia mediterranea [101], *Astrephomene gubernaculifera* [10], *Brachiomonas submarina* (M) [21], *Eremosphaera viridis* [103], *H. zimbabwiensis* [89], *Pyramimonas inconstans* (M) [91], *Stephanosphaera pluvialis* (M) [21], *Volvox globator* (M) [90].	R	R	0

* R = required, 0 = not required;

** The lettering C, L. or M placed in brackets behind the name of the species characterize its growth specificity toward B_{12}-analogues as represented by test organisms: C, for *E. coli* type of specificity; L, *Lactobaccillus* type and M, mammalian type (tested by *O. malhamensis*).

pendence on external vitamin is more frequent among eucariotic species. There is little evidence, so far, that all of the diverse algal species defined as autotrophs utilize vitamin B_{12} as a cofactor in enzyme systems; some of the algae might have followed the evolutionary path of higher plants by developing alternative metabolic pathways not requiring vitamin B_{12}. As described in Chapter 4 the biosynthesis of cobamides in bacteria is a complex process in which approximately 40 enzymes are involved. The various precursors are synthetized in independent paths. Two paths in particular, may be distinguished: the synthesis of the corrinoid-containing part (cobinamide also named Factor B), and the synthesis of the α-5,6-dimethylbenzimidazole nucleotide or its adenine counterpart. These two parts are joined together, in a multistep conversion, to form the biologically active cobalamin or other cobamides. Some bacteria and also algae have the capacity to excise the nucleotide (or nucleoside) moiety of the cobamide and introduce another one. The interconversion of cobamides already observed in bacteria thirty years ago and in algae by Neujahr and Fries [136, 137] in the early sixties, are not fully understood on the biochemical level. Publications on B_{12}-biosynthesis in algae are scarce. It is generally assumed that the synthesis mainly follows the bacterial path. The regulation of the synthesis of such a complex compound in nature seems to present a difficult task. In algae [136] and bacteria [141] appreciable quantities of incomplete forms (or intermediates) of corrinoids – such as cobinamide – are found, which apparently is an indication that the biosynthesis does not run smoothly [136–141]. This multistep process is also easily disturbed by blocking through mutations of any of the numerous enzymes. These features of the B_{12}-biosynthetic path might explain the high incidence of algal mutants. Most auxotrophs, if not all, may be assumed to be such. There is, however, one enzyme system in the biosynthetic path which must not be disturbed even in auxotrophs. It is the conversion of cobamides into their coenzymatic form. Algal mutants, depending on the site of the block, may show various degrees of auxotrophy. Some of them, containing the finishing part of the B_{12}-synthetizing machinery still intact, may utilize incomplete forms of vitamin B_{12} such as cobinamides (Cbi's) [37, 50, 140, 142, 143], as is the case in *Thalassiosira pseudonana* clone 13-1 [37, 140, 142]. Apparently, this clone has also retained the capacity to convert pseudovitamin B_{12} into cobalamin. As may be concluded from the specific requirement of another clone of this species, *T. pseudonana* 3H, this alga utilizes cobalamin exclusively in its enzyme systems [37, 140, 142]. It would not be surprising to find symbiotic relations between two auxotrophs complementing each other by synthetizing the lacking parts of the vitamin. To what extent various forms of corrinoids are utilized by the algal enzymes as cofactors cannot be answered yet, since so far only one enzyme – cobalamin dependent ribonucleotide reductase – has been isolated from a few algal species [144–146]. Its specificity, however, toward various cobamides has not been characterized. For example, the mammalian enzymes, unlike the microbial ones, are highly specific toward cobalamin. In order to ensure the absence of other inactive corrinoids in the cell, mammals developed a selective absorption system discriminating against the uptake of other corrinoids, such as pseudovitamin B_{12}. *Ochromonas malhamensis,* the favourite alga for specific assay of cobalamin (also in the presence of other corrinoids), is also equipped with a specific absorption mechanism for the uptake of cobalamin [147]. The absorption proceeds in two distinct phases: rapid adsorption on the cell surface and slow transport across the membrane. Only in the second phase, energy is required and the transport is specific for cobalamin. Similar uptake systems of various specificity seem to operate in other algae. Algae may accumulate large amounts of vitamin B_{12}, exceeding several thousand times the minimal requirement of the organism [147-150]. This is done by adsorption of the corrinoids on the cell surface as in *O. malhamensis*, or by deposition in the chloroplast fraction, as observed in *Euglena gracilis* [149, 150]. These two algae are auxotrophs with respect to vitamin B_{12} and the large deposit of the vitamin, which reaches in *E. gracilis* [149] and *O. malhamensis* [147] 1.6×10^6 and 0.5×10^6 molecules per cell respectively, is a rich dower for the daughter cells. Even algae which produce the vitamin have the capacity to take it up from the environment. Thus high concentration of vitamin B_{12} and other corrinoids assayed in the alga collected from natural water does not necessarily point to its efficient production in the cell. Corrinoids have been found in brown, red, green and blue-green algae in quantities 0.1-1 μg/g of fresh cells [47, 48, 80, 91, 111, 124, 125, 127, 129-134, 138, 149-165, 174] (Table 6-3 C). These values are higher than the minimal requirement of algae (2-18 molecules/μm^3 of the cell) [38, 147, 166] and much higher than the vitamin content in natural water (0.1-30 ng/l, see section 6.2.2). Rough estimates, expressed in molar concentration, range from 0.1 to 20×10^{-12}M in water, 3 to 27×10^{-9}M in the algal cell as minimal for growth and 0.8 to 8×10^{-7}M actually found in various algae. The latter concentration is approximately 10^5-10^6 times higher than in natural water. The growth limit for auxotrophic algae may be evaluated based on the minimal requirement of algae and the concentration of B_{12} in water.

Table 6-3 C Occurrence of vitamin B_{12} in algae and plankton

Source	Concentration of Vitamin B_{12}		Test organism	Reference
	μg/100 g dry algae	ng·l^{-1} particulate matter suspended		
Brown algae mean of 10 species	7		*E. coli, L. lactis Dorner*	[138]
Red algae mean of 10 species	27		*E. coli, L. lactis Dorner*	[138]
Green algae mean of 4 species	35		*E. coli, L. lactis Dorner*	[138]
Vaucheria dichotoma	280		*E. coli, L. lactis Dorner*	[138]
Scenedesmus obliquus	2-170		*E. coli* 113-3	[171]
Plankton from Lake Tsukui, Japan:			*L. leichmannii*	[176]
winter		*1.9*		
summer		*10-20*		
(Fragilaria crotonensis)				
Anabaena flos-aquae		16.5		[159]
Gelidium capillaceum	5.5		*L. leichmannii*	[158]
Cystoseira crinita	0.0098		*L. leichmannii*	[157]

Since the vitamin occurs under natural conditions in water at such low concentrations algae developed a very effective system, for its uptake. This system is based on B_{12}-receptor proteins with a very high affinity to corrinoids, which are built into the membrane of the cell. Studies with algae used as test organisms for B_{12}-assay provided quantitative data as to the affinity of the receptors to the vitamin characterized in terms of the concentration needed for half the maximal growth rate. The values for several algae expressed in $ng \cdot l^{-1}$ (in brackets) are: *Ochromonas malhamensis* [167] (13), *Monochrysis lutheri* [170] (2.8), *Isochrysis galbana* [168] (1.7), *Skeletonema costatum* [170] (2), *Thalassiosira pseudonana* clone 3H [168] (0.39), *T. pseudonana* clone 13-1 (0.2–0.4), *Bellerochea polymorpha* clone 675-d [143] (~0.2). Studies on the uptake of [^{60}Co] Cbl by *O. malhamensis* revealed Michaelis-Menten type kinetics [147] with a K_m of 4nM and V_{max} of 5×10^4 B_{12}-molecules$\cdot sec^{-1} \cdot cell^{-1}$. Employing this data one can compute the velocity (v) of B_{12} uptake at a given concentration (S) by using the known equation: $v = V_{max} \times S \times (K_m + S)^{-1}$. One of the crucial limiting factors for vitamin B_{12} synthesis by autotrophic algae is the concentration of cobalt in the water [172, 173]. Little consideration has been given in the recent literature to the occurrence, availability and circulation of cobalt in natural water. As can be concluded from the considerable concentration of vitamin B_{12} at greater depth, cobalt incorporated into the vitamin sinks with particulate matter and dead cells beyond the euphotic zone. It seems that under various local conditions the capacity for B_{12}-synthesis of autotrophic algae may not be exploited due to the lack of this element in water; they are forced to compete for the vitamin with the auxotrophs. Many of the algal species excrete significant amounts of vitamin B_{12}-binding proteins into the water. These proteins increase appreciably the density of the binding sites for B_{12} and compete with growing algae for the free vitamin. Unlike the binding on the cell surface, the binding of B_{12} by the free proteins, however firm, is not irreversible, thus a flow of vitamin B_{12} from the protein-complex to the algal cell, via dissociation of the complex, is possible. Interest in the purification and properties of these proteins increased recently and the elucidation of their biological function might follow (for comparison see mammalian B_{12}-binding proteins in Chapter 8).

References

[1] Albritton, E. C., Standard values in nutrition and metabolism, W. B. Saunders, Philadelphia, 1954.

[2] Aldrich, D. V., Photoautotrophy in *Gymnodinium breve* Davis, Science, N. Y. **137**, 988–90, 1962.

[2] Allen, M. B., The cultivation of *Myxophyceae*, Arch. Mikrobiol **17**, 34–53, 1952.

[4] Allen, M. B., Studies with *Cyanidium caldarium*, anomalously pigmented chlorophyte, Arch. Mikrobiol. **32**, 270–7, 1959,

[5] Antia, N. J., Cheng, J. Y. & Taylor, F. J. R., The heterotrophic growth of a marine photosynthetic cryptomonad *(Chroomonas salina)*, Proc. VI Int. Seaweed Symp. **6**, 17–29. Direccion General Pesca Maritima, Madrid 1969.

[6] Baker, A. F. & Bold, H. C., Phycological studies. X. Taxonomic studies on the *Oscillatoriaceae*, Univ. Texas Publ. No. 7004., 1970.

[7] Belcher, J. H. & Fogg, G. E., Studies on the growth of Xanthophyceae in pure culture, *Tribonema aequale*. Arch. Mikrobiol. **30**, 17–22, 1958

[8] Belcher, J. H. & Miller, J. D. A., Studies on the growth of Xanthophyceae in pure culture. IV. Nutritional types amongst the X-anthophyceae, Arch. Mikrobiol. **36**, 219–28, 1960.

[9] Boalch, G.T., Studies on *Ectocarpus* in culture. II. Growth and nutrition of a bacteria-free culture, J. mar. biol. Ass., U.K. **41,** 287-304, 1961.

[10] Brooks, A.E., Physiology and genetics of *Astrephomene gubernaculifera,* Ph.D. Thesis, Indiana University, Bloomington, Indiana, 1965.

[11] Carefoot J.R., Nutrition of *Volvulina,* J. Protozool. **14,** 15-18, 1967.

[12] Carefoot, J.R., Culture and heterotrophy of the freshwater dinoflagellate *Peridinium cinctum* fa *ovoplanum,* J. Phycol. **4,** 129-31, 1968.

[13] Carlucci, A., unpubl.

[14] Cox, E.R. & Bold, H.C., Phycological studies. VII. Taxonomic investigations of *Stigeoclonium,* Univ. Texas Publ. No. 6618, 167 pp, 1966.

[15] De, P.K., The role of blue-green algae in nitrogen fixation in rice-fields, Proc. R. Soc. Lond., B. **127,** 121-39, 1939.

[16] Droop, M.R., Studies on the phytoflagellate *Haematococcus pluvialis,* Thesis, Cambridge Univ., England, 1954.

[17] Droop, M.R., A pelagic marine diatom requiring cobalamin, J. mar. biol. Assoc., U.K. **34,** 229-31, 1955.

[18] Droop, M.R., Auxotrophy and organic compounds in the nutrition of marine phytoplankton, J. gen. Microbiol. **16,** 286-93, 1957.

[19] Droop, M.R., Requirement for thiamine among some marine and supralittoral protista, J. mar. biol. Ass., U.K. **37,** 323-9, 1958.

[20] Droop, M.R., Water-soluble factors in the nutrition of *Oxyrrhis marina,* J. mar. biol. Ass., U.K. **38,** 605-20, 1959.

[21] Droop, M.R., *Haematococcus pluvialis* and its allies. III. Organic Nutrition, Rev. Algol. **5,** 247-59, 1961.

[22] Droop, M.R., Organic nutrients, in Physiology and Biochemistry of Algae, ed. Lewin, R.A., pp. 141-59, Academic Press, New York & London, 1962.

[23] Droop, M.R., McLaughlin, J.J.A., Pintern, I.J. & Provasoli, L., Specificity of some protophytes toward vitamin B_{12}-like compounds, Preprints Intern. Oceanog. Congr. 916-18. A.A.A.S., 1959.

[24] Dusi, H., La pyrimidine et le thiazol facteurs de croissance pour le flagellé à chlorophylle, *Euglena pisciformis,* C. r. Séanc. Soc. Biol. **130,** 419-22, 1939.

[25] Dusi, H., Culture bacteriologiquement pure et nutrition autotrophe d'*Eudorina elegans.* Rôle du fer pour la formation des colonies, Ann. Inst. Pasteur, **64,** 340-3, 1940.

[26] Eppley, R.W. & Maciasr, F.M., Rapid growth of sewage lagoon *Chlamydomonas* with acetate, Physiologia Pl. **15,** 72-9, 1962.

[27] Forsberg, C., Nutritional studies of *Chara* in axenic culture, Physiologia Pl. **18,** 275-90, 1965.

[28] Fries, L., A marine red alga requiring vitamin B_{12}, Nature, Lond. **183,** 558-9, 1959.

[29] Fries, L., Vitamin requirements of *Nemalion multifidum,* Experientia **17,** 75-6 1961.

[30] Fries, L., *Polysiphonia urceolata* in axenic culture, Nature, Lond. **202,** 110, 1964.

[31] Fries, L. & Pettersson, H., On the physiology of the red alga *Asterocytis ramosa* in axenic culture, Br. Phycol. Bull **3,** 417-22, 1968.

[32] Gerloff, G.C., Fitzgerald, G.P. & Skoog, F., The isolation, purification, and culture of blue-green algae, Am. J. Bot. **37,** 215-18, 1950.

[33] Gibor, A., The culture of brine algae, Biol. Bull., mar. biol. Lab., Woods Hole **111,** 223-9, 1956.

[34] Gold, K., Aspects of marine dinoflagellate nutrition measured by ^{14}C assimilation, J. Protozool. **11,** 85-9, 1964.

[35] Groover, R.D. & Bold, H.C., Phycological studies. VIII. The taxonomy and comparative physiology of the Chlorosarcinales and certain other edaphic algae, Univ. Texas Publ. No. 6907, 165 pp., 1969.

[36] Guillard, R.R.L., Organic sources of nitrogen for marine centric diatoms, Symp. Mar. Microbiol. ed. Oppenheimer, C.H., pp. 93-104, Thomas Co., Springfield, Ill. 1963.

[37] Guillard, R.R.L., B_{12} specificity of marine centric diatoms, J. Phycol. **4,** 59-64, 1968.

[38] Guillard, R.R.L. & Cassie, V., Minimum cyanocobalamin requirements of some marine diatoms, Limnol. Oceanog. **8,** 161-5, 1963.

[39] Guillard, R.R.L. & Ryther, J.H., Studies of marine planktonic diatoms. I. *Cyclotella nana* Hustedt, and *Detonula confervacea* (Cleve), Can. J. Microbiol. **8,** 229-39, 1962.

[40] Hargraves, P.E. & Guillard, R.R.L., Personal communication.

[41] Harris, D.O., Nutrition of *Platydorina caudata,* J. Phycol. **5,** 205-10, 1969.

[42] Haxo, F.T. & Sweeney, B.M., Bioluminescence in *Gonyaulax polyedra,* In The Luminescence of Biological Systems, ed. Johnson, F.H., pp. 415-20, A.A.A.S., 1955.

[43] Heinrich, H.C., Der B-Vitamin-Bedarf der Chrysophyceen, *Ochromonas danica* nom. provis. Pringsheim und *Ochromonas malha-*

mensis Pringsheim, Naturwissenschaften **14,** 418, 1955.

[44] Hoham, R., Laboratory and field studies on snow algae of the Pacific Northwest, Ph. D. Thesis, Univ. of Washington, Seattle, 1971.

[45] Holz, G. G., The oxidative metabolism of a cryptomonad flagellate, *Chilomonas paramecium,* J. Protozool. **1,** 114-20, 1954.

[46] Hutner, S. H., Essentiality of constituents of sea water for growth of a marine diatom, Trans. N. Y. Acad. Sci. **10,** 136-41, 1948.

[47] Hutner, S. H. & Provasoli, L., The phytoflagellates, in Biochemistry and Physiology of Protozoa, ed. Lwoff, A., **1,** 27-128. Academic Press, New York & London, 1951.

[48] Hutner, S. H. & Provasoli, L., A pigmented marine diatom requiring vitamin B_{12} and uracil, News Bull. Phycol. Soc. Amer. **6,** 7-8, 1953.

[49] Hutner, S. H., Provosoli, L. & Filfus, J., Nutrition of some phagotrophic fresh-water chrysomonads, Ann. N. Y. Acad. Sci. **56,** 852-62, 1963.

[50] Iwasaki, H., Nutritional studies of the edible seaweed *Porphyra tenera.* I. The influence of different B_{12} analogues, plant hormones, purines and pyrimidines on the growth of *Conchocelis.,* Pl. Cell Physiol., Tokyo, **6,** 325-36, 1965.

[51] Iwasaki, H., Studies on the red tide dinoflagellates. III. On *Peridinium hangoei* Schiller appeared in Gokasho Bay, Shima Peninsula. Bull. Blankton Soc., Japan, **16,** 132-9, 1969.

[52] Iwasaki, H., Studies on the red tide dinoflagellates. V. On *Polykrikos schwartzi* Butschli, Bull. Jap. Soc. Sci. Fish. **37,** 606-9, 1971.

[53] Iwasaki, H., Studies on the red tide dinoflagellates. VI. *Eutreptiella* sp. and *Exuviaella* sp., J. Ocean. Soc., Japan, **27,** 152-7, 1971.

[54] Iwasaki, H., Fujiyama, T. & Yamashita, E., Studies on the red tide dinoflagellates. I. On *Entomosigma* sp. appeared in coastal area of Fukuyama, J. Fac. Fish. & Anim. Husb. Hiroshima Univ. **7,** 259-67, 1968.

[55] Iwasaki, H., Okaka, A. & Tanake, S., Studies on the red tide dinoflagellates. IV. On *Rhodomonas ovalis* Nygaard appeared in coastal area of Fukuyama. Bull. Plankton Soc., Japan, **16,** 140-4, 1969.

[56] Iwasaki, H. & Sasada, K., Studies on the red tide dinoflagellates. II. On *Heterosigma inlandica* appeared in Gokasho Bay, Shima Peninsula, Bull. Jap. Soc. Sci. Fish. **35,** 943-7, 1969.

[57] Kain, J. M. & Fogg, G. E., Studies on the growth of marine phytoplankton. III. *Prorocentrum micans* Ehrenberg, J. mar. biol. Ass., U. K. **39,** 33-50, 1960.

[58] Korn, R. W., Nutrition of *Cosmarium turpini,* Physiologia Pl. **22,** 1158-1165, 1969.

[59] Kratz, W. A. & Myers, J., Nutrition and growth of several blue-green algae, Am. J. Bot. **42,** 282-7, 1955.

[60] Lewin, J. C., Heterotrophy in diatoms, J. gen Microbiol. **9,** 305-13, 1953.

[61] Lewin, J. C., Heterotrophy in marine diatoms, In Symp. Marine Microbiol., ed. Oppenheimer, C. H., pp. 229-35, Thomas Co., Springfield, Ill, 1963.

[62] Lewin, J. C., The thiamine requirement of a marine diatom. Phycologia, **4,** 142-4, 1965.

[63] Lewin, J. C., & Lewin, R. A., Auxotrophy and heterotrophy in marine littoral diatoms, in Preprints Int. Oceanog. Congress, A. A. A. S. pp. 928-9, 1959.

[64] Lewin, J. C. & Lewin, R. A., Auxotrophy and heterotrophy in marine littoral diatoms, Can. J. Microbiol **6,** 127-34, 1960.

[65] Lewin, J. C. & Lewin, R. A., Culture and nutrition of some apochlorotic diatoms of the genus *Nitzschia,* J. gen. Microbiol. **46,** 361-7, 1967.

[66] Lewin, R. A., Vitamin requirements in the Chlorococcales, Phycol. News Bull. **5,** 21-2, 1952.

[67] Lewin, R. A., A marine *Stichococcus* sp. which requires vitamin B_{12}, J. gen. Microbiol. **10,** 93-6, 1954.

[68] Lewin R. A., Vitamin-bezonoj de algoj, in Sciencaj Studoj, ed. Neergaard, P., pp. 187-92, Modersmaalet, Haderslev, Copenhagen, 1958.

[69] Loeblich, A. R. III, Nutrition of the marine dinoflagellate *Cachonina niei,* J. Protozool. **16,** (Suppl.), p. 20, 1969.

[70] Lwoff, A. & Dusi, H., Culture de divers flagelles leucophytes en milieu synthetique, C. r. Seanc. Soc. Bio. **127,** 53-6, 1938.

[71] McLaughlin, J. J. A. & Provosoli, L., Nutritional requirements and toxicity of two marine amphidinium, J. Protozool. **4** (Suppl.), p. 7, 1957.

[72] McLaughlin, J. J. A. & Zahl, P. A., Vitamin requirements in symbiotic algae, In Preprints Int. Oceanog. Congress, A. A. A. S. pp. 930-1, 1959.

[73] McLaughlin, J. J. A. & Zahl, P. A., Axenic zooxanthellae from various invertebrate hosts, Ann. N. Y. Acad. Sci. **77,** 55-72, 1959.

[74] Miller, J. D. A. & Fogg, G. E., Studies on the growth of Xanthopyceae in pure culture. I. The mineral nutrition of *Monodus subterraneus* Petersen, Arch. Mikrobiol. **28,** 1-17, 1957.

[75] Miller, J. D. A. & Fogg, G. E., Studies on the growth of Xanthophyceae in pure culture. II. The relations of *Monodus subterraneus* to organic substances, Arch. Mirkobiol **30,** 1-16, 1958.

[76] Nakahara, H., Alternation of generations of some brown algae in unialgal and axenic cultures, Ph. D. Thesis, Hokkaido University, Sapporo, Japan, 1972.

[77] Osterud, K. L., Trophic potentialities of the green flagellate *Lobomonas piriformis* Pringsheim, Physiol. Zool. **19,** 19-34, 1946.

[78] Palmer, E. G. & Starr, R. C., Nutrition of *Pandorina morum,* J. Phycol **7,** 85-9, 1971.

[79] Parker, B. C., Bold, H. C. & Deason, T. R., Facultative heterotrophy in some chlorococcacean algae, Science, N. Y. **133,** 761-3, 1961.

[80] Pedersén, M., Marine brown algae requiring vitamin B_{12}, Physiologia Pl. **22,** 977-83, 1969.

[81] Pincemin, J. M., Action de facteurs physiques, chimiques et biotiques sur quelques dinoflagellés et diatomeés en culture, Ph. D. Thesis, Université D'Aix-Marseille, 1971.

[82] Pintner, I. J. & Provasoli, L., Artificial cultivation of a red-pigmented marine blue-green alga *Phormidium persicinum,* J. gen. Microbiol. **18,** 190-7. 1958.

[83] Pintner, I. J. & Provasoli, L., The nutrition of *Volvox globator* and *V. tertius,* Proc. IX Int. Bot. Congr. Montreal, Abstr. pp. 300-1, 1959.

[84] Pintner, I. J. & Provasoli, L., Nutritional chracteristics of some chrysomonads, in Symp. Marine Microbiol., ed. Oppenheimer, C. H., pp. 114-21. Thomas Co., Springfield, Ill, 1963.

[85] Pintner, I. J. & Provasoli, L., Heterotrophy in subdued light of 3 *Chrysochromulina* species, Bull. Misaki Mar. Biol. Inst. Kyoto Univ. **12,** 25-31, 1968,

[86] Pringsheim, E. G., Kleine Mitteilungen über Flagellaten und Algen. III. Über *Ochromonas denica* n. sp. und andere Arten der Gattung, Arch. Microbiol. **23,** 181-92, 1955.

[87] Pringsheim, E. G., Über Mixotrophie bei Flagellaten, Planta, **52,** 405-30, 1958.

[88] Pringsheim, E. G., *Chlamydomonas pallens,* a new organism proposed for assays of vitamin B_{12}, Nature, Lond. **195,** 604, 1962.

[89] Pringsheim, E. G., Nutritional requirements of *Haematococcus pluvialis* and related species, J. Phycol. **2,** 1-7, 1966.

[90] Pringsheim, E. G., (1970) Identification and cultivation of European *Volvox* sp., Antonie van Leuwenhoek **36,** 33-43, 1970.

[91] Provasoli, L., Nutrition and ecology of protozoa and algae, A. Rev. Microbiol. **12,** 279-308, 1958.

[92] Provasoli, L., Growing marine seaweeds, in Proc. IV Int. Seaweed Symp., ed. De Virville, A. D. & Feldman, J., pp. 9-17, Pergamon Press, Oxford, London, New York & Paris, 1964.

[93] Provasoli, L. & Gold, K., Nutrition of the American strain of *Gyrodinium cohni,* Arch. Mikrobiol. **42,** 196-203, 1962.

[94] Provasoli, L. & McLaughlin, J. J. A., Auxotrophy in some marine and brackish dinoflagellates, J. Protozool. **2** (Suppl.), 10, 1955.

[95] Provasoli, L. & McLaughlin, J. J. A., Limited heterotrophy of some photosynthetic dinoflagellates, in Symp. Marine Microbiol., ed. Oppenheimer, C. H., pp. 105-13, Thomas Co. Springfield, Ill, 1963.

[96] Provasoli, L. & Pintner, I. J., Ecclogical implications of *in vitro* nutritional requirements of algal flagellates, Ann. N. Y. Acad. Sci. **56,** 839-51, 1953.

[97] Provasoli, L. & Pintner, I. J., Culture of *Trachelomonas* in chemically defined media, Phycol. News Bull. **8,** 7, 1955.

[98] Rayburn, W. R., Morphology, Nutrition and physiology of sexual reproduction of *Pandorina unicocca* sp. nov., Ph. D. Thesis, Indiana Univ., Bloomington, Ind, 1971.

[99] Saito, S., Growth of *Gonium multicoccum* in synthetic media, J. Phycol. **8,** 169-75, 1972.

[100] Samejima, H. & Myers, J., On the heterotrophic growth of *Chlorella pyrenoidosa,* J. gen Microbiol. **18,** 107-15, 1958.

[101] Shepard, D. C., Axenic culture of *Acetabularia* in synthetic medium, in Methods in Cell Physiology IV, pp. 46-69, Academic Press, New York & London, 1969.

[102] Shihira, I. & Krauss, R. W., *Chlorella:* physiology and taxonomy of 41 isolates, 97 pp., Port City Press, Baltimore, Md, 1965.

[103] Smith, R. L. & Bold, H. C., Phycological studies. IV. Investigations on the algal genera *Eremosphaera* and *Oocystis,* 121 pp., Univ. Texas Publ. no. 6612, 1966.

[104] Stein, J. R., Growth and mating of *Gonium pectorale* (Volvocales) in defined media, J. Phycol. **2,** 27-8, 1966,

[105] Stewart, W. D. P., Fixation of elemental nitro-

gen by marine blue-green algae, Ann. Botany, London, N.S. **26**, 439-45, 1962.

[106] Storm, J. & Hutner, S.H., Nutrition of *Peranema*, Ann. N.Y. Acad. Sci. **56**, 901-9, 1953.

[107] Sweeney, B.M., *Gymnodinium splendens*, a marine dinoflagellate requiring vitamin B_{12}, Am. J. Bot. **41**, 821-4, 1954.

[108] Swift, M.J. & McLaughlin, J.J.A., Some nutritional, physiological, and morphological studies on *Melosira juergensii* and *Heterocapsa kollmeriana*, Ann. N.Y. Acad. Sci. **175**, 577-600, 1970.

[109] Tassigny, M., Observations sur les bésoins en vitamines des Desmidiées (Chlorphycées-Zygnematales), J. Phycol. **7**, 213-15, 1971.

[110] Tatewaki, M., Personal communication.

[111] Tatewaki, M. & Provasoli, L., Vitamin requirements of three species of *Antithamnion*, Botanica mar. **6**, 193-203, 1964.

[112] Thomas, W.H., Heterotrophic nutrition and respiration of *Gonyaulax polyedra*, J. Protozool. **2** (Suppl.), 2-3, 1955.

[113] Thomas, W.H., Effects of temperature and illuminance on cell division rates of three species of tropical oceanic phytoplankton, J. Phycol. **2**, 17-22, 1966 and personal communication.

[114] Turner, M.F., A note on the nutrition of *Rhodella*, Br. Phycol. J. **5**, 15-18, 1970.

[115] Van Baalen, C., Studies on marine blue-green algae, Botanica mar. **4**, 129-39, 1962.

[116] Watanabe, A., Production in cultural solution of some amino acids by the atmospheric nitrogen-fixing blue-green algae, Archs. Biochem. Biophys. **34**, 50-5, 1951.

[117] Wetherell, D.F., Obligate phototrophy in *Chlamydomonas eugametos*, Physiologia Pl. **11**, 260-74, 1958.

[118] Wetherell, D.F. & Krauss, R.W., X-Ray induced mutations in *Chlamydomonas eugametos*, Am. J. Bot. **44**, 609-19, 1957.

[119] Williams A.E. & Burris, R.H., Nitrogen fixation by blue-green algae and their nitrogenous composition, Am. J. Bot. **39**, 340-2, 1952.

[120] Wilson, W.B., The suitability of seawater for the survival and growth of *Gymnodinium breve* and some effects of phosphorus and nitrogen on its growth, Florida Board Conservation, Professional Papers, Series **7**, 42 pp, 1966.

[121] Staub, R., Ernährungsphysiologisch-autokologische Untersuchungen an der planktischen Blaualge *Oscillatoria rubescens* DC, Schweizerische Zeitschrift für Hydrologie **23**, 82-198 a, 1961.

[122] Leedale, G.F., Meeuse, B.J.D. & Pringsheim, E.G., Structure and physiology of *Euglena spirogyra*. III-VI., Arch. Microbiol. **50**, 133-55, 1965.

[123] Pringsheim, E.G., Chlorophyllarme Algen. II. *Lobomonas sphaerica* nov. spec., Arch. Mikrobiol. **46**, 227-37, 1964.

[124] Scott, R., Ericson, L.E., Some aspects of cobalt metabolism by *Rhodymenia palmata* with particular reference to vitamin B_{12} content, J. Exp. Bot. **6**, 348, 1955.

[125] Printner, I.J., and Altmeyer, V.L., Vitamin B_{12}-binder and other algal inhibitors, J. Phycol. **15**, 391-398, 1979.

[126] Swift, D.G., and Guillard, R.R.L., Unexpected response to vitamin B_{12} of dominant centric diatoms from the spring bloom in the Gulf of Maine, J. Phycol. **14**, 377-386, 1978.

[127] Fries, L., Requirements for organic substances in seaweeds, Bot. Mar. **16**, 19-31, 1973.

[128] Conrad, H.M., and Eppley, R., Interaction of gibberellic acid and allyltrimethylammonium bromide upon growth of *Ulothrix*. Plant Physiol. **36**, 685-687, 1961.

[129] Loiseanx, S., and Rozier, C., Axenic culture of *Pylaiella littoralis*, Rev. Algol. **13**, 333-340, 1978.

[130] Messina, D.S., and Baker, A.L., Interspecific growth regulation in species succession through vitamin B_{12} competitive inhibition. J. Plankton Res. **4**, 41-46, 1982.

[131] Ohki, K., Fujita Yoshiniko, Laboratory culture of the pelagic blue-green alga *Trichodesmium thiebautii*: Conditions for unialgal culture, Mar. Ecol. Prog. Ser. **7**, 185-190, 1982.

[132] Tsukidate, J., Some notes on *Grinnelia americana* in culture, Bull. Jap. Soc. Sci Fish. **36**, 1109-1114, 1970.

[133 a] Fiala, M., Vitamin B_{12} and phytoplankton at the thermocline level in the Northwest Mediterranean, Oceanol. Acta **5**, 339-348, 1982.

[133 b] Fiala, M., and Oriol, L., Vitamin B_{12} and phytoplankton in the Antarctic Ocean: Distribution and experimental approach, Mar. Biol. Berl **79**, 325-332, 1984.

[134] Holt, J.R., Pfiester, L.A., A survey of auxotrophy in 5 freshwater dinoflagellates *(Pyrrophyta)* J. Phycol. **17**, 415-416, 1981.

[135] Antia, N.J., McAlister, C.D., Parsons, T.R., Stephens, V., and Strickland, J.D.H., Further measurements of primary production using a large-volume plastic sphere, Limnol. Oceanogr. **8**, 166-183, 1963.

[136] Neujahr, H.Y., and Fries, L., On the occur-

rence of light-sensitive corrinoids in axenic cultures of unicellular algae, Acta Chem. Scand **20,** 347-360, 1966.

[137] Fries, L., Influence of different B_{12} analogues on the growth of *Goniotrichum elegans* (Chauv), Physiol. Plantarum **13,** 264, 1969 and: *Polysiphonia urceolata* in axenic culture, Nature, **202,** 110, 1964.

[138] Ericson, L. E., and Lewis, L., Occurrence of vitamin B_{12} factors in marine algae, Arkiv Kemi **6,** 427-442, 1953.

[139] Provasoli, L., Carlucci, A. F., Vitamins and growth regulators, in Algal Physiology and Biochemistry, Ed. W. D. P. Stewart, Blackwell Scientific Publications, Oxford 1974, pp. 741-787.

[140] Swift, G., Vitamins and phytoplankton growth, in The Physiological Ecology of Phytoplankton, Ed. J. Morris, Blackwell Scientific Publications, Oxford 1980, pp. 329-386.

[141] For occurrence of vitamin B_{12} analogues in bacteria see text in sections 6.1 and 6.4 in this Chapter and Tables 6-1, 6-13, 6-14. Formulae of the analogues are given in Table 3-1.

[142] Carlucci, A. F., Silbernagel, S. B., Bioassay of seawater. I. A ^{14}C uptake method for determination of concentrations of vitamin B_{12} in seawater. Can. J. Microbiol. **12,** 175-183, 1966.

[143] Swift, G., Guillard, R. R. L., Diatoms as tools for assay of total B_{12} activity and cyanocobalamin activity in sea water, J. Mar. Res. **35,** 309-320, 1977.

[144] Gleason, F. K., Frick, T. D., Adenosylcobalamin-dependent ribonucleotide reductase from the blue-green alga, *Anabaena* sp., J. Biol. Chem. **255,** 7728-7733, 1980.

[145] Gleason, F. K., and Hogenkamp, H. P. C., Ribonucleotide reductase from *E Euglena gracilis* a deoxyadenosylcobalamin-dependent enzyme, J. Biol. Chem. **245,** 4894-4899, 1970.

[146] Gleason, F. K., Control of ribonucleotide reductase in *Synechococcus* sp. a unicellular Cyanobacterium, Arch. Microbiol. **123,** 15-21, 1979.

[147] Bradbeer, C., Transport of vitamin B_{12} in *Ochromonas malhamensis,* Arch. Biochem. Biophys. **144,** 184-192, 1971.

[148] Reeves, R. B., and Fay, F. S., Cyanocobalamin (vitamin B_{12}) uptake by *Ochromonas malhamensis,* Am. J. Physiol. **210,** 1273, 1966.

[149] Sarhan, F., Houde, M., Cheneval, J. P., The role of vitamin B_{12} binding in the uptake of the vitamin *Euglena gracilis,* J. Protozool, **27,** 235-238, 1980.

[150] Varma, T. N. S., Abraham, A., and Hansen, J. A., Accumulation of ^{58}Co-Vitamin B_{12} by *Euglena gracilis,* J. Protozool. **8,** 212-216, 1961.

[151] Brown, F., Cuthberston, W. F. J. and Fogg, G. E., Vitamin B_{12} activity in *Chlorella vulgaris* Beij and *Anabena cylindrica Lemm.,* Nature **177,** 188, 1956.

[152] Hashimoto, Y., J. Vitaminol. (Osaka) **1,** 49-54, 1954.

[153] Prasad, P. V. D., and Chowdary, Y. B. K., Growth and mineral nutrition of *Gloeotaenium loitlesbergerianum(Chlorococcales, Chlorophyceae).* Indian. J. Bot. **2,** 166-170, 1979.

[154] Baker, E. R., McLaughlin, J. J. A., Hutner, S. H., Deangelis, B., Feingold, S., Frank, O., Baker, H., Water-soluble vitamins in cells and spent culture supernatants of *Poteriochromas stipitata, Euglena gracilis* and *Tetrahymena thermophila,* Arch. Microbiol. **129,** 310-313, 1981.

[155] Bunt, J. S., Uptake of cobalt and vitamin B_{12} by tropical marine macroalgae, J. Phycol. **6,** 339-343, 1970.

[156] Carlucci, A. F., and Bowes, P. M., Production of vitamin B_{12}, thiamine and biotin by phytoplankton, J. Phycol. **6,** 351-357, 1970.

[157] Güven, K. C., Hakyemez, T., and Güner, H., Biochemical studies on *Cystoseira crinita* Bory, Bot. Mar. **23,** 201-201, 1980.

[158] Güven, K. C., Güler, E., and Yücel, A., Vitamin B_{12} content of *Gelidium capillaceum* Kütz, Bot. Mar. **19,** 395-396, 1976

[159] Grieco, E., and Desrochers, R., Production of vitamin B_{12} by a blue alga. Can. J. Microbiol. **24,** 1562-1566, 1978.

[160] Nishijima, T., Shiozaki, R., and Hata, Y., Production of vitamin B_{12} thiamine, and biotin by freshwater phytoplankton, Bull. Jpn. Soc. Sci. Fish. **45,** 199-204, 1979.

[161] Kanazawa, A., Vitamins in algae, Bull. Jap. Soc. Sci. Fish. **29,** 713-731, 1963.

[162] Cowey, C. B., A preliminary investigation of the variation of vitamin B_{12} in oceanic and coastal waters, J. Mar. Biol. Ass. England **35,** 609, 1956.

[163] Thomas, W. H., Nutrient requirements and utilization: Algae, in Metabolism, Eds. P. L. Altman and D. S. Ditmer, Fed. Am. Soc. Exp. Biol. 1968, pp. 210-228.

[164] Carlucci, A. F., and Bowes, P. M., Vitamin production and utilization by phytoplankton

in mixed culture, J. Phycol. **6,** 393-400, 1970 and Vitamin B_{12}, thiamine and biotin contents of marine phytoplankton, J. Phycol. **8,** 133-137, 1982.
[165] Aaronson, S., Deangelis, B., Frank, O., and Baker, H., Secretion of vitamins and amino-acids into the environment by *Ochromonas danica,* J. Phycol. **7,** 215-218, 1971.
[166] Droop, M.R., Vitamin B_{12} in marine ecology, Nature **180,** 1041-1042, 1957.
[167] Ford, J.E., B_{12} vitamins and growth of the flagellate *Ochromonas malhamensis,* J. Gen. Microbiol. **19,** 162-172, 1958.
[168] Swift, D.G., and Taylor, W.R., Growth of vitamin B_{12}-limited cultures: *Thalassiosira pseudonana, Monochrysis lutheri* and *Isochrysis galbana,* J. Phycol. **10,** 385-391, 1974.
[169] Carlucci, A.F. and Silbernagel, S.B., Effect of vitamin concentrations on growth and development of vitamin-requiring algae, J. Phycol. **5,** 64-67, 1969.
[170] Droop, M.R., Vitamin B_{12} and marine ecology. 5. Continuous culture as an approach to nutrional kinetics. Helgoländer Wiss. Meeresunters. **20,** 629-636, 1970.
[171] Fink, G., Herold, J., and Lundin, H., Z. Naturforsch. **13,** 605, 1958, cited after Friedrich, W., in Fermente, Hormone, Vitamine, vol. III, Eds. R. Ammon and W. Discherl, Georg Thieme Verlag, Stuttgart 1975, pp. 169-174.
[172] Holm-Hansen, O., Gerloff, G.G., and Skoog, F., Cobalt as an essential element for blue-green algae, Physiol. Plant. **7,** 665, 1954.
[173] Buddhari, W., Cobalt as an essential element for blue-green algae, Ph. D. Diss. Univ. California, Berkley, 1960.
[174] Gumares, R.P., Rosenberg, J.A., Costa, F.A., Vitamin content of *Scenedesmus quadricauda:* 2. Vitamin B_{12}, Rev. Latinoam Microbiol. **25,** 275-280, 1983.
[174] Kristensen, H.P.O., A vitamin B_{12}-binding factor found in cultures of *Euglena gracilis* var. bacillaris, Acta Physiol. Scand. **37,** 8-13, 1956.
[175] Droop, M.R., Vitamin B_{12} and marine ecology, 4. The kinetics of uptake growth and inhibition in *Monochrysis lutheri,* J. Mar. Biol. Ass. U.K. **48,** 689-733, 1968.
[176] Pitner, iJ., and Altmeyer, V.L., Vitamin B_{12}-binder and other algal inhibitors, J. Phycol. **15,** 391-398, 1979.
[177] Daisley, K.W., The occurrence and nature of *Euglena gracilis* proteins that bind vitamin B_{12}, Int. J. Biochem. **1,** 561-574, 1970.
[178] Ohwada, K., Otsuhata, M., and Taga, N., Seasonal cycles of vitamin B_{12}, thiamin and biotin in the surface water of Lake Tsukui, Bull. Jap. Soc. Sci. Fish. **38,** 817-823, 1972.

6.3.2 Higher plants

The microbiological tests done by various authors in the course of fifties did not indicate measurable quantities of vitamin B_{12} in the green parts of the higher plants [1]; thus in the following years there was little research efforts on this problem. The recent discovery that plants contain a B_{12} dependent enzyme [2] may bring to light again the problem of B_{12} in plants. So far there is little evidence on physiological responses of plants to vitamin B_{12} treatment [3]. A great deal of work has been done, however, on physiological effects of cobalt in plants [4-21, 25, 47-49]. The forms of cobalt causing the particular physiological effects are often not determined; thus, it is possible that some of them are corrinoids. The presence of cobalt ions in the green parts of the plants is of great ecological and agricultural importance. The plant eating animals, in particular the ruminants, which synthetize in the rumen relatively large quantities of various corrinoids depend on cobalt ions from plants (sheep require 0.5-1 mg Co per day). The cobalt is not evenly distributed among plants, it occurs in the concentration range of 0.02-0.5 μg per g dry weight (0.02-0.5 ppm) [22-24, 28-46]. There are plants, however, which cumulate Co like: *Crotolaria cobalticola* [44] (500-800 ppm) and *Nyssa sylvatica* [45] (up to 1000 ppm). The latter is used as test plant for occurrence of Co in soil in forms adequate as plant nutrition. The concentration of Co in *Nyssa sylvati-*

ca below 5 ppm is an indication for cobalt poor soil. In general aquatic plants contain relatively the highest- while grasses the lowest level of this element. It tends to concentrate in the generative organs of plants. Its concentration in the plant depends also very much on what is available in soil.

Another problem is the occurrence of vitamin B_{12} in roots. Since soil contains often relatively high concentration of corrinoids it can not be assumed with certainty that the corrinoids assayed in roots with the sensitive microbiological assays are not there due to contamination. On the other hand, the root nodules of the symbiotic plants contain appreciable concentration of corrinoids which are produced by symbiotic bacteria (bacteroids) [1, 26–33]. Tables 6-4 and 6-5 indicate the quantities of vitamin B_{12} found in roots of symbiotic plants. The effectiveness of symbiotic bacteria to fix dinitrogen in particular those of *Rhizobium* species is correlated with the capacities of various strains to synthetize cobalamin (Table 6-6). Cobalt appears also in root nodules in the ionic form. The supernatant solution from homogenized root nodules separated from bacteroids contains 50% cobalt in the ionic form and only 19% is build into corrinoids [35, 47]. The presence of vitamin B_{12} in root nodules was subject of

Table 6-4 Vitamin B_{12} content in leguminous plants, in mustard, and in soil at three stages of growth: young plants (I), phase of flowerage (II) and 10 days after flowerage (III) assayed with *Euglena gracilis* (from Duda J., Pędziwilk Z., and Zodrow K. [8])

Plant	stage	Vitamin B_{12} content, picomoles per 1 g dry matter			
		root nodules	roots	leaves	soil
Lupin - *Lupinus angustifolus*	I	82.0	16.0	0.5	5.7
	II	81.5	3.7	0.0	3.8
	III	42.0	7.5	0.0	4.9
Pea - *Pisum sativum*	I	615.0	58.0	0.4	8.1
	II	500.0	10.0	0.0	4.4
	III	960.0	168.0	0.0	4.4
Horse bean - *Vicia faba minor*	I	947.0	12.5	0.0	6.8
	II	316.0	5.0	0.0	4.3
	III	374.0	17.0	0.0	4.5
Clover - *trifolium pratense*	I	369.0	5.1	0.0	4.4
	II	518.0	14.7	0.0	4.1
	III	128.0	11.9	0.0	5.2
Bird's trifoil - *Lotus corniculatus*	I	297.0	12.0	0.0	12.7
	II	19.0	4.0	0.0	4.6
	III	71.0	11.0	0.0	7.8
Alfalfa - *Medicago sativa*	I	435.0	1.2	0.0	4.6
	II	230.0	3.2	0.0	4.4
	III	–	2.7	0.0	6.1
Mustard - *Sinapis alba*	I	–	9.9	0.3	5.9
	II	–	1.9	0.0	4.5
	III	–	1.1	0.0	5.0

Table 6-5 B_{12} coenzyme contents of the nodules from various plants and of *Rhizobium meliloti* (from Kliewer and Evans [34]. Reprinted by permission from Nature **195**, 4843, 1962. Copyright 1962 Macmillan Journals Ltd.)

Name	age (days)	B_{12}-coenzyme nmol/g fresh tissue[a]
Medicago sativa	78	64
Glycine max	39	39
Glycine max	73	30
Pisum sativum	26	47
Pisum sativum	45	16
Trifolium pratense	81	40
Alnus oregona	2 years	61
Rhizobium meliloti	36 hours	77

[a] The values were obtained by an enzyme test and pertain to total contents of all cobamide coenzymes active as cofactors in the test enzyme used. The enzyme assay is less specific than the *Ochromonas* test which is specific to cobalamin only (see the data in Table 6-13)

Table 6-6 B_{12} coenzyme contents of effective and ineffective strains of *R. meliloti and R. japonicum*[a] (from Kliewer and Evans [38]). The enzyme test used by the authors is not specific for cobalamin coenzyme. The data pertain to total contents of all cobamide coenzymes active as cofactors (see also footnote to Table 6-5)

Rizobium species and strain[b]	Nitrogen fixing efficiency	B_{12} coenzyme in fresh cells[c] nmol/g
R. meliloti		
F29	Effective	25.5
3DOal	Ineffective	6.7
3DOol	Ineffective	15.4
3DOa26	Parasitic	9.4
R. japonicum		
b	Effective	15.0
3I1b4	Ineffective	6.7
3I1b69	Ineffective	8.0
3I1b24	Parastic	6.7

[a] All values reported are means of two replicates
[b] All bacteria were grown with 50 ppb Co in the culture medium
[c] Coenzyme contents are expressed as nanomoles of 5,6-dimethylbenzimidazolylcobamide coenzyme per gram of fresh cells.

speculation whether corrinoids are directly involved in reduction of N_2 to ammonia. Such function seemed to be possible since the reduced form of vitamin B_{12} is able to evolve hydrogen even from water (see B_{12s} in section 3.1.3). B_{12s} could function also as protector of nitrogenase from being inactivated by oxygen. Some cobalt coordination compounds bind N_2 [50]. On the other hand neither vitamin B_{12} nor cobalt have been reported in nitrogenases purified from various sources. It seems rather that the correlation of dinitrogen fixation with B_{12} production in Rhizobium (Table 6-6) is due to

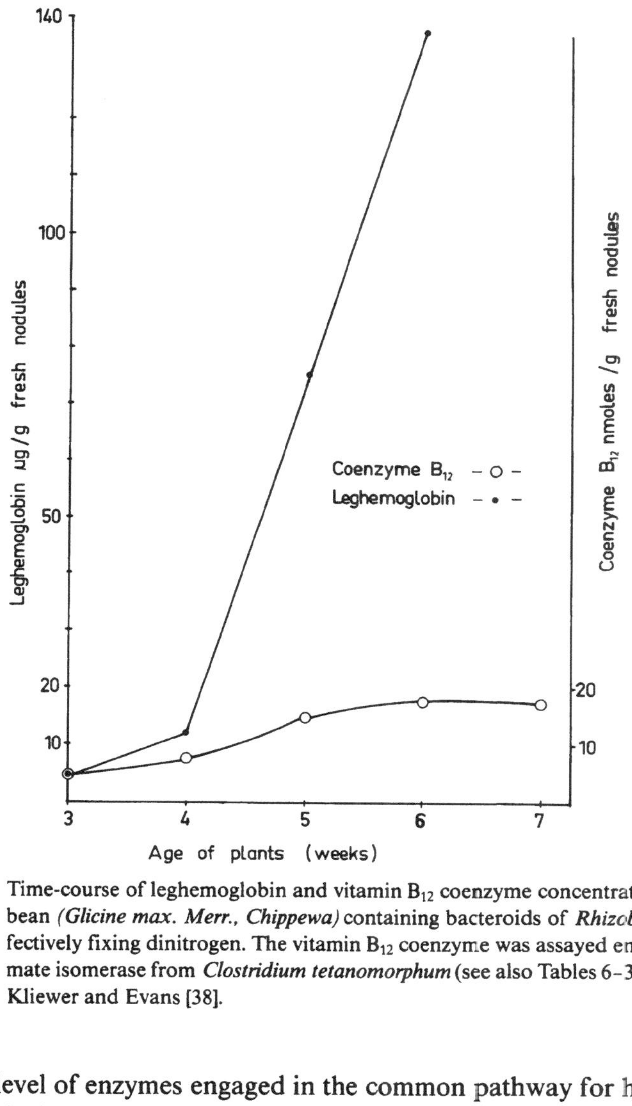

Fig. 6-1 Time-course of leghemoglobin and vitamin B_{12} coenzyme concentration in root nodules of soybean *(Glicine max. Merr., Chippewa)* containing bacteroids of *Rhizobium japonicum,* a strain effectively fixing dinitrogen. The vitamin B_{12} coenzyme was assayed enzymatically by using glutamate isomerase from *Clostridium tetanomorphum* (see also Tables 6-3 and 6-4). Data taken from Kliewer and Evans [38].

higher level of enzymes engaged in the common pathway for heme- (the nonprotein part of leghemoglobin) and vitamin B_{12} biosynthesis (for details see Figs. 4-1 and 4-2). The complete form of vitamin B_{12} inhibits, by feedback action, further formation of the corrin ring making effective the conversion of its precursor (uro' gen III) to heme. Other authors suggest that coenzyme B_{12} is engaged in nitrogen fixation through synthesis of the heme precursor [49].

References

[1] Duda, J., Pędziwilk, Z., Zodrow, K., Studies on the vitamin B_{12} content of the leguminous plants, Acta Microbiol. Polon. **6**, 233-238, 1967.

[2] Peston, J. M., Leucine 2,3-aminomutase: A cobalamin-dependent enzyme present in bean seedlings, Science **195**, 301-302, 1977.

[3] Spiess, L. D., Lippincott, B. B., Lippincott,

J. A., Effect of hormones and vitamin B_{12} on gametophore development in the moss *Pylaisiella selwynii,* Amer. J. Botany **60,** 708, 1973.
[4] Busse, M., Über die Wirkung von Kobalt auf die Zellwand bei Avenokoleoptilen, Planta **53,** 25, 1959.
[5] Shkolnik, M., Trace elements and plant growth regulators, Referativnyi Zhurnal **18,** 839, 1976.
[6] Lipskaya, G. A., Molecular aspects of investigation of the photosynthetic apparatus at chloroplast level and the productivity of plants given different cobalt rates, Referativnyi Zhurnal, Biologiya **1,** 1978.
[7] Foy, C. D., Chaney, R. L., White, M. C., The physiology of metal toxicity in plants, Annual Review of Plant Physiology, **29,** 511-566, 1978.
[8] Lipskaya, G. A., Ivanov, N. P., Dolgorukova, T. N., Leaf plastid apparatus and grain yield of barley as affected by application of cobalt to newly reclaimed peat-bog soil of transitional type, Agrokhimya, **2,** 122-127, 1978.
[9] Herich, R., Bobak, M., Influence of exogenously added cobalt on the submicroscopical structure of the proplastids, Acta Facultatis Rerum Naturalium Universitatis Comenianae Physiologia Plantarum, **13,** 3-8, 1977.
[10] Wallace, A., Alexander, G. V., Chadhry, F. M., Phytotoxicity of cobalt, vanadium, titanum, silver, and chromium, Comunications in Soil Science and Plant Analysis, **8,** (9), 751-756, 1977.
[11] Wallace, A., Romney, E. M., Kinnear, J., Alexander, G. V., Single and multiple trace metal excess effects on three different plant species, J. of Plant Nutr. **2,** 11-23, 1980.
[12] Ormrod, D. P., Proctor, J. T. A., Hofstra, G., Phillips, M. L., Air pollution effects on agricultural crops in Ontario: a review, Canadian Journal of Plant Science, **60,** (3), 1023-1030, 1980.
[13] Purves, D., Trace-element contamination of the environment, Elsevier Scientific Publishing 1977, p 260.
[14] Commonwealth Bureau of Soils, Cobalt in plant nutrition and plant metabolism, Annotated Biblography, Commonwealth Bureau of Soils 1977. This contains copies of relevant abstracts from Soils and Fertilizers and cover 1956-76.
[15] Lau, O. L. and Yang, S. F., Inhibition of ethylene production by cobaltous ions, Plant Physiol. **58,** 114-117, 1976.
[16] Wallace, A., Romney, E., Kinnear, J. E., Moeller, R. T., Metal ratios as an index of plant species differences on uptake characteristics of different trace metals, J. of Plant Nutr., **2,** 25-34, 1980.
[17] Lipskaya, G. A., Skurko, I. E., On the relationship of cobalt with biosynthesis of pigments, Fiziologiya Biokhimya Kulturnykh Rastenii **13,** 173-179, 1981.
[18] Lipskaya, G. A., Chlorphyll heterogeneity and activity of Hill reaction under the residual effect of cobalt, Fiziologiya i Biochkhimiya Kulturnykh Rastenii **7,** 522-525, 1975.
[19] Lipskaya, G. A., Effect of cobalt and heteroauxin on the morphology and structure of barley leaves, Biyalagichnykh Navuk No **5,** 121-123, 1974.
[20] Tandon, P. K., Awasthi, C. P., Effect of cobalt toxicity on the germination of barley seeds with special reference to catalase and peroxidase activity, Indian Journal of Agricultural Chemistry, **12,** 103-106, 1979.
[21] Austenfeld, F. A., Phytotoxicity of Ni and Co to *Phaseolus vulgaris* grown in solution culture, Zeitschrift für Pflanzenernährung und Bodenkunde **142,** 786-791, 1979.
[22] Saito, Y., Takahashi, K., Studies on heavy metal pollution in agricultural land. 5. Interaction between heavy metals in zinc, nickel, cobalt and copper absorption and translocation in rice plants, Agricultural Experiment Station **33,** 23-31, 1979.
[23] Application of the spectrophotometric method with 2,2-dipyridyl cetoxime to the determination of cobalt in plants, Anais da Escola Superior de Agricultura "Luiz de Queiroz" **33,** 55-66, 1976.
[24] Yadav, D. V., Khanna, S. S., Chaudhary, M. L., Cobalt content in different plant species as affected by available cobalt status of Haryana soils, Haryana Agricultural Univ. Journal of Research **6,** 12-16, 1976.
[25] Pfau, A. and Kallistratos, G., Aufnahme und Verteilung von ^{60}Co-Vitamin B_{12} und $^{60}CoCl_2$ in *Vicia taba,* in Vitamin B_{12} und Intrinsic Factor 2. Europäisches Symposium, Ed. H. C. Heinrich, Enke, Stuttgart 1962, pp 432-443.
[26] Kliewer, M., Evans, H. J., B_{12} coenzyme content of the nodules from legumes, alder and of *Rhizobium meliloti,* Nature **194,** 108-109, 1962.
[27] Kliewer, M., Evans, H. J., Identification of cobamide coenzyme in nodules of symbionts, isolation of the B_{12} coenzyme from *Rhizobium meliloti,* Plant. Physiol. **38,** 55-59, 1962.
[28] Kliewer, M., Evans, H. J., Effect of cobalt defi-

ciency on the B_{12} coenzyme content of *Rhizobium meliloti,* Arch. Biochem. Biophys. **97,** 428-429, 1962.

[29] Kliewer, M., Evans, H.J., B_{12} coenzyme content of the nodules from legumes, alder and of *Rhizobium meliloti,* Nature **195,** 828, 1962.

[30] Evans, H.J., Kliewer, M., Ann. N.Y. Acad Sci. **112,** 735, 1964.

[31] Fries, L., Vitamin B_{12} in *Pisum sativum,* Physiol. Plant. **15,** 566, 1962.

[32] Shemahanova, N.M., Bunko, J.P., Content of cobalamin and riboflavin in nodules of *Phaseolus vulgaris* infected with active and less active nodule forming bacteria species (Russian), Proc. Acad. Sci Sov. Union **47,** 24, 1969.

[33a] Kliewer, M., Evans, H.J., Cobamide coenzyme contents of soybean nodules and nitrogen fixing bacteria in relation to physiological conditions, Plant. Physiol. **38,** 99-104, 1963.

[33b] Troitskaya, G.N., Vitamin B_{12} content in soybean root nodules as affected by plant growth conditions, Fiziol. Rast. Mosc. **30,** 158-164, 1983.

[34] Ramenskaya, M.L., Microelements in plants in the Far North. Referativnyi Zhurnal, Biologiya **1,** 1975.

[35] Wilson, S.B., Hallsworth, E.G., The distribution of cobalt in T. subterranean, Plant Soil **23,** 60, 1965.

[36] Ahmed, S., Evans, H.Y., Cobalt: A micronutrient element for the growth of soybean plants under symbiotic conditions, Soil. Sci. **90,** 205, 1960.

[37] Wilson, J.B., Reisenauer, H.M., Cobalt requirement of symbiotically grown alfalfa, Planta Soil **11,** 364, 1963.

[38] Bond, G., Hewitt, E.J., Cobalt and fixation of nitrogen by root nodules of alnus and caguarina, Nature **195,** 94, 1962.

[39] Ahmed, S., Evans, H.Y., The essentiality of cobalt for soybean plants grown under symbiotic conditions, Proc. Nat. Acad. Sci. USA **47,** 24-36, 1961.

[40] Shemahanova, N.M., Bunko, J.P., Content of cobalamin and riboflavin in nodules of *Phaseolus vulgaris* infected with active and little active species, Izv. A.N. USSR (Russian) **3,** 413, 1969.

[41] Gulati, S.L., Nodulation and yield of soybean and berseem crops as affected by culture methods of preparation of *Rhizobium* inoculants in different media, Seeds Microbiol. **7,** 39-43, 1981.

[42] Robson, A.D., Dilworth, M.J., Chatel, D.W., Cobalt and nitrogen fixation in *Lupinus angustifolius.* 1. growth, nitrogen concentrations and cobalt distribution, New Phytologist **83,** 53-62, 1979.

[43] Dilworht, M.J., Robson, A.D., Chatel, D.L., Cobalt and nitrogen fixation in *Lupinus angustifolius L.* 2. Nodule formation and function, New Phytologist **83,** 63-79, 1979.

[44] Vanselow, A.P., Cobalt, in Criteria for plants and soils, ed. H.D. Chapman, University of California, Div. of Agric. Sciences, 1966, pp. 142-156.

[45] Mengel, K., and Kirkby, E.A., Principles of plant nutrition, International Potash Institute, Worblaufen-Bern, Switzerland 1980.

[46] Duvigneaud, P., and Denaeyer-de Smet, S., Effect of some heavy metals in the soil (copper, cobalt, manganese, uranium) on the vegetation in Upper Katanga Ier Colloq. Soc. Bot. France 1959, p. 121.

[47] Wilson, S.B., Hallsworth, E.G., Studies on nutrition of the forage legumes. IV. The effect of cobalt on the growth of nodulated and non-nodulated *Trifolium subterraneum L.,* Plant, Soil **22,** 260, 1967.

[48] Wilson, S.B., Nicholas, D.J.D., A cobalt requirement for non-nodulated legumes and for wheat, Phytochemistry **6,** 1057-1060, 1967.

[49] Evans, H.J., Russell, S.A., Physiological chemistry of symbiotic nitrogen fixation by legumes, in The chemistry and biochemistry of nitrogen fixation, ed. J.R. Postgate, Plenum Publishing Co, 1971, pp. 191-244.

[50] Bottomley, F., Dinitrogen complexes of the transition metals, in A treatise on dinitrogen fixation, eds. R.W.F. Hardy, F. Botomley, R.C. Burns, Willey and Sons, New York 1979, pp. 109-167.

6.4 Cobalamin in food and feeding stuffs

The corrinoids present in food derived from animal tissues consist entirely of various forms of cobalamin; while food obtained by fermentation may also contain other corrinoids inactive for man and animals. Due to very specific cobalamin absorption mechanism of man's digestive system the B_{12}-analogues present in food in reasonable quantities are not harmful for man. However the B_{12}-analogues, present in some sources in quantities exceeding the true vitamin B_{12}, may interfere with the microbiological assays. For that reason, whenever the contamination of non-cobalamin corrinoids is suspected, a specific assay for cobalamin has to be applied e.g. the *Euglena*- or *Ochromonas* tests. For the assay of cobalamin in animal derived food not exposed to fermentation the more convenient *E. coli* test is adequate (for details of microbiological assays see section 5.3.1). The amounts of vitamin B_{12} found in food by various authors differ [1-14]. The discrepancies arise for two main reasons: (i) the values obtained by microbiological assay depend on the organism used; the *Ochromonas* test produces the lowest-, while the *E. coli* plate test the highest values (compare the analytical data obtained by various tests in Tables 6-13, 6-14); (ii) the same food items obtained from different batches or in different geographic areas vary greatly in cobalamin content. The amounts of vitamin B_{12} in meat, particularly in liver, milk [5, 8-12] and eggs, depend very much on the animals diet, its age and other environmental factors (see occurrence of vitamin B_{12} in ruminants in section 6.5). Thus the values of B_{12} in food items presented in Table 6-7 A are only roughly adequate to that what we expect to find in products eaten everyday. Appreciable losses of vitamin B_{12} are observed in processed food [3, 7, 11, 13]. Vitamin B_{12}, however resistant to boiling if in pure aqueous solutions, is quite labile in presence of proteins in particular those containing thiol residues. Various forms of cobalamin show different stability: Ado-Cbl and Me-Cbl are more stable than OH-Cbl. The cyanocobalamin (CN-Cbl) used in medicine is most stable to heat treatment, but unless not supplemented, it does not occur (or occurs in negligible quantities) in raw food. So far dietitians do not specify what proportion of the cobalamin forms in the diet is optimal for man. Since the concentration of Me-Cbl in human tissues is decreasing with age (see section 6.6) conclusions concerning cobalamin forms desired in the diet can be drawn already from this fact. Early estimates did not distinguish cobalamin forms since the analyzed food samples were treated, prior to assay, with cyanide which results in conversion of all forms to CN-Cbl. In the course of seventies, efforts were undertaken to distinguish the particular cobalamins in food [4]. Due to technical difficulties approximative data are available only (Table 6-7 B). Since higher plants do not contain appreciable quantities vitamin B_{12} in the green parts [15] (see occurence of vitamin B_{12} in plants, Table 6-4 and text in section 6.3.2), no cobalamin has been found in plant derived food such as: bread, wheat flour, rice, fruits, vegetables etc [6, 15-19]. Algae and other plankton, which are assumed to stand at the beginning of the food chain, contain appreciable quantities of cobalamin (see Table 6-3 c and text in section 6.3.1).

Table 6-7 A Occurrence of vitamin B_{12} in food and other biological materials

Source	μg of vit. B_{12} per 100 g (or 100 ml)		Vitamin B_{12} forms			Test organism	Reference author
	Fresh weight	Dry weight	Ado-Cbl	Me-Cbl	OH-Cbl		
1	2	3	4	5	6	7	8
Raw food:							
Beef muscle	2–8					not indicated	Smith [1]
Veal muscle	2					"	"
Beef kidney	20–50					"	"
" "		85					Friedrich [2]
Chicken kidney	14					*L. leichmannii*	"
Rabbit kidney	165					not indicated	"
Beef heart	25					"	Smith [1]
Beef liver	50–130					"	"
" "		214					Friedrich [2]
Chicken liver	8					not indicated	Smith [1]
" "	13					*L. leichmannii*	Friedrich [2]
Herring liver	34					not indicated	Smith [1]
Rabbit liver	113					"	Friedrich [2]
Pork	0.1–5					not indicated	Smith [1]
Egg (whole)	0.8–1.25 μg/egg					"	Friedrich [2]
Egg yolk	1.2 μg per yolk		+	+	+		Adams
Egg white	0					not indicated	Smith [1]
Herring (whole)	11					"	"
Herring (fillet)	13					"	"
Milk (cow)	0.2–0.6		+	+	+		Adams et al. [3]
" "	0.36	2.97				*L. leichmannii*	Friedrich [2]
" "	0.23–0.3					*E. gracilis*	"
" "	0.26–0.33					*E. coli 113-3*	Friedrich [2]
" "	0.65					*E. gracilis*	Adams et al. [3]
Milk (goat)	0.01					not indicated	Smith [1]
Milk (human)	0–0.24					*Lactobacillus*	Neujahr [5]
Milk (human colostrum)	0.243						Samson et al. [14]
Prepared food:							
Beef	0.8–1.39		+	o	+	*E. gracilis*	Adams et al. [3, 4]
Chicken	0.25–0.55					"	"
Ham	0.21–0.52		+	o	+	"	"

Table 6-7 A continued

Source	µg of vit. B_{12} per 100 g (or 100 ml)		Vitamin B_{12} forms			Test organism	Reference author
	Fresh weight	Dry weight	Ado-Cbl	Me-Cbl	OH-Cbl		
1	2	3	4	5	6	7	8
Kidney (sheep)	79.1					"	"
Kidney (ox)	5.8					"	"
Lamb	1.35					"	"
Liver (ox)	75,9					"	"
Luncheon meat canned	0.79					"	"
Pork	0.4–0.45					"	"
Sausages (beef and pork)	0.4–0.92					"	"
Veal fricassee	0.91–1.44					"	"
Anchovy fillets (canned)	21.84					"	"
Brisling	11.0					"	"
Crab (canned)	0.53					"	"
Herring	5.3–8.5					"	"
Prawns (canned)	0.55					"	"
Sardines (canned)	12.4		+	o	+	"	"
Shrimps (canned)	0.79					"	"
Tuna	0.46					"	"
Cheese: Brie	1.29					"	"
Cheddar	0.77		+	+	+	"	"
Cheshire	0.89					"	"
Danish blue	1.18					*E. gracilis*	Adams et al [3, 4]
Edam	1.25–2.0					"	Friedrich [2]
Edam	1.7–2.6					*E. coli 113-3*	"
Milk (whole pasteurized)	0.39		o	o	+	*E. gracilis*	Adams et al. [3, 4]
" (" evaporated)	0.15					"	"
Egg: boiled	1.03					"	"
fried	0.61					"	"
scrambled	0.30					"	"
Feeds:							
Fish meal	10–25					not indicated	Smith [1]
" "		19–150				*L. leichmannii*	Friedrich [2]
Fish solubles	15–40					not indicated	Smith [1]
" "	29–37					*L. leichmannii*	Friedrich [2]
Silage	0.1–2					not indicated	Smith [1]

Table 6-7B Relative amounts of the different forms of vitamin B_{12} found in food (from Farquharson and Adams [4])

Food and method of preparation	Vitamin B_{12} present[a] as:				
	Adenosyl-cobalamin	Hydroxo-cobalamin	Cyano-cobalamin	Methyl-cobalamin	Sulphito-cobalamin
	Meat and poultry				
Bacon, grilled	+	+	O	+	O
Beef, brisket, boiled	++	++	O	O	O
Beef, rump, braised	+++	++	O	O	O
Beef, corned, canned (St. Michael; Marks Spencer Ltd, Baker Street, London)	O	++−	O	O	−+
Beef, spread (Princes; Princes Foods Ltd, Liverpool)	++	++−	O	++	O
Beef, spread (Shippams; Shippams Ltd, East Walls, Chichester)	++	++	O	+	O
Chicken, casserole	+++	++	O	++	O
Gammon, canned (Danoxa; S. Daniels Co. Ltd, Wilec House, City Road, London)	+	++	O	O	O
Ham and chicken roll, canned (Cross Blackwell; Nestle Ltd, St. Georges House, Croydon)	+	+++	O	O	O
Lamb, gigot chop, braised	+	++	O	O	O
Liver, lamb, raw	+++	+++	O	++	O
Tongue, ox, canned (De Haan (Foods) Ltd, Park Lane, Abram, Wigan, Lancs)	O	+++	O	O	++
	Fish				
Haddock, boiled	+++	++	+	+	O
Salmon, canned (John West; John West Foods Ltd, Stanley Street, Liverpool)	++	+++	O	O	++
Salmon, spread (Princes; Princes Foods Ltd)	++	+++	O	+	O
Sardines, canned (Joy to Eat; Joy to Eat Foods Ltd, 50 Mark Lane, London)	++	+++	O	O	++
	Dairy produce				
Cheese, cheddar	++	++	O	++	+
Cheese, spread (Kraft; Kraft Foods Ltd, Braeview, East Kilbride)	++	++	O	+++	O
Egg yolk, raw	+	++	O	+++	+
Milk, cows', whole, fresh	+	++	O	+	O
Milk, cows', pasteurized	O	++	O	O	O
Milk, cows', evaporated (Carnation; Carnation Foods Ltd, Carnation House, II High Road, London)	++	O	O	++	O

[a] Detected and identified by thin-layer chromatography and bioautography and results scored as follows: O, cobalamin absent from extract; +, cobalamin present in undiluted extract; ++, coablamin present in undiluted extract and in 1:4-dilution extract; +++, cobalamin present in undiluted extract and in both 1:4- and 1:16-dilution extracts.

References

[1] Smith, E.L., Vitamin B_{12}, Methuen, London, 3rd edn, 1965.

[2] Friedrich, W., Vitamin B_{12} und verwandte Corrinoide, in Fermente, Hormone, Vitamine, Band III/2, Eds. Ammon, R. and Dirscherl, W., Georg Thieme Verlag, Stuttgart 1975, pp.169-186.

[3] Adams, J.F., McEvan, F., Wilson, A., The vitamin B_{12} content of meals and items of diet, Brit. J. Nutr. **29**, 65-72, 1973.

[4] Farquharson, J., Adams, J.F., The forms of vitamin B_{12} in foods, Brit. J. Nutr. **36**, 127-136, 1976.

[5] Neujahr, H.Y., Jonasson, M., Lundin, H., On vitamin B_{12} and its occurrence in milk, Milk Dairy Research, Alnarp, Sweden Report No.59, 1960 (see also references 27-30 in section 6.5).

[6] McCance, R.A., Widdowson, E.M., The composition of foods, Spec. Rep. Ser. Med Res. Coun., 297, 1960.

[7] Molonon, B.R., Bowers, J.A., and Dayton, A.D., Vitamin B_{12} and folic acid content of raw and cooked turkey muscle, Poult. Sci. **59**, 303-307, 1980.

[8] Craft, J.L., Matthews, D.M., Linnell, J.C., J. Clin. Path. **24**, 449, 1971.

[9] Obrusiewicz, K., Roczniki Instytutu Przemyslu Mleczarskicgo (Polish) **11**, 81, 1960.

[10] Rosenthal, M.L., in The Vitamins, Vol.2, Eds. H.Sebrell and L.Harris, Academic Press, New York 1968, p 86.

[11] Kilshaw, P.J., Heppell, L.M.J. and Ford, J.E., Effects of heat treatment of cow milk and whey on the nutritional quality and antigenic properties, Arch. Dis. Child. **57**, (11), 842-847, 1982.

[12] Samson, R.R. and McClelland, D.B.L., Vitamin B_{12} in human colostrum and milk. Acta Paediatr. Scand. **69**, 93-100, 1980.

[13] Bennink, M.R. and Ono, K., Vitamin B_{12}, vitamin E and vitamin D content of raw and cooked beef, J. Food Sci. **47**, 1786-1792, 1982.

[14] Casey, P.J., Speckman, K.R., Ebert, F.J. and Hobbs, W.E., Radioisotope dilation technique for determination of vitamin B_{12} in foods, J. Assoc. Off. Anal. Chem. **65**, 85-88, 1982.

[15] Janicki, J., Pędziwilk, F., Occurrence of vitamin B_{12} in seeds of leguminous plants, Acta Physiol. Polon. **5**, 295-297, 1958 (see also Table 6-4 and the reference to this Table).

[16] Rao, D.R., Reddy, A.V., Pulusani, S.R., Cornwell, P.E., Biosynthesis and utilization of folic acid and vit. B_{12} by lactic cultures in skin milk, J. Dairy., Sci. **67**, 1168-1174, 1984

[17] Scott, K.J., Bishop, D.R., Zechalko, A., Edwards-Webb, J.D., Nutrient content of liquid milk. 2. Content of vitamin C, riboflavin, folic acid thiamin, vitamin B_{12} and B_6 in pasteurized milk as delivered to the home and after storage in the domestic refrigerator, J.Dairy. Res. **51**, 51-58, 1984

[18] Syltie, P.W., Dahnke, W.C., The vit. B_1, vit. B_2 vit. B_6, vit. B_{12} and vit. E. Contents of hard red spring wheat Triticum aestivum at influenced by fertilization and cultivar. Qual. Plant. Plant Foods Hum. Nutr. **32** 51-58, 1983

[19] Murdock, F.A. and Fields, M.L., B-vitamin content of natural lactic acid fermented corn meal. J. Food. Sci **49**, 373-375, 1984

6.5 Vitamin B_{12} in animals with special reference to ruminants

All animals seem to need vitamin B_{12} for growth and health. Since the microorganisms are the primary source of the vitamin the animals developed, depending on species, various routs of its uptake and transfer into the body. Man and a number of animals acquire the vitamin B_{12} from the diet; ruminants produce it by fermentation in the rumen; other animals gain it from food polluted with soil or from the faeces by coprophagy [1-4]. The latter source might be significant for some mammals as: rat, rabbit and dog and also for fowl. Appreciable quantities of corrinoids are produced by the intestinal flora of fish [5b], earth worms [5c] and insects [5d]. However, little is known to what extent they are utilized by the animal organisms. There is considerable synthesis of cobalamin and other cobamides in the caecum and large gut [5a]. This source, however, is not available, since vitamin B_{12} is neither absorbed from the large gut nor possible is the form of the vitamin in intact bacteria available for absorption. The animals take up daily relatively small amounts of the vitamin in the range approx. 0.05–0.5 μg/kg body weight. Its concentration in the diet of most animals is also very low. In order to absorb it with high efficiency the animals developed very effective absorption mechanisms based on proteins with high affinity to B_{12} and resistance to digestion and microbial degradation (see Chapter 8). Due to high storage capacity of the animal organs (mainly liver and kidneys) the animals maintain in good health for a longer period of time even when the vitamin B_{12} is absent in the diet or is insufficiently absorbed for various reasons. But when the vitamin B_{12} level in the animals body declines below a certain limit, several more or less specific symptoms appear [3 b, 6–17 a, b] e.g. loss of appetite and body weight (see ruminants below); growth retardation and infertility; and abnormal high concentration of methylmalonic acid in urine. B_{12}-deficient animals do not show typical for man abnormal hematological or neurological changes (see Chapter 9). Vitamin B_{12}-depleted young animals respond with growth to diet supplemented with vitamin B_{12} [15 a, b]. This feature has been utilized in B_{12}-assay by using chicks, rats or piglets as test animals (see Chapter 5). In general the distribution of the vitamin in organs of animals, in particular in baboon, is similar to those of man (compare Tables 6–8, 6–9, 6–10 and 6–16) [18–26]. The liver contains by far its greatest amount followed by kidneys and the walls of alimentary tract. Differences of mean B_{12} contents in organs observed amongst species are frequently smaller than those of normal-and vitamin B_{12}-deficient animals within one species. The concentration of cobalamin in milk of various species differs markedly (Table 6–12) [4, 27–30]. It is correlated with B_{12} concentration in serum [71] (Fig. 6–12). Concerning the nutritional value, milk contains a protein [27 b] which strongly enhances the absorption of the vitamin B_{12} by suckling piglets [28, 29]. Since there is a broad range of vitamin B_{12} content in animal tissues considered to be normal, the nutritional value of meat and its products may show undesired variations in

Table 6-8 Distribution of tracer dose of vitamin B_{12} in animals (per cent total radioactivity in body; compiled by Chanarin [4]).

	Rooster [65]	Calf [65]	Pig [65]	Mouse [20]	Hamster [20]	Guinea pig [19]	Rat [19]	Dog [21]
Liver	27	32	39	29.5	21.6	36.2	16	47
Kidney	9	2	6	7.8	8.7	2.9	12	6.4
Pancreas								3
Spleen				1.8	0.4	0.3	1	4
Brain							0.8	5.9
Gonad				1.9		0.4	2.6	0.8
Heart				0.5	0.5	0.3	2.6	8.2
Muscle	12	14	12.5	20.0	9.3			
Lung							2.4	3.8
Marrow								
Alimentary tract							11	18.5
Bone	4.5	12.5	10					
Skin	5	4	5					
Carcass				32.5	52		54	

Table 6-9 Distribution of ^{58}Co in tissues and alimentary tract contents after subcutaneous injection of 500 μg [^{58}Co]cyanocobalamin (13 μCi) over 20 days into a vitamin B_{12}-deficient sheep (from Smith and Marston [58]).

Organ, tissue or section of alimentary tract	Tissues[a]			Contents of alimentary tract[b]			
	Wet weight g	Specific activity nCi/g wet	Percent-age of total administered dose	Wet weight g	Dry matter %	Specific activity nCi/g dry	Percentage of total administered dose
Liver	401	5.35	16.5	–	–	–	–
Kidneys	81	8.42	5.3	–	–	–	–
Spleen	55	1.45	0.6	–	–	–	–
Lungs	385	1.03	3.1	–	–	–	–
Brain	95	0.40	0.3	–	–	–	–
Heart	158	0.35	0.4	–	–	–	–
Skeletal muscle	–	0.13	–	–	–	–	–
Bone marrow	–	0.05	–	–	–	–	–
Blood	–	0.08	–	–	–	–	–
Gall bladder	2	0.72	–	–	–	–	–
Gall bladder contents	2.4	0.59	–	–	–	–	–
Oesophagus	37	0.27	0.1	–	–	–	–
Rumen	472	2.90	10.5	3150	16.4	0.12	0.5
Reticulum	119	2.05	1.9	330	12.6	0.05	0.1
Omasum	133	3.06	3.2	161	17.2	0.33	0.1
Abomasum	178	0.81	1.2	286	10.4	0.54	0.1
Duodenum	9	1.18	0.1	12	6.2	3.06	0.1
Ileum 1	40	2.24	3.8 (Ileum 1–8)	69	10.5	1.81	0.8 (Ileum 1–8)
2	43	1.21		98	10.3	2.19	
3	20	1.42		73	10.9	1.47	
4	25	1.49		75	7.6	2.19	
5	75	1.71		140	8.0	2.22	
6	33	1.10		93	7.5	1.42	
7	41	1.53		109	6.6	0.44	
8	44	1.51		92	7.0	0.41	
Caecum	144	0.82	0.9	580	11.4	0.64	0.4
Colon 1	58	0.71	1.3 (Colon 1–4)	75	14.5	0.55	0.3 (Colon 1–4)
2	53	0.63		58	18.2	0.68	
3	58	0.86		50	23.5	0.86	
4	80	0.58		44	21.7	0.68	
Rectum	89	0.40	0.3	34	27.4	0.51	0.1

[a] It appears from this experiment that the liver and the rumen wall accumulate appreciable quantities of vitamin B_{12} amounting 16.5% and 10.5% of the total administered dose respectively. In this experiment the vitamin reaches the rumen wall tissues via blood stream. So far there is no evidence for absorption of the vitamin B_{12} from the rumen content.

[b] The rise of specific radioactivity of the duodenum content seems to be an indication for secretion of the radioactivity into the duodenum presumably with the bile. The decline of specific radioactivity in sections 6–8 of the ileum can be interpreted as absorption of the radioactivity. The biliary secretion of vitamin B_{12} and its reabsorption in the ileum is observed in man (see Fig. 8–23).

Table 6-10 Vitamin B_{12} concentrations (µg/g) in tissues from two baboons *(Papio cynocephalus anubis)*. Total vitamin B_{12} (µg/organ) is given in parentheses (from Siddons and Jacob [2 a]).

	Baboon no.	
	129	64
Liver	0.720 (153.4)	0.392 (125.8)
Pituitary	0.192	0.084
Kidney	0.145 (5.2)	0.069 (3.1)
Heart	0.125 (8.5)	0.062 (6.6)
Spleen	0.090 (1.3)	0.110 (2.2)
Pancreas	0.075 (1.0)	0.081 (1.5)
Lymph nodes	0.070	0.045
Stomach	0.057	0.036
Duodenum	0.055	0.023
Ileum	0.047	0.049
Brain	0.034 (5.0)	0.016 (2.6)
Colon	0.029	0.031
Caecum	0.026	0.034
Thyroid	0.025	0.014
Tongue	0.025	0.006
Spinal cord	0.023	–
Testis	–	0.013 (1.0)
Bladder	0.014	0.011
Lung	0.011	0.019
Muscle	0.008	0.002
Skin	0.001	0.001

Table 6-11 Various forms of cobalamin in rat and cow (from Brazenas et al. [18]).

Organ or tissue	Cobalamins in ng/g or ng/ml			
	Total Cbl	Me-Cbl	Ado-Cbl	OH-Cbl
Rat:				
Liver	133.8	traces	85	47
Kidney	473.8	48.7	200	224
Spleen	63.7	4.2	34.7	24.8
Plasma	0.6	0.18	0.36	0.05
Cow:				
Liver	1161.0	traces	836.0	748.0

this respect. The B_{12}-content in animals is of serious concern of dietitians. In particular ruminants are the most significant source of cobalamin for man. Thus proper diet and farming of those animals is of great nutritional importance.

Ruminants. Unlike other mammals the ruminants produce a great part of their food by fermentation in the rumen. In particular relatively large quantities of volatile fatty acids such as acetic- and propionic acid are produced by ruminous fermentation [31,

Table 6-12 Vitamin B_{12} contents in milk of various species (Compiled by Chanarin [4]).

Source of milk	B_{12} content µg/l
human	0.1 - 1.5
cow	3.2 - 12.4
ewe	1.0 - 6.0
goat	0.07 - 0.18
sow	0.03 - 2.7
rat	11.0 - 95.0
dog	0.7 - 13.0
horse	0.02

32]. The absorption and metabolism of these acids is disturbed in vitamin B_{12}-deficient animals [32-44a-d]. The microorganisms in the rumen which do the fermentation require the corrinoids for growth as well. They produce the vitamin themselves or take it from other ruminous bacteria [68]. Among various forms of corrinoids synthetized in course of ruminous fermentation a small part only (3-15%) is in the form of cobalamin [45-51]. Under natural conditions the synthesis of corrinoids is limited by concentration of cobalt in the diet which occurs there frequently in quantities not sufficient for optimal requirement of the animal [52-71]. Cobalt depleted diet, mainly grass, below a certain limit (0,08 ppm) [54] can be the cause of severe losses in population of domestic animals [69, 70] and may lead to vitamin B_{12} deficiency in man. This problem has been a major concern in Australia where the cobalt deficiency in sheep known as bush sickness was first observed half a century ago. Thus most extensive studies on dietary cobalt and B_{12}-production and absorption have been done using sheep. Sheep fed experimentally with cobalt deficient diet lose apetite and almost one third of its weight and finally die [58, 59]. The deficiency of cobalt in the diet affects markedly the level of cobalamin in liver and serum. It has been shown that a daily supplement of fixed amounts of cobalt given to cobalt depleted diet over a longer period of time results in a certain balance between the vitamin B_{12} level in serum and liver and the quantity of dietary cobalt (Figs. 6-2A, 6-2B, 6-3) [58, 59]. Sheep require daily approximately 1 mg of Co in the diet in order to produce and absorb cobalamin in quantities sufficient to saturate the storage capacity of the liver. On the other hand, a daily intake of 20-50 µg of Co is adequate for the maintenance of body weight and health [58, 59]. The transfer of sheep from cobalt supplemented- (1 mgCo/d) to cobalt poor diet results in dramatic decline of liver- and serum B_{12}-level (10-14 fold). Presumably all other tissues and body fluids tend to adjust vitamin content accordingly to the B_{12}-level of the liver. Thus the concentration of vitamin B_{12} in meat, milk and other products derived from ruminants depends essentially on dietary cobalt. The data in Table 6-7A indicating the vitamin B_{12} content in animal derived food may not be representative to meat or its products from ruminants fed with cobalt poor feed. The minimal daily requirement of sheep for vitamin B_{12} is approximately 11 µg; while an absorption of 34 µg is sufficient to keep the liver saturated with the vitamin

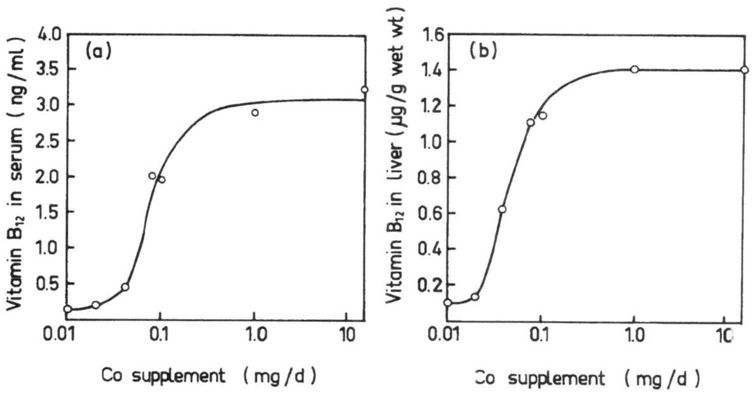

Fig. 6-2 Relationship between daily supplements of cobalt administered per os to sheep and the mean concentrations of vitamin B_{12} in (a) the blood sera and (b) the livers of the appropriate groups. The sheep previously depleted of vitamin B_{12} by feeding cobalt poor diet were next fed a diet with the cobalt supplements indicated in the Figures. After 71 weeks of cobalt administration liver samples taken by aspiration biopsy and serum of the animals were analyzed for vitamin B_{12} content (from Marston [58]).

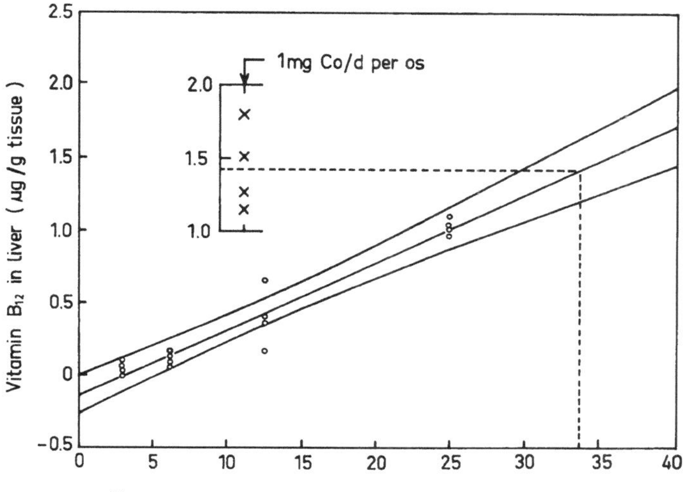

Fig. 6-3 Relationship in sheep between amount of vitamin B_{12} administered parenterally and concentration of the vitamin in the liver after 50 weeks of treatment. Prior treatment the animals were depleted of vitamin B_{12} by feeding cobalt poor diet (ca. 30 weeks). Inset: concentrations of vitamin B_{12} in the livers of four sheep fed a diet supplemented with 1 mg Co/d. The broken lines indicate the average concentration in the liver and the amount of vitamin B_{12} estimated to be absorbed each day from the gut. The curved lines on either side of the linear regression indicate fiducial limits at $P = 0.01$. One can conclude from the regression that intramuscular injection of 6.25 µg vitamin B_{12}/d provided close to the minimum supplement necessary (see also Fig. 6-2) to maintain the physiological requirements of sheep kept on Co-deficient diet (from Marston [58]).

[58, 59] (Fig. 6-3). Other authors found higher values [48, 49, 61]. The above indicated quantities of the vitamin correspond to 0.48 μg and 1.5 μg of cobalt respectively. Since much greater quantitiy of cobalt is required for maintenance of growth and health it appears that the bulk of it is not utilized directly by the sheep. The efficiency of utilization of cobalt is higher when its concentration in the diet is lower (Table 6-13). One of the main causes of the low efficiency is the production of relatively large quantities of other corrinoids by ruminous microorganisms which are not absorbed-, or not utilized, by the animal, such as: cobinamide (Cbi), pseudovitamin B_{12} (Ade-Cba), Factor A (2-MeAde-Cba) or Factor III (5-OHBza-Cba) [45]. Cultures of bacteria isolated from rumen synthetize these inactive corrinoids in quantities exceeding up to 30 times that of cobalamin (Table 6-14). The amount of B_{12} analogues indicated in rumen is somewhat lower but still accounts approximately for 90% of all corrinoids (Table 6-13). These corrinoids, however, are growth factors for microorganisms and the nucleotide containing forms are active in bacterial enzyme systems. They seem to play its role in ruminous fermentation (see B_{12}-dependent enzymes in Chapter 7). The low efficiency of cobalamin production in rumen is also due to short fermentation period and anaerobic conditions unfavourable for the synthesis of cobalamin precursor, the base 5,6-dimethylbenzimidazole (see Chapter 4). Indeed a supplement of the base to sheep's diet causes an immediate rise of cobalamin synthesis in the rumen at the expense of other forms and a greater uptake of the vitamin by the animal [49]. The addition of benzimidazole to feed seems to be a solution for more effective utilization of cobalt. Its application might be useful in cobalt poor areas. The supplementation of the diet of ruminants with cobalamin has not been found useful [58]. The ruminous microorganisms are converting the bulk of supplemented cobalamin into inactive forms of corrinoids before it can be absorbed by the animal [58, 59]. It seems that interconversion of various forms of corrinoids, based mainly on substitution of the nucleotide moiety by another one, is very vigorous in the rumen and other parts of the alimentary tract. This metabolism of corrinoids observed already in microorganisms three decades ago is little understood on biochemical level (see Chapter 4). The distribution of cobalamin in ruminants is similar to that of other animals (compare Tables 6-8 and 6-9). As shown by labelling studies the sheep (and may be all ruminants) are distinguished, however, by high content of cobalamin in the tissues of the rumen and other organs of the alimentary tract. The tissues of the alimentary tract contain approximately 23% of total cobalamin, following by liver, 16.5%, and kidneys, 5.3% (Table 6-9). As mentioned above the quantities of cobalamin produced in the rumen and present in the contents of the alimentary tract and in the faeces and urine depend essentially on the intake of cobalt, although a part of cobalamin may be excreted into the alimentary tract with the bile (see B_{12} content in duodenum in Table 6-9). Under cobalt deficient conditions sheep produce in the rumen approximately 110 μg of cobalamin per day with 15% efficiency; while a daily supplement of 1 mg Co to diet results in production of 730 μg of cobalamin per day with 3% efficiency [58]. Other authors providing special diet and controlled quantities of cobalt could predict the production of ruminal cobalamin in the range of 113 to 2137 μg daily [55]. The vitamin B_{12} deficient sheep excrete in faeces 5-44 μg of cobalamin per day [58]; while animals fed with Co supplemented diet (1 mg Co/d) excrete 86-300 μg per day [58]. The excretion in urine depends on the rate of absorption and the B_{12}-status of the

Table 6-13 Vitamin B_{12} activities in rumen contents of sheep during 8 h after feeding (from Smith and Marston [58]).

Animal no.	Treatment	Time after feeding (h)	Food intake over preceding 2 h (g)	Lignin content of dry matter in rumen (%)	Vitamin B_{12} activity[a]			
					Ochromonas (µg/g) (dry)	*E. coli* (plate) (µg/g) (dry)	*E. coli* (tube) (µg/g) (dry)	*L. leichmanii* (µg/g) (dry)
193	1 mg cobalt/d	0	0	17.9	1.23	13.9	2.5	–
		2	686	16.3	0.86	9.8	1.8	–
		4	364	16.6	0.73	9.0	1.6	–
		6	0	15.8	0.85	10.0	1.9	–
		8	0	16.0	1.01	12.6	2.1	–
193	Rumen depleted of Co	0	0	19.7	0.102	0.61	0.33	0.27
		2	756	17.3	0.078	0.79	0.30	0.23
		4	294	16.9	0.097	1.03	0.37	0.31
		6	0	15.6	0.090	1.16	0.41	0.34
		8	0	15.8	0.108	1.25	0.45	0.35
110	1 mg Co/d	0	0	19.5	0.66	13.4	2.9	2.1
		2	548	16.2	0.43	7.5	1.9	1.4
		4	349	15.8	0.39	6.8	1.6	1.2
		6	133	16.0	0.46	7.6	1.8	1.5
		8	16	17.5	0.59	9.6	2.3	1.9
110	Rumen depleted of Co	0	0	19.0	0.039	0.30	0.12	–
		2	499	17.2	0.051	0.51	0.20	–
		4	345	15.6	0.074	0.96	0.37	–
		6	169	15.9	0.071	0.89	0.33	–
		8	13	16.1	0.082	1.22	0.43	–

[a] The *Ochromonas* assay is assumed to be most specific for cobalamin since this organism responds very little (or not) to other corrinoids; while *E. coli* responds by growth to a number of vitamin B_{12} analogues, such as: Ade-Cba, 2-MeAde-Cba, Chi, Chi-GDP and 5-OHBza-Cba, which seem to dominate in the rumen. Thus the values obtained by *E. coli* assay are much higher (compare also the production of corrinoids by microorganism isolated from cow's rumen listed in Table 6-14.

Table 6-14 Comparative production of vitamin B_{12} and its analogues by synthetizing strains of ruminal bacteria from cow (from Dryden et al. [45]. Reprinted by permission from Nature **195**, 201, 1962. Copyright © 1962 Macmillan Journals Ltd.).

species, strain[a]	*E. coli*[b] plate assay ng/ml	*E. coli*[b] tube assay ng/ml	*O. malhamensis*[b] assay ng/ml
Butyrivibrio fibrisolvens			
49	3.0 ± 0.3 (6)	1.5 ± 0.1 (3)	0.14 ± 0.04 (3)
Peptostreptococcus elsdenii			
B 159	8.8 ± 1.2 (5)	6.3 ± 0.4 (5)	0.24 ± 0.00 (3)
T 81	4.7 ± 0.4 (5)	2.1 ± 0.2 (6)	0.26 ± 0.02 (2)
Selenomonas ruminantium			
GA 192	35.8 ± 3.6 (6)	15.6 ± 2.3 (5)	0.56 ± 0.07 (3)
HD_4	78.8 ± 7.5 (7)	62.4 ± 6.8 (6)	1.26 ± 0.08 (5)
PC 18	47.8 ± 5.3 (7)	12.3 ± 0.6 (3)	0.31 ± 0.02 (2)
GA 31	52.6 ± 9.3(5)	18.9 ± 2.9 (6)	0.15 ± 0.01 (3)
Unnamed group			
B 385-1	19.5 ± 2.4 (8)	10.1 ± 2.1 (4)	0.41 ± 0.01 (3)

[a] The microorganisms listed have been found the main producers of corrinoids among 21 species isolated from cow's rumen

[b] *O. malhamensis* responds by growth mainly to cobalamin and thus the assay with the use of this organism is most specific, while *E. coli* responds by growth to an number of noncobalamin corrinoids, such as: Ade-Cba, 2-MeAde-Cba, and Cbi, which seem to dominate in the assay medium. It should be emphasized, however, that the *E. coli* plate assay is not reliable if applied to a mixture of corrinoids containing predominantly cobinamide and other derivatives (see Table 5-4). Diffused growth zones appear, which if evaluated against cobalamin standard (with definite growth zones) produce values up to several fold higher than the true total concentration of corrinoids. On the other hand, underestimation is observed if such a mixture is assessed by the *E. coli* tube test (see in reference 6, section 6.1, critical discussion of the assay of corrinoid mixtures by *E. coli*).

whole organism. The values for sheep receiving cobalt depleted- and supplemented diets are in the ranges of 0.07-0.17 μg B_{12}/d and 0.76-3.6 μg B_{12}/d respectively [58].

References

[1a] Rosenthal, H. L., and Cravitz L., Organ, urine and faeces vitamin B_{12} content of normal and starved rabbits, J. Nutr. **64**, 281, 1958.

[1b] Limsuwan, T., and Lovell, R. T., Intestinal synthesis and absorption of vitamin B-12 in channel catfish *Ictalurus punctatus*, J. Nutr. **111**, 2125-2132, 1981.

[2a] Siddons, R. C., and Jacob, F., Vitamin B_{12} nutrition and metabolism in the baboon, Br. J. Nutr. **33**, 415-424, 1975.

[2b] Cornwell, P. E., and Boots, L. R., Selected coenzymes and vitamin values in the female baboon *(Papio cynocephalus)*. J. Med. Primatol. **10**, 284-286, 1981.

[3a] Tonder, S. V., Metz J., and Green R., Vitamin B_{12} metabolism in the fruit bat *(Rousettus aegyptiacus)*. The induction of vitamin B_{12} deficiency and its effect on folate levels, Br. J. Nutr. **34**, 397-410, 1975.

[3b] Van der Westhuyzen, J., Cantrill, R. C., Fernandes-Costa, F., Metz, J., Effect of a vitamin B_{12}-deficient diet on lipid and fatty acid com-

position of spinal cord myelin in the fruit bat Rousettus aegyptiacus, J. Nutr. **113,** 531-537, 1983.

[4] Chanarin, I., in The megaloblastic anaemias, Blackwell Scientific Publications, Oxford 1969.

[5a] Kaipainen, W. J., Wallen, S., and Totterman, G., On the vitamin B_{12} content in faeces, Scand. J. Clin. Lab. Inv. **6,** 85, 1954.

[5c] Lovell, R. T., and Limsuwan, T., Intestinal synthesis and dietary non-essentiality of vitamin B_{12} for Tilapia nilotica, Trans. Am. Fish. Soc. **111,** 485-490, 1982.

[5e] Atlavinyte, O., Galvelis, A., Dacinlyte, J., and Luganskas, A., Effects of entobacterin Bacillus thuringiensis var. galleriae on earthworm activity, Pedobiologia, **23,** 372-379, 1982.

[5d] Wakayama, E. J., Dillwith, J. W., Howard, R. W., and Blomquist, G. J., Vitamin B_{12} levels in selected insects, Insect. Biochem. **14,** 175-180, 1984.

[6] Rosenblum, C., Reizenstein, P. G., Cronkite, E. P., and Meriwether, H. T., Tissue distribution and storage forms of vitamin B_{12} injected and orally administered to the dog, Proc. Soc. Exp. Biol. Med. **112,** 262, 1963.

[7] Nishino, K., and Itokawa, Y., Thiamin metabolism in vitamin B_6 or vitamin B_{12}-deficient rats, J. Nutr. **107,** 775-782, 1977.

[8] Chiao, F., and Stokstad, E. L. R., Effect of methionine on the metabolism of formate and histidine by rats fed folate vitamin B_{12}-methionine-deficient diet, Biochim. Biophys. Acta **497,** 225-233, 1977.

[9] Tseng, R. Y. L., Cohen, N. L., Reyes, P. S., and Briggs, G. M., Metabolic changes in golden hamsters fed vitamin B_{12}-deficient diets, J. Nutr. **106,** 77-85, 1976.

[10] Ryś, R., and Koreleski, J., The effect of dietary propionic acid on the requirement of chicks for vitamin B_{12}, Br. J. Nutr. **31,** 143-146, 1974.

[11] Davidson, G. E., Weir, D. G., and Scott, J. M., The metabolic consequences of vitamin B_{12}-methionine deficiency in rats, Biochim. Biophys. Acta **392,** 207-215, 1975.

[12] Williams, D. L., and Spray, G. M., Metabolism of propionate in the vitamin B_{12}-deficient rat, Br. J. Nutr. **28,** 263-268, 1972.

[13] Cullen, R. W., and Oace, S. M., Methylmalonic acid and vitamin B_{12} excretion of rats consuming diets varying in cellulose and pectin, J. Nutr. **108,** 640-647, 1978.

[14] Oxnard, C. E., Some variation in the amount of vitamin B_{12} in the serum of the rhesus monkey, Nature **201,** 1188, 1964.

[15a] Johnson, B. C., Studies on the mechnism of action of vitamin B_{12} in animal nutrition, Am. J. Clin. Nutr. **6,** 18, 1958.

[15b] Thacker, P. A., Bowland, J. P., and Fenton, M., Effects of vitamin B_{12} on serum lipids and lipoproteins of pigs fed diets supplemented with propionic acid or calcium propionate, Can. J. Anim. Sci. **62,** 527-536, 1982.

[16a] Barness, L. A., Vitamin B_{12} deficiency with emphasis on methylmalonic acid as a diagnostic aid, Am. J. Clin. Nutr. **20,** 573, 1967.

[16b] Peters, J. P., and Elliot, J. M., Effect of cobalt or hydroxocobalamin supplementation on vitamin B_{12} content and S methylmalonyl coenzyme A mutase (EC 5.4.99.2) activity of tissue from cobalt-depleted sheep, J. Nutr. **114,** 660-670, 1984.

[16c] Peters, J. P., and Elliot, J. M., Effect of vitamin B_{12} Status on performance of the lactating ewe and gluconeogenesis from propionate, J. Dairy Sci. **66,** 1917-1925, 1983.

[17a] McGing, P. G., and Scott, J. M., Evidence that the decreased liver folate status following vitamin B_{12}-inactivation in the mouse is due to increased loss rather than impaired uptake, Biochem. Biophys. Acta **673,** 594-597, 1981.

[17b] Burns, R. A., and Jackson, N., The metabolism of tritium-labeled pteroylglutamic acid by chicken liver in vivo, Effects of folate deficiency and vitamin B_{12} supplementation. Int. J. Vitam. Nutr. Res. **52,** 149-158, 1982.

[18] Brazenas, G. R., Bartkeviciene, V. V., Aleksiene, A. M., Alioniene, A. J., Konopkaite, S. J., and Zoryte, V. F., On some properties of the effect and metabolism of different cobalamins in tumorbearing rats, in Vitamin B_{12}, Eds. B. Zagalak and W. Friedrich, Walter de Gruyter, Berlin 1979, pp. 1113-1118.

[19] Booth, M. A., and Spray, G. H., The excretion and distribution in the tissues of partenterally administered ^{58}Co-vitamin B_{12} in normal and gastrectomized rats, Br. J. Haemat. **8,** 114, 1962.

[20] Miller, A., Gaull, G., Ross, J. F., and Lemon, H. M., Tissue distribution of parenteral ^{60}Co-vitamin B_{12} in mouse, hamster, rat and guinea pig, Proc. Soc. Exp. Biol. Med. **93,** 33, 1956.

[21] Cooperman, J. M., Luhby, A. L., Teller, D. N., and Morley, J. F., Distribution of radioactive and nonradioactive vitamin B_{12} in the dog, J. Biol. Chem. **235,** 191, 1960.

[22] Mattews, D. M., Distribution of cobalamins in the animal body, in Vitamin B_{12}, Eds. B. Zagalak and W. Friedrich, Walter de Gruyter, Berlin 1979, pp. 681-694.

[23] Boots, L. R., Cornwell, P. E., and Beck, L. R., Diurnal and menstrual cycle variations in serum vitamin B_{12} and folacin in baboons, J. Nutr. **105,** 571-576, 1975.

[24] Strength, D. R., Alexander, W. F., and Wack, J. P., Intracellular distribution of vit. B_{12}-Co^{60} in liver and kidney of B_{12} deficient and normal rats, Proc. Soc. Exp. Biol. Med. **102,** 15, 1959.

[25] Yacowitz, H., Hill, C. H., Norris, C., and Heuser, G. F., Distribution of vitamin B_{12} in the organs and tissues of the chick, Proc. Soc. Exp. Biol. Med. **79,** 279, 1952.

[26] Fischer, N. A., Benson, E. M., and Swendscid, M. E., The distribution of vitamin B_{12} in the developing chick egg, Arch. Biochem. Biophys. **74,** 458, 1958.

[27a] Collins, R. A., Harper, A. E., Schreiber, M., and Elvehjem, C. A., The folic acid and vitamin B_{12} content of the milk of various species, J. Nutr. **43,** 313, 1951.

[27b] Sandberg, D. P., Begley, J. A., and Hall, C. A., The content, binding and forms of vitamin B_{12} in milk, Am. J. Clin. Nutr. **34,** 1717-1724, 1981.

[28] Ford, J. E., Some observations on the possible nutritional significance of vitamin B_{12}- and folate-binding proteins in milk, Br. J. Nutr. **31,** 243-257, 1974.

[29] Ford, J. E., Scott, K. J., Sansom, B. F., and Taylor, P. J., Some observations on the possible nutritional significance of vitamin B_{12}- and folate-binding proteins in milk. Absorption of [^{58}Co]cyanocobalamin by suckling piglets, Br. J. Nutr. **34,** 469-491, 1975.

[30a] Collins, R. A., Schreiber, M., Elvehjem, C. A., and Hart, E. B., Vitamin B_{12} and folic acid in milk diets, J. Nutr. **49,** 485, 1953.

[30b] Collins, R. A., Boldt, R. E., Elvehjem, C. A., and Hart, E. B., Further studies on the folic acid and vitamin B_{12} content of cow's milk, J. Dairy Sci. **36,** 24, 1953.

[30c] Gregory, M. E., Ford, J. E., and Kon, S. K., A vitamin B_{12}-binding factor in sow's milk, Biochem. J. **51,** 29, 1952.

[30d] Gregory, M. E., The microbiological assay of vitamin B_{12} in the milk of different animal species, Brit. J. Nutr. **8,** 340, 1954.

[30e] Gregory, M. E., and Holdsworth, E. S., A cyanocobalamin-protein complex from sow's milk and desiccated pig stomach, Nature **173,** 830, 1954.

[30f] Hartman, A. M., Dryden, L. P., and Riedel, G. H., Vitamin B_{12} content of milk and milk products as determined by rat assay, J. Nutrition, **59,** 77, 1956.

[30g] Karlin, R., La vitamine B_{12} dans le lait humain et bovin. Vitamin B_{12} in human milk and cow's milk, Compt. Rend. Soc. Biol. **148,** 371, 1954.

[30h] Van Koetsveld, E. E., Differences in the vitamin B_{12} content of cow's milk during the last month indoors and the first week at pasture, Nature **171,** 483, 1953.

[30i] Neujahr, H. Y., On vitamins in sewage sludge. IV. Isolation of new vitamin B_{12}-like factors. Acta Chem. Scand. **10,** 917, 1956.

[30j] Nurmikko, V., and Virtanen, A. J., Der Vitamin-B_{12}-(Cobalamin)-Gehalt der Kuhmilch. Milchwiss. **11,** 192, 1956.

[30k] Sreenivasamurthy, V., Nambudripad, V. K. N., and Iya, K. K., Further studies on the vitamin B_{12} activity of milk, Indian J. Dairy Sci. **6,** 105, 1953.

[30l] Callieri, D. A., Studies on the vitamin B_{12}, cyanocobalamin-binding capacity, desocyribsides, and methionine in some commercial milk products and cheese. Acta Chem. Scand. **13,** 737, 1959.

[31] Bauman, D. E., Davis, C. L., and Bucholtz, H. F., Propionate production in the rumen of cows fed either a control or high-grain, low-fiber diet, J. Dairy Sci. **54,** 1282, 1971.

[32] Menke, K. H., Salewski, A., and Ehrensvärd, U., Methylmalonsäure and Vitamin B_{12}-Ausscheidung im Harn des ketotischen Rindes, in Energy metabolism of farm animals, Eds. A. Schurch and C. Wenk, European Ass. Anim. Prod. Publ. no 13, Switzerland, 1970, p. 77.

[33] Smith, R. M., and Marston, R. H., Metabolism of propionate by pair-fed vitamin B_{12} deficient and vitamin B_{12} treated sheep, Br. J. Nutr. **26,** 41, 1971.

[34] Smith, R. M., and Osborne-White, W. S., Folic acid metabolism in vitamin B_{12} deficient sheep. Depletion of liver folates, Biochem. J. **136,** 279, 1973.

[35] Smith, R. M., Osborne-White, W. S., and Russel, G. R., Methylmalonic acid and coenzyme A concentrations in the livers of pair-fed vitamin B_{12} deficient and vitamin B_{12} treated sheep, Biochem. J. 1969.

[36] Somers, M., Volatile fatty acid clearance studies in relation to vitamin B_{12} deficiency in sheep, Austr. J. Biol. Med. Sci. **47,** 219, 1969.

[37] Cox, E. V., and White, A. M., Methylmalonic acid excretion. An index of vitamin B_{12} deficiency, Lancet **2,** 853, 1962.

[38] Frenkel, E. P., Kitchens, R. L., and Johston, J. M., The effect of vitamin B_{12} deprivation on

the enzymes of fatty acid synthesis, J. Biol. Chem. **248,** 7540, 1973.

[39] Garton, G.A., Hovell, F.D.D., and Duncan, W.R.H., Influence of dietary volatile fatty acids on the fatty acid composition of lamb triglycerides, with special reference to the effect of propionate on the presence of branched-chain components, Br. J. Nutr. **28,** 409, 1972.

[40] Frobish, R.A., and Davis, C.L., Theory involving propionate and vitamin B_{12} in the low-milk fat syndrome, J. Dairy Sci. **60,** 268, 1977.

[41] Walker, C.K., and Elliot, J.M., Lactational trends in vitamin B_{12} status on conventional and restricted-roughage rations, J. Dairy Sci. **55,** 474, 1972.

[42] Burton, W.C., and Frenkel, E.P., Effect of vitamin B_{12}-deficiency on the hepatic tissue concentration of acyl carnitines, Biochim. Biophys. Acta **398,** 217-223, 1975.

[43] Elliot, J.M., Propionate metabolism and vitamin B_{12}, Eds. Y. Ruckebusch and P. Thivend, in Digestive Physiology and Metabolism in Ruminants, MTP Press Ltd., Lancaster, England 1980, pp 485-503.

[44a] Marston, H.R., Allen, S.H., and Smith, R.M., Production within the rumen and removal from the blood-stream of volatile fatty acids in sheep given a diet deficient in cobalt, Br. J. Nutr. **27,** 147-157, 1972.

[44b] Peters, P., Bergman, E.N., and Elliot, J.M., Influence of vitamin B-12 status on hepatic propionic acid uptake in sheep, J. Nutr. **113,** 1221-1228, 1983.

[44c] Peters, J.P., Bergman, E.M., and Elliot, J.M., Changes of glucose, insulin and glucagon associated with propionate infusion and vitamin B_{12} status in sheep, J. Nutr. **113,** 1229-1240, 1983.

[44d] Tsugoshi, I., Abe, N., Takasaki, H., Ohmiya, M., and Kancho, K., Occurrence of cobalt-vitamin B_{12} deficiency in sheep in the Konsen district of Hokkaido, Japan, Bull. Fac. Agric Tamagawa Univ. **21,** 16-35, 1981.

[45] Dryden, L.P., Hartman, A.M., Bryant, M.P., Robinson, J.M., and Moore, L.A., Production of vitamin B_{12} analogues by pure cultures of ruminal bacteria, Nature **195,** 201-202, 1962.

[46] Bigger, G.W., Elliot, J.M., and Rickard, T.R., Estimated ruminal production of pseudovitamin B_{12}, factor A and factor B in sheep, J. Anim. Sci. **43,** 1077, 1976.

[47] Dawborn, M.C., Hine, D.C., and Smith, J., The determination of vitamin B_{12}-activity in the organs and excreta of sheep. 4. The separation of vitamin B_{12}-active factors from rumen contents by paper ionophoresis, Australian J. Exp. Biol. **35,** 97, 1957.

[48] Rickard, T.R., and Elliot, J.M., Absorption of vitamin B_{12} and factor B from the intestine of sheep, J. of Animal Sci. **46,** 304-308, 1978.

[49] Rickard, T.R., Bigger, G.W., and Elliot, M.J., Effects of 5,6-dimethylbenzimidazole, adenine and riboflavin on ruminal vitamin B_{12} synthesis, J. of Animal Sci. **40,** 1199-1204, 1977.

[50] Ford, J.E., The microbiological assay of vitamin B_{12}. The specificity of the requirement of *Ochromonas malhamensis* for cyanocobalamin, Brit. J. Nutr. **7,** 299, 1953.

[51] Rickard, T.R., and Elliot, J.M., Effect of factor B on vitamin B_{12} status and propionate metabolism in sheep, J. Animal Sci. **55,** 168-173, 1982.

[52] Kercher, C.J., and Smith, S.E., The response of cobalt-deficient lambs to orally administered vitamin B_{12}, J. Anim. Sci. **14,** 458, 1955.

[53] Gelman, A.L., Determination of cobalt in plant material by atomic absorption, J. Sci. Food Agr. **23,** 299, 1972.

[54] Underwood, E.J., in Trace elements in human and animal nutrition. Academic Press, New York 1971.

[55] Hedrich, M.F., Elliot, J.M., and Lowe, J.E., Response in vitamin B_{12} production and absorption to increasing cobalt intake in the sheep, J. Nutr. **103,** 1646, 1973.

[56] Wootten, J., The production and absorption of vitamin B_{12} in sheep in relation to diet. M.S. thesis, Univ. Aberdeen, Scotland, 1972.

[57] Eliot, J.M., Kay, R.N.B., and Goodall, E.D., Production and absorption of vitamin B_{12} in the sheep – a preliminary study, Life Sciences **10,** 647, 1971.

[58] Smith, R.M., and Marston, H.R., Production absorption, distribution and excretion of vitamin B_{12} in sheep, Br. J. Nutr. **24,** 857-867, 1970.

[59] Marston, H.R., The requirement of sheep for cobalt or for vitamin B_{12}, Br. J. Nutr. **24,** 615, 1970.

[60] Maro, J.K., and Kategile, J.A., Studies on copper and cobalt in dairy calves, Br. J. Nutr. **44,** 25-31, 1980.

[61] Sutton, A.L., and Elliot, J.M., Effect of ratio of roughage to concentrate and level of feed intake on ovine ruminal vitamin B_{12} production, J. Nutr. **102,** 1341, 1972.

[62] Marston, M.R., The cobalt story. Med. J. Australia **2,** 105, 1959.

[63] Gawthorne, J.M., The effect of cobalt intake on the cobamide and cobinamide composition of the rumen contents and blood plasma of sheep, Austr. J. Exp. Biol. Med. Sci. **48**, 285, 1970.
[64] Gawthorne, J.M., *In vitro* studies of the factors affecting the metabolism of cobamides and cobinamides by sheep rumen micro-organisms, Austr. J. Exp. Biol. Med. Sci. **48**, 293, 1970.
[65] Reizenstein, P.G., Body distribution, turnover rate, and radiation dose after parenteral administration of radiovitamin B_{12}, Acta Med. Scand. **165**, 467, 1959.
[66] Macpherson, A., and Dixon, J., Effect of copper and cobalt treatment of finishing steers, Anim. Prod. **30** (3), 373-382, 1980.
[67] Burdelev, T.E., Kokorina, E.K., and Ivanova, L.Ya., The growth and development of calves in addition of vitamins and polysalt trace elements to the feed, Izv. Timiryazev. S-Kh. Akad. **0** (5), 137-144, 1979.
[68] Bonhomme, A., Durand, M., Quintana, C., and Halpern, S., Influence of cobalt and vitamin B_{12} on growth and survival of rumen cilliates in vitro according to bacterial population, Reprod. Nutr. Dev. **22**, 107-122, 1982.
[69] Ishii, T., Abe, N., Takasaki, H., Ohmiya, M., and Kaneko, K., Occurrence of cobalt-vitamin B_{12} deficiency in sheep in the Konsen District of Hokkaido, Japan, Bull. Fac. Agric. Tamagawa Univ. **0** (21), 16-35, 1981.
[70] Duncan, W.R.H., Morrison, E.R., and Garton, G.A., Effects of cobalt deficieny in pregnant and post-parturient ewes and their lambs, Br. J. Nutr. **46**, 337-344, 1981.
[71] Mykkanen, H.M., and Korpela, H., Serum vitamin B_{12} levels in dairy cows before and after parturition, Zentralbl. Veterinaermed. Reihe A **28**, 526-528, 1981.

6.6 Cobalamin in man

6.6.1 Concentration of cobalamin in organs and body fluids [1-34]

The corrinoids occurring in man and animals consist mainly of cobalamin[a]. This is mainly due to specific absorption system. The distribution of cobalamin varies greatly depending on tissue (Tables 6-15, 6-16). A man's body contains totally 3-5 mg of cobalamin from which a great part (approx. 1.5 mg) is present in the liver [25]. Among other rich in cobalamin organs is the kidney while brain [53b] and spleen are the poorest. Three main forms of cobalamin are found under natural conditions in man and animals: methylcobalamin (Me-Cbl) adenosylcobalamin (Ado-Cbl) and hydroxocobalamin (OH-Cbl) [1]. Cyanocobalamin (CN-Cbl), an artificial product, is used in pharmaceuticals due to its greater stability. It is converted in the organism into these three forms. The conversion of CN-Cbl to other forms takes several days-thus appreciable quantities of cyanocobalamin may be found in human tissues in result of medical treatment. On the other hand CN-Cbl may be formed in the body from OH-Cbl whenever a small amount of cyanide is available. Cyanide may reach the tissues from cigarette smoke [1] or cyanogenic diet [43, 57, 58]. It may be produced also within the body e.g. in course of phagocytosis [44]. The share of the three forms in total cobalamin content is different in various organs and vary in course of aging. Some diseases affect significantly the total cobalamin in tissues and serum as well as the percentage of these three forms. Vegetarians have low cobalamin level in serum (< 200 pg/ml) and presumably also in the whole body [1a-c].

---

[a] A vitamin B_{12} analog has been recently discovered in man. This not closer defined compound contributes appreciably (up to 40%) to the whole corrinoid content in serum of people with vitamin B_{12} deficiency. It is inactive in some test organisms and probably also in man. The analog interferes, however, in the radioisotope dilution assay (see section 5.3.3 and the references 35a-c in that section).

Table 6-15 Cobalamins in human tissues and body fluids

Tissue or fluid	Total Cbl ± S. D. or range		Forms of cobalamin (%)				References
	ng/g or ng/ml	nmol/kg or nmol/l	Me-Cbl	Ado-Cbl	OH-Cbl	CN-Cbl	
Liver:							
biopsies (n = 7)	1050	772	1.0	61	38	0	[1, 2]
autopsy samples	555	408	3	84	13	0	[1, 2]
Kidney	134	99	27	52	23	0	[1, 2]
Spleen	63	46	37	43	20	0	[1, 2]
Brain	81	60	11	61	28	0	[1, 2]
Spinal cord (ventral horn)	99	73	2.6	17	24	56	[3]
Putuitary	230	169	21	48	31	0	[1, 2]
Bone marrow	12-13	9-10	13	55	30	2	[1, 2]
Leucocytes	4	3	20	48	28	4	[1, 2]
	38 pg/10^7 cells	28 femtomoles/10^7 cells					[4, 5]
Erythrocytes	0.2	0.15	15	53	26	6	[2]
	0.11-0.5	0.08-0.4	< 15	> 50	~25	–	[6]
	0.209	0.154					[7]
Plasma	0.4	0.3	65	23	10	2	[2]
	0.44	0.32	78	6	12	4	[45]
Serum	0.2-0.9	0.1-0.7					[8]
	0.60 ± 0.12	0.44 ± 0.09					[8]
	0.89 ± 0.21	0.66 ± 0.15					[8]
Cerebrospinal fluid (CSF)	0.017-0.032	0.013-0.24					[9, 10, 11]
	0.02	0.015	8	74	8	10	[2]
Aqueous humor	0.012-0.060	0.009-0.044	0	predominating forms			[2]
Bile	1.1-11.5	0.8-8.5					[12]
	4.5-8.8	3.3-6.5					[13]
	18	13	4	50	39	7	[2]
Bile excretion	0.5-5 μg/day	0.4-4 nmol/day	–	–	–	–	[15]
	43 μg/day	32 nmol/day	–	–	–	–	[15]
Milk	0.1-1.5	0.07-1.1	–	–	–	–	[16, 17]
	0.26-0.66	0.19-0.49	60-80	–	–	–	[18]
Milk: B_{12} output	0.12-0.16 μg/day	0.09-0.12 nmol/day	–	–	–	–	[18]
Urine B_{12} output:	31-43 ng/day	23-32 pmol/day	–	–	–	–	[19]
	110-240 ng/day	81-176 pmol/day	–	–	–	–	[20]
	50-250 ng/day	37-184 pmol/day	–	–	–	–	[21]
non smokers	60 ± 7.9 ng/day	44 ± 5.8 pmol/day	–	–	–	–	[22]
smokers	81 ± 8.7 ng/day	60 ± 6.4 pmol/day	–	–	–	–	
Amniotic fluid	0.12		–	–	–	–	[23, 24]
Whole body	2-5 mg	1.5-3.7 μmol					[25]

Table 6-16 The distribution of a tracer dose of vitamin B_{12} in man (compiled by Chanarin [30])

Authors	[a]Doscherholmen, Hagen [26] values at 115 days	[b]Adams [27] values at 4 days	[b]Cronkite [28] values at 180 days
liver	70	47.2	58.6
Kidney	8	1.2	0.23
Pancreas	11	1.2	0.25
Spleen	3	1.1	3.60
Brain		0.9	5.9
Adrenals			0.04
Gonads			0.0
Heart	5	0.9	1.2
Muscle			8.56
Lungs	1	1.8	1.16
Marrow			4.4
Alimentary tract			4.46

[a] Expressed as percentage of retained radioactivity
[b] Expressed as percentage of administered tracer dose

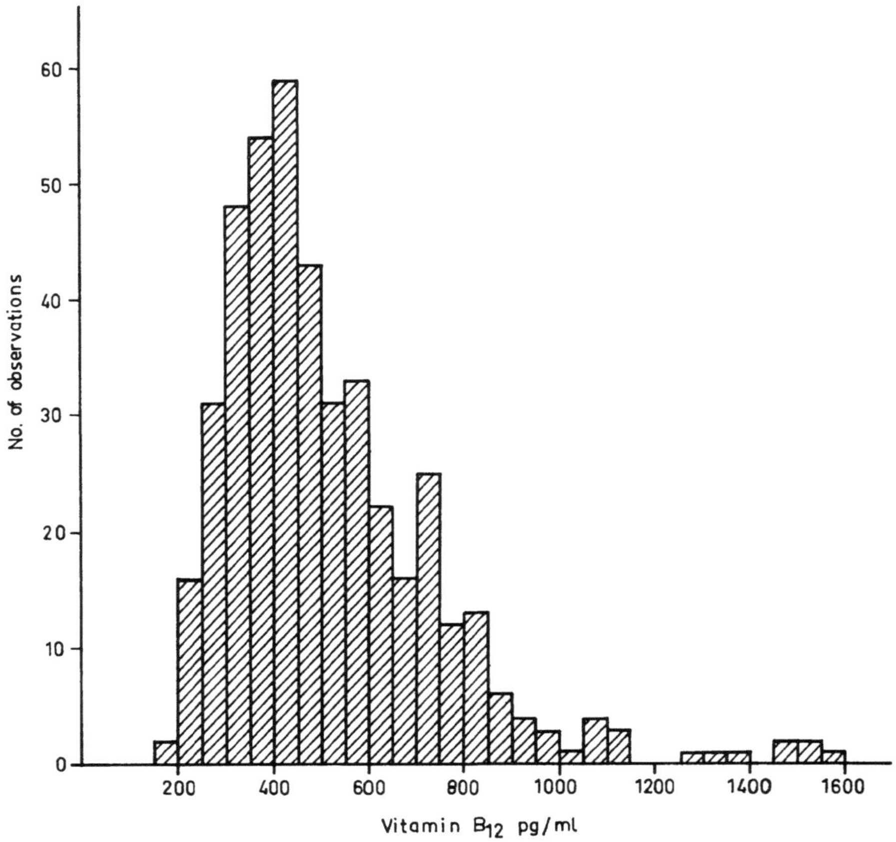

Fig. 6-4 The distribution of serum vitamin B_{12} levels in 434 normal subjects (with permission of Doctors W. O. Maror and J. L. Mould from Royal Hampshire County Hospital, England). See also reference [54].

6.6.2 Dynamic aspects

The total cobalamin content in serum and liver [46–48] and presumably also in other organs changes with age (Table 6–17) [1]. The content of liver B_{12} in man increases with age in terms of mean value, however, a broader range of liver cobalamin concentration is observed in the population of older people. This means that among aged people a significant percentage have very low liver B_{12} level. The share of the three forms of cobalamin: Ado-Cbl, CH_3-Cbl, OH-Cbl changes with age as well [58, 60, 62, 63]. This is evident particularly in case of Me-Cbl in liver and brain. A dramatic decrease of Me-Cbl is observed in liver in the age range 1–15 years (Fig. 6–5). The retention of two forms of cobalamin CN-Cbl and OH-Cbl in serum is different [30]. As shown in Figure 6–6 the retention of OH-Cbl in serum is greater than CN-Cbl. The faster removal of CN-Cbl from blood serum ist also depicted in Fig. 6–7 showing that smokers, whose blood contains appreciable quantities of CN-Cbl due to smoking, have lower serum B_{12} levels [22] (compare also B_{12} output in urine by smokers and non-smokers in Table 6–15). The rate of interconversion of the various forms of cobalamin into each other seems to be the basic factor responsible for the share of Ado-Cbl, Me-Cbl and OH-Cbl in total cobalamin contents in various tissues [35, 59]. As shown in Figs. 6–8, 6–9, 6–10 the cobalamin which is produced most readily from other forms is Ado-Cbl. This cobalamin, however, is converted with a slower rate into Me-Cbl. A significant increase of Me-Cbl is observed in serum of vegetarians after application of 1 μg of cobalamin (Fig. 6–11) [36].

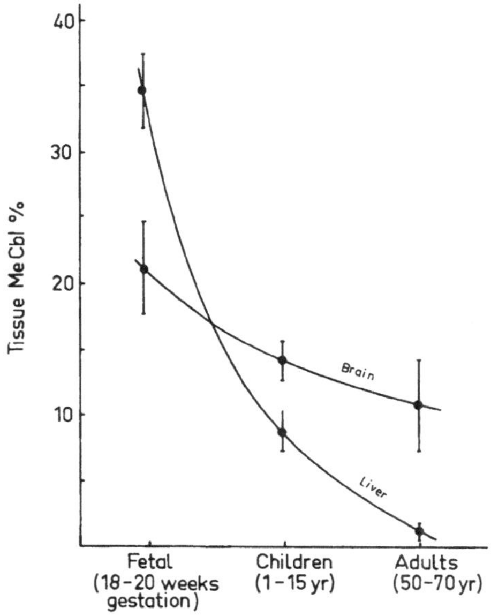

Fig. 6–5 Age-related changes in the proportion of MeCbl in liver and brain (mean ± SAM) (from Linnell [1] Copyright 1974 John Wiley & Sons, Inc. Reprinted with permission of John Wiley & Sons, Inc.).

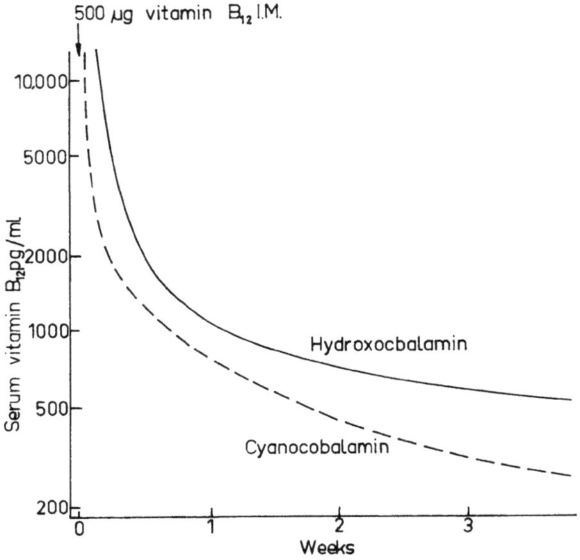

Fig. 6-6 The serum vitamin B_{12} levels following a single parenteral injection of 500 µg of hydroxocobalamin and cyanocobalamin (from Chanarin [30]).

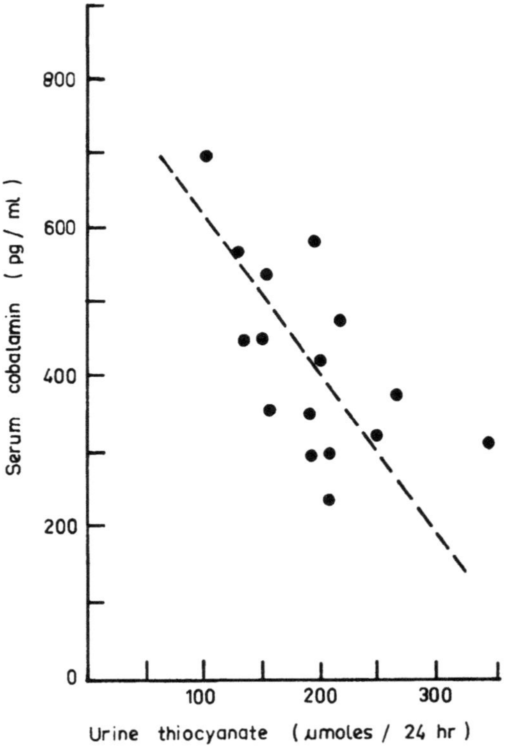

Fig. 6-7 Correlation between serum total cobalamin and urine thiocyanate excretion in healthy smokers. The increased cyanide intake from smoking (the cyanide is converted by the organism into thiocyanate which is excreted in the urine) negatively affects cobalamin concentration in serum (see also clearance of CN-Cbl from plasma Fig. 6-3 and Table 6-15) (from Linnell et al., Brit. Med. J. **2**, 215, 1968).

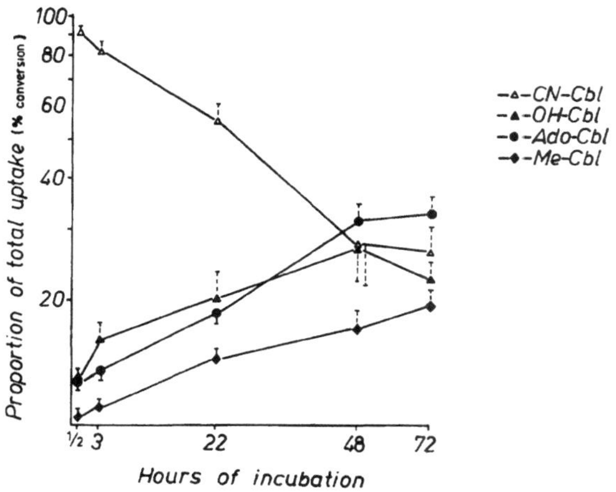

Fig. 6-8 Formation of labelled cobalamins from [^{57}Co]cyanocobalamin by human lymphocytes in *vitro* (from Quadros et al. [35]).

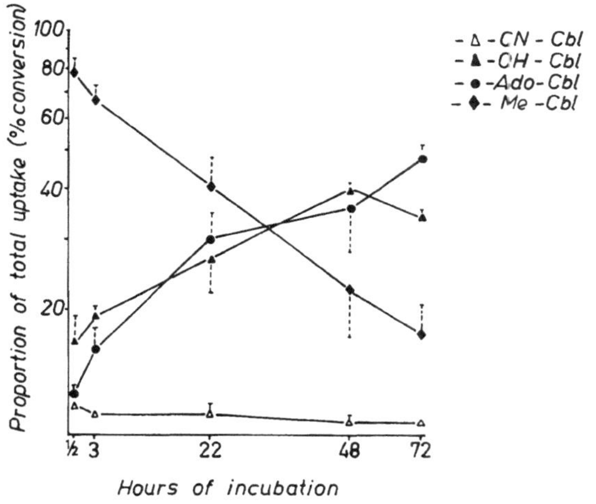

Fig. 6-9 Formation of labelled cobalamins from [^{57}Co]methylcobalamin by human lymphocytes in *vitro* (from Quadros et al. [35]).

Table 6-17 Liver total cobalamin concentration at different ages (from Hsu et al., reprintet by permission from *Nature*, **210,** 1264 (1966). Copyright (c) 1966 Macmillan Journals Limited).

Age	Number of Cases	Total Cobalamin (ng/g wet wt)
Stillborn	5	56 ± 28[a]
4-48 h	3	108 ± 16
0.5-2 yr	2	383 ± 120
20-40 yr	5	310 ± 111
40-60 yr	6	598 ± 168
60-85 yr	6	775 ± 611

[a] Mean ± SD

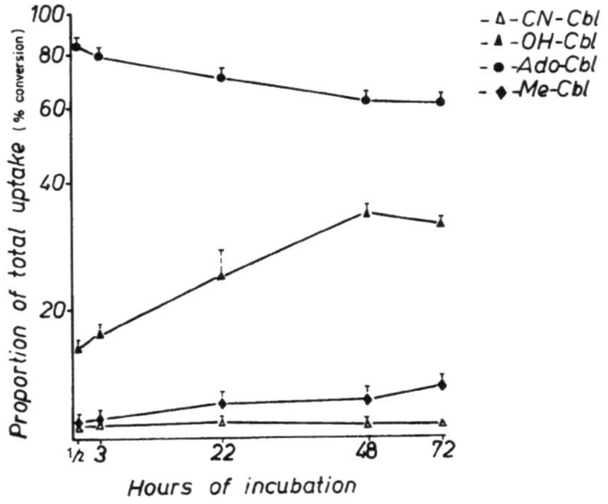

Fig. 6-10 Formation of labelled cobalamins from [^{57}Co]adenosylcobalamin by human lymphocytes in *vitro* (from Quadros et al. [35]).

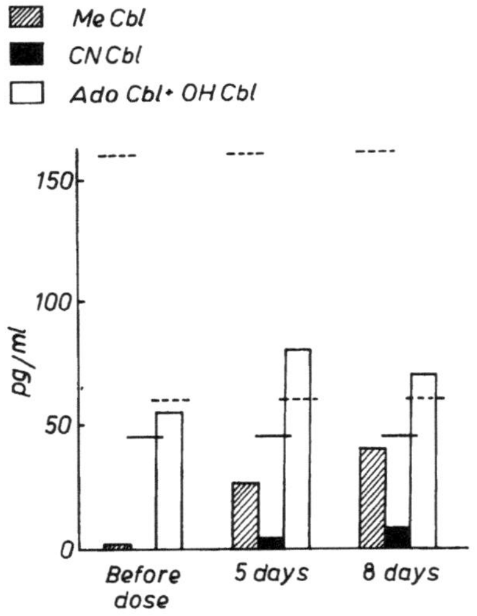

Fig. 6-11 Plasma cobalamins in a cobalamin-deficient vegan given 1 μg CN-Cbl daily (from Linnell [36]).
- - - = lower limit of normal
—— = upper limit of normal

6.6.3 Loss of vitamin B_{12}

Vitamin B_{12} may be lost by excretion in the urine and faeces. The quantitative calculations are based on experiments with radioactive vitamin B_{12} (mainly [^{60}Co]CN-Cbl). After an injection of a tracer dose of vitamin B_{12} an initial rapid clearance from the plasma and some loss in the urine is followed by a phase of slower plasma clearance and redistribution of the dose within the body. The second phase apears after 250–300 days [31] when the equilibrium of the tracer dose with the body pool is attained. On the basis of clearance in the second phase the half time for hepatic vitamin B_{12} has been found 330–420 days [26] and for the whole body by other investigators 1360 days [49]. A daily loss of 0.05% of the total vitamin B_{12} was estimated [30, 31]. On the basis of 5 mg B_{12} value for whole body vitamin B_{12} a daily loss of 2.5 µg can be calculated. Other authors cite 2.0, 4.8 and 6.0 µg loss of vitamin B_{12} each day [24, 30]. These data are in rough agreement with the estimates of daily optimal requirement of vitamin B_{12} by man, found by various authors, which are in the range 2–5 µg/day [1, 13, 15, 26, 49, 50]. Minimal daily requirement has been found in the range 0.6–1.0 µg/day [56]. Smoking seems to increase the loss of vitamin B_{12} as judged by lower serum B_{12} level of smokers proportional to the intensity of this habit (Fig. 6-7). Women lose appreciable quantities of cobalamin in course of menstruation, for supplying the fetus and during lactation. The serum B_{12} level is an indication for vitamin B_{12} status in the whole body (see Fig. 6-4). In pregnant women the serum B_{12} falls by about 25% (from 413 to 307 pg/ml of mean value) [25]. Particularly high losses of Me-Cbl in serum are observed during pregnancy. The fetal serum B_{12} level is similar to the serum B_{12} level of the mother [24]. The amniotic fluid contains about 120 pg/ml of cobalamin (Table 6-15) [23]. There is also a strong correlation between serum B_{12} and B_{12} in milk (Fig. 6-12). Injection of CN-Cbl to the lactating woman is immediately followed by a dramatic output of cobalamin into the milk in unchanged form (Fig. 6-13) [37, 52].

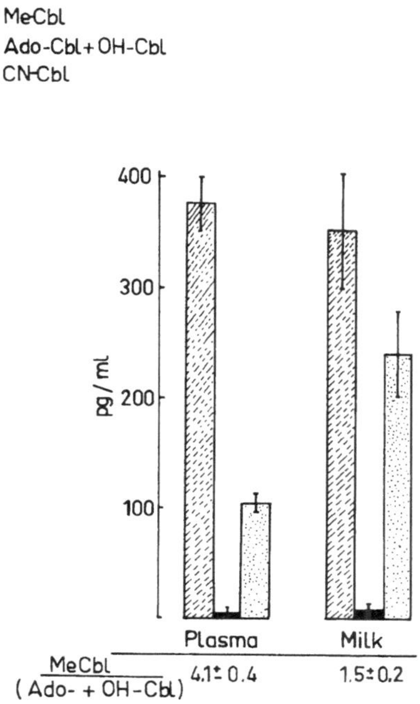

Fig. 6-12 Plasma- and milk cobalamins in 10 healthy mothers 4 to 8 days post partum (mean values SEM. Plasma: total Cbl 490 ± 59 pg/ml; milk: total Cbl 605 ± 185 pg/ml) (from Craft et al. [37]).

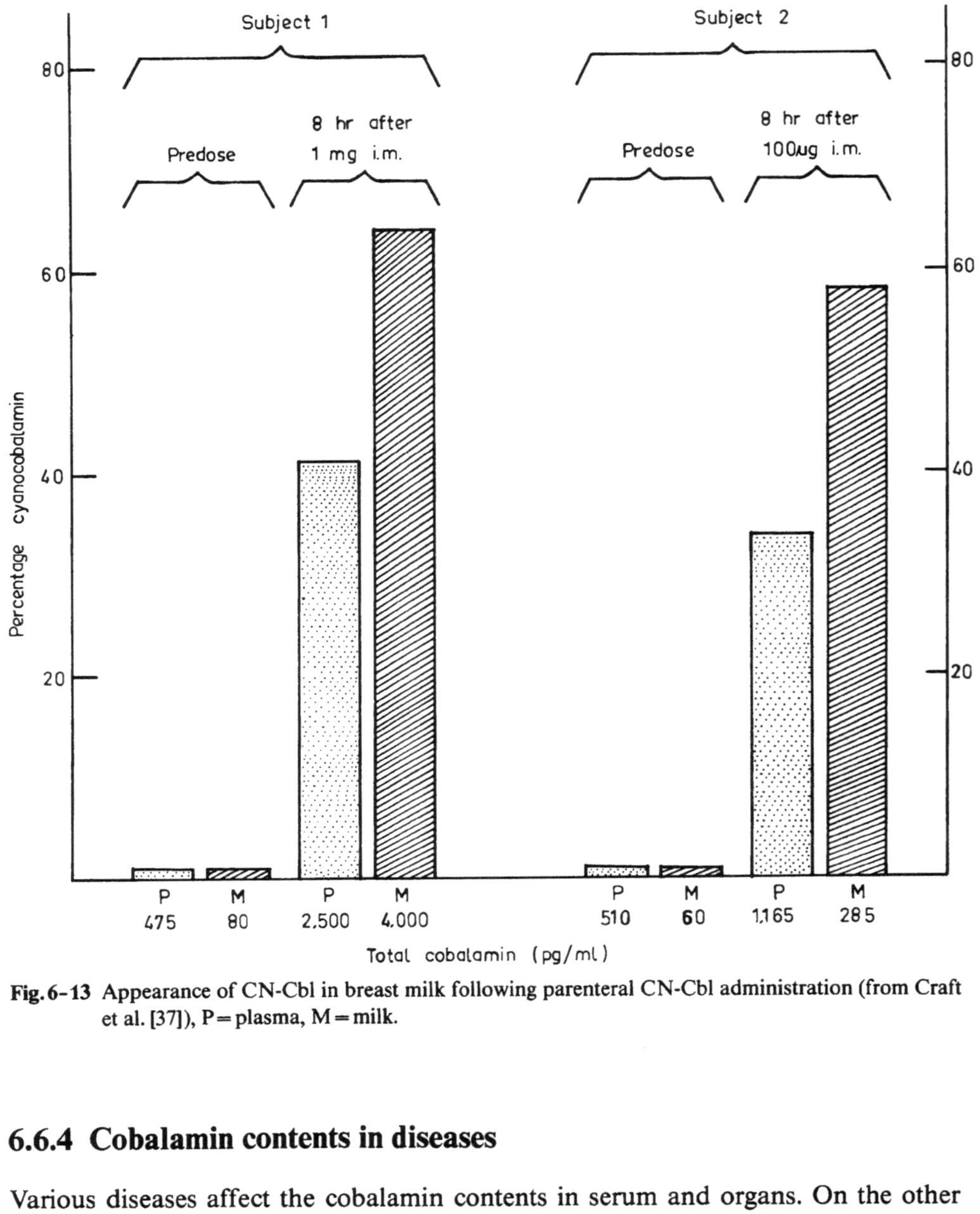

Fig. 6-13 Appearance of CN-Cbl in breast milk following parenteral CN-Cbl administration (from Craft et al. [37]), P = plasma, M = milk.

6.6.4 Cobalamin contents in diseases

Various diseases affect the cobalamin contents in serum and organs. On the other hand some diseases like pernicious anaemia and neurological disorders are caused by deficiency of cobalamin (see chapter 9). Since liver is the main depot of cobalamin in man, the liver diseases, in particular virus infections, are manifested by dramatic cobalamin losses from this organ [40, 41]. Usually the serum B_{12} level increases in the first week of infection (Table 6-18). Table 6-18 also shows that various forms of leukemia and cancer are accompanied by higher serum-B_{12} levels. See also Table 6-19 illustrating contents of various forms of cobalamin in tumors of rat organs and plasma and text in section 9.2.3.

Table 6-18 Cobalamin in human tissues and body fluids in diseases.

Organ or tissue	Disease	Total Cbl mean ± SD or range		Form of Cbl				References
		pg/ml or pg/g	p mol/l or p mol/kg					
Plasma	Pernicious anaemia	88 ± 11	65 ± 8					[38]
Erythrocytes		202 ± 25	149 ± 18					[38]
Leucocytes		2575 ± 1467	1893 ± 1079					[38]
Bone marrow		6450 ± 1404	4743 ± 1032					[38]
Cerebro spinal fluid (CSF)		20		Ado-Cbl 85%				[38]
urine		20000–30000 pg/day	14706–22059	Ado-Cbl and OH-Cbl				[1]
		8000 pg/day	5882	Me-Cbl				[21]
Plasma	Partial gastrectomy	255	188	Me-Cbl				[1]
	Acute hepatitis	2000	1471					[39]
Serum	Virus hepatitis: controls	484 ± 99	356 ± 73					[40, 41]
	1st week	1044 ± 453	767 ± 333					[40, 41]
	3rd week	510 ± 263	376 ± 193					[40, 41]
	6st week	312 ± 134	229 ± 99					[40, 41]
	Chronic active hepatitis (n = 19)	415 ± 200	305 ± 147					[40, 41]
	Liver cirrhosis (n = 40)	1195 ± 522	879 ± 384					[40, 41]
	Hepatic coma (n = 5)	4408 ± 1981	3241 ± 1457					[40, 41]
	Chronic myeloid leukemia (CML)	1200 – 8000	882 – 5882					[8]
		3210 ± 1105	2360 ± 813					
	Acute myeloid leukemia (AML)	400 – 900	294 – 662					[8]
	Acute promyeloid leukemia (APL)	2000 – 3500	1471 – 2574					[8]
		4890 ± 1940	3596 ± 1426					[8]
	Acute leukemia	3048 ± 1419	2241 ± 1043					[8]
	Polycytemia Vera	860 ± 290	632 ± 213					[8]
				Me-Cbl, %	Ado-Cbl, %	OH-Cbl, %	CN-Cbl %	
Plasma	Primary hepatoma	104	76	74	26	–	0	[42]
Erythrocytes	Primary hepatoma	180	132	76	13	12	0	[42]
Ascitic fluid	Primary hepatoma	223	164	46	43	9	4	[42]
Liver	Primary hepatoma	344×10^3	253×10^3	21	65	13	0	[42]
Tumor	Primary hepatoma	86×10^3	63×10^3	24	64	12	0	[42]
Kidney	Primary hepatoma	54×10^3	40×10^3	26	46	28	0	[42]
Spleen	Primary hepatoma	24×10^3	18×10^3	8	83	8	0	[42]
Brain	Primary hepatoma	3×10^3	2×10^3	37	30	33	0	[42]
Spinal cord (ventral horn)	Amyotrophic lateral sclerosis (ALS)	16.2×10^3	12×10^3	3.3	6.5	1.6	88.6	[3]

Table 6-19 Contents of various forms of cobalamin in tumors of rat organs and blood plasma (from Brazenas et al. [29]).

Cobalamins	Healthy	ZAH[1]	AC-RS-1[2]	S-45[3]	KS-W[4]
Liver ng/g					
Total	133.8	71.6	111.6	195.7	191.6
Me-Cbl	Traces	7.8	Traces	Traces	Traces
OH-Cbl	47.0	23.7	48.0	60.0	60.0
Ado-Cbl	85.0	40.1	63.0	135.0	131.0
Kidney ng/g					
Total	473.3	134.6	499.0	974.8	423.3
Me-Cbl	48.7	23.0	45.4	146.2	72.4
OH-Cbl	200.2	69.4	178.6	250.5	149.4
Ado-Cbl	224.3	142.2	275.0	578.1	201.5
Spleen ng/g					
Total	63.7	31.6		43.1	
Me-Cbl	4.2	9.0		2.8	
OH-Cbl	24.8	6.4		13.8	
Ado-Cbl	34.7	16.2		26.5	
Blood plasma pg/ml					
Total	600.0	260.0		750.0	
Me-Cbl	180.0	100.0		170.0	
OH-Cbl	50.0	80.0		170.0	
Ado-Cbl	350.0	80.0		410.0	

1 ZAH = Zajdela ascite hepatoma
2 AC-RS-1 = alveolar carcinoma RS-1
3 S-45 = sarcoma 45
4 KS-W = Walker carcinosarcoma

Various diseases affecting the alimentary tract are the cause of vitamin B_{12} deficiency [61] (see section 8.1.3 and the listed references). Patients after surgery may have lower serum B_{12} levels [56, 57]. Populations of elder people, suffering frequently from various mild defects of the alimentary tract for a longer period of time, may show lower serum B_{12} values which are difficult to correct by appropriate diet [60, 62, 63].

References

[1 a] Linnel, J.C., The fate of cobalamin *in vivo,* in cobalamin-/biochemistry and pathophysiology, Ed. Babior, B.M., John Wiley and Sons, New York 1974, pp. 288-333.

[1 b] Dong, A., Scott, S.C., Serum vitamin B_{12} and blood cell values in vegetarians, Ann Nutr. Metabol. **26,** 209-216, 1982.

[1 c] Shinwell, E.D. and Gorodischer, R., Totally vegetarian diets and infant nutrition, Pediatrics **70,** 582-586, 1982.

[2] Matthews, D.M., Distribution of cobalamins in the animal body, in Vitamin B_{12}, Eds. B. Zagalak and W. Friedrich, Walter de Gruyter, Berlin 1979, pp. 681-694.

[3] Inada, M., Kameyama, M., and Toyoshima, M., The significance of vitamin B_{12} binders in the central nervous system, in Vitamin B_{12}, Eds. B. Zagalak and W. Friedrich, Walter de Gruyter, Berlin 1979, pp. 821-826.

[4] Thomas, J.W., and Anderson, B.B., Brit. J. Hematol. **2,** 41, 1956.

[5] Kidd, H.M., and Thomas, J.W., Brit. J. Hematol. **8,** 64, 1962.

[6] Biggs, J.C., Mason, S.L.A., and Spray, G.H.,

Vitamin B_{12} activity in red cells, Brit. J. Hematol. **10,** 36, 1964.
[7] Kelly, A., and Herbert, V., Blood **29,** 139, 1967.
[8] Rachmilewitz, B., Schlesinger, M., Rabinowitz, R., and Rachmilewitz, M., The origin and clinical implications of vitamin B_{12} binders – the transcobalamins, in Vitamin B_{12}, Eds. B. Zagalak, and W. Friedrich, Walter de Gruyter, Berlin 1979, pp. 765-776.
[9] Linnel, J. C., Hussein, A. A. A., Matthews, D. M., Patterns of plasma cobalamins in control subjects and in cases of vitamin B_{12} deficiency, J. Clin. Pathol. **22,** 545, 1969.
[10] Herbert, V., and Zalusky, R., Fed. Proc. **20,** 453, 1961.
[11] Basil, W., Brown, J. K., and Matthews, D., J. Clin. Pathol. **18,** 317, 1965.
[12] Ardeman, S., Chanarin, I., and Berry, V., Studies on human gastric intrinsic factor. Observations on the possible absorption and entero-hepatic circulation, Br. J. Haemat. **11,** 11, 1965.
[13] Okuda, K., Gräsbeck, R., and Chow, B. F., J. Lab. Clin. Med. **51,** 17, 1958.
[14] Herbert, V., Am. J. Clin. Nutr. **21,** 743, 1968.
[15] Gräsbeck, R., Nyberg, W., and Reizenstein, P., Bilary and faecal vit. B_{12} excretion in man. An isotope study, Proc. Soc. Exp. Biol. Med. **97,** 780, 1958.
[16] Neujahr, H. Y., Jonasson, M., Lundin, H., On vitamin B_{12} and its occurrence in milk, Milk and Dairy Research, Alnarp, Sweden, Report No **59,** 1960. (see also references 30a-301 in section 6.5)
[17] Collins, R. A., Harper, A. E., Schreiber, N., and Elvehjem, C. A., J. Nutr. **43,** 313, 1951.
[18] Karlin, R., Proc. Soc. Biol. **148,** 371, 1954.
[19] Register, U. D., and Sarett, H. P., Proc. Soc. Exp. Biol. Med. **77,** 837, 1951.
[20] Mollin, D. L., and Ross, G. I. M., J. Clin. Pathol. **5,** 129, 1952.
[21] Heinrich, H. C., Metabolic basis of diagnosis and therapy of vitamin B_{12} deficiency, - in Seminaries Hematology, Ed. C. P. A. Miescher, **1,** 1964, pp. 199-249.
[22] Linnell, J. C., Smith, A. D. M., Smith, C. L., Wilson, J. and Matthews, D. M., Brit. Med. J. **2,** 215, 1968.
[23] Clarke, Int. Vitam. Nutr. Res. **43,** 324, 1973.
[24] Friedrich, W., in Fermente-Hormone-Vitamine, vol III/2, Vitamin B_{12} und verwandte Corrinoide, Eds. R. Ammon and W. Dirscherl, Georg Thieme Verlag, Stuttgart 1975, pp. 319-327.
[25] Friedrich, W., in Fermente-Hormone-Vitamine, Vol III/2, Vitamin B_{12} und verwandte Corrinoide, Eds. R. Ammon and W. Dirscherl, Georg Thieme Verlag, Stuttgart 1975, pp. 316-318.
[26] Doscherholmen, A., Hagen, P. S., Kinetics of ingested radiocyanocobalamin in man, Vitamin B_{12} und Intrinsic Factor, 2. Europäisches Symposium, Hamburg 1961, Ed. H. C. Heinrich, Ferdinand Enke, Stuttgart 1962, pp. 381-396.
[27] Adams, J. F., The measurement of the total assayable vitamin B_{12} in the body, Vitamin B_{12} und Intrinsic Factor, 2. Europäisches Symposium, Hamburg 1961, Ed. H. C. Heinrich, Ferdinand Enke, Stuttgart 1962, pp. 397-43.
[28] Cronkite, E. P., Henley, E., Driscoll, D. H., Meyer, L. M., Dohan, I. C., Rubini, R. J., Wolins, W., Studies on the kinetics of intravenously injected ^{60}Co-labelled vitamin B_{12} in man, Haemat. Lat. **2,** 265, 1959.
[29] Brazenas, G. R., Bartkeviciene, V. V., Aleksiene, A. M., Alioniene, A. J., Kanopkaite, S. J., and Zovyte, V. F., On some properties of the effect and metabolism of different cobalamins in tumor-bearing rats, in Vitamin B_{12}, Eds. B. Zagalak and W. Friedrich, Walter de Gruyter, Berlin 1979, pp. 1113-1118.
[30] Chanarin, I., in The Megaloblastic Anaemias, Blackwell Scientific Publications, Oxford 1969.
[31 a] Reizenstein, P. G., Body distribution, turnover rate, and radiation dose after parenteral administration of radiovitamin B_{12}, Acta Med. Scand. **165,** 467, 1959.
[31 b] Reizenstein, P., Robertson, J. S., and Cronkite, E. P., Relations of the turnover of tracer vitamin B_{12} to that of unlabelled B_{12} in the body stores, in Vitamin B_{12} und Intrinsic Factor, 2. Europäisches Symposium, Hamburg 1961, Ed. H. C. Heinrich, Ferdinand Enke, Stuttgart 1962, pp. 404-410.
[32] Miller, A., Gaull, G., Ross, J. F., Lemon, H. M., Tissue distribution of parenteral ^{60}Co-vitamin B_{12} in mouse, hamster, rat and guinea pig. Proc. Soc. Exp. Biol. Med. **93,** 33, 1956.
[33] Booth, A., Spray, G. H., The excretion and distribution in the tissues of parenterally administered ^{58}Co-vitamin B_{12} in normal and gastrectomized rats, J. Br. J. Haemat. **8,** 114, 1962.
[34] Cooperman, J. M., Luhby, A. L., Teller, D. N., Morley, J. F., Distribution of radioactive and nonradioactive vitamin B_{12} in the dog, J. Biol. Chem. **235,** 191, 1960.

[35] Quadros, E. V., Jackson, B., Hoffbrand, A. V., Linnell, J. C., Interconversion of cobalamins in human lymphocytes in vitro and the influence of nitrous oxide on synthesis of cobalamin coenzymes, in Vitamin B_{12}, Eds. B. Zagalak, and W. Friedrich, Walter de Gruyter, Berlin 1979, pp. 1045-1054.
[36] Linnell, J. C., in The Cobalamins, Eds. Arnstein, H. R. V. and Wrighton, R. J., Churchill-Livingstone, Edinburg 1971.
[37] Craft, I. L., Matthews, D. M., and Linnell, J. C., J. Clin. Pathol. **24**, 449, 1971.
[38] Linnell, J. C., Mackenzie, H. M., Wilson, J., Matthews, D. M., Clin. Sci. Mol. Med. **46**, 163, 1974.
[39] Matthews, D. M., and Linnell, J. C., in The Cobalamins. A Glaxo Symposium, Eds. Arnstein, H. R. V., and Wrighton, R. J., Churchill-Livingstone, London 1971, p. 23.
[40] Wierusz-Wysocka, B., Vitamin B_{12} binding by serum proteins in Virus Hepatitis, Pol. Arch. Med. Wewn. **56**, 445-456, 1976.
[41] Wierusz-Wysocka, B., Binding of vitamin B_{12} by serum proteins in chronic liver disease, Pol. Arch. Med. Wewn. **58**, 139-147, 1977.
[42] Linnell, J. C., and Matthews, D. M., Recent advances in cobalamin metabolism: Abnormalities in coenzyme distribution in tumor development and inherited metabolic disease, in Vitamin B_{12}, Eds. B. Zagalak and W. Friedrich, Walter de Gruyter, Berlin 1979, pp. 1101-1111.
[43] Cooke, R. D., and Coursey, D. G., Cassava: a major cyanide-containing food crop, and also in this book: Nartey, F., Cyanogenesis in tropical feeds and food stuffs, in Cyanide in Biology, Eds. B. Vennesland et al., Academic Press, New York 1981, pp. 93-132.
[44] Stelmaszynska, T., and Zgliczynski, J. M., The role of myeloperoxidase in phagocytosis, with special regard to HCN formation, in Cyanide in Biology, Eds. B. Vennesland et al., Academic Press, New York 1981, pp. 373-383.
[45] Gimsing, P., Nexø, E., and Hippe, E., Determination of cobalamins in biological material, Anal. Biochem. **129**, 296-304, 1983.
[46] Hershko, C., Levy, S., Matzner, Y., Grossowicz, N., and Izak, G., Prevalence and causes of anaemia in the elderly in Kiryat Shmoneh, Israel. Gerontology **25**, 42-48, 1979.
[47] Landahl, S., Jagenburg, R., and Svanborg, A., Blood components in a 70 years old population, Clin. Chim. Acta **112**, 301-314, 1981.
[48] Magnus, E. M., Bache-Wiig, J. E., Aanderson, T. R., and Melbostad, E., Folate and vitamin B_{12} blood levels in elderly persons in geriatric homes, Scand. J. Haematol. **28**, 360-366, 1982.
[49] Heinrich, H. C., and Pfau, A. A., Mit Hilfe eines Gesamtkörper-Radioaktivitäts-Detektors durchgeführte Untersuchungen zur Resorption, biologischen Halbwertzeit und Umsatzrate des Vitamin B_{12} im menschliohen Organismus, in Vitamin und Intrinsic Factor, 2. Europäisches Symposion, Hamburg 1961, Ed. H. C. Heinrich, Enke, Stuttgart 1962, pp. 351-381.
[50] Okuda, K., Yashima, K., Takara, I., Kitazaki, T., Kurashiga, M., and Takamatsu, M., Vitamins (Japan) **40**, 224, 1969.
[51] FAO Nutrition Meetings Report Series No **47**, Rom 1970.
[52] Thomas, M. R., Kawamoto, J., Sneed, S. M., Eakin, R., The effects of vitamin C, vitamin B_6 and vitamin B_{12} supplementation on the breast milk and maternal status of well-nourished women, Am. J. Clin. Nutr. **32**, 1679-1683, 1979.
[53] Baker, H., Frank, O., and Jaslow, S. P., Oral vs intramuscular vitamin supplementation for hypovitaminosis in the elderly, J. Am. Geriatr. Soc. **28**, 42-45, 1980.
[54] Fleming, A. F., Ogunfumnilade, Y. A., and Carmel, R., Serum vitamin B_{12} levels and vitamin B_{12}-binding proteins of serum and saliva of healthy Nigerians and Europeans. Am. J. Clin. Nutr. **31**, 1732-1738, 1978.
[55] Dong, A. and Scott, S. C., Serum vitamin B_{12} and blood cell values in vegetarians. Ann. Nutr. Metab. **26**, 209-216, 1982.
[56] Baker, S. J. and Mathan, V. J., Evidence regarding the minimal daily requirement of dietary vitamin B_{12}, Am. J. Clin. Nutr. **34**, 2423-2433, 1981.
[57] Monekosso, G. L., Wilson, J., Plasma thiocyanate and vitamin B_{12} in Nigerian patients with degenerative neurological disease, Lancet **1**, 1062-1064, 1966.
[58] Osuntokun, B. O., Matthews, D. M., Hussein, H. A. A., Wise, J. J., Linnell, J. C., Plasma and hepatic cobalamins in tropical ataxic neuropathy, Clin. Sci. Mol. Med. **46**, 563-567, 1974.
[59] Baker, H., Frank, O., Chen, T., Feingold, S., Dengelis, B. and Baker, E., Vitamin content of some normal human brain segments. J. Neurosci Res. **11**, 419-436, 1984.
[60] Muir, M., Chanarin, J., Separation of cobalamin analogs in human sera binding to intrinsic factor and to R-type vitamin B_{12} binders, Br. J. Haematol. **54**, 613-622, 1983.

[61] Nilsson, L.O., Myrvold, H.E., Swolin, B., Ojerskog, B., Vitamin B_{12} in plasma in patients with continent ileostomy and long observation time, Scand. J. Gastroenterol. **19,** 369-374, 1984.

[62] Bambach, Ch.P. and Hill, G.L., Long-term nutritional effects of extensive resection of the small intestine. Anst. N.Z.J Surg **52,** 500-506, 1982.

[63] Hutchins, D.A. and Ball, P.E., Changes in the tissue concentration of folate and vitamin B_{12} during maturation and aging. Neurochem. Int. **5,** 231-236, 1983.

[64] Gimsing, P., Hippe, E., Helleberg-Rasmussen, I., Moesgaard, M., Lanng Nielsen, J., Bastrup-Madsen, P., Berlin, R., Hansen, T., Cobalamin forms in plasma and tissue during treatment of vitamin B_{12} deficiency. Scand, J. Haematol. **29,** 311-318, 1982.

[65] Hooper, P.L., Hooper, E.M., Hunt, W.C., Carry, P.J., Goodwin, J.S., Vitamins, lipids and lipoproteins in a healthy elderly population. Int. J. Vitam. Nutr. Res. **53,** 412-419, 1983.

[66] Batt, R.M., Bush, B.M., Subcellular biochemical studies of a naturally occurring enteropathy in the dog resembling chronic tropical sprue in humans, Am. J. Vet. Res. **44,** 1492-1496, 1983.

[67] Magnus, E.M., Bache-Wiig, J.E. Aanderson, F.R., and Melbostad, E., Folate and vitamin B_{12} cobalamin blood levels in elderly persons in geriatric homes, Scand. J. Haematol. **28,** 360-366, 1982.

[68] Goodwin, J.S., Goodwin, J.M., Garry, P.J., Association between nutritional status and cognitive functioning in a healthy elderly population, Jama J. Am. Med Assa **249,** 2917-2921, 1983.

7 Cobamide Dependent Enzymes

7.1 Introduction

In this chapter, enzymatic reactions dependent on cobamides are presented. These reactions can be divided into two groups, which require two types of cobamides: the adenosylcobamides and the methylcobamides (Fig. 7-1.)

The first group includes:

1) Reactions of the rearrangements involving the transfer of hydrogen atom from one carbon atom to an adjacent carbon atom and its replacement by another adjacent group, Fig. 7-1 (A.1).

A.1.

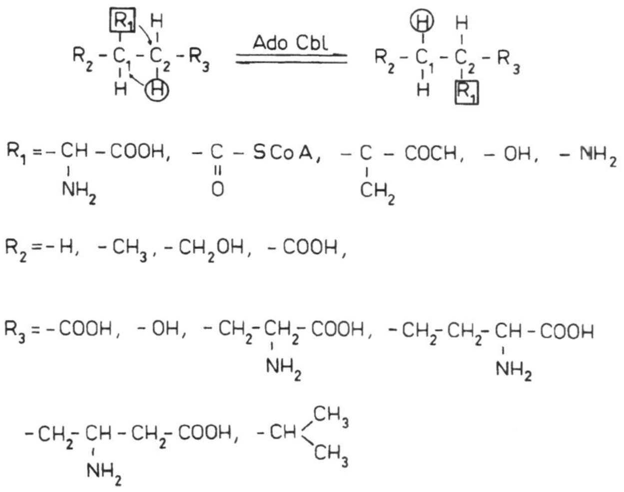

2) Reaction of reduction of ribunucleotide triphosphate to 2′-deoxyribonucleotide triphosphate, Fig. 7-1 (A.2).

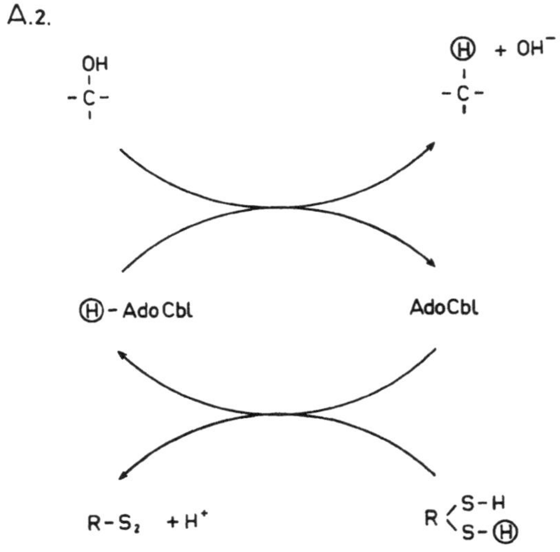

The second group contains reactions requiring the methylcobamides engaged as co-factors in methyl transfer, (Fig. 7-1 (B).

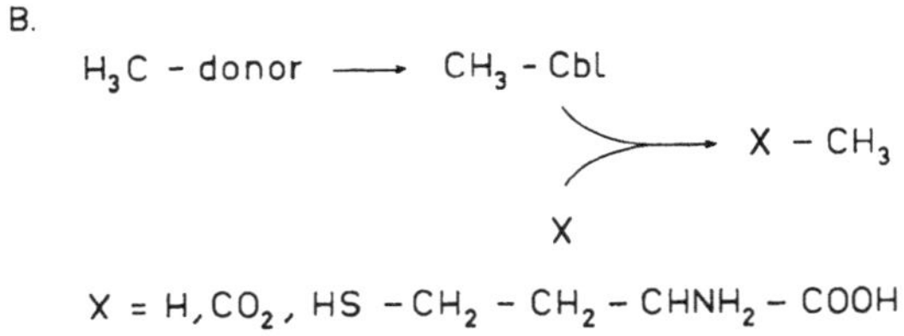

Fig. 7-1 Enzymatic reactions dependent on: A. adenosylcobamide B. methylcobamide

These reactions were reviewed by Barker [1], Hogenkamp [2], Stadtman [3], and other authors [4-6].

7.2 Adenosylocobamide - dependent reactions

by A. Stroinski

7.2.1 Glutamate Mutase (L-threo-3-Methylaspartate carboxy-aminomethylmutase EC 5.4.99.1)

Sources: *Clostridium tetanomorphum* [7, 8]
Propionibacterium arabinosum [9]

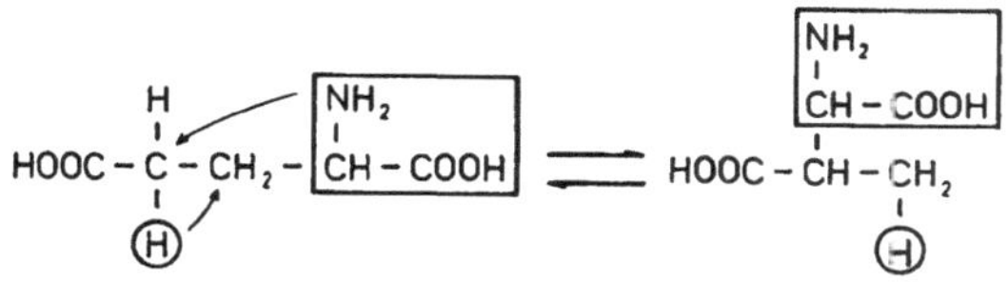

B_{12} dependence

Active forms [9]:	K_M ($\times 10^7$ M)
AdoCbl	128.0
(2,6-diaminopurinyl)AdoCba	16.4
(Ade)AdoCba	12.4
(5(6)Nitro-Bza)AdoCba	5.3
(5(6)-Amino-Bza)AdoCba	4.6
(5(6)-Me-Bza)AdoCba	4.3
(5(6)-Trifluoro-Me-Bza)AdoCba	4.0
(Bza)AdoCba	2.3

Inactive forms:
AdoCbi, MeCbl, Carboxy MeCbl, Carboxy EtCbl, UrdCbl [10, 11].

Molecular structure [12–15]
Subunits: E = 128 000, binds AdoCbl (1 mol/1 mol E)
S = 17 000, contains 5 SH groups
Apoenzym: (E + S) binds 2 mol AdoCbl/mol apoenzyme

Catalytic properties [15]
$K_{eq} = 107$ for glutamate

Substrate: K_M values:
L-Glutamate $= 1.5 \times 10^{-5}$ M
β-Methylaspartate $= 0.5 \times 10^{-3}$ M

Temperature optimum: at 40 °C for 30′ incubation period.
pH optimum: 8.5
$\Delta F° = -5{,}99$ kJ/mol for methylaspartate →glutamate

Subunits: E and S separately inactive

Assay system: In the presence of β-methylaspartase, β-methylaspartate is deaminated to 2-methylfumarate, which is estimated spectrophotometrically [15].

Stability and Storage [15]: components S and E are most stable in neutral phosphate buffer at −10 °C. Mercaptoethanol decreases the stability.

7.2.2 **Methylmalonyl-CoA mutase** (2-methylmalonyl CoA: CoA-carbonyl mutase, EC 5.4.99.2)

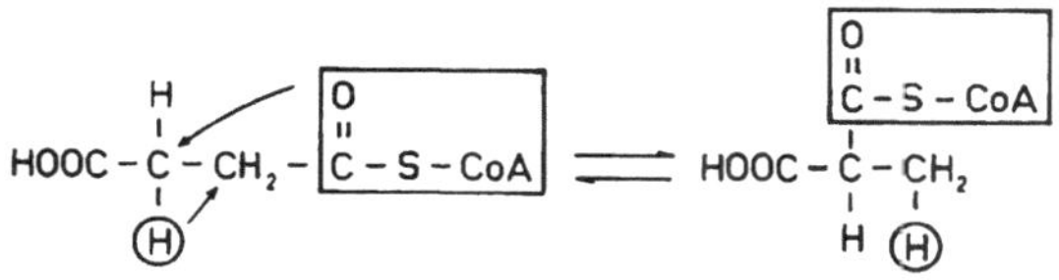

Sources: *Propionibacterium shermanii* [16, 17, 19]
Protaminbacter ruber [20]
Sheep liver [22, 23]

B_{12}-dependence:
animal enzyme [24, 25]
active forms: AdoCbl, (Bza)AdoCba
inactive forms: (Ade)AdoCba

Molecular strukture:
bacterial enzyme [19]: 124000
subunits: 61000 and 66000
animal enzyme [22]: 165000

Catalytic properties:

	bacterial enzyme [16, 17, 19, 21]	animal enzyme [22]
K_{eq} (for succinyl-CoA)	23.5	18.6
K_M values:		
AdoCbl	3.5×10^{-8} M	2.1×10^{-8} M
L-methylmalonyl-CoA	8.0×10^{-5} M	2.4×10^{-4} M
succinyl-CoA	3.45×10^{-5} M	6.2×10^{-5} M
number of active sites	–	2

Substrate: specific for L-enantiomer [28]
Requirements
Bacterial enzyme [16]: no metal ions required, EDTA, N-ethylmaleimide and p-hydroxymercurobenzoate have no effect on the activity.

Assay method [21, 23]: the rate of conversion of L-methylmalonyl-CoA to succinyl-CoA can be assayed spectrophotometrically by means of a series of coupled reactions.

Inactivation: The holoenzyme undergoes irreversible destruction of coenzyme with the formation of 5′-deoxyadenosine [32].

Mechanism:
1. No carboxyl transfer occurs [26]
2. The movement of the thioester group is intramolecular [27].
3. There is a direct hydrogen transfer from the methyl carbon of methylmalonyl-CoA to C-3 of succinyl-CoA [18],
4. The configurations are retained for carbon atom receiving the migrating hydrogen atom [28] and for the carbon receiving the migrating acyl group [29],
5. Adenosylcobalamin is the intermediate hydrogen carrier [30–32].

Stability: Animal enzyme can be stored as ammonium sulfate suspension at −12 °C [22].

7.2.3 α-Methyleneglutarate mutase

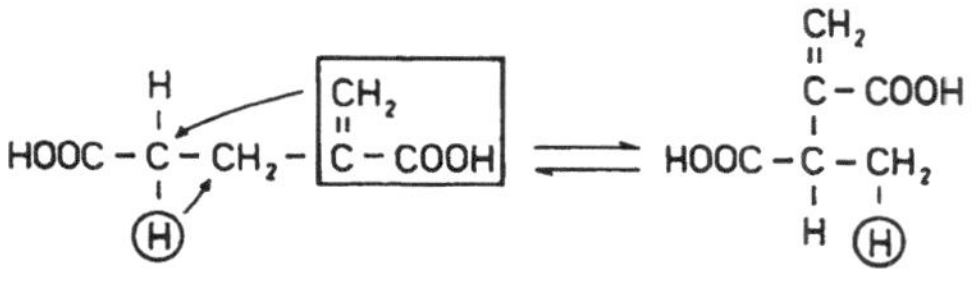

Source: *Clostridium barkeri* [33]

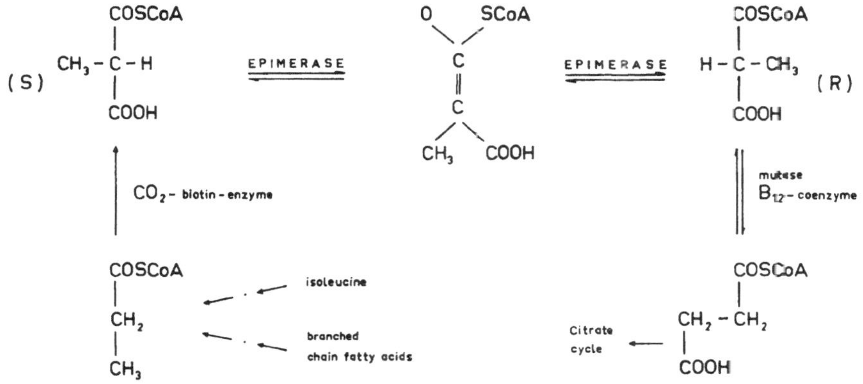

Fig. 7-2 Metabolic function of methylmalonyl-CoA mutase (reprinted from Retey [32a])

B_{12}-dependence

active forms [34]	K_M
AdoCbl	7.3×10^{-8} M
(Bza)AdoCba	3.0×10^{-7} M
(Ade)AdoCba	1.25×10^{-6} M

Molecular weight: 170000 [33]

Catalytic properties [34]:
$K_{eq} = 0.23$ (methylitaconate → α-methyleneglutarate)
pH optimum 7.3
cofactor requirements: does not require metallic ions nor sulfhydryl reagents
$\Delta F° = +3{,}72$ kJ/mol (methyleneglutarate → methylitaconate, 34 °C, pH 7.9)
Substrate: K_M for α-methyleneglutarate $= 7.1 \times 10^{-3}$ M
Inhibition:

Competitive	K_i (mM)
Itaconate	1.2
Mesaconate	1.2
Succinate	1.8
1-Me-1,2-trans-cyclo-propanedicarboxylate	3.3
L-Malate	7.0
Maleate	–

Non-competitive	
Glutaconate	4.5
1-Me-1,2-cis-cyclo-propanedicarboxylate	38
Citrate	–

Assay system: methylitaconate is isomerized to dimethylmaleate, which is estimated colorimetrically [34].

Reaction mechanism: AdoCbl acts as a hydrogen carrier as evidenced by the incorporation of tritium into substrate and product from [5′-^{3}H]AdoCbl, and the proton transfer does not involve protons from the medium [35].

7.2.4 Dioldehydratase (Propanediol dehydratase, 1,2-Propanediol hydro-lyase; EC 4.2.1.28)

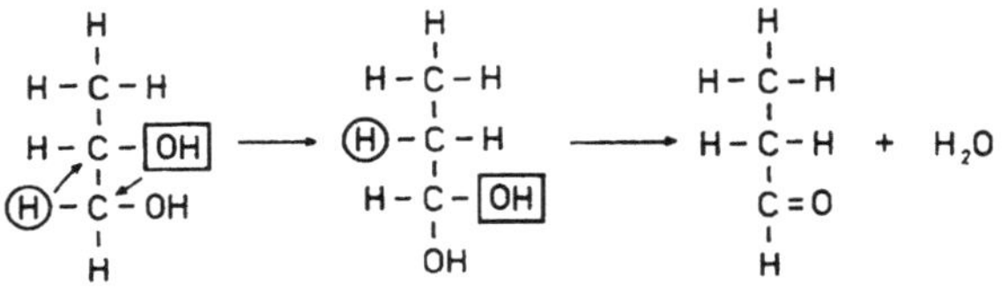

Sources: *Klebsiella pneumonia* ATCC 8724 (*Aerobacter aerogenes* ATCC 8724) grown anaerobically on glycerol [36–39], or on glycerol plus 1,2-propanediol [45], or on 1,2-propanediol [46].

Klebsiella pneumonia ATCC 25955 (*Aerobacter aerogenes* PZH 572 Warsaw) grown in a 1,2-propanediol or a glycerol medium, ATCC 13882 *(Klebsiella aerogenes)* and

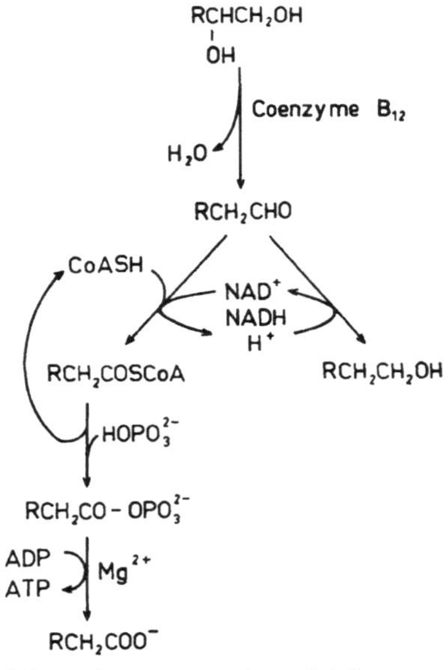

Fig. 7-3 Schematic representation of 1,2-propanediol and 1,2-ethanediol fermentation by *Klebsiella pneumoniae* ATCC 8724. R = CH_3 or H (from Fukui and Toraya [49a])

ATCC 8308 *(Aerobacter aerogenes)*, Citrobacter freundii (Escherichia freundii) AKU 0009, and *Citrobacter intermedium (Escherichia intermedia)* ATCC 8454 and AKU 0010 grown in a 1,2-propanediol medium [46, 46a]. *Klebsiella sp.* No 101 and *Micrococyclus eburneus* [46b].

Activity of coenzyme B_{12} analogues:

***Active* [55, 60]**	K_M (µM)	***Inactive*** [37, 55]	K_i (µM)
AdoCbl	0.8	CH_3-Cbl	0.73
(Ara-Ado) Cbl	1.38	CN-Cbl	1.87
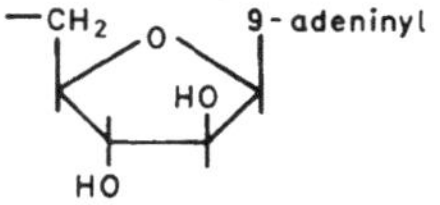		(L-Ado)Cbl	1.17
		Inosyl-Cbl	22.2
		CydCbl	6.05
Aristeromycyl-Cbl	1.05	(Bza)-ribosyl-Cbl	3.25
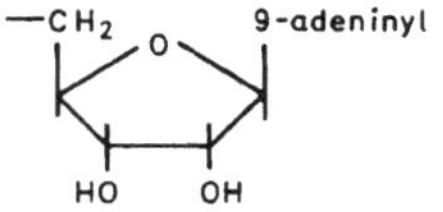		Adeninylpentyl-Cbl	0.27
		Adeninbutyl-Cbl	0.47
3-Izo-Ado Cbl	1.35	Adeninylpropyl-Cbl	–
Nebularyl-Cbl	1.67	Adeninylethyl-Cbl	–
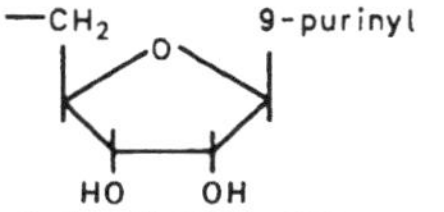		Adenosylethyl-Cbl	–
		OH-Cbl	–
AdoCbl(10-Cl)	0.6		

Molecular structure:

apoenzyme: 250000 [47],
230000 [48]

subunits [48–50]: F = 30000
S = 200000

S subunit consists of [48]: $S_1 = 60000$
$S_2 = 23000$
$S_3 = 15500$
$S_4 = 14000$

Apoenzyme possesses one active site [47], which has mercurial-sensitive sulfhydryl group(s) involved in binding of cobalamin [52, 55].

Catalytic properties:

K_{eq} - irreversible

Activity of substrate analogues:

1,2-Butanediol and styrene glycol are catalytically inactive and are weak competitive inhibitors [54]. 3-fluoro-1,2-propanediol is an active analog of substrate [56]. Glycerol is a substrate and inactivator [57, 59]. Apoenzyme distinguishes between R and S binding conformation, the enzyme-(R)-glycerol complex is responsible for the catalytic reaction, while the enzyme-(S)-glycerol is more favorable for inactivation (Table 7-1) [57].

pH optimum: 6–10 [45].

Table 7-1 Parameters of substrates and inactivators of diol dehydratase (Reprinted from Bachovchin et al., Biochemistry **16,** 1082, 1977 with permission. Copyright 1977 American Chemical Society)

Substrate	$K_M \times 10^4$	$k_p(S^{-1})$	$k_i(min^{-1})$
S-1,2-Propanediol	0.186	191	≈0
RS-1,2-Propanediol	0.30	250	≈0
R-1,2-Propanediol	0.60	340	≈0
Glycerol	16.0	412	1.30
1-Chloro-2,3-propanediol	4.1	>0	0.36
Trifluoropropanediol	1.24(K_i)	>0	0.016
2,3-Butanediol(meso + D)	2.0(K_i)	4.0	0.068
1,2-Butanediol	5.9(K_i)	>0	0.054
Isobutylene glycol	8.8	18	0.47
Ethylene glycol	6.6	150	0.098
Thioglycerol	0.1	~0	1.74

Monovalent ions: K^+ ion is absolutely required for product formation [37] because this ion plays an essential role in the interaction of the apoenzyme with AdoCbl [52, 53].

Table 7-2 Effects of monovalent cations on propanediol dehydratase activity (reprinted with permission from Toraya et al., Biochemistry **10,** 3475, 1971. Copyright 1971 American Chemical Society)

Monovalent Cation	V_{max} (μmol/min)	Relative cofactor activity (%)[b]	K_M(M)
None	0.00	0	
Na^+	6.58×10^{-2}	57	1.92×10^{-2}
K^+	1.15×10^{-1}	(100)	6.67×10^{-4}
NH_4^+	1.22×10^{-1}	106	1.18×10^{-4}
Rb^+	1.08×10^{-1}	96	1.82×10^{-3}
$CH_3NH_3^+$	1.06×10^{-1}	92	8.34×10^{-3}
Ca^+	5.06×10^{-2}	44	5.00×10^{-3}

[b] Expressed in relation to the activity with K^+ taken as 100

Holoenzyme dissociates in the absence of an appropriate monovalent cation [52-54]. Ions: Li^+, $(CH_3)_2NH_2^+$, $(CH_3)_3NH^+$, $C_2H_5NH_3^+$, $(C_2H_5)_2NH_2^+$ and $(C_2H_5)_3NH^+$ have no cofactor activity [52].

Inhibition: Adenosine is inhibitor of the reaction [39].

Inactivation of apoenzyme complexes: AdoCbl [37, 58], (Ara) Ado Cbl-, and 3-Izo-Ado-Cbl-apoenzyme complexes [51] are inactivated irreversibly in the absence of 1,2-propanediol and in the presence of oxygen. In contrast, (Aristeromycyl-Cbl)- and (Nebular-Cbl)-apoenzyme complexes are not inactivated under identical conditions [51]. In the presence of glycerol and in the absence of 1,2-propanediol and oxygen, apoenzyme-AdoCbl complex undergoes irreversible inactivation [57-59]. The rate of inactivation is a first order reaction with respect to enzyme. 1,2-propanediol is a competitive inhibitor of this process. Enzyme-(S)-glycerol complex is responsible for inactivation.

Reaction mechanism [63 a, 63 b]: The reaction involves the displacement of the secondary hydroxyl group by one of the C-1 hydrogen atoms of the diol. The hydrogen transfer runs in two steps: 1. the transfer of hydrogen from C-1 of the substrate to C-5′ of active complex, enzyme-AdoCbl [61], 2. the transfer of hydrogen from C-5′ of the complex to C-2 of the substrate:

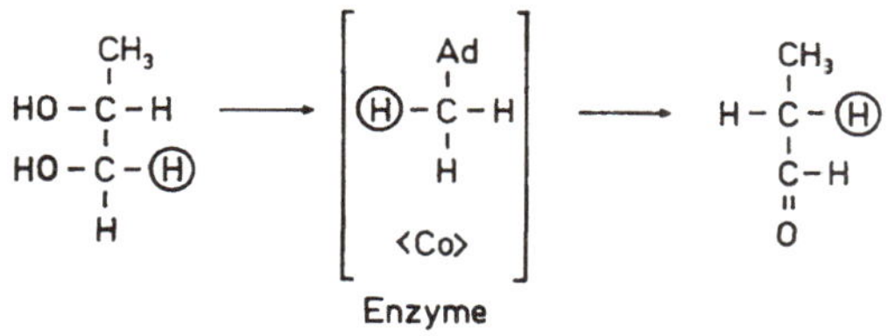

The reaction involves the transfer of the oxygen atom from C-2 to C-1 with the formation of propane-1,1-diol followed by dehydration to propionaldehyde [40–45, 47, 58, 62]. These transformations require the homolytic cleavage of the carbon-cobalt bond of coenzymes, during interaction between apoenzyme and coenzyme [54, 58]. Interactions of both adenine and the sugar moiety of the β ligand with apoenzyme play an essential role in the activation of the Co-C bond [51]. The first hydrogen transfer step is the rate-limiting step [57].

Assay systems: The product of the reaction can be estimated:
1. colorimetrically with 2,4-nitrophenyl-hydrazine [37],
2. by 3-methyl-2-benzothiazolinone hydrazone method [51],
3. spectrophotometrically by coupling to a NAD-alcohol dehydrogenase system [57].

7.2.5 **Glycerol dehydratase** (Glycerol hydro-lyase, EC 4.2.1.30)

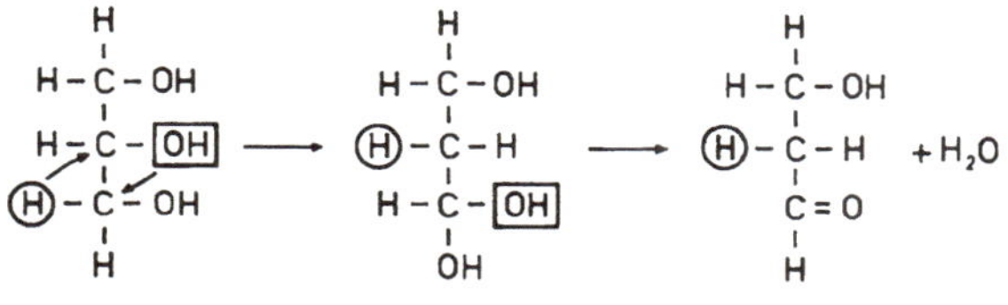

Sources: *Lactobacillus sp* [65, 66],
Klebsiella pneumoniae ATCC 25955 (Formerly named *Aerobacter aerogenes*, PZH 572, Warsaw) [67, 68],
Klebsiella pneumoniae ATCC 13882 *(Klebsiella aerogenes)* [69],
Klebsiella pneumoniae ATCC 8308 *(Aerobacter aerogenes)* [69],
Citrobacter freundii (Escherichia freundii) AKU 0009 [69], *Citrobacter intermedium (Escherichia intermedia)* ATCC 8454 and AKU 0010 [69].

Induction of the enzyme in bacteria: Glycerol, 1,2-propanediol, ethyleneglycol can induce glycerol dehydratase [76]. α-phosphoglycerol, β-phosphoglycerol, 1,3-propanediol and 2,3-butanediol are not inducers [76]. Glucose lowers both intracellular cAMP and the activity of enzyme. Extracellular cAMP derepresses the enzyme [86, 86 a]. In opposition to glucose, cAMP enhances the uptake of glycerol, and reduces the uptake of glucose to cells [86 a].

Scheme of anaerobic catabolism of glycerol by *Klebsiella sp.* [46 a]

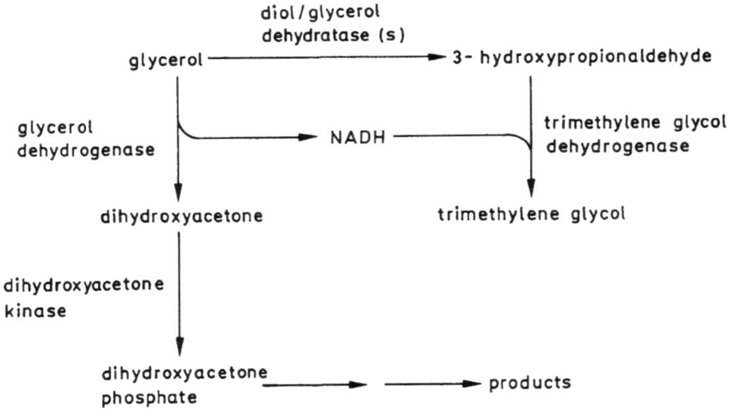

Glycerol is attacked by coenzyme B_{12}-dependent diol dehydratase [36 a, 59, 59 a] and the product, 3-hydroxypropionaldehyde, is reduced by an NADH-linked reductase to trimethylene glycol (1,3-propanediol). 3-hydroxypropionaldehyde may spontaneously form acrolein by intramolecular dehydration [65 a].

Activity of coenzyme B_{12} analogues [68, 70–72, 74, 93]

Active forms:	K_M(M)
AdoCbl	2×10^{-8}
(Ade)AdoCba	–
(6-exo-N-MeAdo)Cbl	$1{,}8 \times 10^{-7}$
(6-exo-N-BuAdo)Cbl	$2{,}9 \times 10^{-7}$
InoCbl	$1{,}5 \times 10^{-7}$
(3,5,6,-Me_3Bza)AdoCba	$4{,}4 \times 10^{-7}$
(3-benzyl,5,6,Me_2Bza)AdoCba	$4{,}0 \times 10^{-7}$
AdoCbl(a, b, c, d, g, -pentaamide)	$5{,}0 \times 10^{-7}$
AdoCbl(e-ethoxy-a, b, c, d, g, -pentaamide)	$1{,}8 \times 10^{-7}$
2′dAdoCbl	$4{,}1 \times 10^{-8}$
3′dAdoCbl	$4{,}4 \times 10^{-8}$
IsoAdoCbl	$5{,}3 \times 10^{-8}$
AdoCbl-5′P	active
Inactive forms:	K_i(M)
MeCbl	1×10^{-7}
(Ade)MeCba	$1{,}5 \times 10^{-6}$
(Ade)AdoCba	$8{,}0 \times 10^{-7}$
(CN)-Cbl	$1{,}0 \times 10^{-8}$

(3,5,6-Me_3Bza)AdoCba	$6{,}0\times10^{-5}$
CydCbl	$4{,}0\times10^{-7}$
UrdCbl	$2{,}8\times10^{-7}$
(Ade)UrdCba	$1{,}4\times10^{-5}$
(3,5,6-Me_3Bza)UrdCba	$1{,}4\times10^{-5}$
(6-exo-N-benzoyl-Ado)Cbl	$4{,}5\times10^{-8}$
GuaCbl	$3{,}3\times10^{-7}$
(1-N-oxide)AdoCbl	$3{,}6\times10^{-8}$
AdoCbi	$1{,}8\times10^{-6}$
CN-Cbl (ethoxy-a, b, c, d, g-pentamide)	$3{,}5\times10^{-7}$
OH-Cbl	–
Cyd-Cba	–
(3,5,6,-Me_3Bza)CydCba	–
1-MeAdoCbl	–
(OH)MeCbi	–
(OH)EtCbi	–
(OH)PrCbi	–
(OH)BuCbi	–
(OH)(Bu-2)Cbi	–
(OH)PeCbi	–
(OH)OctylCbi	–
(OH)DecylCbi	–
(OH)CarboethoxyCbi	–
(2′,3′-Isopropylidene)AdoCbl	–
InoCbl	–

Molecular structure: Apoenzyme-HO-Cbl complex 188000 (by gradient centrifigation [75]).

Apoenzyme ($\alpha_2\beta_2$) = 195000 ± 15000 (by gel filtration [76, 78, 79])

Subunits A (α_2) = 22000 [78].
α = 12000 [78].
B (β_2) = 189000 ± 22000 [80].
β = 90000 ± 25000 [80].
Subunit β has binding sites for glycerol, AdoCbl and UMP [85].

Catalytic Properties:
K_{eq} - irreversible

Substrate:
K_M: glycerol = 1,3 mM [76,8], 1,5 mM [74].
1,2-propanediol = $0{,}8\times10^{-4}$ M [74].
ethyleneglycol = 0,66 ± 0,05 mM [82].
1,2-Butanediol, 2,3-butanediol, 1,3-butanediol are inactive [84].

Monovalent ion requirements: K^+, or Rb^+, or NH_4^+ [65, 67, 69, 76, 80, 81, 83], or Tl^+ [83] are essential for enzymic activity.

Monovalent cations are involved in the binding of AdoCbl or MeCbl by apoenzyme [78, 87].

K_M values: $K^+ = 1{,}0 \times 10^{-2}$ M [79], and 7×10^{-3} M [83],
$NH_4^+ = 4{,}0 \times 10^{-3}$ M [83],
$Tl^+ = 1{,}0 \times 10^{-3}$ M [83],

Table 7-3 Kinetic parameters for glycerol dehydratase substrates and their analogues (reprinted from Poznanskaya and Korsova [86c])

Substrates and analogs	K_M, mM	$V_{rel.}$	K_D, mM	k_{in}, min^{-1}
Glycerol	1.30	1		0.35
Ethylene glycol	0.60	0.4		0.19
(RS)-1,2-propanediol	0.080	0.5		0.04
(R)-1,2-propanediol	0.096	0.75		
(S)-1,2-propanediol	0.04	0.4		
1,2-butanediol			2.2	0.02
3-chloro-1,2-propanediol			2.1	0.54
Glycerol methyl ether			4.0	0.05
Glycerol ethyl ether			3.9	0.07
1,3-propanediol			7.0	4.3
1,3-butanediol			150	1.1
1,4-butanediol			110	0.03

Na^+ is the noncompetitive inhibitor and Cs^+ is the competitive inhibitor with K^+ [76, 85].

$(CH_3)_4N^+$, $(C_2H_5)_4N^+$ and $(HOCH_2)_3N^+CH_3$ have no cofactor activity [80].

Allosteric properties: Positive cooperativity in glycine buffer and negative cooperativity in ethanolamine buffer for glycerol binding sites [76, 80]. Phosphate [80], K^+ and UMP [76], are allosteric activators. UTP is allosteric inhibitor [76].

pH Optimum: 8.6 [78].

Temperature Optimum: 30–35 °C [78, 79].

Inhibition: N-ethylmaleimide (1 mM) completely blocks the reaction [73].

Assay System: The reaction product can be estimated:
1. tryptophan method [66],
2. 2,4-dinitrophenylhydrazine method [89],
3. MBTH method with 3-methyl-2-benzothiazolinone hydrazone [80],
4. Coupling with an alcohol dehydragenase reaction, which is measured spectrophotometrically or by a pH-metric method [82].

The activity of glycerol dehydratase may be estimated in presence of diol dehydratase using the 67-fold difference in their K_H values for coenzyme B_{12} [86b].

Holoenzyme stability: The holoenzyme (enzyme-AdoCbl) undergoes rapid, irreversible inactivation, under aerobic condition, faster in the absence of glycerol [79, 80]. The inactivation is accompanied with the cleavage of the coenzyme at the carbon-cobalt bond [78, 87], and with the formation of a new covalent bond between enzyme and cobalt [87]. The binding of HO-Cbl to apoenzyme is accompanied by a change in the cobalamin spectrum (Fig. 7-4) [75].
CN-Cbl-5'-P, CN-Cbl (monocarboxylic), (aq,CN)Cbi, (aq,CN)Cbi-P, (aq,CN)Cbi-GDP do not inactivate the apoenzyme [93].

Effects of heating: The apoenzyme is very unstable to heat treatment. Four minutes heating at 46 °C results in 50% destruction of apoenzyme. Complex, enzyme-Cbl, is much more stable, no destruction occurs after 4 min heating at 70 °C and 50% destruction only after 4 min at 82 °C [78, 81].

Thermal inactivation of the apoenzyme is due to the thermal lability of subunit A (α_2). Subunit B (β_2) is stable up to 70 °C (Fig. 7-5) [78]. The heating of enzyme-Cbl complex over 70 °C produces a still active subunit B an inactivate subunit A and HO-Cbl. Li^+ and Na^+ ions promote this thermal dissociation, while K^+ ions inhibit it [83]. The inactivation of subunit A at 70 °C is reversible in the presence of ions: Na^+, Rb^+, K^+ and Cs^+, but not Li^+ [92].

Mechanism of reaction [89, 90]:

1. Interaction of the coenzyme with the apoenzyme induces homolytic fission of cobalt-carbon bond of the coenzyme and formation of Cob(II)alamin and the 5′-deoxyadenosyl radical (Fig. 7-6) [90].
2. The hydrogen atom of the substrate at C-1 is transferred to C-5′ of deoxyadenosyl radical.
3. The hydrogen from C-5′ of the cobalamin is next transfered to C-2 of the substrate by a mechanism which involves:

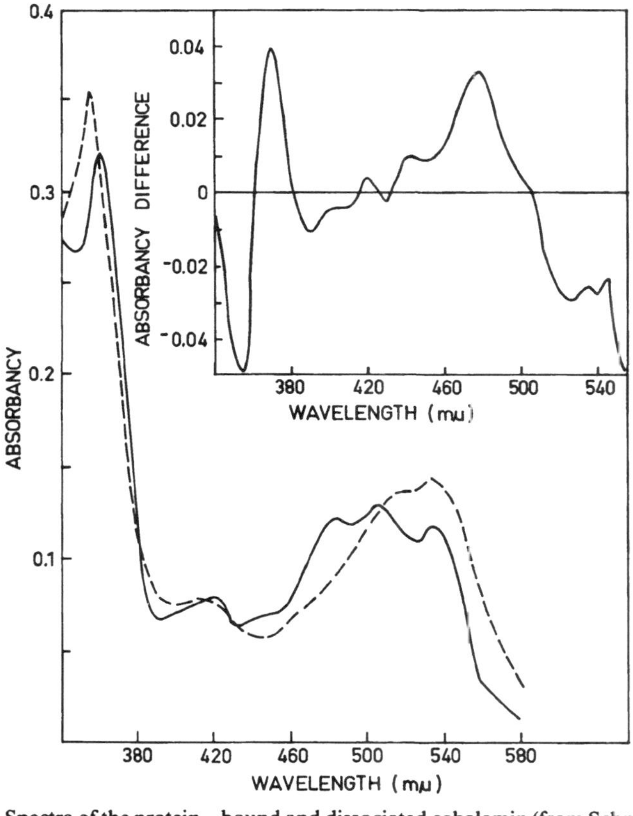

Fig. 7-4 Spectra of the protein - bound and dissociated cobalamin (from Schneider et al. [75]).

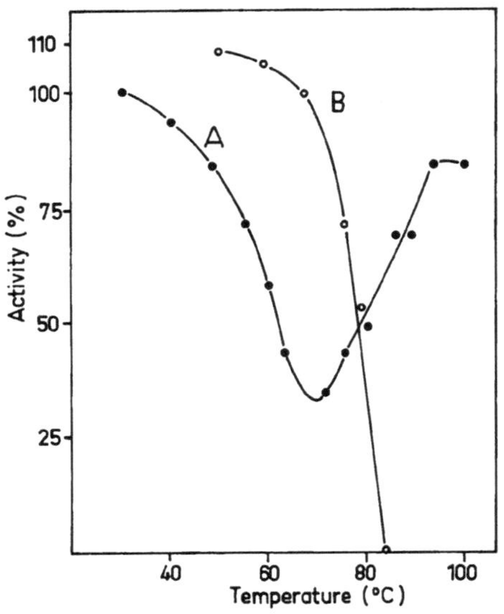

Fig. 7-5 Effect of temperature on the activity of proteins A and B (from Schneider and Pawełkiewicz [78]).

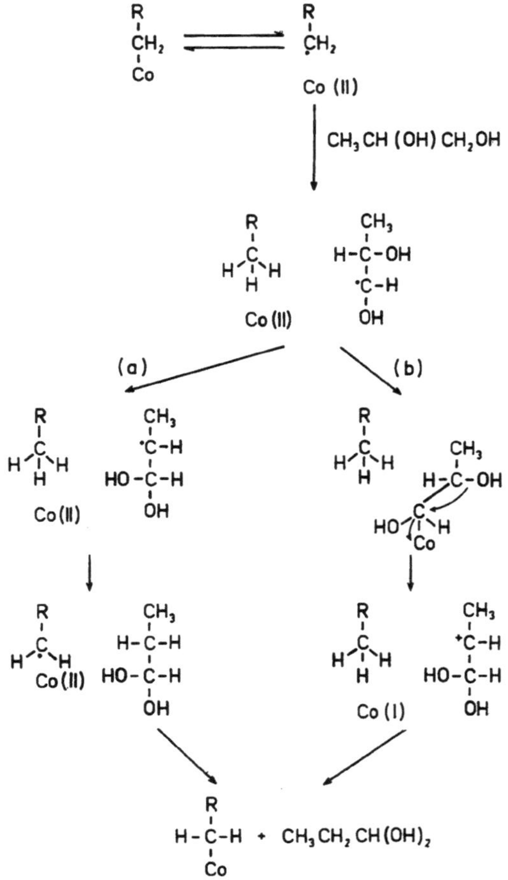

Fig. 7-6 Possible mechanisms for conversion of propane-1,2-diol to propionaldehyde by glyceroldehydratase (reprinted with permission from Cockle et al., J. Chem. Soc. **94**, 275, 1972. Copyright 1972 American Chemical Society)

a) The shift of OH radical
or b) the formation of bond between Co and C-1 of substrate.
4. The transfer of hydrogen is intermolecular [89].

Storage conditions: The enzyme preparations dried in Sephadex gels (G-50) proved to be very stable and could be stored at room temperature for several years without change [91 a, b]. The full activity of enzyme is still preserved after two years of storage at −5 °C in 50% glycerol [86 d].

Methods of enzyme purification:
– using gel filtrations [80, 81],
– by means of hydrophobic interaction chromatography on Octyl-Sepharose [86 d].

7.2.6 Ethanolamine deaminase (Ethanolamine ammonia-lyase, EC 4.3.1.7)

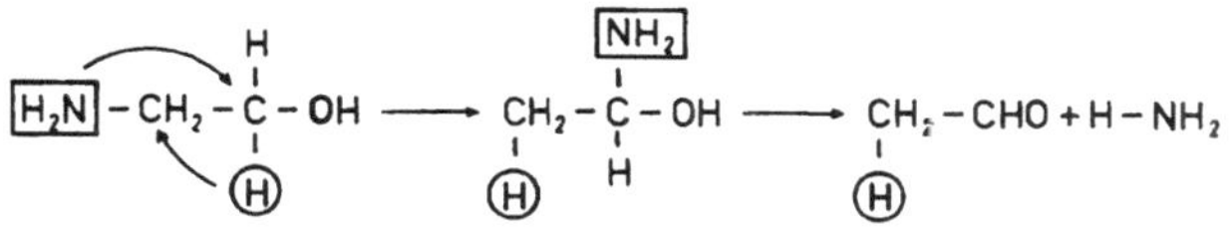

Sources: *Clostridium sp.* [94],
Salmonella typhimurium [109],
Klebsiella aerogenes [110].

Activity of coenzyme B_{12} forms:

active forms [94, 95, 103	***inactive forms*** [94, 103, 104]
AdoCbl	OH-Cbl
(OH)AdoCbi [1, 2]*	CN-Cbl
(Bza)AdoCba	MeCbl
(2′-deoxy Ado)Cbl	β-OH EtCbl
	β-(2-Tetrahydropyryloxy)EtCbl
	(1-Me Ado)Cbl
	δ-(9-Ade)BuCbl
	InoCbl
	UrdCbl

K_M Values:

AdoCbl	$= 8.2 \times 10^{-7}$ M [94],
	1.5×10^{-6} M [95],
(Bza)AdoCba	$= 4.0 \times 10^{-7}$ M [94],
	1.9×10^{-7} M [95],
(Ade)AdoCba	$= 7.7 \times 10^{-7}$ M [95],

* This is the only known example, in which an analog without α-nucleoside is active.

Molecular structure:
apoenzyme 520000 (sedimentation) [96],
subunits 51000
36000 (SDS-acrylamide gel electrophoresis)
active site number 2 [102, 103, 104].

Catalytic properties:

Reversibility: none with ethanolamine and slight with propanolamine [112].

Turnover number: (molecules of product/molecule of enzyme)
for ethanolamine $= 100\ sec^{-1}$ [102]
for propanolamine $= 1\text{-}2\ sec^{-1}$ [112]

Substrates: ethanolamine and D- and L-2-aminopropanol [95, 97, 98]
K_i values of analogues of ethanolamine [95]:

DL-2-Amino-1-propanol	7.1×10^{-6} M,
L-2-Amino-1-propanol	6.3×10^{-6} M,
1-Amino-2-propanol	2.0×10^{-5} M,
2-Amino-1-butanol	2.7×10^{-5} M,
3-Amino-propanol	1.6×10^{-2} M
2(Methylamino)-ethanol	3.9×10^{-3} M,
2(Dimethylamino)-ethanol	1.4×10^{-2} M

K_M values: ethanolamine $= 2.9 \times 10^{-5}$ M [94],
1.0×10^{-5} M [95, 111].

Monovalent ions: K^+, or NH_4^+, Rb^+ are required, Na^+ and Li^+ are competitive inhibitors to the appropriate ions [95].
K_M for $K^+ = 4.7 \times 10^{-4}$ M [95].

pH optimum: 6.8–8.2 [95].

Inactivation: by intrinsic factor [94], by sulphydryl reagents [95, 108 d], and by incubation of the apoenzyme with AdoCbl in the absence of substrate [107].

Mechanism of reaction [100 a, 100 b, 108 e]:
1. The carbinol carbon of the substrate becomes the carbinyl carbon of the product [96, 99].
2. The alcoholic oxygen atom is retained in the product [99].
3. The hydrogen from the alcohol carbon is transfered to C-5 of apoenzyme bound AdoCbl and next to the amine carbon [98–101, 107, 111, 113].
4. The carbon-cobalt bond of AdoCbl is cleaved homolytically [108 a, b, 112] or heterolytically [114–116].

Assay systems:
1. Spectrophotometric method, coupled to yeast alcohol dehydrogenase, which reduces the acetaldehyde to ethanol [95],
2. Chromatographic method using [^{14}C]ethanolamine [102].

Storage conditions: The enzyme is stable at −70 °C or in liquid N_2 and changes slowly at −20 °C [99].

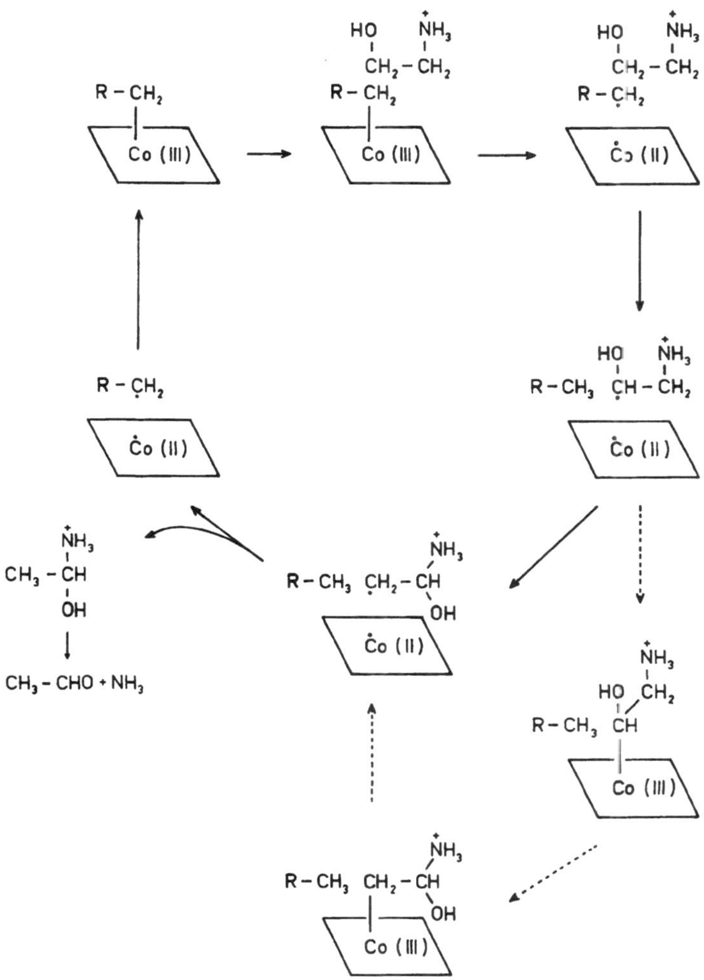

Fig. 7-7 Mechanistic scheme for the rearrangement of 2-aminoethanol catalysed by EAL. $RCH_2 = 5'$-deoxy-5'-adenosylcobalamin; the catalytic steps of the reaction are believed to proceed by mechanism which may (stippled arrows) or may not involve substrate-metal binding during rearrangement. A similar scheme can, in principle, be drawn for L-2-aminopropanol as substrate (reprinted from Hollaway et al. [100b]).

7.2.7 D-α-Lysine mutase

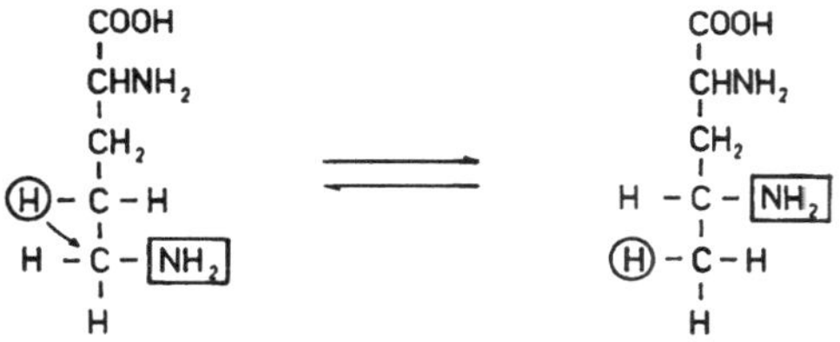

Source: *Clostridium sticklandii* [117].

B_{12} dependence:
K_M for AdoCbl $= 5.0 \times 10^{-6}$ M [118].

Molecular structure:
molecular weight: 250000 [119],
two subunits [119]: E_1 - containing AdoCbl and pyridoxal phosphate,
E_2 - sulfhydryl protein.

Catalytic properties:
K_{eq} (for D-α-Lysine) $= 1.9$ [119].

Substrate:
K_M values: for D-α-Lysine $= 1.5 \times 10^{-3}$ M, and 0.2×10^{-3} M in presence of ATP [118].

pH optimum: 8.5–9.0 [118].

Cofactor requirements: pyridoxal phosphate, K^+, Mg^{2+}, ATP and sulfhydryl reagent [118].

Allosteric activator: ATP [118].

Inhibition: light and alkylating reagents, Na^+ and Li^+ [118].

Reaction mechanism:
1. Transfer of hydrogen from C-5 to C-6, AdoCbl mediated [120].
2. Transfer of $-NH_2$ from C-6 to C-5 probably pyridoxal phosphate mediated [118, 120].

Assay systems:
1. with acidic ninhydrin reagent [118],
2. measurement of radioactive product separated by TLC [118].

7.2.8 L-β-Lysine mutase

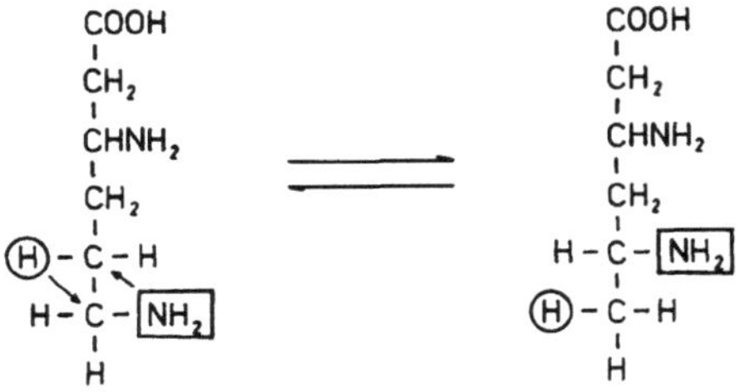

Source: *Clostridium stichlandii* [121, 122, 125].

B_{12}-dependence:
(Ade)AdoCba-native cofactor of E_1 subunit [121],
AdoCbl - active cofactor [121], $K_M = 2 \times 10^{-6}$ M [123],
HO-Cbl - inhibitor [121].

Molecular structure:
Two subunits: E_1 ($\alpha_2\beta_2$) = 170000,
a cobamide protein [121, 123],
α = 32000
β = 52000
E_2 = sulfhydryl protein = 60000 [121, 123].

Catalytic properties:
K_{eq} - reversible [121]
K_M values: β-lysine = 3×10^{-4} M
pyridoxal-5'-P = 2×10^{-7} M [123].

pH optimum: 7.8 with buffer system $N(EtOH)_3$-HCL [123].

Temperature optimum: 37 °C [123].

Cation requirement: monovalent ions: K^+, Li^+, Rb^+, Na^+, NH_4^+, and Cs^+ stimulate the reaction [123]. Divalent ions: Mg^{2+}, Mn^{2+}, Ca^{2+}, Zn^{2+} and Cu^{2+} activate the enzyme [123].

other requirements: enzyme exhibits absolute dependency on pyridoxal phosphate [123] and on mercaptan, pyruvate and ATP [126]. Dithiothreitol is better than BAL or mercaptoethanol, pyruvate can be partially replaced by α-ketobutyrate, but not by α-ketoglutarate. ATP can be replaced by 2'-dATP, but not by GTP, CTP, ITP, UTP, DPN, TPN, inorganic phosphate, pyridoxal phosphate, 3',5'-cAMP, acetyl-CoA, PRPP, formyl phosphate, acetyl phosphate and phosphoenolpyruvate [126].

Reaction mechanism:
1. There is a stereospecific hydrogen migration from C-5 to C-6 of substrate [124],
2. The hydrogen transfer from 5'-H-coenzyme B_{12} to C-5 of substrate and C-6 of product occurs [124].
3. The migrating amino group does not exchange with ammonia nitrogen [122].

Assay systems: The product is estimated:
1. enzymatically with 3,5-diaminohexaconate dehydrogenase [123]
2. Using the Chinard acid ninhydrin [126].

Stability: The purified enzym can be stored at −22 °C for 3 months without loss of activity [126].

7.2.9 Ornithine mutase

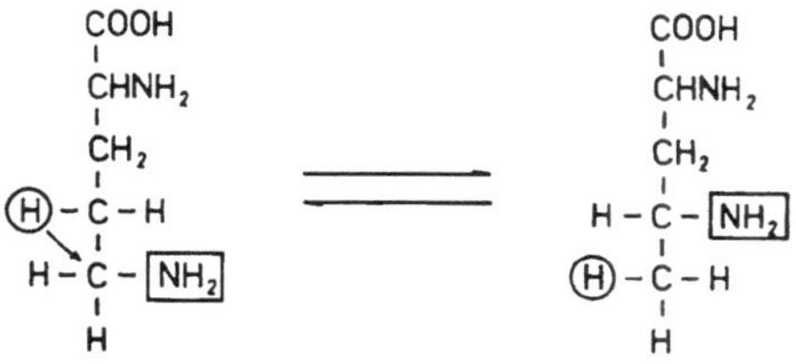

Source: *Clostridium sticklandii* [127, 128]

B_{12}-dependence:
K_M for AdoCbl = 3×10^{-6} M [129]

Molecular structure:
apoenzyme (α_2): 180000 [129].

Catalytic properties:
$K_{eq} = 1$

Substrate: specific for D-α-ornithine

analogues of substrate	% inhibition [129]
L-α-ornithine	57
D,L-α-Lysine	62
β-L-Lysine	56
δ-amino-n-valeric acid	9
cadaverine	0
putrescine	0
N-α-acetyl-L-ornithine	0

K_M for D-α-ornithine $= 0.44\text{-}6.7 \times 10^{-6}$ M [129].

Other cofactor: pyridoxal phosphate [128],
$K_M = 3.6 \times 10^{-7}$ M (coupled assay [129]).
1.8×10^{-6} M (radiochemical assay [129]).

Temperature optimum: 37 °C [129].

pH optimum: 9.0 [129].

Assay systems: The product of the reaction can be estimated: 1. by coupling with 2,4-diaminopentanoic acid dehydrogenase, which is measured spectrophotometrically [128], 2. by ionophoresis using [^{14}C]ornithine [128].

7.2.10 Leucine 2,3-Aminomutase

Sources: *Clostridium sporogenes* [130],
Clostridium leuputrescens [130],
Rat liver [130],
Rat kidney [130],
Sheep liver [130],
Rhesus monkey liver [130],
African green monkey liver [130],
Human leukocytes [130],
Phaseolus vulgaris [131],
Lolium ssp. [131],
Solanum tuberosum [132]

B_{12}-dependence: the enzyme requires AdoCbl [130].

Catalytic properties:

K_{eq} - reversible [130]

Inhibition: by intrinsic factor which is overcome by the addition of AdoCbl [130].

Assay methods:

1. α-leucine is measured in amino acid analyzer or by gas chromatography [130].
2. β-leucine is estimated chromatographically [130].

7.2.11 Ribonucleotide reductase

Sources [133]: *Lactobacillus leichmanii* [134–136, 138, 139]
Lactobacillus acidophillus [145],
Clostridium tetanomorphum [158],
Clostridium sticklandii [158],
Rizobium melioti [147, 147 a–b]
Euglena gracilis [143, 175]
Propionibacterium shermannii [174]

Activity of coenzyme B_{12} analogues

active forms [137, 139, 149]	K_M (M)	Substrate
AdoCbl	8×10^{-7}	GTP
(Bza)AdoCba	8×10^{-7}	GTP
(Ade)AdoCba	2.8×10^{-6}	GTP
2′-dAdoCbl	v. low activity	GTP
AdoCbl	5.5×10^{-6}	CTP
AdoCbl	8.3×10^{-6}	ATP

K_i values [149]: 2′-dAdoCbl $= 2 \times 10^{-5}$ M, 2′,3′-Isopropylidene AdoCbl $=$ 2–3 $\times 10^{-5}$ M.

Inactive forms [134, 137, 139, 149]: CydCbl, UrdCbl, MeCbl, EtCbl, CN-Cbl, HOCbl, CN-Cbl-(ethylamide), CN-Cbl-(anilide), CN-Cbl-(monocarboxylic acid), CN-Cbl-(dicarboxylic acid), (CN,OH)-Cbi.

Molecular structure: Monomeric, single polypeptide chain contains 8 methionine residues [140].

M. W. = 76000 [140],
110000 [142],
145000 [143],

Thioredoxin: 12000 [148].

Thioredoxin reductase: 55000 [148].

Catalytic properties:

K_{eq} - irreversible

Substrates: nucleoside triphosphates [141-145].

substrate	K_M (M)	
	L. leichmanii [149]	*Rizobium meliloti* [147]
GTP	2.4×10^{-4}	7.6×10^{-4}
ATP	8×10^{-4}	2.3×10^{-4}
CTP	3.6×10^{-4}	1.6×10^{-3}

Table 7-4 Activity of analogues of adenosylcobalamin in ribonucleotide reduction (reprinted from Hogenkamp [169])

Cbl[d]	Nucleosidyl ligand[d] (β ligand)	Rate of ATP[a] reduction	Apparent K_I	
			reduction	exchange
		units per mg	$M \times 10^5$	$M \times 10^5$
I	Adenosyl	100	–	0.46 ±0.05
II	3-Isoadenosyl	59	–	0.83 ±0.08
III	Nebularyl	13	–	2.4 ±0.5
IV	Tubercidyl	3	0.88 ±0.15	1.7 ±0.3
V	Diaminopuryl[b]	77	0.58 ±0.05	ND
VI	Inosyl	0.3	1.7 ±0.4	3.5 ±0.6
VII	Toyocamycyl[b]	65	2.2 ±0.44	ND
VIII	Uridyl	0.3	3.3 ±0.7	4.3 ±1.2
IX	Cytidyl	0.3	2.3 ±0.4	4.5 ±1.1
X	Benzimidazole ribosyl	0.3	0.54 ±0.09	2.7 ±0.4
XI	2′-Deoxyadenosyl	5	–	0.54 ±0.5
XII	Ara-adenosyl	0.6	3.4 ±0.4	2.3 ±0.4
XIII	Adenosylethyl	0.3	0.53 ±0.05	0.55 ±0.08
XIV	Isopropylidene adenosyl	1.3	2.8 ±0.7	1.7 ±1.4
XV	L-Adenosyl	1.1	3.6 ±0.5	3.9 ±0.9
XVI	Aristeromycyl	0.3	1.2 ±0.1	2.0 ±0.2
XVII	Thymidyl	0.3	1.7 ±0.4	2.1 ±0.4
XVIII	Adenylethyl[c]	0.3	1.6 ±0.2	1.2 ±0.2
	Adenylpropyl	0.3	7.7 ±1.0	7.9 ±1.3
	Adenylbutyl	0.3	1.6 ±0.3	1.1 ±0.2
	Adenylpentyl	0.3	0.024 ±0.004	0.017 ±0.002
	Adenylhexyl	0.3	0.45 ±0.08	0.63 ±0.3

[a] Analogues were present at a concentration of approximately 50 μM. Initial velocities are expressed relative to that obtained with AdoCbl (100 units per mg). For I-III, 0.025 mg of enzyme and for IV-XVIII 0.20 mg of enzyme was used [167].

[b] Activity in the ribonucleotide reduction system of *Corynebacterium nephridii* (% of activity and K_m)

[c] See reference [168]

[d] See Figs. 7-8 and 7-9 for structure of nucleosides

Table 7-5 Coenzyme properties of analogues of adenosylcobalamin (reprinted from Hogenkamp [169])

Cbl[b]	Nucleosidyl ligand[b] (β ligand)	Rate of ATP reduction	Equilibrium concentration of rapid intermediate[a]	Rate of degradation[a]
		units per mg	Δ A525	cm
I	Adenosyl	100	0.121	9.4
II	3-Isoadenosyl	59	0.104	11.9
III	Nebularyl	13	0.006	8.8
IV	Tubercidyl	3	0	7.3
VI	Inosyl	0.3	0	0
VIII	Uridyl	0.3	0	11.6
IX	Cytidyl	0.3	0	4.5
X	Benzimidazole ribosyl	0.3	0	1.7
XI	2'-Deoxyadenosyl	5	0.004	17.3
XII	Ara-adenosyl	0.6	0	12.6
XIII	Adenosylethyl	0.3	0	0
XIV	Isopropylidene adenosyl	1.3	0	0
XV	L-Adenosyl	1.1	0	0
XVI	Aristeromycyl	0.3	0	12.8
XVII	Thymidyl	0.3	0	2.2

[a] See reference [167]

[b] See Figs. 7-8 and 7-9 for structure of nucleosides

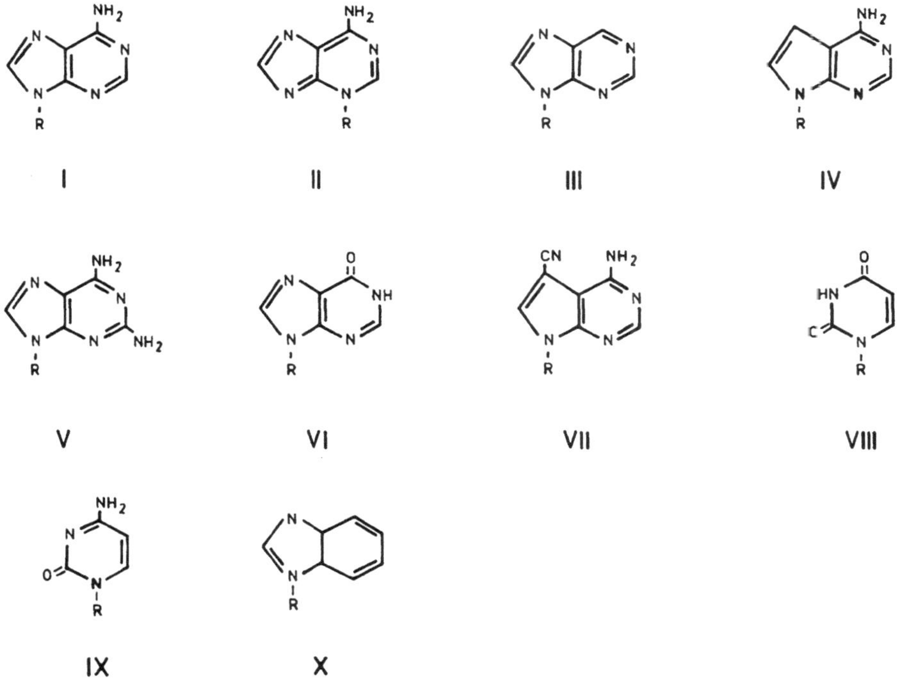

Fig. 7-8 Structures of the nucleoside axial ligands of adenosylcobalamin analogs, modifications in the base moiety (reprinted from Hogenkamp [169], see Tables 7-4 and 7-5).

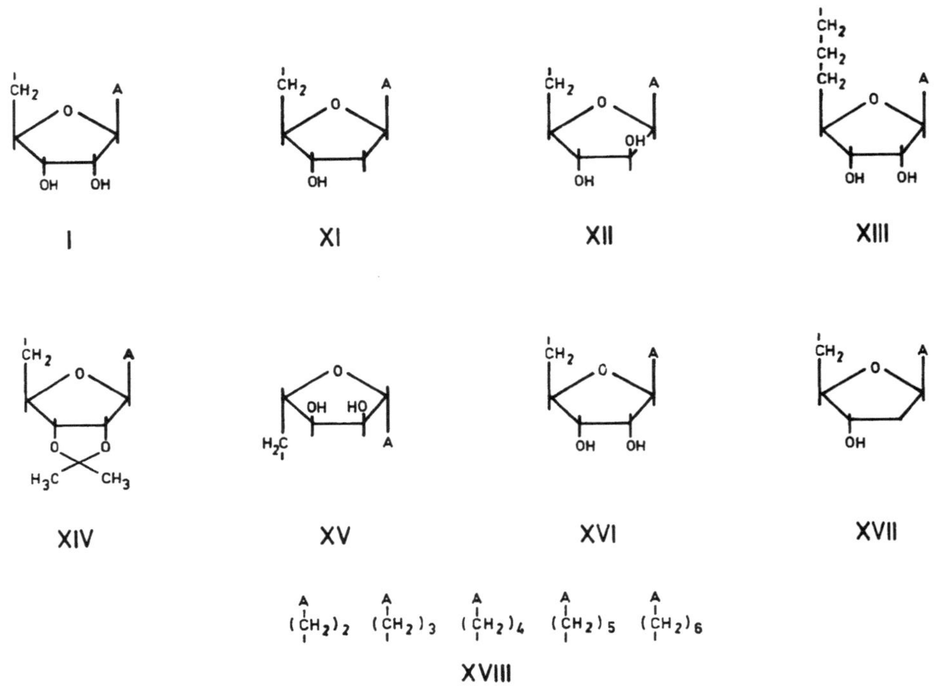

Fig. 7-9 Structures of the nucleoside axial ligands of adenosylcobalamin analogs, modifications in the sugar moiety (reprinted from Hogenkamp [169], see Tables 7-4 and 7-5).

Table 7-6 Comparison of the properties of ribonucleotide reductase from the different organisms (reprinted from Sato et al. [173])

	E. coli.	*L. leichmannii*	*R. melilati*
Substrate	CDP	CTP	CDP
Hydrogen transport system	NADPH →$(SH)_2$	NADPH →$(SH)_2$ →AdoCbl(H)	NADPH →$(SH)_2$ →AdoCbl(H)
Metal requirement	Mg^{2+}	(Mg^{2+})*	(Mg^{2+})*
Activator	ATP	dATP	dATP
Inhibitor	ATP	–	–
Subunits	$B_1 + B_2$	Single polypeptide	$3S_1 + S_2$
Molecular weight	B_1 160000 B_2 78000 ca. 240000	76000	S_1 130000 S_2 110000 ca. 500000 (pH 9.0) ca. 1000000 (pH 7.5 aggregated)
Optimal pH	~7.5	7.5~8.0	9.0~10.0

* Conditional requirement

Allosteric activators [142, 144, 147 a, 149]: dGTP for ATP reduction, dATP and ATP for CTP reduction - K_M change for CTP from 9.5 to 1.7×10^{-3} M [142].

Competitive inhibitors:
for AdoCbl [156]:
5′deoxyadenosine: $K_i = 1.4 \pm 0.22 \times 10^{-5}$ M.
cob(II)alamin: $K_i = 3.7 \pm 0.2 \times 10^{-5}$ M.
3′-deoxyadenosine, 2′-deoxyadenosine, and 4′,5′-didehydro-5′-deoxyadenosine have no significant inhibitory effect.

Hydrogen donors:
Monothiols are inactive [146].
Dithiols are active [139, 146].

Dithiols	% activity
dihydrolipoic acid	100
dihydrolipoyl-glycine	45
dihydrolipoyl-glycyl-glycine	55
dihydrolipoyl-L-lysine	12
dithioerythrol	55
dithiothreitol	54
1,3-dithiopropane	16
1,3-dithiopropan-2-ol	55
2,3-dithiopropanol	–

The natural reducing agent is thioredoxin [146, 148].
K_M for thioredoxin $= 3 \times 10^{-6}$ M [148].

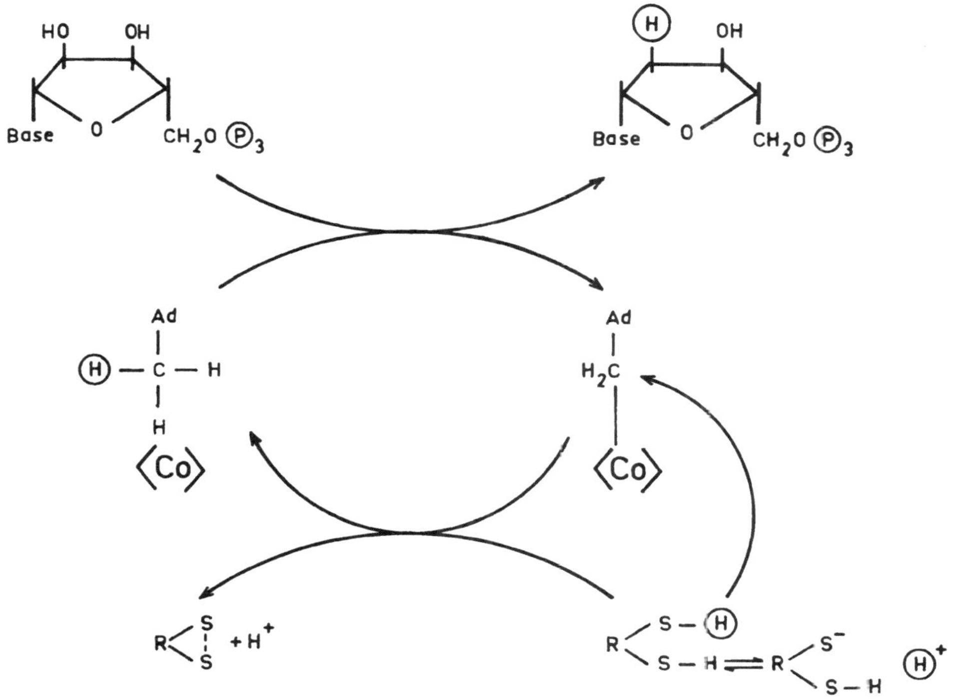

pH optimum: 7.3-7.5 [137, 142].

Inactivation: In the presence of dihydrolipoate and dGTP, AdoCbl is degraded to cob(II)alamin but the absence of the reducing substrate and exposure to air produces aqCbl and 5′-deoxyadenosine bound to the active center [156]. A slow rearrangement of the adenosyl radical-thiol system can be responsible for the irreversible degradation of AdoCbl [165].

Reaction mechanism:

1. The specific exchange of hydrogen between (5′5′-H_2) AdoCbl and water via a thiol group and the replacement of the hydroxyl at C-2′ of substrate by hydrogen from C-5 of AdoCbl with retention of the configuration [150-155, 158-161].
2. Interactions AdoCbl/enzyme/dithiols and allosteric site/active center cause homolytic fission of the cobalt-carbon bond and formation of cob(II)alamin and a deoxyadenosyl radical [157, 162, 163, 165, 166].
3. Dithiols compounds participate in the cleavage of Co-C bond [4]:
   - reduced thioredoxin reduces a disulfide group on the enzyme,
   - homolytic fission of Co-C,
   - the transfer of hydrogen atom from one of the sulfhydryl groups to deoxyadenosyl radical produces 5′-deoxyadenosine.

Assay:

1. radioactive methods
2. colorimetric assay

Storage conditions: full activity after 1 year at −20 °C in 50% glycerol.

References

[1] Barker, H. A., Biochemical functions of corrinoid compounds, Biochem. J. **105**, 1-14, 1967.

[2] Hogenkamp, H. P. C., Enzymatic reactions involving corrinoids, Ann. Rev. Biochem. **37**, 225-245, 1968.

[3] Stadtman, T. C., Vitamin B_{12}, Science **171**, 859-867, 1971.

[4] Babior, B. M., Cobamides as cofactors. Adenosylcobamide-dependent reactions, in Cobalamin. Biochemistry and Pathophysiology, A Wiley - Interscience Publication, New York 1975, pp. 141-212.

[5] Poston, J. M., and Stadtman, T. C., Cobamides as cofactors. Methylcobamides and the synthesis of methionine, methane and acetate, in Cobalamin. Biochemistry and Pathophysiology. Wiley - Interscience Publication, New York, 1975, pp. 111-140.

[6] Friedrich, W., Vitamin B_{12} und verwandte Corrinoide, in Fermente, Hormone, Vitamine, Band III/2, Georg Thieme Verlag, Stuttgart 1975, pp. 194-243.

[6a] Zagalak, B., Vitamin B_{12} als biologisch aktive Modellsubstanz, Naturwissenschaften **69**, 63-74, 1982.

[6b] Lexa, D., Saveant, J. M., The Electrochemistry of Vitamin B_{12}, Acc. Chem. Res. **16**, 235-243, 1983.

[6c] Brown, K., Vitamin B_{12} in The Chemistry of Enzyme Action (Page, M. I. Ed.) Elsevier Science Publishers B. V. 1984, pp. 433-459.

[6d] Down, P., Trivedi, B. K., On the mechanism of action of vitamin B_{12}. Model studies directed toward the hydrogen abstraction reaction, J. Org. Chem. **50**, 206-217, 1985.

[6e] Halpern, J., Mechanisms of coenzyme B_{12}-dependent rearrangements, Science **227**, 869-875, 1985.

[7] Barker, H. A., Weisbach, H., and Smyth, R. D., A Coenzyme containing pseudovitamin B_{12}, Proc. Nat. Acad. Sci. **44**, 1039-1097, 1958.

[8] Barker, H. A., Fermentation of nitrogenous compounds, in The Bacteria: Vol. II, Chapter 3, Eds. Gunsalus, I. C., and Stainier, R. S., Academic Press, New York 1961, 161-168.

[9] Tochey, J. I., Perlman, D., Barker, H. A., Purification and properties of cobamide coenzymes obtained from *Propionibacterium arabinosum*, J. Biol. Chem. **236**, 2119-2127, 1961.

[10] Barker, M. A., Suzuki, F., Iodise, A., and Rooze, V., Glutamate mutase reaction, Ann. N. Y. Acad. Sci. **112**, 644-654, 1964.

[11] Barker, H. A., Chemistry and Biology of vitamin B_{12} coenzymes in „Vitamin B_{12} and Intrinsic Factor", Ed., Heinrich, H. C., F. Enke, Stuttgart, 1962.

[12] Barker, H. A., Rooze, V., Suzuki, F., and Iodice, A., The glutamate mutase system (Assays and Properties), J. Biol. Chem. **239**, 3260-3266, 1964.

[13] Suzuki, F., and Barker, H. A., Purification and properties of component E of glutamate mutase, J. Biol. Chem. **241**, 878-888, 1966.

[14] Switzer, R. L., and Barker, H. A., Purification and characterization of component S of glutamate mutase, J. Biol. Chem. **242**, 2658-2674, 1967.

[15] Barker, H. A., Glutamate mutase, Meth. Enzymol. **13**, 319-330, 1969.

[16] Kellermeyer, R. W., Allen, S. H. C., Stjernholm, R., and Wood, H. G., Methylmalonyl isomerase. IV. Purification and properties of the enzyme from propionibacteria. J. Biol. Chem. **239**, 2562-2569, 1964.

[17] Wood, G. H., Kellermeyer, H. W., Stjernholm, R., and Allen, S. H. C., Metabolism of Methylmalonyl-CoA and the role of biotin and B_{12}-Coenzymes, Ann. N. Y. Acad. Sci. **122**, 661-679, 1964.

[18] Erfle, J. D., Clarc, J. M., Jr., Nystrom, R. F., and Johnson, B. C., Direct hydrogen transfer by methylmalonyl coenzyme A mutase, J. Biol. Chem. **239**, 1920-1921, 1964.

[19] Zagalak, B., Rétey, J., and Sund, H., Studies on methylmalonyl-CoA mutase from *Propionbacterium shermanii*, Eur. J. Biochem. **44**, 529-535, 1974.

[20] Sato, K., Ueda, S., and Shimizu, S., Methylmalonyl-CoA mutase in methanol utilizing bacterium, *Protaminbacter ruber*, FEBS Letters **71**, 248-251, 1976.

[21] Kellermeyer, R. W., and Wood, H. G., 2-Methylmalonyl-CoA mutase from *Propionibacterium shermanii* (Methylmalonyl-CoA isomerase), Methods Enzymol. **13**, 207-215, 1969.

[22] Cannata, J. J. B., Focesi, A., Mazumder, R., Warner, R. C., and Ochoa, S., Metabolism of propionic acid in animal tissues. XII. Properties of mammalian methylmalonyl coenzyme. A mutase, J. Biol. Chem. **240**, 3249-3257, 1965.

[23] Mazumder, R., and Ochoa, S., Methylmalonyl-CoA mutase from sheep liver, Methods Enzymol. **13**, 198-207, 1967.

[24] Mazumder, R., Sasakawa, T., and Ochoa, S., Metabolism of propionic acid in animal tissues. Methylmalonyl coenzyme A mutase holoenzyme, J. Biol. Chem. **238**, 50-53, 1963.

[25] Lengyel, P., Mazumder, R., and Ochoa, S., Mammalian methylmalonyl isomerase and vitamin B_{12} coenzymes. Proc. Nat Acad. Sci. **16**, 1313-1018, 1980.

[26] Eggerer, H., Overach, P., Lynen, F., and Stadtman, E. R., On the mechanism of the cobamide coenzyme dependent isomerization of methylmalonyl-CoA to succinyl acid, J. Ann. Chem. Soc. **82**, 2643-2644, 1960.

[27] Kellermeyer, R. W., and Wood, H. G., Methylmalonyl isomerase: A study of the mechanism of isomerization, Biochemistry **1**, 1124-1131, 1962.

[28] Sprecher, M., Clarc, M. J., and Sprinson, D. B., The absolute configuration of methylmalonyl coenzyme A and stereochemistry of the methylmalonyl coenzyme A mutase reaction, J. Biol. Chem. **241**, 872-877, 1966.

[29] Rétey, J., Zagalak, B., Stereochemie der coenzym B_{12}-abhängigen Methylmalonyl-CoA-Mutase-Reaktion. Untersuchung mit Äthylmalonyl-CoA, Angew. Chem. **85**, 721-722, 1973.

[30] Rétey, J., Arigoni, D., Coenzyme B_{12} als gemeinsamer Wasserstoffüberträger der Dioldehydrase- und der Methylmalonyl-CoA-Mutase-Reaktion, Experienta (Basel) **22**, 783-784, 1966.

[31] Cardinale, G. J., Abeles, R. H., Mechanistic similarities in the reactions catalyzed by dioldehydrase and methylmalonyl-CoA mutase, Biochim. Biophys. Acta **132**, 517-518, 1967.

[32] Babior, B. M., Woodams, A. D., and Brodie, J. D., Cleavage of coenzyme B_{12} by methylmalonyl coenzyme A mutase, J. Biol. Chem. **248**, 1445-1450, 1973.

[32a] Retey, J., The mechanism of action of methylmalonyl-CoA mutase as studied with isotope labelling and synthetic models. In Vitamin B_{12}, Eds. Zagalak, B., Friedrich, W., Walter de Gruyter, Berlin 1979, pp. 439-460.

[32b] Rask, H., Schou-Olesen, A., Mortensen, J. Z., Freund, L. G., Nitrons oxide and urine methylmalonic acid in man, Scand. J. Haematol **31**, 45-48, 1983.

[33] Kung, H.F., Cederbaum, S., Tsai, L., and Stadtman, T.C., Nicotinic acid metabolism, V. A cobamide coenzyme-dependent conversion of *β*-methyleneglutaric acid to dimethylmaleic acid, Proc. Nat. Sci. **65**, 978-984, 1970.

[34] Kung, H.F., and Stadtman, T.C., Nicotinic acid metabolism. VI. Purification and properties of *β*-methyleneglutarate mutase (B_{12}-dependent) and methylitaconate isomerase, J. Biol. Chem. **246**, 3378-3388, 1971.

[35] Kung, H.F., and Tsai, L., Nicotinic acid metabolism. VII. Mechanisms of action of clostridial *β*-methyleneglutarate mutase (B_{12}-dependent) and methylitaconate isomerase, J. Biol. Chem. **246**, 6436-6443, 1971.

[36] Abeles, R.H., and Lee, H.A., An intramolecular oxidation-reduction requiring a cobamide coenzyme, J. Biol. Chem. **236**, 2347-2350, 1961.

[36a] Abeles, R.H., Brownstein, A.M., Randles, C.H., *β*-Hydroxypropionaldehyde, an intermediate in the formation of 1,3-propanediol by *Aerobacter aerogenes*. Biochim. Biophys. Acta **41**, 530-531; 1960.

[37] Lee, H.A. Jr., and Abeles, R.H., Purification and properties of dioldehydrase, an enzyme requiring a cobamide coenzyme, J. Biol. Chem. **238**, 2367-2373, 1963.

[38] Abeles, R.H., Lee, H.A., Studies of the mechanism of action of cobamide coenzymes, Ann. N.Y., Acad. Sci. **112**, 695-702, 1964.

[39] Yamane, T., Shimizu, S., Fukui, S., Effect of light-irradiation and adenosine on the activity of a dioldehydratase requiring cobamide coenzyme, Biochim. Biophys. Acta, **110**, 616-618, 1965.

[40] Brownstein, A.M., Abeles, R.H., The mechanism of conversion of 1,2-propanediol to propionaldehyde = a reaction requiring a vitamin B_{12} coenzyme, J. Biol. Chem. **236**, 1199-1200, 1961.

[41] Zagalak, B., Frey, G.L., Karabatsos, G.L., and Abeles, R.H., The stereochemistry of the conversion of D- and L-1,2-propanediols to propionaldehyde, J. Biol. Chem. **241**, 3028-3035, 1966.

[42] Abeles, R.H., Zagalak, B.J., The Nature of the hydrogen transfer in the dimethylbenzimidazolylcobamide coenzyme catalyzed conversion of 1,2-propanediol to propionaldehyde, J. Biol. Chem. **241**, 1245-1246, 1966.

[43] Frey, P.A., Essenberg, M.R., and Abeles, R.H., Studies on the mechanism of hydrogen transfer in the cobamide coenzyme-dependent dioldehydratase reaction, J. Biol. Chem. **242**, 5369-5377, 1967.

[44] Frey, P.A., Abeles, R.H., The role of the B_{12}-coenzyme in the conversion of 1,2-propanediol to propionaldehyde, J. Biol. Chem. **241**, 2732-2733, 1966.

[45] Abeles, R.H., Dioldehydrase, in Methods in Enzymology **9**, 686-689, 1966.

[46] Toraya, T., and Fukui, S., Immunochemical evidence for the difference between coenzyme-B_{12}-dependent diol dehydratase and glycerol dehydratase, Eur. J. Biochem. **76**, 285-289, 1977.

[46a] Forage, R.G., Foster, M.A., Glycerol fermentation in *Klebsiella pneumoniae:* Functions of the coenzyme B_{12}-dependent glycerol and diol dehydratases. J. Bacteriol. **149**, 413-419, 1982.

[46b] Nishio, N., Kawagishi, T., Matsuno, R., Kamikubo, T., Metabolism of 1,2-propanediol by methanol utilizing bacteria and some properties of 1,2-propendiol dehydrogenating enzyme. Agric. Biol. Chem. **42**, 1095-1100, 1978.

[47] Essenberg, M.K., Frey, P.A., and Abeles, R.H., Studies on the mechanism of hydrogen transfer in the coenzyme B_{12}-dependent dioldehydrase reaction, J. Ann. Chem. Soc. **93**, 1242-1251, 1971.

[48] Poznanskaya, A.A., Tanizawa, K., S.K. Toraya, T., and Fukui, S. Propanediol dehydratase from *Aerobacter aerogenes:* Purification enzymatic properties and molecular structure. Bioorg. Chim. **3**, 111-119, 1977.

[49] Toraya, T., Vesaba, M., Kondo, M., and Fukui, S., Dissociation of diol dehydrase into two different protein components. Biochem. Biophys. Res. Commun. **52**, 350-355, 1973.

[49a] Fukui, S., Toaraya, T., Coenzyme B_{12}-dependent diol dehydratase distribution and metabolic role in *Enterobacteriaceae,* enzymological properties and interaction with coenzyme B_{12}. in Vitamin B_{12}, Eds. Zagalak, B., Friedrich, W., Walter de Gruyter, Berlin 1979, pp. 413-430.

[50] Toraya, T., Uesaka, M., and Fukui, S., Coenzyme B_{12} dependent diol dehydrase system. Dissociation of the enzyme into two different protein components and some properties of the components, Biochemistry **13**, 3895-3899, 1974.

[51] Toraya, T., Ushio, K., Fukui, S., and Hogenkamp, H.P.C., Studies on the mechanism of

the adenosylcobalamin-dependent dehydratase reaction by the use of analogs of the coenzyme, J. Biol. Chem. **252,** 963-970, 1977.

[52] Toraya, T., Sugimoto, Y., Tamao, Y., Shimizu, S., and Fukui, S., Propanediol dehydratase system. Role of monovalent cations in binding of vitamin B_{12} coenzyme or its analogs to apoenzyme, Biochemistry **10,** 3475-3484, 1971.

[53] Toraya, T., Sugimoto, Y., Tamao, Y., Shimizu, S., and Fukui, S., Potassium ion as an essential factor for binding of vitamin B_{12} coenzyme to apopropanediol dehydratase, Biochem. Biophys. Res. Commun. **41,** 1314-1320, 1970.

[54] Toraya, T., and Fukui, Y., Coenzyme B_{12}-dependent propandiol dehydratase system. Ternary Complex between apoenzyme, coenzyme and substrate analog, Biochim. Biophys. Acta **284,** 536-548, 1972.

[55] Toraya, T., Kondo, M., Isemura, Y., and Fukui, S., Coenzyme B_{12}-dependent propanediol dehydratase system. Nature of cobalamin binding and some properties of apoenzyme-coenzyme B_{12} analog complexes, Biochemistry **11,** 2599-2606, 1972.

[56] Eagar, R.G. Jr., Bacovchin, W.W., and Richards, J.H., Mechanism of action of adenosylcobalamin: 3-Fluoro-1,2-propanediol as substrate for propanediol dchydrase-mechanistic implications, Biochemistry **14,** 5523-5528, 1975.

[57] Bachovchin, W.W., Eagar, R.G., Jr., Moor, K.W., and Richards, J.H., Mechanism of action of adenosylcobalamin: glycerol and other substrate analogues as substrates and inactivators for propanediol dehydratase - kinetics, stereospecificity and mechanism, Biochemistry **16,** 1082-1092, 1977.

[58] Wagner, O.W., Lee, H.A., Jr., Frey, P.A., and Abeles, R.H., Studies on the mechanism of action of cobamide coenyzmes. Chemical properties of the enzyme-coenzyme complex, J. Biol. Chem. **241,** 1751-1762, 1966.

[59] Toraya, T., Shirakashi, T., Kosuga, T., and Fukui, S., Substrate specificity of coenzyme B_{12}-dependent diol dehydratase: glycerol as both a good substrate and a potent inactivator, Biochem. Biophys. Res. Commun. **69,** 475-480, 1976.

[59a] Forage, R.G., Foster, M.A., Resolution of the coenzyme B_{12}-dependent dehydratases of *Klebsiella sp.* and *Citrobacter freundii.* Biochim. Biophys. Acta **569,** 249-258, 1979.

[60] Tamao, Y., Morikawa, Y., Shimizu, S., and Fukui, S., Coenzyme activity of 5'-deoxyadenosyl-10-chlorobalamin (10-CL-DBCC) in propanediol dehydratase system, B.B.R.C., **28,** Biochem. Biophys. Res. Commun. **28,** 692-698, 1967.

[61] Retey, J., Uniani-Ronchi, A., Arigoni, D., Zur Stereochemie der Propandioldehydrase-Reaktion, Experienta **22,** 72-73, 1966.

[62] Retey, J., Uniani-Ronchi, A., Seibl, J., Arigoni, D., Zum Mechanismus der Propandioldehydrase-Reaktion, Experienta **22,** 502-503, 1966.

[63] Finlay, T.H., Valinsky, J., Sato, K., and Abeles, R.H., Studies on the mechanism of action of coenzyme B_{12}. The formation of 5'-deoxyadenosine and B_{12} in the reaction of dioldehydrase with chloroacetaldehyde, J. Biol. Chem. **247,** 4197-4207, 1972.

[63a] Arigoni, D., A stereochemical approach to the diol dehydratase reaction. In Vitamin B_{12} Eds. Zagalak, B., Friedrich, W., Walter de Gruyter, Berlin 1979, pp. 389-411.

[63b] Abeles, R.H., Current status of the mechanism of action of B_{12}-coenzyme. In Vitamin B_{12} Eds. Zagalak, B., Friedrich, W., Walter de Gruyter, Berlin 1979, pp. 372-388.

[64] Valinsky, J.E., Abeles, R.H., and Mildvan, A.S., Electron spin resonance studies with dioldehydrase. II. The hyperfine structure of proposed radical intermediates derived from isotopic substitution in 2-chloroacetaldehyde, J. Biol. Chem. **249,** 2751-2755, 1974.

[65] Smiley, K.L., and Sobolov, M., Some properties of the glycerol dehydratase system from a species of *Lactobacillus,* Ann. N.Y. Acad. Sci. **112,** 706-712, 1964.

[65a] Sobolov, M., Smiley, K.L., Metabolism of glycerol by an acrolein-forming *Lactobacillus.* J. Bacteriol. **79,** 261-266, 1960.

[66] Smiley, K.L., and Sobolov, M., A cobamide-requiring glycerol dehydratase from an acrolein-forming *Lactobacillus,* Arch. Biochem. Biophys. **97,** 538-543, 1962.

[67] Zagalak, B., and Pawełkiewicz, J., An enzymic conversion of ethylene glycol into acetaldol requiring corrinoid coenzymes, Life Sci. 395-399, 1962.

[68] Pawełkiewicz, J., and Zagalak, B., A conversion of 1,2-diols into corresponding deoxyaldehydes by an enzymatic system from *Aerobacter aerogenes* (PZH.572), Ann. N.Y. Acad. Sci. **112,** 703-705, 1964.

[69] Toraya, T., and Fukui, S., Immunochemical evidence for the difference between coenzyme-B_{12}-dependent diol dehydratase and

glycerol dehydratase, Eur. J. Biochem. **76,** 285-289, 1977.

[70] Zagalak, B., and Pawełkiewicz, J., Synthesis and properties of some analogues of the corrin coenzymes, Acta Biochim. Polon. **11,** 49-59, 1964.

[71] Zagalak, B., and Pawełkiewicz, J., Synthesis and properties of analogies of coenzyme B_{12} methylated in the adenosyl group, Acta Biochim. Polon. **12,** 103-114, 1965.

[72] Zagalak, B., and Pawełkiewicz, J., Synthesis and properties of Co-adenine nucleoside analogues of coenzyme B_{12}, Acta Biochim. Polon. **12,** 219-228, 1965.

[73] Pawełkiewicz, J., and Zagalak, B., Enzymic conversion of glycerol into β-hydroxypropionaldehyde in a cell-free extract from *Aerobacter aerogenes,* Acta Biochim. Polon. **12,** 207-218, 1965.

[74] Yakusheva, M.I., Poznanskaya, A.A., Pospelova, T.A., Rudakova, I.P., Yurkevich, A.M., and Yakovlev, V.A., Study on the mechanism of action of adenosylcobalamin-dependent glycerol dehydratase from *Aerobacter Aerogenes.* I. Role of Structural components of adenosylcobalamin in the formation of the active site of glycerol dehydratase, Biochim. Biophys. Acta **484,** 216-235, 1977.

[75] Schneider, Z., Larsen, E.G., Jacobson, G., Johnson, B.C., and Pawełkiewicz, J., Purification and properties of glycerol dehydratase, J. Biol. Chem. **245,** 3388-3396, 1970.

[76] Stroiński, A., Properties of glycerol dehydratase from *Aerobacter aerogenes,* Ph.D. Thesis, Academy of Agricultural, Poznań, 1970.

[77] Stroiński, A., and Pawełkiewicz, J., Transition of homotropic effect and of stability of vitamin B_{12}-dependent glycerol dehydrase, Fed. Proc. **31,** Meeting Abstr. **911,** 1972.

[78] Schneider, Z., and Pawełkiewicz, J., The Properties of glycerol dehydratase isolated from *Aerobacter aerogenes* and the properties of the apoenzyme subunits, Acta Biochim. Polon. **13,** 311-328, 1966.

[79] Schneider, Z., Pech, K., and Pawełkiewicz, J., Enzymic transformation of glycerol to β-hydroxypropionic aldehyde. II. Dissociation of the enzyme into two protein fragments. Bull. Acad. Polon. Ser. Sci. Biol. **14,** 7-12, 1966.

[80] Stroiński, A., Pawełkiewicz, J., and Johnson, B.C., Allosteric interactions in glycerol dehydratase. Purification of enzyme and effects of positive and negative cooperativity for glycerol. Arch. Biochem. Bioph. **162,** 321-330, 1974.

[81] Johnson, B.C., Stroiński, A., and Schneider, Z., Glycerol dehydratase from *Aerobacter aerogenes,* Methods in Enzymology **42,** Part C, 315-323, 1975.

[82] Yakusheva, M.I., Malahov, A.A., Poznanskaya, A.A., and Yakovlev, V.A., Determination of glycerol dehydratase activity by the coupled enzymic method. Anal. Biochem. **60,** 293-301, 1974.

[83] Korsova, T.L., Levina, T.S., Poznanskaya, A.A., and Yakovlev, V.A., Role of monovalent cations in reactions catalyzed by glycerol dehydratase from *Aerobacter aerogenes,* Biochimija **40,** 559-565, 1975.

[84] Stroiński, A., (Unpublished experiments).

[85] Stroiński, A., Regulatory properties of vitamin B_{12}-dependent glycerol dehydratase, Abstracts of X. Intern. Con. Biochem. Hamburg, Abs. 07-5-158, 403, 1976.

[86] Stroiński, A., Regulatory role of cAMP in glycerol dehydratase synthesis in *Aerobacter aerogenes,* Abstracts of XV. Meeting of PTBioch., Ps - 27, 264, 1977.

[86 a] Stroiński, A., Participation of cAMP in regulation of coenzyme B_{12}-dependent glycerol dehydratase synthesis in *Klebsiella pneumoniae* ATCC 25955, in Vitamin B_{12} (Eds. Zagalak, B., Friedrich, W.), Walter de Gruyter, Berlin 1979, pp. 1029-1032.

[86 b] Forage, R.G., Foster, M.A., Resolution of the coenzyme B-12-dependent dehydratases of *Klebsiella sp.* and *Citrobacter freundii.* Biochim. Bioph. Acta **569** 249-258, 1979.

[86 c] Poznanskaya, A.A., Korsova, T.L., Adenosylcobalamin-dependent glycerol dehydratase interaction with substrates and their analogs, in Vitamin B_{12}, Eds. Zagalak, B., Friedrich, W., Walter de Gruyter, Berlin 1979, pp. 431-436.

[86 d] Stroiński, A., Wieczorek-Floryszak, J., In preparation.

[87] Stroiński, A., Schneider, Z., and Pawełkiewicz, J., Coenzyme B_{12}-dependent glycerol dehydratase from *Aerobacter aerogenes,* Properties of self-inactivated holoenzyme, Bull. Acad. Polon. Sci. Ser. Sci. Biol. **15,** 727-731, 1967.

[88] Schneider, Z., and Johnson, B.C., Studies on the thermal dissociation of glycerol dehydratase, Arch. Biochem. Biophys. **144,** 491-495, 1971.

[89] Zagalak, B., Mechanism of action of coenzyme B_{12}-dependent glycerol dehydratase, Bull. Acad. Sci. Ser. Sci. Biol. **16,** 67-71, 1967.

[90] Cockle, S.A., Hill, H.A.O., Williams, R.J.P., Davies, S.P., and Foster, M.A., The detection of intermediates during the conversion of propane-1,2-diol to propionaldehyde by dependent reaction, J. Amer. Chem. Soc. **94,** 275-277, 1972.

[91 a] Schneider, Z., Stroiński, A., and Pawełkiewicz, J., Thermostability of enzyme in the threedimensional network of polysaccharide chains, Bull. Acad. Polon. Sci. Ser. Sci. Biol. **16,** 203-204, 1967.

[91 b] Schneider, Z., and Friedmann, H.C., Stabilization of enzymes in Sephodex - a long-term experiment, J. Appl. Biochem. **3,** 97, 1981.

[92] Schneider, Z., and Johnson, B.C., unpublished data.

[93] Schneider, Z., and Friedmann, H.C., Studies on the enzymatic dephosphorylation of vitamin B_{12} 5'-phosphate, Arch. Biochem. Bioph. **152,** 488-495, 1972.

[94] Bradbeer, C., The clostridial fermentations of choline and ethanolamine. II. Requirement for a cobamide coenzyme by ethanolamine deaminase. J. Biol. Chem. **240,** 4675-4681, 1965.

[95] Kaplan, B.H., and Stadtman, E.R., Ethanolamine deaminase, a cobamide coenzyme-dependent enzyme. Purification, assay, and properties of the enzyme. J. Biol. Chem. **243,** 1787-1793, 1968.

[96] Kaplan, B.H., and Stadtman, E.R., Ethanolamine deaminase, a cobamide-coenzyme-dependent enzyme. II. Physical and chemical properties and interaction with cobamides and ethanolamine. J. Biol. Chem. **243,** 1794-1803, 1968.

[97] Carty, T.J., Babior, B.M., and Abeles, R.H., The Mechanism of action of ethanolamine ammonia-lyase, a B_{12}-dependent enzyme. Evidence for two intermedias in the catalytic process. J. Biol. Chem. **249,** 1683-1688, 1974.

[98] Babior, B.M., Carty, T.J., and Abeles, R.H., The Mechanism of action of ethanolamine ammonia-lyase, a B_{12}-dependent enzyme. The reversible formation of 5'-deoxyadenosine from adenosylcobalamin during the catalytic process. J. Biol. Chem. **249,** 1689-1695, 1974.

[99] Babior, B.M., The mechanism of action of ethanolamine deaminase. Studies with isotopic hydrogen and oxygen. J. Biol. Chem. **244,** 449-456, 1969.

[100] Retey, J., Suchling, C.J., Arigoni, W., and Babior, B.M., The stereochemistry of the reaction catalyzed by ethanolamine ammonia-lyase, an adenosylcobalamin-dependent enzyme. An example of recemisation accompanying substitution. J. Biol. Chem. **249,** 6359-6360, 1974.

[100 a] Babior, B.M., Recent studies on the mechanism of action of ethanolamine ammonia-lyase, in Vitamin B_{12} Eds. Zagalak, B., Friedrich, W., Walter de Gruyter, Berlin 1979, pp. 461-470.

[100 b] Hollaway, M.R., White, H.A., Joblin, K.N., Johnson, A.W., Lappert, M.F., Wallis, O.C., Studies on the mechanism of reactions catalyzed by ethanolamine amonialyase, in Vitamin B_{12}, Eds. Zagalak, B., Friedrich, W., Walter de Gruyter, Berlin 1979, pp. 471-484.

[101] Carty, T.J., Babior, B.M., and Abeles, R.H., The mechanism of action of ethanolamine ammonia-lyase, a B_{12}-dependent enzyme. IX. Interaction of the enzyme-coenzyme complex with reaction products. J. Biol. Chem. **246,** 6313-6317, 1971.

[102] Babior, B.M., and Li, T.K., The binding of cobamides to ethanolamine deaminase. Biochemistry **8,** 154-160, 1969.

[103] Babior, B.M., The mechanism of action of ethanolamine deaminase. III. Inhibition by coenzyme B_{12} analogues. J. Biol. Chem. **244,** 2917-2926, 1969.

[104] Babior, B.M., The mechanism of action of ethanolamine deaminase. IV. Cobamide-dependent binding of substrate to ethanolamine deaminase. J. Biol. Chem. **244,** 2927-2934, 1969.

[105] Law, P.Y., and Wood, J.M., Fed. Proc. **30,** 1088, 1971.

[106] Law, P.Y., Brown, D.G., Lien, E.L., Babior, B.M., Wood, J.M., Synthesis and catalytic activity of spin-labelled cobinamide coenzymes. Biochemistry **10,** 3428-3435, 1971.

[107] Babior, B.M., The mechanism of action of ethanolamine deaminase. VI. Ethylene glycol, a quasi-substrate for ethanolamine deaminase. J. Biol. Chem. **245,** 1755-1766, 1970.

[108] Babior, B.M., The mechanism of action of ethanolamine ammonia-lyase, a B_{12}-dependent enzyme. VII. The mechanism of hydrogen transfer. J. Biol. Chem. **245,** 6125-6133, 1970.

[108 a] Krouwer, J.S., Babior, B.M., Mechanism of action of ethanolamine amonia-lyase, an adenosylcobalamin-dependent enzyme. Interaction between the enzyme and a postulated organocobalamin intermediate. J. Biol. Chem. **252,** 5004-5009, 1977.

[108 b] Krouwer, J.S., Schultz, R.M., Babior, B.M., The mechanism of action of ethanolamine ammonia-lyase, an adenosylcobalamin-dependent enzyme. Reaction of the enzyme cofactor complex with 2-aminoacetaldehyde. J. Biol. Chem. **253,** 1041-1047, 1978.

[108 c] Wallis, O.C., Johnson, A.W., Lappert, M.F., Studies of the structure of the adenosylcobalamin-dependent enzyme ethanolamine ammonia-lyase. FEBS Letters [97], 196-199, 1979.

[108 d] Mauck, L., Babior, B.M., Inactivation of ethanolamine ammonia-lyase by 5,5'-dithiobis (2-nitrobenzoic acid). Further evidence for the involvement of sulfhydryl group in adenosylcobalamin-dependent rearrangements. Biochim. Bioph. Acta **480,** 489-794, 1977.

[108 e] Hollaway, M.R., White, H.A., Joblin, K.N., Johnson, A.W., Lappert, M.F., Wallis, O.C., A spectrometric rapid kinetic study of reaction catalysed by coenzyme-B_{12}-dependent ethanolamine ammonia-lyase. Eur. J. Biochem. **82,** 143-154, 1978.

[109] Mang, G.W., Chang, J.T., Evidence for the B_{12}-dependent enzyme ethanolamine deaminase in *Salmonella*. Nature **254,** 150-151.

[110] Scarlett, F.A., and Turner, J.M., Microbial metabolism of amino alcohols. Ethanolamine catabolism mediated by coenzyme B_{12}-dependent ethanolamine ammonia-lyase in *Escherichia coli* and *Klebsiella aerogenes*. J. Gen. Microbiol. **95,** 173-176, 1976.

[111] Weisblat, D.A., Babior, B.M., The mechanism of action of ethanolamine ammonia-lyase, a B_{12}-dependent enzyme. VIII. Further studies with compounds labelled with isotopes of hydrogen: identification and some properties of the rat-limiting step. J. Biol. Chem. **246,** 6064-6071, 1971.

[112] Babior, B.M., Moss, T.H., The mechanism of Action of ethanolamine ammonia-lyase, a B_{12}-dependent enzyme. The participation of paramagnetic species in the catalytic deamination of 2-aminopropanol, J. Biol. Chem. **249,** 4537-4544, 1974.

[113] Babior, B.M., Coenzyme B_{12} as hydrogen carrier in the ethanolamine deaminase reaction, Biochim. Biophys. Acta **167,** 456-458, 1968.

[114] Schrauzer, G.N., and Sibert, J.W., Coenzyme B_{12} and coenzyme B_{12} model compounds in the catalysis of the dehydration of glycols, J. Am. Chem. Soc. **92,** 1022-1030, 1970.

[115] Schrauzer, G.N., and Windgasses, R.J., On hydroxyalkylcobaloximes and the mechanism of a cobamide-dependent diol dehydrase. J. Am. Chem. Soc. **89,** 143-147, 1967.

[116] Schrauzer, G.H., Holland, R.J., and Seck, J.A., The mechanism of coenzyme B_{12} action in dioldehydrase. J. Am. Chem. Soc. **93,** 1503-1505, 1971.

[117] Stadtman, T.C., and Tsai, L. A cobamide coenzyme dependent migration of the E-amino group of D-lysine, Biochem. Biophys. Res. Commun. **28,** 920-926, 1967.

[118] Morley, C.G.D., and Stadtman, T.C., Studies on the fermentation of D-α-lysine, purification and properties of an adenosine triphosphate regulated B_{12}-coenzyme dependent D-α-lysine mutase complex from *Clostridium sticklandii,* Biochemistry **9,** 4890-4900, 1970.

[119] Stadtman, T.C., and Grant, M.A., D-α-lysine mutase of *Clostridium,* Methods Enzymol. **17 B,** 211-215, 1971.

[120] Morley, C.G.D., and Stadtman, T.C., Studies of the fermentation of D-α-lysine. On the hydrogen shift catalyzed by the B_{12} coenzyme-dependent D-α–lysine mutase, Biochemistry **10,** 2325-2329, 1971.

[121] Stadtman, T.C., and Renz, P., Anaerobic degradation of lysine. V. Some properties of the cobamide coenzym dependent β-lysine mutase of *Clostridium sticklandii*. Arch. Biochem. Biophys. **125,** 226-239, 1968.

[122] Bray, R.C., and Stadtman, T.C., Anaerobic degradation of lysine. III. ^{15}N Studies on the conversion of lysine to 3,5-diaminohexonoate. J. Biol. Chem. **243,** 381-385, 1968.

[123] Baker, J.J., van der Drift, C., and Stadtman, T.C., Purification and properties of β-lysine mutase, a pyridoxal phosphate and B_{12} coenzyme dependent enzyme. Biochemistry **12,** 1054-1063, 1973.

[124] Retay, J., Kunz, F., Stadtman, T.C., and Arigoni, D., Zum Mechanismus der β-Lysin-Mutase-Reaktion, Experientia **25,** 801-802, 1969.

[125] Tsai, L., and Stadtman, T.C., Anaerobic degradation of lysine. IV. Cobamide coenzymedependent migration of an amino group from carbon 6 of β-lysine (3,6-diaminohexanoate) to carbon 5 forming a new naturally occurring amino acid, 3,5-diaminohexanoate. Arch. Biochem. Biophys. **125,** 210-225, 1968.

[126] Stadtman, T.C., and Grant, M.A., L-β-lysine mutase *Clostridium sticklandii,* Methods Enzymol. **17 B,** 206-211, 1971.

[127] Dyer, J. K., and Costilow, R. N., 2,4-Diaminovaleric acid: an intermediate in the anaerobic oxidation of ornithine by *Clostridium sticklandii*, J. Bact. **101**, 77-83, 1970.

[128] Tsuda, Y., Friedmann, H.C., Ornithine to 2-amino-4-ketopentanoic acid via 2,4-diamino-pentanoic acid, participation of B_{12} coenzyme, pyridoxal phosphate and pyridine nucleotide. J. Biol. Chem. **245**, 5914-5926, 1970.

[129] Somack, R., Costilow, R., Purification and properties of a pyridoxal phosphate and coenzyme B_{12}-dependent D-α-ornithine 5,4-aminomutase. Biochemistry **12**, 2597-2604, 1973.

[130] Poston, J. M., Leucine 2,3-aminomutase, an enzyme of leucine catabolism, J. Biol. Chem. **251**, 1859-1863, 1976.

[131] Poston, J. M., Leucine 2,3-aminomutase: A cobalamin-dependent enzyme present in bean seadlings, Science **195**, 301-302, 1977.

[132] Poston, J. M., Coenzyme B_{12}-dependent enzymes in potatoes: leucine 2,3-aminomutase and methylmalonyl-CoA mutase, Phytochemistry **17**, 401-402, 1978.

[133] Reichard, P., The Biosynthesis of deoxyribonucleotides, Eur. J. Biochem. **3**, 259-266, 1968.

[134] Blakley, R. L., Cobamides and ribonucleotide reduction. I. Cobamide stimulation of ribonucleotide reduction in extracts of *Lactobacillus leichmanii*, J. Biol. Chem. **240**, 2173-2180, 1965.

[135] Beck, W. S., and Hardy, J., Fed. Proc. **24**, 421, 1965.

[136] Downing, M., Adams, A., and Helleuga, L., Enzymatic conversion of ribonucleotides to deoxynucleotides, Biochem. Biophys. Acta **108**, 233-242, 1965.

[137] Beck, W. S., and Hardy, J., Requirement of ribonucleotide reductase for cobamide coenzyme, a product of ribosomal activity. Proc. Natl. Acad. Sci. USA **54**, 286-293, 1965.

[138] Blakley, R. L., and Barker, H. A., Cobamide stimulation of the reduction of ribotides to deoxyribotides in *Lactobacillus leichmannii*, Biochem. Biophys. Res. Commun. **16**, 391-397, 1964.

[139] Abrams, R., and Duraiswami, S., Deoxycytidylate formation from cytidylate without glycoside cleavage in *Lactobacillus leichmannii* extract containing vitamin B_{12} coenzyme, Biochem. Biophys. Res. Commun. **18**, 409-414, 1965.

[140] Panagon, D., Orr, M. D., Dunstone, J. R., and Blakley, R. L., A monomeric, allosteric enzyme with a single polypeptide chain. Ribonucleotide reductase of *Lactobacillus leichmannii*. Biochemistry **1**, 2378-2383, 1972.

[141] Abrams, R., Cytidine 5'-triphosphate as the precursor of deoxycytidylate in *Lactobacillus leichmannii*, J. Biol. Chem. **240**, PC 3697, 1965.

[142] Goulian, M., and Beck, W. S., Purification and properties of cobamide dependent ribonucleotide reductase from *Lactobacillus leichmannii*, J. Biol. Chem. **241**, 4233-4242, 1966.

[143] Gleason, F. K., and Hogenkamp, H. P. C., Ribonucleotide reductase from *Euglena gracilis* a deoxyadenosylcobalamin-dependent enzyme. J. Biol. Chem. **245**, 4394-4899, 1970.

[144] Beck, W. S., Regulation of cobamide-dependent ribonucleotide reductase by allosteric effectors and divalent cations, J. Biol. Chem. **242**, 3148-3158, 1967.

[145] Blakley, R. L., Ghambeer, R. K., Nixon, P., and Vitols, E., The cobamide-dependent ribonucleoside triphosphate reductase of Lactobacilli, Biochem. Bioph. Res. Commun. **20**, 439-445, 1965.

[146] Vitols, E., and Blakley, R. L., Hydrogen-donor specificity of ribonucleoside triphosphate. Reductase from *Lactobacillus leichmannii*, Biochem. Biophys. Res. Commun., **2**, 466-472, 1965.

[147] Cowles, J. R., and Evans, H. J., Some properties of the ribonucleotide reductase from *Rhizobium meliloti*, Arch. Biochem. Biophys. **127**, 770-778, 1968.

[147 a] Inukai, S., Sato, K., Shimizu, S., Regulation of vitamin B_{12}-dependent ribonucleotide reductase from *Rhizobium meliloti* by allosteric effectors, Agric. Biol. Chem. **44**, 1105-1109, 1980.

[147 b] Inukai, S., Sato, K., Shimizu, S., Regulation of vitamin B_{12}-dependent ribonucleotide reductase from *Rhizobium meliloti* by allosteric effectors, Agric. Biol. Chem. **44**, 1105-1110, 1980.

[148] Orr, M. D., and Vitols, E., Thioredoxin from *Lactobacillus leichmannii* and its role as hydrogen donor for ribonucleoside triphosphate reductase, Biochem. Biophys. Res. Commun. **25**, 109-115, 1966.

[149] Vitols, E., Brownson, C., Gardiner, W., and Blakley, R. L., Cobamides and ribonucleotide reduction. A kinetic study of the ribonucleoside triphosphate reductase of *Lactoba-*

cillus leichmannii, J. Biol. Chem. **242,** 3025-3041, 1967.

[150] Vitols, E., Hogenkamp, H. P. C., Brownson, C., Blakley, R. L., and Connellan, G., Reduction of a disulphide bond of ribonucleotide reductase by the dithiol substrate, Biochem. J. **104,** 58C-60C, 1967.

[151] Blakley, R. L., Ghambeer, R. K., Batterham, T. J., and Brownson, C., Studies with hydrogen isotopes on the mechanism of action of cobamide-dependent ribonucleotide reductase, Biochem. Biophys. Res. Commun., **24,** 418-426, 1966.

[152] Gottesman, M. M., and Beck, W. S., Transfer of hydrogen in the cobamide-dependent ribonucleotide reductase reaction, Biochem. Biophys. Res. Commun. **24,** 353-359, 1966.

[153] Batterham, T. J., Ghambeer, R. K., Blakley, R. L., and Brownson, C., Cobamides and ribonucleotide reduction. IV. Stereochemistry of hydrogen transfer to the deoxyribonucleotide, Biochemistry **6,** 1203-1208, 1967.

[154] Griffin, C. E., Hamilton, F. D., Hopper, S. P., and Abrams, R., Stereospecificity of deoxycytidine triphosphate synthesis with the ribonucleotide reductase of *Lactobacillus leichmannii,* Arch. Biochem. Biophys. **126,** 905-911, 1968.

[155] Follman, H., and Hogenkamp, H. P. C., Ribonucleotide reductase. Studies with ^{18}O-labelled substrates, Biochemistry **8,** 4372-4375, 1969.

[156] Yamada, R., Tamao, Y., and Blakley, R. L., Degradation of 5'-deoxyadenosylcobalamin by ribonucleoside triphosphate reductase and binding of degradation products to the active center, Biochemistry **10,** 3959-3968, 1971.

[157] Hamilton, J. A., Yamada, R., Blakley, R. L., Hogenkamp, H. P. C., Looney, F. D., and Winfield, M. E., Cobamides and ribonucleotide reduction. VII. Cob(II)alamin as a sensitive probe for the active center of ribonucleotide reductase. Biochemistry **10,** 347-355, 1971.

[158] Abeles, R. H., and Beck, W. S., The mechanism of action of cobamide coenzyme in the ribonucleotide reductase reaction, J. Biol. Chem. **242,** 3589-3593, 1967.

[159] Hogenkamp, H. P. C., Ghambeer, R. K., Brownson, C., and Blakley, R. L., The relation of hydrogen exchange between deoxyadenosylcobalamin and water to the role of the coenzyme in ribonucleotide reduction. Biochem. J. **103,** 50-70, 1967.

[160] Beck, W. S., Abeles, R. H., and Robinson, W. G., Transfer of hydrogen from cobamide coenzyme to water during enzymatic ribonucleotide reduction, Biochem. Biophys. Res. Commun. **25,** 421-425, 1966.

[161] Hogenkamp, H. P. C., Ghambeer, R. K., Brownson, C., Blakley, R. L., and Vitols, E., Cobamides and ribonucleotide reduction. VI. Enzyme-catalyzed hydrogen exchange between water and deoxyadenosylcobalamin. J. Biol. Chem. **243,** 799-808, 1968.

[162] Hamilton, J. A., Blakley, R. L., Looney, F. D., and Winfield, M. E., Formation of a cobamide containing divalent cobalt by the ribonucleotide reductase of *Lactobacillus leichmannii,* Biochem. Biophys. Acta **177,** 374-376, 1969.

[163] Hamilton, J. A., Tamao, Y., Blakley, R. L., and Coffman, R. E., Electron paramagnetic resonance studies on cobalamin-dependent ribonucleotide reduction, Biochemistry **11,** 4696-4705, 1972.

[164] Hamilton, J. A., and Blakley, R. L., Electron spin resonance studies of ribonucleotide reduction catalyzed by the ribonucleotide reductase of *Lactobacillus leichmannii,* Biochim. Biophys. Acta **84,** 224-226, 1969.

[165] Tamao, Y., and Blakley, R. L., Direct spectrophotometric observation of an intermediate formed from deoxyadenosylcobalamin in ribonucleotide reduction, Biochemistry **12,** 24-34, 1973.

[166] Orme-Johnson, W. H., Beinert, H., and Blakley, R. L., Cobamides and ribonucleotide reduction. XII. The electron paramagnetic resonance spectrum of active coenzyme B_{12}, J. Biol. Chem. **249,** 2338-2343, 1974.

[167] Sando, G. N., Blakley, R. L., Hogenkamp, H. P. C., Hoffman, P. J., J. Biol. Chem. **250,** 8774, 1975.

[168] Sando, G. N., Grant, M. E., Hogenkamp, H. P. C. Biochim. Biophys. Acta **428,** 228, 1976.

[169] Hogenkamp, H. P. C., The chemical synthesis and coenzymatic properties of analogs of adenosylcobalamin in the ribonucleotide reductase reaction, in vitamin B_{12}, Eds. Zagalak, B., Friedrich, W., Walter de Gruyter, Berlin 1979, pp. 489-503.

[170] Inukai, S., Sato, K., Shimizu, S., Agric. Biol. Chem. **44,** (3), 1979.

[171] Thelander, L., Sjöberg, B. M., Erikson, S., Methods in Enzymol. **51,** 227-237, 1978.

[172] Blakley, R. L., Methods in Enzymol. **51,** 246-259, 1978.

[173] Sato, K., Inukai, S., Veda, S., Seki, T., Shimizu, S., Formation and role of vitamin B_{12} in *Protaminobacter ruber* and *Rhizobium meliloti*, in Vitamin B_{12}, Eds. Zagalak, B., Friedrich, W., Walter de Gruyter, Berlin 1979, pp. 361-372.
[174] Jordan, E.P. and Vorobieva, L.J., Localization of vitamin B_{12}-dependent ribonucleotide reductase in *Propionibacterium shermanii* cells, Mikrobiologiya **50**, 736-738, 1981.
[175] Carell, E.F., and Seeger, J.W., Ribonucleotide reductase activity in vitamin B_{12} deficient Euglena gracilis, Biochem. J. **188**, 573-576, 1980.

7.3 Methylcobamide – dependent reactions

by A. Stroinski and Z. Schneider

7.3.1 **Methionine synthetase,** (Homocysteine: N^5-Methyltetrahydrofolate Methyltransferase, EC 2.1.1.13

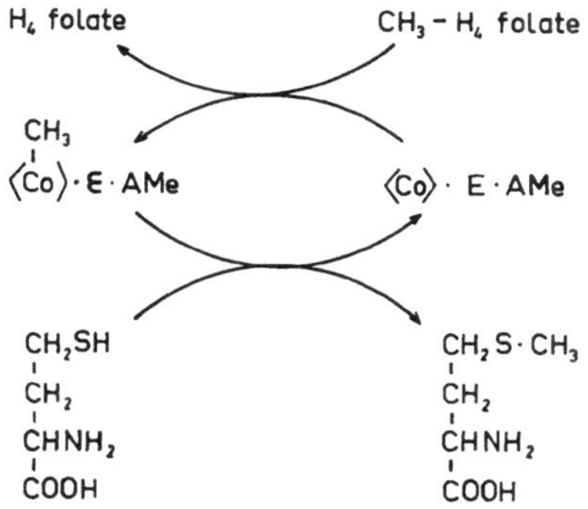

Sources: *Escherichia coli* [1],
Aerobacter aerogenes [12],
Salmonella typhimurium [3],
Cyrnebacterium simplex [4],
Rhodopseudomonas spheroides [5],
Pseudomonas denitrificans [6],
Streptomyces divacens [7],
Ochromonas malhamensis [8, 9],
Mamalian liver [10],
Mamalian cells in tissue culture [11],
Avian liver [12].
The cobalamin dependent enzyme has not been found in yeast and in higher plants [13-18].

Repression: Several enzymes involved in methionine biosynthesis are repressed by methionine or by cobalamin [19-22].

B_{12}-dependence [23, 27, 28, 35–40, 42]:
active form: MeCbl, $K_M < 10^{-8}M$ [43]
inactive form: PrCbl,

Molecular structure:
apoenzyme: 140000 to 150000 (sedimentation equilibrium and gel filtration) [23–25],
186000 (SDS poliacrylamide gel electrophoresis) [43],
holoenzyme: 255000 (gel filtration) [44],
apoenzyme subunits [26]:
M = about 125000 (cobalamin containing protein),
S = 3000 (sulfhydryl protein),
two forms of enzyme: methylated and demethylated depending on the presence of S-adenosyl-methionine [44].

Catalytic properties:
K_{eq} for methionine $= 7 \times 10^6$ [45].

Substrates:
K_M for D, L-Homocysteine $= 3 \times 10^{-3}M$ [24], for D, L-5-Methyltetrahydrofolate $= 6.7 \times 10^{-4}M$ [24].

Activator: S-Adenosylmethionine
$K_M = 1 \times 10^{-7}M$ [46, 47].

Specific activity: 3 μmoles of methionine/mg protein · min [43].

Turnover number: 600 to 800 [43].

pH optimum: 7.0–7.2 [24, 26].

Reaction mechanism:
1. the methyl group of N^5-methyltetrahydrofolate is transferred to the enzyme bound cobalamin to form MeCbl [1, 29–34],
2. the transfer of the methyl group from MeCbl to homocysteine does not require the presence of S-adenosylmethionine [32, 33],
3. the reaction is reversible but the equilibrium lies far to the right, to the formation of methionine [34, 45].

Assay systems:
1. microbiological assay
2. radioisotopic assay using [5-^{14}C]methyltetrahydrofolate [41].

Storage conditions: The enzyme is very unstable, if it is stored frozen during ½ h its activity reduces by a half, but may be stabilized by storage at −20 °C in 50% glycerol [43, 48].

7.3.2 Methane synthetase

$$\begin{matrix} CO_2 \\ HCOOH \\ H_2CO \\ H_3COH \\ CH_3COOH \\ CH_3COCOOH \end{matrix} \longrightarrow CH_3-H_4-\text{folate} \rightarrow CH_3Cbl \rightarrow CH_4$$

Sources: *Methanosarcina barkeri* [49-54],
Methanobacillus omelianskii [55-58, 69],
Methanobacillus kuzneceovii [64],
Methanobacterium (Metanobacillus omelianskii strain M.O.H. [66, 68]).

B_{12}-dependence:
active forms [62]: MeCbl, (Ade) MeCba and (5-OH-Bza) MeCba

Molecular structure:
Active enzyme system contains [54, 57]:
1. high molecular weight B_{12}-enzyme,
2. ferrodoxin,
3. an unknown protein
4. coenzyme M [66, 67]

Catalytic properties:

Requirements: ATP, Mg^{2+} or Mn^{2+} [57].

Inhibition: Enzyme is inhibited by intrinsic factor [53], propyl iodide [62], $CHCL_3$, CCl_4, CH_2Cl_2 [60], competitive inhibitors to MeCbl: $CFCl_3$, CF_2Cl_2, CF_3Cl.

Temperature optimum: 40 °C [49].

pH optimum: 7.0-7.2 [49].

Reaction mechanism [67]:

$$^{14}CH_3\text{-}B_{12} + \text{coenzyme M} \xrightarrow[\text{methyltransferase}]{\text{electron donor}} {}^{14}CH_3\text{-coenzyme M} + B_{12}$$

$$^{14}CH_3\text{-CoM} \xrightarrow[\text{methanase}]{H_2,ATP,Mg^{2+}} CH_4 + \text{coenzyme M}$$

coenzyme M = 2,2-dithiodiethanesulfonic acid,
CH_3-coenzyme = 2(methylthio)ethanesulfonic acid

Assay system: The product is estimated by gas chromatography [57], by flame ionization method [49] and by application of radioactive substrates [49].

7.3.3 Acetate synthetase

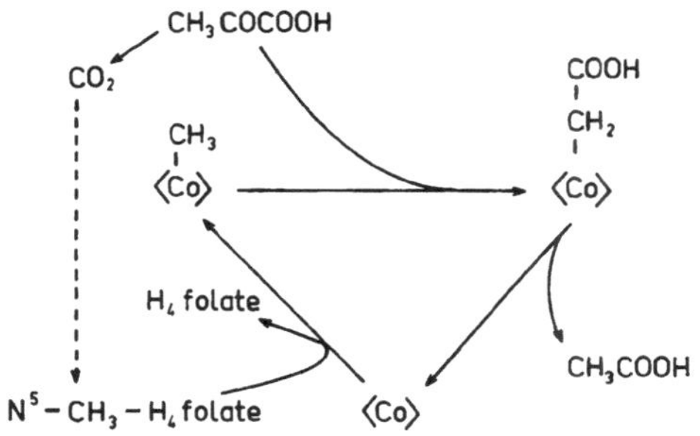

⟨Co⟩ - corrinoid enzyme complex

Sources: *Clostridium thermoaceticum* [70], *Clostridium formicoaceticum* [71], *Acetobacterium woodii* [72],

B_{12}-dependance:
active forms: MeCbl, (Ade) MeCba, (5-methoxy-Bza) MeCba [73]

Molecular structure: Three protein fractions (DEAE-cellulose) are required to convert MeCbl to acetate [70, 74]:
1. A-protein containing corrinoid [75],
2. B-protein catalyzes the NADPH-dependent cleavage of Co-carboxymethylcobalamine, Mr 27000. It contains 1 mole (5-methoxy-Bza) MeCba per mole of protein [76],
3. ferredoxin

The "corrinoid" enzyme is a tetramer [83, 84]:
160000 (sedimentation equilibrium)
subunit: 40000 (SDS-electrophoresis)

Catalytic properties:
Substrate: pyruvate can be replaced by α-ketobutyric acid but not by oxamic acid nor tiglic acid [70].

Requirements: coenzyme A ($K_M = 1.6 \times 10^{-4}$M) [70].

Inhibition: by sulfhydryl poisons, arsenite and cadmium [70], propyl iodide, $CHCL_3$ and CCl_4 [77, 86–92].

pH optimum: 8.0–9.0 [70].

Temperature optimum: *57°C [70].*

Assay system: Acetate can be measured titrimetrically following Celite column chromatography and distillation or colorimetrically following conversion to the hydroxamate by the acetate kinase method, or radioactive method with [Co-$^{14}CH_3$] cobalamin [70].

Reaction mechanism:
1. the methyl group is transferred from the N^5-Methyltetrahydrofolate to the corrinoid [77],
2. the conversion of MeCbl to acetate has been explained by means of three pathways:
   - the carboxylation of the MeCbl followed by reductive cleavage of the (HOO C CH_2-)Cbl [4] (equation 7-16)
   - MeCbl as a biological Grignard reagent [78, 80]
   - the reductive production of a methyl carbanion which then reacts with CO_2 [79],
3. the carboxycarbon of acetate derives from the carboxyl carbon of α-ketoacids [81, 82].

7.3.4 DNA-methylase

The enzyme is methylcobalamin-dependent. It catalyzes specifically the methylation of cytosine residue in DNA isolated from B_{12}-deficient *Propionibacterium shermanii.*

SAM is the methyl group donor. The investigations are at an early stage. Only partially purified enzyme preparations from *P. shermanii* are available [85].

References

[1] Guest, J. R., Helleiner, C. W., Gross, M. J., and Woods, D. D., Cobalamin and the synthesis of methionine by ultrasonic extracts of *Escherichia coli*, Biochem. J. **76**, 396-405, 1960.

[2] Morningster, J. F., and Kisliuk, R. L., Interrelations between two pathways of methionine biosynthesis in *Aerobacter aerogenes*, J. Gen. Microbiol. **39**, 43-51, 1965.

[3] Cauthen, S. E., Foster, M. A., and Woods, D. D., Methionine synthesis by extracts of *Salmonella typhimurium*, Biochem. J. **98**, 630-635, 1966.

[4] Fujii, K., Takeuchi, H., Shimizu, S., and Fukui, S., Agric. Biol. Chem. **36**, 2323, 1972.

[5] Cauthey, S. E., Pattison, J. R., and Lascelles, J., Vitamin B_{12} in photosynthesis bacteria and methionine synthesis by *Rhodopseudomonas spheroides*, Biochem. J. **102**, 474, 1967.

[6] Lago, B. D., and Demain, A. L., Alternate requirement for vitamin B_{12} or methionine in mutants of *Pseudomonas denitrificans*, a vitamin B_{12}-producing bacterium. J. Bacteriol. **99**, 347-349, 1969.

[7] Ohmori, H., Sato, K., Shimizu, S., and Fukui, S., Conformation of a cobalamin-dependent methionine synthesizing system in *Streptomyces olivaceus*, Agric. Biol. Chem. **35**, 338-343, 1971.

[8] Johnson, B. C., Holdsworth, E. S., Porter, J. W. G., and Kon, S. K., Vitamin B_{12} and methyl-group synthesis, Brit. J. Nutr. **11**, 313-323, 1957.

[9] Griffiths, J. M., and Daniel, L. J., Methionine biosynthesis in *Ochromonas malhamensis*, Arch. Biochem. Biophys. **134**, 463-472, 1969.

[10] Dickerman, H., Redfield, B. G., Bievi, J., and Weissbach, H., The role of vitamin B_{12} in methionine biosynthesis in avian liver. J. Biol. Chem. **239**, 2545-2552, 1964.

[11] Langhlin, R. E., Elford, H. L., and Buchanan, J. M., Enzymatic synthesis of the methyl group of methionine, VII. Isolation of a cobalamin-containing transmethylase 5-methyltetrahydrofolate-homocysteine from mammalian, J. Biol. Chem. **239**, 2888-2895, 1964.

[12] Mangum, J. H., Murray, B. K., and North, J. A., Vitamin B_{12}-dependent methionine biosynthesis in cultured mammalian cells, Biochemistry **8**, 3496-3499, 1969.

[13] Burton, E. G., and Sakami, W., The formation of methionine from the monoglutamate form of methyltetrahygrofolate by higher plants, Biochem. Biophys. Res. Commun. **36**, 228-234, 1969.

[14] Burton, E., Selhub, J., and Sakami, W., The substrate specificity of 5-methyltetrahydropteroytriglutamate-homocysteine methyltransferase, Biochem. J. **111**, 793-795, 1969.

[15] Salem, A. R., Foster, M. A., and Wilson, R. H., Methylation of homocysteine in *Coprinus lagopus*, Biochem. J. **118**, 16-17, 1970.

[16] Wilson, R. H., The regulation by methionine of three inducible enzymes in *Coprinus lagopus*, Biochem. J. **118**, 16, 1970.

[17] Botsford, J. L., and Parlas, L. W., Role of S-Adenosylmethionine in methionine biosynthesis in yeast, J. Bacteriol. **94**, 966-971, 1967.

[18] Salem, A. R., Pattison, J. R., and Foster, M. A., Folic acid and the methylation of homocysteine by *Bacillus subtilis*, Biochem. J. **126**, 993-1004, 1972.

[19] Katzen, H. M., and Buchanan, J. M., J. Biol. Chem. **240**, 825, 1965.

[20] Milner, L., Whitfield, C., and Weissbach, H., Effect of L-methionine and vitamin B_{12} on methionine biosynthesis in *Escherichia coli*, Arch. Biochem. Biophys. **133**, 413-419, 1969.

[21] Kung, H. F., Spears, C., Greene, R. C., and Weissbach, H., Regulation of the thermal reactions in methionine biosynthesis by vitamin B_{12} and methionine, Arch. Biochem. Biophys. **150**, 23-31, 1972.

[22] Greene, R. C., Williams, R., Kung, H. F., Spears, C., and Weissbach, H., Effects of methionine and vitamin B_{12} on the activities of methionine biosynthetic enzymes in met J mutants of *Escherichia coli* K 12, Arch. Bioch. Biophys. **158**, 249-256, 1973.

[23] Taylor, R. T., and Weissbach, H., N^5-Methyltetrahydrofolate-homocysteine transmethylase, J. Biol. Chem. **242**, 1502-1508, 1967.

[24] Stavrianopoulos, J., and Jaenicke, L., Reaktionsschritte der Methionin-Synthese bei *Escherichia coli*, Eur. J. Biochem. **3**, 95-106, 1967.

[25] Taylor, R. T., *Escherichia coli* B N^5-Methyltetrahydrofolate-homocysteine cobalamin methyltransferase: gel filtration behavior of apo-

enzyme and holoenzymes, Biochim. Biophys. Acta **242**, 355-364, 1971.

[26] Galivan, J., and Huennekens, F. M., Resolution of the methionine synthetase system from *Escherichia coli* K-12, Biochem. Biophys. Res. Commun. **38**, 46-51, 1970.

[27] Guest, J. R., Friedman, S., Woods, D. D., and Smith, E. L., A methyl analogue of cobamides coenzyme in relation to methionine synthesis by bacteria, Nature **195**, 340-342, 1962.

[28] Weissbach, H., Redfield, B. G., Dickerman, H., and Brot, N., Studies on methionine biosynthesis effect of alkylcobamide derivatives on the formation of holoenzyme, J. Biol. Chem. **240**, 856-862, 1965.

[29] Brot, N., and Weissbach, H., The role of cobamides in methionine synthesis. Enzymatic formation of holoenzyme, J. Biol. Chem. **241**, 2024-2028, 1966.

[30] Brodic, J. D., Origin of photolabile methyl groups in methionine biosynthesis, Biochem. Biophys. Res. Commun. **26**, 261-264, 1967.

[31] Taylor, R. T., and Weissbach, H., Isolation of methyl-B_{12} from *Escherichia coli* B N^5-methyl-H_4-folate-homocysteine vitamin B_{12} transmethylase, Biochem. Biophys. Res. Commun. **27**, 398-404, 1967.

[32] Taylor, R. T., and Weissbach, H., Enzymic synthesis of methionine: formation of a radioactive cobamide enzyme with N^5-methyl^{14}C-tetrahydrofolate, Arch. Biochem. Biophys. **119**, 572-579, 1967.

[33] Taylor, R. T., and Weissbach, H., *Escherichia coli* B N^5-methyltetrahydrofolate-homocysteine vitamin B_{12} transmethylase: formation and photolability of a methylcobalamin enzyme, Arch. Biochem. Biophys. **123**, 109-126, 1968.

[34] Taylor, R. T., and Weissbach, H., *Escherichia coli B* N^5-methyltetrahydrofolate-homocysteine methyltransferase. Sequential formation of bound methylcobalamin with S-Adenosyl-L-methionine and N^5-methyltetrahydrofolate. Arch. Biochem. Biophys. **129**, 728-744, 1969.

[35] Weissbach, H., Redfield, B., and Dickerman, Cobamide-dependent synthesis of methionine: light reactivation of an inhibited enzyme, Biochem. Biophys. Res. Commun. **17**, 17-22, 1964.

[36] Brot, N., Redfield, B., and Weissbach, H., Reduction and alkylation of the cobamide prosthetic group in the enzymatic synthesis of methionine, Biochem. Biophys. Res. Commun. **18**, 18-23, 1965.

[37] Weissbach, H., and Taylor, R., Role of vitamin B_{12} in methionine synthesis, Fed. Proc. **25**, 1649-1656, 1967.

[38] Taylor, R. T., Whitfield, C., and Weissbach, H., Chemical propylation of vitamin B_{12} transmethylase: anomalous behaviour of S-adenosyl-L-methionine, Arch. Biochem. Biophys. **125**, 240-252, 1968.

[39] Burke, G. T., Mangum, J. H., and Brodie, J. D., Mechanism of mammalian cobalamin-dependent methionine biosynthesis, Biochemistry **10**, 3079-3085, 1971.

[40] Weissbach, H., Redfield, B. G., and Dickerman, H., Effect of vitamin B_{12} analogues on methionine formation from N^5-methyltetrahydrofolic acid, J. Biol. Chem. **239**, 146-148, 1964.

[41] Weissbach, H., Peterkofsky, A., Redfield, B. G., and Dickerman, H., Studies on the terminal reaction in the biosynthesis of methionine, J. Biol. Chem. **238**, 3318-3324, 1963.

[42] Taylor, R. T., Methylcobalamin as a substrate at a separate site on *Escherichia coli* B N^5-methyltetrahydrofolate-homocysteine cobalamin methyltransferase, Arch. Biochem. Biophys. **144**, 352-362, 1971.

[43] Jaenicke, L., unpublished data.

[44] Rüdiger, H., and Jaenicke, L., Methionine synthetase: existence and intercovnersion of two enzyme species, Eur. J. Biochem. **16**, 92-95, 1970.

[45] Rüdiger, H., and Jaenicke, L., Methionine synthesis: demonstration of the reversibility of the reaction, FEBS Letters **4**, 316-318, 1969.

[46] Rüdiger, H., and Jaenicke, L., On the role of S-adenosylmethionine in the vitamin B_{12} dependent methionine biosynthesis, Eur. J. Biochem. **10**, 557-560, 1969.

[47] Taylor, R. T., and Hanna, M. L., *Escherichia coli* B 5-methyltetrahydrofolate – homocysteine cobalamin methyltransferase: catalysis by a reconstituted methyl-^{14}C-cobalamin holoenzyme and the function of S-adenosyl-L-methionine, Arch. Biochem. Biophys. **137**, 453-459, 1970.

[47 a] Rask, H., Schou, Olesen, A., Mortensen, J. Z., Freund, L. G., Nitrous oxide and urine methylmalonic acid in man, Scand. J. Haematol. **31**, 45-48, 1983.

[48] Rüdiger, H., and Jaenicke, L., The biosynthesis of methionine, Mol. Cell. Biochem. **1**, 157-168, 1973.

[49] Blaylock, B. A., and Stadtman, T. C., Biosynthesis of methane from the methyl moiety of methylcobalamin, Biochem. Biophys. Res. Commun. **11**, 34-38, 1963.

[50] Blaylock, B.A., and Stadtman, T.C., Biosynthesis of methane from the methyl moiety of methylcobalamin in extracts of *methanosarcina barkerii,* Ann. N. Y. Acad. Sci. **112,** 799-803, 1964.

[51] Gezius, A., and Barker, H.A., Biochemistry **4,** 510, 1965.

[52] Blaylock, B.A., and Stadtman, T.C., Enzymic formation of methylcobalamin in *Methanosarcina barkerii* extracts, Biochem. Biophys. Res. Commun. **17,** 475-480, 1964.

[53] Blaylock, B.A., and Stadtman, T.C., Methane biosynthesis by *Methanosarcina barkerii.* Properties of the soluble enzyme system, Arch. Biochem. Biophys. **116,** 138-158, 1966.

[54] Blaylock, B.A., Cobamide-dependent methanol cyanocob(I)alamin methyltransferase of *Methanosarcina barkerii,* Arch. Biochem. Biophys. **124,** 314-324, 1968.

[55] Wolin, M.J., Wolin, E.A., and Wolfe, R.S., ATP-dependent formation of methane from methylcobalamin by extracts of *Methanobacillus omelianskii,* Biochem. Biophys. Res. Commun. **12,** 464, 1963.

[56] Wolin, M.J., Wolin, E.A., and Wolfe, R.S., The cobalamin product of the conversion of methylcobalamin to CH_4 by extracts of *Methanobacillus omelianskii,* Biochem. Biophys. Res. Commun. **15,** 420-423, 1964.

[57] Wood, J.M., and Wolfe, R.S., Components required for the formation of CH_4 from methylcobalamin by extracts of *Methanobacillus omelianskii,* J. Bacter. **92,** 696-700, 1966.

[58] Wood, J.M., and Wolfe, R.S., Alkylation of an enzyme in the methane-forming system of *Methanobacillus omelianskii,* Biochem. Biophys. Res. Commun. **22,** 119-123, 1966.

[59] Wood, J.M., and Wolfe, R.S., Propylation and purification of a B_{12} enzyme involved in methane formation, Biochemistry **5,** 3598-3603, 1966.

[60] Bauckop, T., Inhibition of rumen methanogenesis by methane analogues, Bacteriol. J. **94,** 171-175, 1967.

[61] Penley, M.W., Brown, D.G., and Wood, J.M., Chemical and biological studies with fluoroalkylcobalamins, Biochemistry **9,** 4302-4310, 1970.

[62] Wood, J.M., Wolin, M.J., and Wolfe, R.S., Formation of methane from methyl factor B and methyl factor III by cell-free extracts of *Methanobacillus omelianskii,* Biochemistry **5,** 2381-2384, 1966.

[63] Siebert, J.W., and Schrauzer, G.N., Enzymatic and nonenzymatic demethylation of methylcobalamin and of abiogenic cobaloxime model substrates methane biosynthesis, by *Methanobacillus omelianskii* J. Amer. Chem. Soc. **92,** 1421-1423, 1970.

[64] Pantskhava, E.S., Pchelkina, V.V., and Bukin, V.N., Formation of methane and acetic acid from methylcobalamine by cell-free extracts of *Methanobacillus kuzneceov,* Biochimija **38,** 507-514, 1973.

[65] Stadtman, T.C., Vitamin B_{12}, Science, **171,** 859-861, 1971.

[66] Mc Bride, B.C., Wolfe, R.S., A new coenzyme of methyl transfer, coenzyme M, Biochemistry **10,** 2317-1971.

[67] Taylor, C.D., Wolfe, R.S., Structure and methylation of coenzyme M, a methyltransfer coenzyme, Fed. Proc. **32,** 589 Abs., 1973.

[68] Bryant, M.P., Mc Bride, B.C., and Wolfe, R.S., Hydrogen-oxidizing methane bacteria. I. Cultivation and methanogenesis, J. Bacteriol. **95,** 1118-1123, 1968.

[69] Bryant, M.P., McBride, B.C., and Wolfe, R.S., Hydrogen-oxidizing methane bacteria. 1. Culturation and methanogenesis, J. Bacteriol. **95,** 1118-1123, 1968.

[70] Poston, J.M., Kuratomi, K., and Stadtman, E.R., The conversion of carbon dioxide to acetate. I. The use of cobalt-methylcobalamin as a source of methyl groups for the synthesis of acetate by cell-free extracts of *Clostridium thermoaceticum,* J. Biol. Chem. **241,** 4209-4216, 1966.

[71] O'Brien, W.E., and Ljungdahl, L.G., Fermentation of fructose and synthesis of acetate from carbon dioxide by *Clostridium formicoaceticum,* J. Bacteriol. **109,** 626-632, 1972.

[72] Tanner, R.S., Wolfe, R.S., and Ljungdahl, L.G., Levels of tetrahydrofolate enzymes in *Acetobacterium woodii* and their implication in the synthesis of acetate from CO_2, J. Bacteriol. **134,** 668-670, 1978.

[73] Ljungdahl, L., Irion, E., and Wood, H.G., Total synthesis of acetate from CO_2. I. Co-methylcobyric acid and Co-methyl-5-methoxybenzimidazolylcobamide as intermediates with *Clostridium thermoaceticum,* Biochemistry **4,** 2771-2780, 1965.

[74] Poston, J.M., and Stadtman, E.R., The Conversion of carbon dioxide to acetate. III. Demonstration of ferrodoxin in the system converting Co-$^{14}CH_3$- cobalamin to acetate, Biochem. Biophys. Res. Commun. **26,** 550-555, 1967.

[75] Ljungdahl, L., Glatzle, D., Goodyear, J., and Wood, H.G., Total synthesis of acetate from

CO_2 by *Clostridium thermoaceticum,* Bacteriol. Proc. Abstr., **128,** 1967.

[76] Ljungdahl, L.G., Le Gall, J., and Lee, J.P., Isolation of a protein containing tighthy bound 5-methoxybenzimidazolylcobamide factor IIIm from *Clostridium thermoaceticum,* Biochemistry **12,** 1802-1808, 1973.

[77] Ghambeer, R.K., Wood, H.G., Schulman, M., and Ljungdahl, L., Total synthesis of acetate from CO_2. III. Inhibition by alkylhalides of the synthesis from CO_2, methyltetrahydrofolate, and methyl-B_{12} by *Clostridium thermoaceticum,* Arch. Biochem. Biophys. **143,** 471-484, 1971.

[78] Ingraham, L.L., Ann. N. Y. Acad. Sci. **112,** 713, 1964.

[79] Schrauzer, G.N., and Sibert, J.W., Acetate synthesis from carbon dioxide and methylcorrinoids. Simulation of the microbial carbon dioxide fixation reaction in a model system, J. Am. Chem. Soc. **92,** 3509, 1970.

[80] Parker, D.J., Wood, H.G., Ghambeer, R.K., and Ljungdahl, L., Total synthesis of acetate from carbon dioxide. Retention of deuterium during carboxylation of trideuteriomethyltetrahydrofolate or tridenteriomethylcobalamin, Biochemistry **11,** 3074, 1972.

[81] Schulman, M., Ghambeer, R.K., Ljungdahl, L.G., Wood, H.G., Synthesis of acetate from methyl-H_4 Folate by transcarboxylation from a keto acid in *Clostridium thermoaceticum,* Fed. Proc. **32,** 627, 1973.

[82] Schulman, M., Ghambeer, R.K., Ljungdahl, L.G., and Wood, H.G., Total synthesis of acetate from CO_2. VII. Evidence with *Clostridium thermoaceticum* that the carboxyl of acetate is derived from the carboxyl of pyruvate by transcarboxylation and not by fixation of CO_2, J. Biol. Chem. **248,** 6255, 1973.

[83] Welty, F.K., Wood, H.G., Purification of the "corrinoid" enzyme involved in the synthesis of acetate by *Clostridium thermoaceticum,* J. Biol. Chem. **253,** 5832-5838, 1978.

[84] Wood, H.G., The role of corrinoids in the total synthesis of acetate from CO_2, in Vitamin B_{12}, Eds. Zagalak, B., Friedrich, W., Walter de Gruyter, Berlin, 1979, pp 529-538.

[85] Antoshkina, N.V., Vorobieva, L.J., and Buryanov, Ya.J., Methylcobalamin-dependent enzymatic DNA-methylation in the cell-free extract of *Propionibacterium shermanii* (in Russian), Mikrobiologiya **50,** 631-635, 1981.

[86] Ragsdale, S.W., Ljungdahl, L.G., and Vartanian V. der, D., Isolation of carbon Monoxide dehydrogenase from acetobacterium woodii and comparison of its properties with those of the clostridium thermoaceticum enzyme. J. Bact. **155,** 1224-1237, 1983.

[87] Clark, J.E., Ragsdale, S.W., Ljungdahl, L.G. and Wiegel, J., Levels of enzymes involved in the synthesis of acetate from CO_2 in Clostridium thermoautotrophieum. J. Bacteriol. **151,** 507-509, 1982.

[88] Ljungdahl, L.G., and Wood, H.G., Acetate biosynthesis, In D. Dolphin ed, B_{12} vol. 2. John Wiley and Sons Inc., New York. 1982, pp.166-202.

[89] Ragsdale, S.W. Clark, J.E., Ljungdahl, L.G., Lundie, L.L. and Drake, H.L., Properties of Purified Carbon monoxide dehydrogenase from clostridium thermoaceticum, a nickel, iron-sulfur protein J. Biol. Chem. **258,** 2364-2369, 1983.

[90] Pezacka, E., and Wood, H.G., The synthesis of acetyl-CoA by clostridium thermoaceticum from carbon dioxide, hydrogen, coenzyme a and methyltetrahydrofolate Arch. Microbiol. **137,** 63-69, 1984.

[91] Hu, S-I., Pezacka, E., and Wood, H.G., Acetate Synthesis from carbon monoxide by clostridium thermoaceticum J. Biol. Chem. **259,** 8892-8897, 1984.

[92] Thauer, R., Nickelenzyme im Stoffwechsel von methanogenen Bakterien Biol. Chem. Hoppe-Seyler **366,** 103-112, 1985.

8 Non-enzymatic Vitamin B_{12} Binding Proteins in Man and Animals

Z. Schneider

The absorption and transport of vitamin B_{12} in man and animals is exclusively mediated by proteins. Three groups of proteins may be distinguished on the basis of their source, biological function, relation to specific membrane acceptors, immunological differences and molecular weight. Historically, *Intrinsic Factor* (IF) was the first protein found at the time vitamin B_{12} was isolated. Two other proteins, *Transcobalamin* (TC) and *Cobalophilin* (CP), were isolated and described in the early sixties. They were named differently by various authors and it was not until 1979 when the names *Transcobalamin* and *Cobalophilin* were commonly accepted. The IF and CP are glycoproteins containing appreciable, but variable amounts of carbohydrates which is the cause of difficulties in the isolation of homogeneous preparations, reproduction of purification procedures and the production of accurate chemo-physical data such as molecular weight, isoelectric point and binding capacity. The absorption of vitamin B_{12} by bacteria is protein-mediated as well. The bacterial binding proteins involved in vitamin B_{12} transport are less characterized, however, in literature. The abbreviation, Cbl (for cobalamin) is used here synonymously for B_{12}. Various forms of Cbl are described in chapters 2 and 3.

8.1 Intrinsic Factor

8.1.1 Abbreviations

IF, IF-Cbl for complex with Cbl

8.1.2 Site of synthesis and output

Parietal cells (stomach) in man, monkey, rabbit, guinea pig, cat and cow [1–4].

Chief cells (stomach) in rat and mouse [4–8].

Intrinsic factor secretion in man.

The intrinsic factor secretion was studied by several investigators (Tables 8-1 and 8-2).

The quantity of secreted IF is expressed in nanogram units which equal nanograms of labelled vitamin B_{12} bound to IF (see Chapter 8.1.5.).

An average secretion of 70000 units/day is assumed [9].

The secretion of IF is stimulated by histamine, betazole, gastrin, pentagastrin and insulin (Tables: 8-3, 8-4, 8-5 and Figures 8-1, 8-2, 8-3).

Table 8-1 Intrinsic factor secretion in man (compiled by Chanarin [9]).

	No	Basal secretion			
		Concentration (units/ml)		Output per hour units	
		Range	Mean	Range	Mean
Ardeman, et. al. [10]					
Medical students	9	9-52	25	900-8300	3600
D.U.	14	10-62	30	600-3200	1900
Hospital control	14	20-124	56	500-5900	2700
Ardeman [15]					
Males	62	3-124	36	100-8300	2600
Females	43	5-125	33	100-3500	1300
Irvine et. al. [16]	11	–	–	640-5230	–
Schipperijn [17]					
Volunteers	11	18-53	36	600-4340	2661
Patients	12-16	10-52	37	520-5979	2478

Table 8-2 Basal intrinsic factor output per hour in relation to age (from Chanarin [9]).

Age (years)	No	Mean basal intrinsic factor output units
19-30	38	2700
31-50	24	2400
>50	28	1300

Table 8-3 Intrinsic factor secretion in man in response to histamine stimulation (compiled by Chanarin [9]).

	No	Concentrations units/ml		Output per hour units	
		Range	Mean	Range	Mean
		Post-histamine secretion (40 μg/kg)			
Ardeman et. al. [10]					
Medical students	9	14-147	56	2200-18.300	8900
D.U.	14	26-83	51	2700-14.000	8600
Hospital control	14	14-111	53	1000-8600	4200
Ardeman [15]					
Males	37	14-111	60		
Females	32	15-105	55	400-14.900	5800
Irvine et. al.	11	–	–	5720-23.300	–
Wangel [18]	5	–	–	6200-16.700	11.800
Røbro et. al. [19]					
Normal	15	60-70	–	5400-25.000	16.200
D.U.	24			12.400-40.500	23.200

Table 8-3 continued

	No	Concentrations units/ml		Output per hour units	
		Range	Mean	Range	Mean
		Post-histamine secretion (0.5 mg)			
Schipperijn [17]	12	22-81	44.3	1288-7707	4545
		Post-histamine secretion (continuous infusion)			
Weir et al. [11] mostly D. U.	10	71-97	82	17.522-37.098	25.568
Dotevall et al. [12] (D. U.)	14	–	52	–	13.825

Table 8-4 Effect of histamine dosage on intrinsic factor output (compiled by Chanarin [9]).

Dose of histamine	Intrinsic factor (nanogram units)	
	Per ml	Per hour
None	36	2600
0.5 mg	44	4500
±2.0 mg	55-60	9000
Infusion	62	14.000

Table 8-5 Effect of age on intrinsic factor secretion after histamine stimulation (from Chanarin [9]).

Age years	No	Mean intrinsic factor output after histamine (nanogram units)
19-30	21	9500
31-50	16	9700
>50	25	5700

The secretion of IF is not stimulated by cholinergic drugs (carbachol, methacholine chloride and bethanecol chlorine) and by glucagon, predisone, secretine and acetazolamide. It is inhibited by colchicine [216].

8.1.3 Biological function

Intrinsic factor is a glycoprotein which promotes the absorption of vitamin B_{12} in human and animal digestion systems. It is capable of binding strongly the tiny amount of vitamin B_{12} (3-5 μg daily in human), released from the digested meal, into a com-

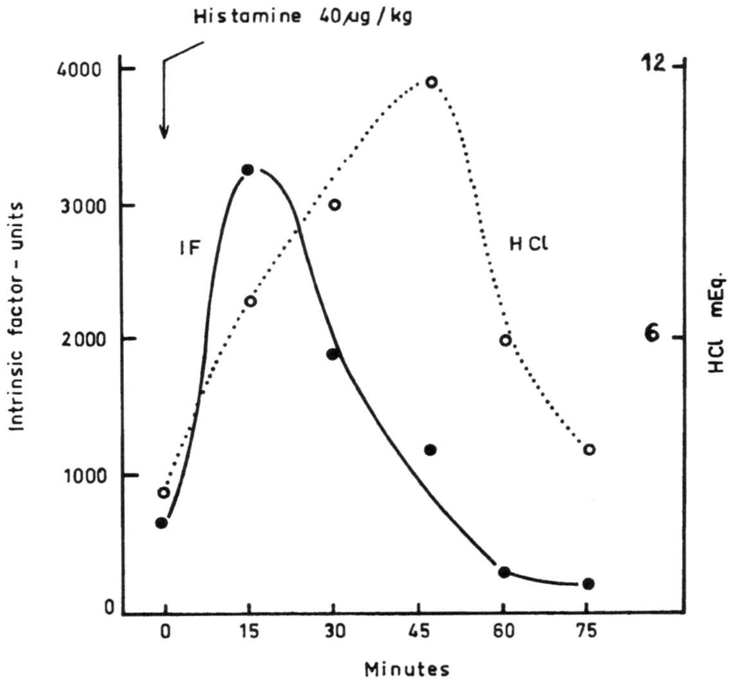

Fig. 8-1 The early rise in intrinsic factor output in man after histamine followed by an increase in hydrochloric acid secretion (from Chanarin [9])

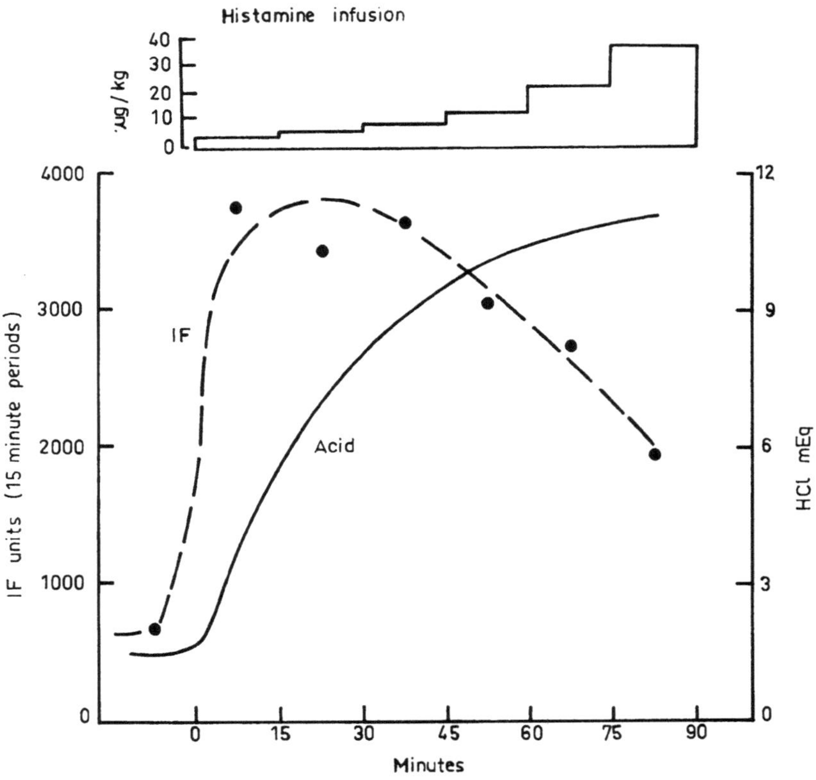

Fig. 8-2 The effect of infusing increasing amounts of histamine on intrinsic factor- and acid secretion in man (from Dotevall et al. [12] ©1964 Munksgaard International Publishers Ltd., Copenhagen, Denmark.

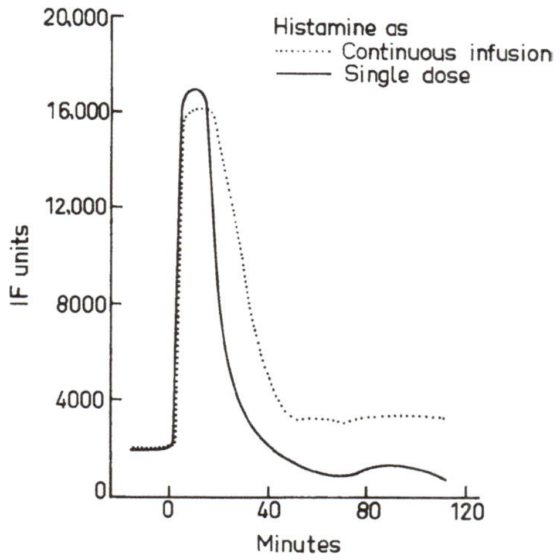

Fig. 8-3 A comparison of the effects of histamine given as a single dose and as a continuous infusion on intrinsic factor output in man (from Weir et al. [11])

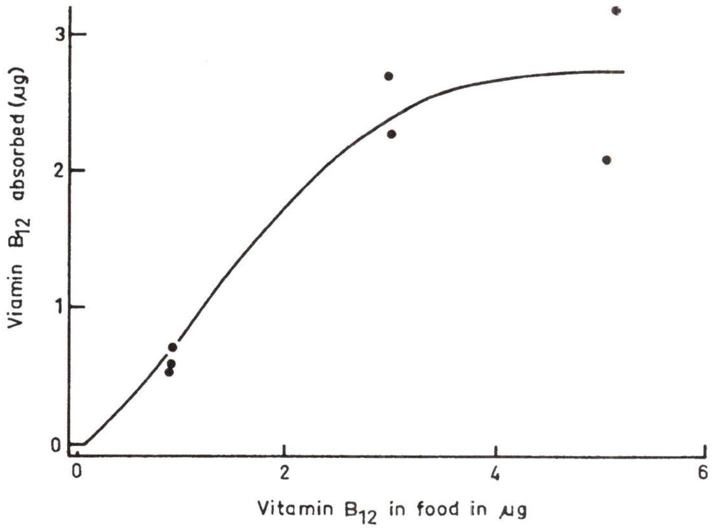

Fig. 8-4 Absorption of vitamin B_{12} from meat (from Heyssel et al. [211] © Am. J. Clin. Nutr. American Society for Clinical Nutrition)

plex resistant to digestion and to absorption by intestinal microorganisms. The IF-Cbl complex binds specifically to a receptor protein in the cell membrane of the terminal ileum (human) [20, 21]. Subsequently the cobalamin is transported by an energy dependent process accross the intestinal epithelium. Within the ileal cell it becomes attached to lysosomes and subsequently to *Transcobalamin* (TC) [218, 220], another B_{12} binder, and transported as TC-Cbl complex into the blood stream [22–24]. This absorption system serves presumably also for salvage of Cbl secreted with the bile.

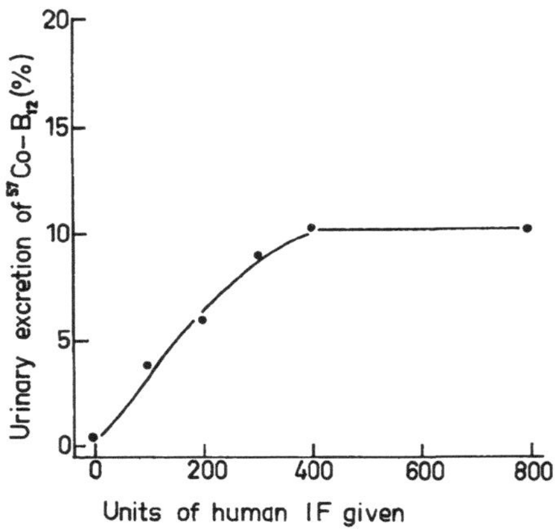

Fig. 8-5 The amount of human intrinsic factor required for maximum absorption of 1.0 μg dose of vitamin B_{12} in pernicious anaemia (from Chanarin [9])

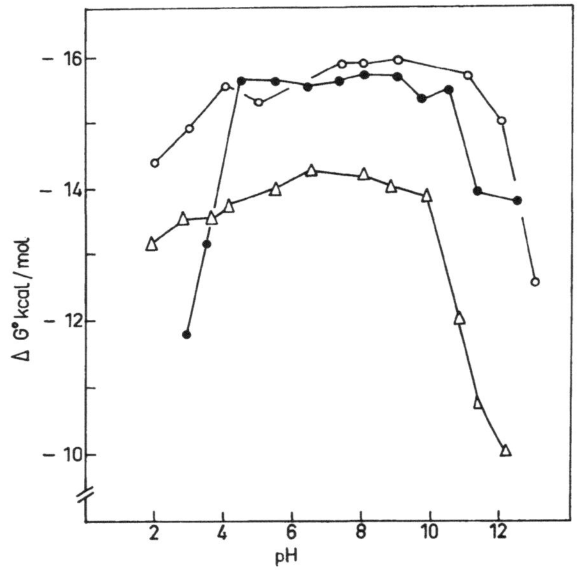

Fig. 8-6 Effect of varying pH on free energy of reaction, ΔG°, for binding of CN-Cbl to human intrinsic factor (Δ) at 22 °C, transcobalamin (●) and cobalophilin (O) at 22 °C (from Hippe and Olesen [125])

Appreciable quantities of Cbl in the digestion system remain bound to *Cobalophilin* (CP) a B_{12}-binding protein which has a much higher affinity to B_{12} than IF, particularly at acidic and neutral pH. This non-IF binding protein does not mediate the absorption of cobalamin. The binding of Cbl to IF is possible after its release from the CP complex by digestion with pancreatic enzymes [217]. Thus the greater part of Cbl is not bound to IF under physiological conditions until in the intestine [25-28] (see Fig. 8-23).

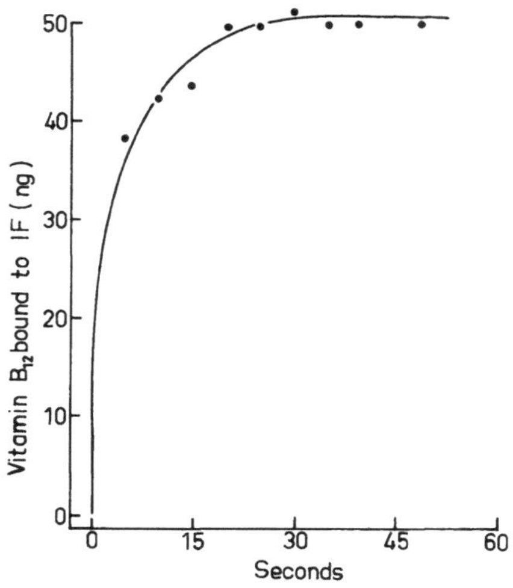

Fig. 8-7 The rate of binding of vitamin B_{12} to human intrinsic factor (from Chanarin [9])

Table 8-6 Species specificity of IF activity in the intestine (from Glass [35]).

IF Donor	IF Recipient										
	Man	Hog	Rat	Guinea pig	Hamster	Rabbit	Monkey	Chicken	Beaver	Dog	Cat
Man	+		±	+	–	±	+	–	+	+	
Hog	+	?	–	±	±		+	–	+	+	
Rat	+		+	+	+	+	+	–	+	+	
Mouse			+								
Guinea pig	–		–	+	±	–	–				
Hamster			–	±	+	–	+				
Rabbit			–	+	+	+					
Monkey	+						+				
Chicken			+					+			
Cow/calf	±								+		
Beaver			+								
Frog							–				
Horse	–									–	
Dog										–	
Cat											– ?

Key: + active, – inactive, ± active slightly or under special conditions, ? controversial.

Species specificity of Cbl absorption system. The IF mediated absorption mechanism of Cbl uptake seems to be common for many animal species. Because of the small concentration of cobalamin available in food the system requires high affinity of the binding protein for the corrinoid as well as for the IF-receptor protein in the ileum mucosa. The interaction of the IF receptor with the IF-Cbl has been found species specific; however, a number of IF proteins from various animals are complementary to the receptor of other species (see Table 8-6). Intrinsic factor preparations from the following animals are active in man [29-34]: monkey, pig, rat, rabbit, hamster, wild boar, silver fox, lion, tiger, leopard, ferret, cow. The effective binding of IF-Cbl to the ileal receptor requires appropriate conditions such as pH, calcium ions and possibly processing of IF by the digestive enzymes. These factors seem to contribute significantly to species specific cobalamin absorption.

Specificity of IF toward corrinoids. Human and animal tissues contain exclusively cobalamin (a corrinoid which contains 5,6-dimethylbenzimidazole in the nucleotide part). This is consistent with the requirement for cobamide coenzymes of the two so far known enzymes in tissues: methionine synthetase and methylmalonyl-CoA mutase (see Chapter 7). Other cobamides, particularly those containing the adenine or its derivatives instead of benzimidazole in the nucleotide, such as (Ade) AdoCba (pseudovitamin B_{12}) and (2-Me Ade) AdoCba or corrinoids without the nucleotide moiety (Cbi's) are inactive as coenzymes in these enzymes (see Chapter 7). The quantity of these corrinoids, inactive for man and animals, ingested with food derived from fermentation or produced by microorganisms in the digestive systems of man and animals, may exceed many times the amount of the biologically active Cbl (see Chapter 6). The intrinsic factor is highly selective with regard to the nucleotide moiety. It binds efficiently cobalamin and other benzimidazole containing cobamides. Cobamides containing purine in the nucleotide and cobinamides are weakly bound by IF [36-39a]. The coordinative linkage between the heterocyclic base and the central metal ion is required (base for strong binding to IF. This was shown with the metal free-, Zn-, Cu- and Rh analogues of cobalamin [39b]. Another barrier discriminating the uptake of the noncobalamin corrinoids seems to be at the IF receptor site. The binding of IF-cobamide complex to the ileal receptor may not be followed by transport of the cobamide into the blood stream. In the light of the recent discovery of non-cobalamin corrinoids in human serum, their uptake will come under more precise studies [40, 41a-d]. The early experiments on binding of cobalamin analogs to intrinsic factor were carried out with nonhomogeneous preparations of the binding protein. Gastric juice also contains cobalophilin, a far less specific but very strong B_{12} binder, which does not mediate the absorption of vitamin B_{12}. So far, the binding affinities of cobalamin analogs to IF were not specified in terms of association constants. Several biologically active cobalamin analogs have been found to compete effectively with cobalamin for the IF binding site [36-38].

Naturally occurring corrinoids (see chapters 3 and 6):
- 5-Hydroxybenzimidazolyl cobamide (5-OH-Bza) Cba
- 5-Methoxybenzimidazolyl cobamide (5-Me-O-Bza) Cba

obtained through guided biosynthesis (see Chapters 3 and 4):
- Benzimidazolylcobamide Bza-Cba
- 5(6)Methylbenzimidazolyl cobamide [5(6)-Me-Bza] Cba
- Naphtimidazolyl cobamide Nza-Cba
- 5-Amino-6-methylbenzimidazolyl cobamide (5-Am-6-Me-Bza) Cba

The IF is less selective with regard to substitutions of (or at) the β ligand [42-44]. The association constants of CN-Cbl, MeCbl, AdoCbl and aqCbl, with IF are of the same order (see section 8.1.6). However the IF mediated uptake of various forms of cobalamin by the ileal mucosa and the transfer into the blood stream proceed at different rates.

Minimal dose of IF active in pernicious anaemia test:
human IF [35] 40 μg
hog IF [45] 40–100 μg

Capacity of the Cbl absorption system. The capacities of IF and IF-receptors are limited and the percentage of the Cbl absorbed decreases drastically by application of larger, not physiological doses (Table 8-7). Vitamin B_{12} given orally in very large doses (500–1000 μg) may reach the blood stream via non-specific absorption.

Table 8-7 The absorption of a single oral dose of cyanocobalamin in normal human subjects (Data from Chanarin [9]).

Oral Dose (μg)	Amount absorbed	
	μg	%
0.1	0.08	80
0.25	0.19	76
0.5	0.35	70
1.0	0.56	56
2.0	0.92	46
5.0	1.4	28
10.0	1.6	16
20.0	1.2	6
50.0	1.5	3

Malabsorption of Cbl related to IF function:
Impaired secretion of acid and pepsin [46, 47]
stomach resection [48–55 a], vagotomy [55 b].
Presence of antibodies directed against IF [56–61]
(see Section 8.1.8).
Congenital IF deficiency [62, 63] or abnormal IF [64].
Damage of the intestinal mucosa by infection [21 a, 22].
Ionizing radiation of the intestine [21 b].
Resection of the small intestine [21 b–f].
Pancreatic insufficiency [27, 65–68 a].
Cystic fibrosis [68 b].
Chelation of divalent ions [67, 69].
Congenital IF-receptor deficiency [70–73].
Celiac disease – steatorrhea [72 a].
Gastric hypersecretion [72 b–c, 73].
Diabetics [219].
Scurvy [223].

Drug induced malabsorption:
paraaminosalicylic acid [74], biguanidines [75a, b], colchicine [76, 216], Neomycyne [77], alcohol [72a, b].

8.1.4 Methods of purification

Precipitation of IF at the isoelectric point [78, 79], precipitation with ammonium sulfate [78, 79], ultracentrifugation [80], electrophoresis [44, 80], proteolytic digestion [45, 80].

Chromatography of human IF: Chromatography on Amberlite XE-64 (CG-50) [80], calcium phosphate gel [80], CM-cellulose [81], DEAE-cellulose [81], DEAE-Sephadex [81], CM-Sephadex [81], Sephadex G-200 [81], immuno adsorption [82], affinity chromatography [83-86], affinity photorelease technique [44, 87, 88].

Chromatography of hog IF: DEAE-cellulose, DEAE-Sephadex [45, 79], affinity chromatography [89].

8.1.5 Methods of IF assay

There are a number of techniques in vivo and in vitro for IF assay. The methods apply suitable cobalt labelled cobalamins (see labelled cobalamins in chapters 3 and 5). The amount of the IF assayed in vitro is expressed usually by the capacity to bind Cbl (e.g. nanogram unit = nanogram of Cbl bound to IF). The in vivo tests measure the capacity of the whole absorption system (IF and IF-receptors) promoting the uptake of Cbl. Some of the methods are applied for medical diagnosis.

The following assays apply whole organisms, organs or tissues:

Urinary excretion test (UET) or Schilling test is based on the measurement of the labelled cobalamin in urine after oral application [90-92a-d].

Whole body counting (WBC) and faecal collection [21, 68, 93].

Stool sample test (SST) employs a double tracer technique with $^{51}CrCl_3$ and labelled Cbl [45, 47, 89].

IF-Cbl adsorption by suitable ileal loops, fragments or homogenates of the ileum [94-104].

Hematological response in pernicious anaemia – is based on the capacity of IF to promote absorption of Cbl in pernicious anaemia (U.S. Pharmacopoeia 1955) [9].

The following methods are based on measurement of Cbl bound to IF:

Enzyme inhibition assay – measures the change of cobalamin-dependent enzyme activity after introduction of a competitive Cbl binder [27, 105, 106]. The method is not selective with regard to other binders.

Radiodilution assay applying hemoglobin coated charcoal to adsorb the excess of free labelled Cbl [19, 107] (for details see Cbl assay in Chapter 5). The method does not distinguish IF from the other binders.

Cobinamide blocking assay – is an improved radiodilution assay due to addition of an excess of cobinamide to the tested crude IF preparation. Cobinamide blocks the non IF binders but not IF itself and thus permits a selective estimation of IF in presence of other B_{12} binders [25, 108, 109].

Zirconium gel adsorption [110]. The method is based on the adsorption of the cobalamin labelled binders by zirconyl phosphate gel. Free Cbl is not being adsorbed. The method is not selective with regard to other binders.

Immunological blocking assay – applies antibodies against IF or against CP [111-118]. This method allows for a selctive assay of IF in gastric juice in presence of other B_{12}-binders.

Gel-filtration assay permits a complete separation of the cobalamin labelled IF-Cbl from other labelled binders and free Cbl [119, 120].

Electrophoretic separation of the cobalamin labelled IF-Cbl from free Cbl and other binders followed by autoradiography [121-123].

Microbial absorption inhibition [124]. The method is based on the resistance of IF-Cbl to digestion and absorption by microorganisms used for Cbl assay. It is not selective with regard to other binders.

8.1.6 Physical and chemical properties

Thermodynamic data for IF-Cbl complex formation [125]

human IF:	$\Delta G°$ (26 °C)	-13.47 kcal/mol (-56.4 kJ/mol)
	$\Delta H°$	-22.73 kcal/mol (-95.17 kJ/mol)
	$\Delta S°$	-30.96 cal/K/mol (-129.63 J/K/mol)

Association constants

human IF [39, 126, 127]				
	IF–CN-Cbl [126]		$1.5 \times 10^{10} M^{-1}$	
	IF–CN-Cbl		$5 \times 10^{10} M^{-1}$	
	IF–aqCbl		$7 \times 10^{10} M^{-1}$	
	IF–CH_3-Cbl		$6 \times 10^{10} M^{-1}$	
	IF–AdoCbl		$4 \times 10^{10} M^{-1}$	
	IF–CN-Cbl [125]	38 °C	$1.4 \times 10^{9} M^{-1}$	0.1 M sodium
		34 °C	$3 \times 10^{9} M^{-1}$	phosphate pH 7.4
		26 °C	$6 \times 10^{9} M^{-1}$	in 1M NaCl
		22 °C	$13 \times 10^{9} M^{-1}$	
hog	IF [126]			
	IF–CN-Cbl		$1.5 \times 10^{10} M^{-1}$	

Isoelectric point (I. P.)

The I. P. values are scattered over a wider range than would result from the experimental error. This is due to variable contents of sialic acid in the carbohydrate part of IF. Sialic acid deficient IF's have their I. P. values shifted toward the alkaline pH. (see carbohydrate content):

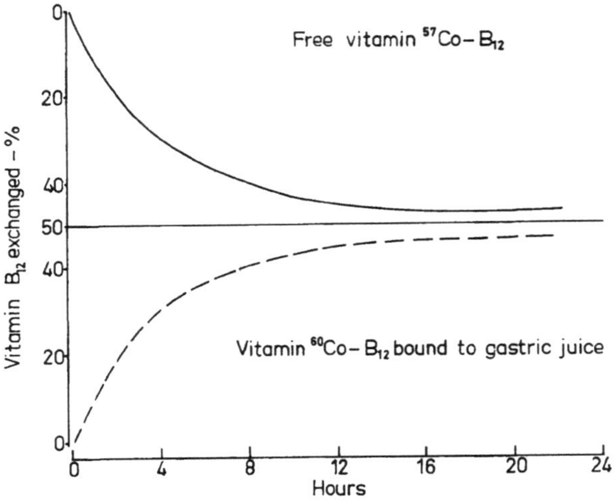

Fig. 8-8 The exchange of vitamin B_{12} in a mixture of ^{57}Co-vitamin B_{12} bound to human gastric juice and an equal amount of free ^{57}Co-vitamin B_{12} (from Donaldson and Katz [212]. Reproduced from The Journal of Clinical Investigation, **42**, 534, 1963, by copyright permission of The American Society for Clinical Investigation)

		I. P.
human:		
	IF – CN-Cbl [128–131]	4.8–5.6
	IF	4.12–4.21
	sialic acid deficient [132]	5.95–6.52
hog:		
	IF	4.35–5.40
	IF – CN-Cbl	4.15–5.20

Electrophoretic mobility:
Hog IF [133] 1.9×10^{-5} cm^2/V/sec at pH 7.8
β_2 globulin like mobility (veronal buffer pH 8.6) [81]

Radiolabelling of IF:
[^{51}Cr] IF [134], [^{125}]IF [135].

Stokes radius: [136, 137]

human	IF	3.66 nm
	IF-Cbl	3.25 nm

Molecular weight

Human IF:
60000 by gel filtration (Gräsbeck et al. 1968 [43])
45000–47700 by ultracentrifugation (Allen and Mehlman 1973 [138])
44200 computed from aminoacid and carbohydrate analysis (Allen and Mehlman 1973 [138])
59000–66000 by SDS polyacrylamide gel electrophoresis (Allen and Mehlman 1973 [138])

60000–73000 (Visuri and Gräsbeck 1973 [139])
54000 by SDS polyacrylamide gel electrophoresis (Jacobsen and Montejano 1979 [88])

Human IF-Cbl:
114000–119000 by ultracentrifugation (Gräsbeck et al. 1966 [81])

Hog IF:
55000 by ultracentrifugation (Holdsworth 1961 [133])
53000 calculated from binding capacity (Bromer and Davisson 1962 [80])
53000 by ultracentrifugation (Higley et al. 1967 [92])
53300–58600 by ultracentrifugation (Allen and Mehlman 1973 [138])
66000–75000 by SDS polyacrylamide electrophoresis (Allen and Mehlman [138])
56800 by SDS polyacrylamide electrophoresis (Jacobsen and Montejano 1979 [88])

Hog IF-Cbl:
100000 by ultracentrifugation (Higley et al. 1967 [140]).

Spectral properties of IF-Cbl

The UV-VIS spectra and the CD spectra of cobalamins bound to binding proteins differ from the corresponding spectra of the unbound Cbl forms (Table 8–8 and Fig-

Table 8–8 Absorbance bands and millimolar absorption coefficients (in brackets) of hog IF-Cbl complexes (data from Nexø [44]).

IF-Cbl complex	γ_1	D	β	α
free aqCbl	351 (26.2)	412 (3.0)	500 (8.7)	526 (9.0)
IF-aqCbl	359 (34.0)	414 (5.6)	510 (9.6)	536 (9.9)
free OH-Cbl	358 (20.2)	420 (4.1)	516 (8.2)	538 (8.7)
IF-OH-Cbl	361 (33.0)	416 (5.8)	510 (10.3)	540 (10.9)
free CN-Cbl	361 (28.0)	408 (3.4)	520 (7.7)	552 (8.2)
IF-CN-Cbl	366 (37.0)	414 (6.1)	519 (10.0)	552 (10.3)

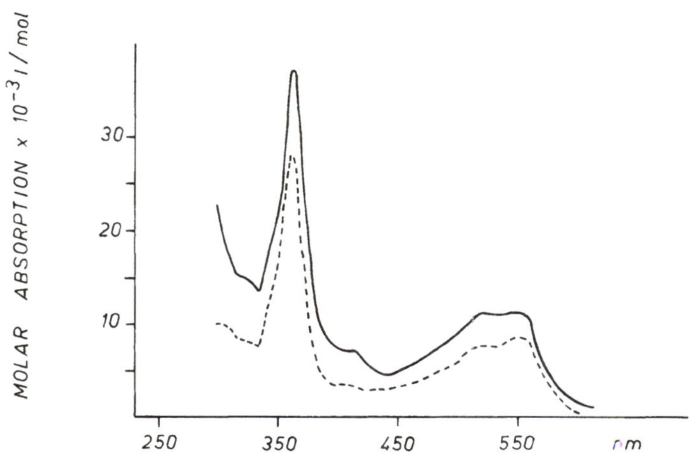

Fig. 8–9 UV-VIS spectra of IF-CN-Cbl complex (——) and free CN-Cbl (- - -) (from Nexø et al. [44])

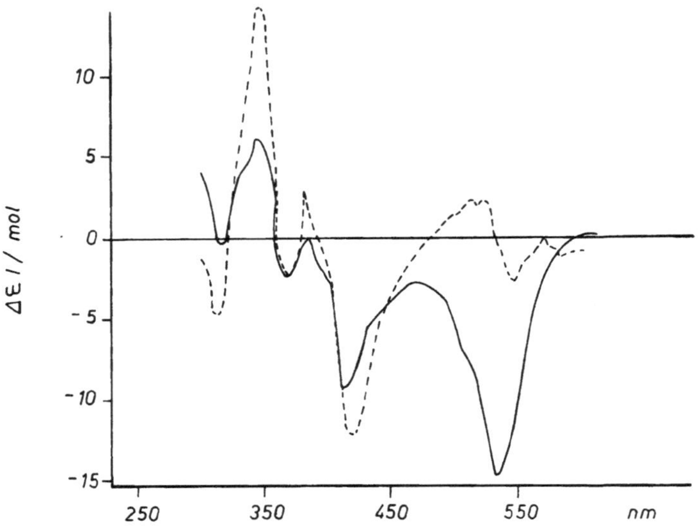

Fig. 8-10 Circular dichrosim spectra of IF-aqCbl complex (——) and free aqCbl (- - -) (from Nexø et al. [44])

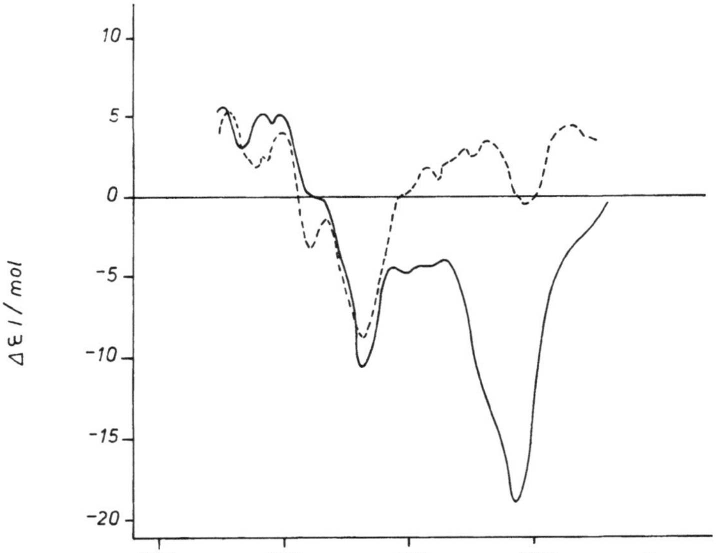

Fig. 8-11 Circular dichroism spectra of IF-OH-Cbl complex (——) and free OH-Cbl (- - -) (from Nexø et al. [44])

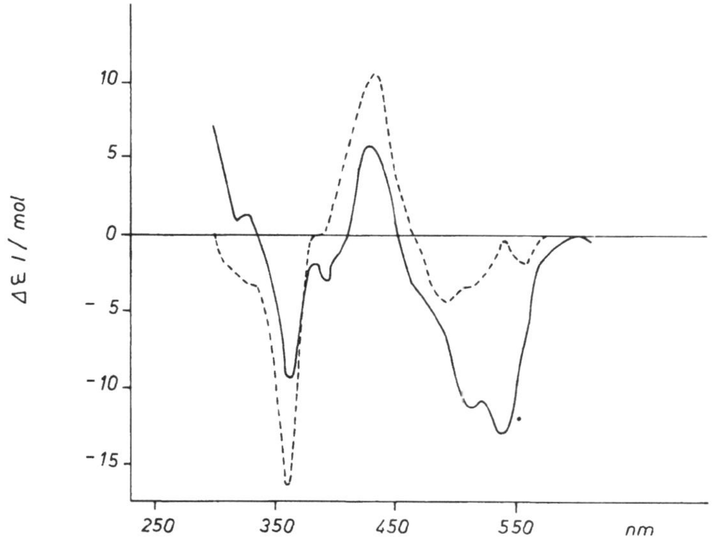

Fig. 8-12 Circular dichroism spectra of IF-CN-Cbl complex (——) and free CN-Cbl (- - -) (from Nexø et al. [44])

ures 8-9, 8-10, 8-11, 8-12). The evident spectral differences of UV spectrum, particularly in the region of the γ band (350-365 nm), are characteristic in this respect. They may result mainly from substitutions of α or β ligands at the cobalt ion in the corrin ring or the change of the Co-valency (see chapter 3). The contribution of the glycoprotein part to the UV-VIS spectra is insignificant for the wavelengths > 300 nm.

Absorption coefficient of IF at 278 nm $E_{1\,cm}^{1\,\%} = 9.1-9.5$ [81].

Effect of temperature and pH on IF stability

IF is resistant to freeze drying [35].

Heating of human IF at neutral pH for 1 hour at 55 °C results in 50% loss of activity [141].

Rat IF-Cbl loses 90% of activity after heating at 50 °C for 30 minutes [142].

IF is stable at neutral pH in room temperature or frozen [35].

IF-Cbl dissociates reversibly at pH > 12.5 [143] (see Fig. 8-6).

IF dried at temperature > 0 °C with Sephadex G-50 may be stored at room temperature for several years (see Cbl assay in chapter 5).

Heat inactivation at neutral pH of IF-Cbl does not cause release of free Cbl [144].

Effect of chemicals and solvents

IF is destroyed by treatments with: iodine and acetic anhydride [125], trichloroacetic acid [145] fluorodinitrobenzene [146], ethanol (conc. 70-95%) at room temperature cause considerable loss of human IF activity [147], 8 M urea for 24 hours [148].

Human and hog IF resists treatments with: saturated ammonium sulfate pH 4.5 [148], ethanol(50-80%) at low temperature [147], nitrous acid [125], p-chloro-mercuribenzoate [149].

Table 8-9 Effect of some specific group blocking agents on vitamin B_{12} binding and IF activity. (compiled by Glass [35] from [146, 150–154]).

Blocking agents or procedures	Conditions of the reactions	Material tested (substrate)	Effect on B_{12} binding	Effect on IF activity
Fluorodinitrobenzene	32 mg/ml at pH 8.6 for 3½ h	DHGJ[a], HIFC[b]	66% inactivation	
Fluorodinitrobenzene	30 mg/ml pH 9.0, 3 h room temp.	RGMH[c]	100% inactivation	100% inactivation[d]
Alcoholic iodine (I_2)	Final concentration 0.025 M at pH 8.6	DHGJ & HIFC	96% inactivation	100% inactivation[e]
Alcoholic iodine (I_2)	Final concentration 0.0005 M, pH 8.6	DHGJ & HIFC	very slight inactivation	
Alcoholic iodine (I_2)	Final concentration 0.008 M 4 mg/ml, 0 °C at pH 9.0 for 5 h	RGMH		100% inactivation[d]
U.V. light	280 nM, 4×10^{19} quanta/16 mg O_2 supplied for 24 h at 4 °C	HIFC	25% inactivation	
p-chloromercuribenzoate	1.7 mg/ml at pH 7.4 + 10 mg substrate for 24 h at 4 °C	HIFC RGMH	None < 10% inactivation	 None[d]
Acetic acid anhydride	0.12 ml/ml in half saturated Na-acetate at 0 °C for 1 h	HIFC	None	
$NaNO_2$	1 M at pH 4.6, 0 °C for up to 82 h	HIFC	None	
di-isopropylfluorophosphate	0.25 mg or 0.5 mg + 10 mg substrate/ml for 3 h	RGMH	< 10% inactivation	None[d]
Fluorodinitrobenzene	0.001 M, at pH 8, 2 h at 20 °C	HIFC	100% inactivation	
Iodine (I_2)	0.01 M I_2 in 0.01 M KI, 2 h at 20 °C	HIFC	100% inactivation	
p-chloromercuribenzoate	0.001 M, 2 h at 20 °C	HIFC	20% inactivation	
Formaldehyde	0.1 M, 2 h at 20 °C, at pH 8.0 0.1 M, 2 h at 20 °C, at pH 11.0	HIFC HIFC	30% inactivation 42% inactivation	
Iodine, sol. in ethanol	3.8 mM	DHGJ		90% inactivation[f]
Iodine, sol. in water	3.8 mM	DHGJ		None
Acetic acid anhydride	(as above)	DHGJ		100% inactivation
Ethanol		DHGJ		55% inactivation
Reduction and alkylation	by 1,4-dithiothreitol in TRIS-HCl buffer, pH 8.6 and by iodoacetamid		Difficult to interpret	

Breaking of alkalilabile linkages	by 0.5 M NaOH, at 4 °C for 72 h	No distinguishable effect
Blockage of histidine residues	by alkylation at pH 5.5 by iodoacetic acid, 57 °C, 6 h	None
Blockage of ε-NH_2 groups of lysine	by succinic anhydride	None
Esterification of free carboxyl groups	in the presence of CMC[g] and diethylammonium chloride	None
Blockage of guanidinic groups	by 1,2-cyclohexanedione at 10 X concentration of arginine at 25 °C for 18 h	100% inactivation

a DHGJ-depesinized human gastric juice
b HIFC-hog intrinsic factor concentrate
c RGMH-rat gastric musoca homogenate
d on rats by faecal excretion test
e on PA patients using urinary excretion test
f on guinea pig intestinal musoca homogenate
g CMC-1-cyclohexyl-3(2-morpholinoethyh-(4)-ethyl) carbodiimide metho-ptoluene sulfonate

Table 8-10 Composition of the carbohydrate moiety and nitrogen content of IF preparations from human and hog stomach (in % wt/wt) (data compiled by Glass [35]).

	IF Preparations								
	From human stomach			From hog stomach					
	Gräsbeck et al. 1966 [81]		Faillard et al. 1962 [159, 160]	Latner et al. 1956 [161]	Glass et al. 1959 [162]	Holdsworth 1961 [146]	Faillard et al. 1962 [159, 160]	Highley et al. 1967 [140]	
	Complex S ($IF-B_{12}$)	Complex I ($IF-B_{12}$)						IF	$IF-B_{12}$
Galactose Mannose	6.0	4.3	8.0–9.0		8.9	7.2 3.2		11.8 3.5	10.4 5.1
Fucose	0.9	0.7			2.4	3.0		4.0	3.8
Galactosamine Glucosamine	4.1	4.4	5.0–6.0	5.8	8.0	2.1[x] 5.0[x]	5.0–6.0	3.6 9.1	2.7 9.1
Sialic acid	1.7	1.7	1.5–1.7		0.9		2.0–2.5	3.7	3.2
Reducing substances						20.0		19.2	19.3
Nitrogen	12.1		13.0	11.2		10.3	11.0–12.0		

x Most or all acetylated.

Effect of group blocking agents

No effect on binding of Cbl to IF is produced by the following agents: Histidine alkylation, lysine-NH_2 groups succinylation, glutamic and aspartic acid carboxyl groups esterification, and SH group blocking [150] (see Table 8-9). The Cbl binding is affected by the following treatments: reduction of di-sulfur bridges followed by alkilation with iodoacetamide [150], blocking guanidine and arginine groups with 1,2-cyclohexanedione [150], pH > 12.5 [120] (see (Table 8-9).

Decomposition of IF by enzymes

Human IF is slowly destroyed by pepsin [155].

Acid gastric juice at 37 °C decreases the binding capacity of IF by 50% after 2 hours [141].

Pancreatic enzymes in the intestine split off carbohydrate residues [130].

Neuramidase inactivates IF [156].

IF-Cbl is resistant toward proteolitic enzymes in vivo [25, 130, 143, 157].

IF-Cbl is digested by neuramidase (sialic acid is being split off) but the activity is retained [156, 158].

Carbohydrate content

The IF preparations from man and hog purified by various authors contained significant but variable amounts of carbohydrates [81, 82, 140, 146, 159-161]. This fact seems to be the major cause of the difficulties in precise determination of physical parameters (see I.P. values and molecular weight of IF preparations in section 8.1.6). Among carbohydrate components sialic acid affects significantly the charge of the IF molecule. Depending on the source: mucosa cells or gastric juice, the IF preparations may contain, besides complete IF molecules, also juvenile populations or partially decomposed forms respectively. One of the major roles of carbohydrates in IF is to prevent gastric and microbial decomposition. Table 8-10 shows the carbohydrate contents in IF preparations from man and hog according to various in vestigators [35].

Aminoacid composition

Human IF has been sequenced to a total of 84 N-terminal aminoacid residues out of about 340 [131, 163]. The N-terminal aminoacid of human IF is serine [164]. Human IF and hog IF have similar aminoacid sequences [26, 131, 164-167].

8.1.7 Isoproteins of IF

Five isoproteins of hog IF were resolved by the isoelectric focusing method [81, 132, 168-169]. Variable content of carbohydrates (sialic acid) seems to be responsible for this microheterogeneity (see carbohydrate content Table 8-10).

8.1.8 IF antibodies (IF-Abs)

Source:

Sera of rabbit and goat immunized with human-, hog- or rat gastric juices or gastric mucosa preparations [170-180].

Table 8-11 Aminoacid composition of IF preparations from hog gastric mucosa. (Data compiled by Glass [35]).

Amino Acids	IF preparations				
	Latner et al. (1956) [161]	Holdsworth (1961) [146]	Highley et al. (1967) [140]		Inada (1970) [167]
			IF	IF + B_{12}	
	in % protein, dry wt.				
Aspartic acid	+	9.9	5.8	6.5	9.2
Threonine	+	4.7	4.1	4.1	5.8
Serine	+	5.2	3.5	3.6	5.1
Proline	+	4.3	3.7	2.7	4.8
Glutamic acid	+	9.4	6.7	6.7	10.4
Glycine	+	2.7	1.7	1.6	2.7
Alanine	+	4.3	2.4	2.5	4.5
Valine	+	5.6	3.1	3.2	4.7
Cystine	+	1.9	small amount		0.5
Methionine	+	0.9	1.0	0.8	1.7
Isoleucine	+	4.5	2.8	3.1	3.7
Leucine	+	8.2	4.7	5.4	7.5
Tyrosine	2.4	1.5	2.3	2.6	3.2
Phenylalanine	+	2.7	2.2	2.7	4.0
Lysine	+	4.5	2.3	2.9	5.4
Histidine		2.4	1.4	1.4	2.4
Arginine	+	3.0	2.1	2.4	3.5
Tryptophan	1.3	1.9	1.1	1.1	1.6
Ammonia		2.3			1.8
Total		77.6	50.9	53.3	80.9

Sera of pernicious aneamia patients contain auto-IF-Abs [88, 148, 181-183]. (see also section 9.3.1.)

Treatment of Cbl deficient patients with hog IF (per os) induces auto-IF-Abs [88, 182-185].

Types of IF-Abs

Two types of auto-IF-Abs distinguished by binding to different antigenic reactive sites on the IF molecules were found (see Fig. 9-20).

Type I — synonyms: "blocking antibodies"
"combining antibodies"

Type I IF-Abs prevent the binding of Cbl to IF (most common type) [150, 186].

Type II — synonyms: "complex binding antibodies"
"precipitating antibodies"

Type II IF-Abs do not prevent the binding of Cbl to IF. These antibodies have their antigenic combining site on IF at some distance from the Cbl-binding site. The binding of type II IF-Abs to IF prevents the Cbl blocking effect of type I IF-Abs [150, 183, 186-188].

Table 8-12 Classification of vitamin B_{12} binders of human and animals by immunological properties.

antigen	Antiserum		
	Intrinsic Factor	Transcobalamin	Cobalophilin
Intrinsic Factor	+	–	–
Transcobalamin	–	+	–
Cobalophilin	–	–	+

Structure of IF-Abs

Type I and type II IF-Abs have no common origin as concluded from heavy and light chain composition [186].
Ig G – predominant class of IF-Abs in serum [189–191].
Ig A type of IF-Abs was found in gastric juice [192].
Full immunological cross-reactivity between human and hog IF and IF-Abs [5, 44].

Methods of IF-Abs assay

These methods are essentially similar to those for IF assay (see section 8.1.5).

Combinations of the following techniques are applied:

Type I Zirconium gel assay, charcoal radioimmunoassay, gel filtration, neutralization of biological activity in vivo and in vitro [191, 193–194a].

Type II Electrophoretic methods, salting out procedures, zirconium gel assay, gel filtration, inhibition of biological activity in vivo and in vitro [150, 187–188].

8.1.9 IF-receptor

The transport of IF bound Cbl across the intestinal epithelium is mediated by IF-receptor localized in the ileal cell membrane. Injury of the receptor site by ionizing radiation [21a, b] (see section 8.1.3.), giardiasis [21c] or resection of the terminal ileum [21d–f] affects the receptor mediated absorption of Cbl.

Source and site of action

Human: terminal part of ileum [21d–f, 195] (Fig. 8–13)
Hog: intestine [195–199]
Guinea pig: intestine (small bowel) [200].

Purification

Solubilization of the membrane bound receptor with the detergent Triton X-100 [200–202] or digestion with papain [220].
Purification by conventional means and affinity chromatography on IF-Cbl-Sepharose [158, 196–198].

Assay

Cobalamin labelled IF-Cbl is used for binding to the receptor. The quantity of the receptor is expressed in terms of nanograms of Cbl bound. The IF-Cbl-receptor complex is separated from the free IF-Cbl by gel filtration on Sephadex G-200 column [195–200, 203].

Association constant

Guinea pig receptor [131, 200, 202]: $1.2 \times 10^9 M^{-1}$

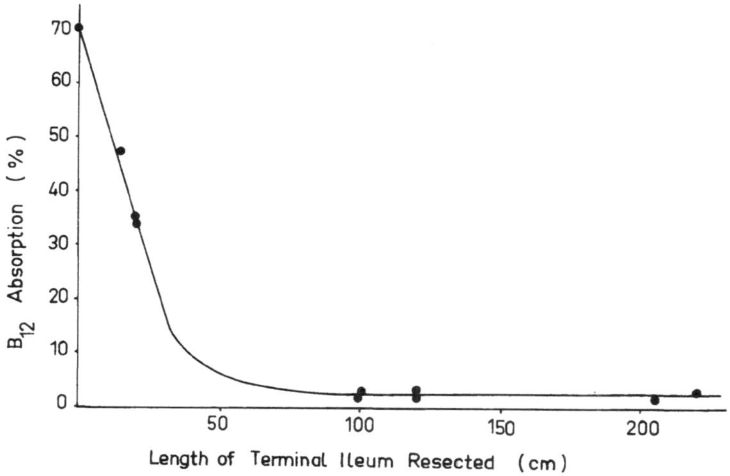

Fig. 8-13 Relation of vitamin B_{12} absorption to the length of terminal ileum resected (from Morishita and Uchino [21])

Binding features

Calcium ions are required for binding of IF-Cbl to the receptor [69, 204, 205 a, b]. IF dissociates from the receptor at pH < 5 [102, 195, 205-206].

IF-Cbl has greater affinity for the receptor than free IF [207]. The capacity of guinea pig intestine to bind IF-Cbl is controlled by hormones (pentagastrin, pregnancy) [158, 171, 200, 208-210].

Molecular weight

SDS polyacrylamide gel electrophoresis of the alkylated receptor protein

two protein subunits [195, 214]	α	β
human	90000	140000
hog	70000	130000

[^{125}I]α-subunit binds Cbl and forms complexes with goat-anti-intrinsic factor antibody and rabbit anti-cobalophilin antibody (gel filtration) [26].

8.1.10 Vitamin B_{12} releasing enzyme

The enzyme causes the release of vitamin B_{12} from IF-B_{12} complex presumably by splitting the IF. It is localized in the ileal brush border. Its function is probably to free the Cbl from IF in the course of transport of the IF-Cbl across the ileal cell membrane. So far only the enzyme from rat has been described [213].

Purification

Ammonium sulphate precipitation, gel filtration and Con. A-Sepharose 4B affinity chromatography

Molecular features

M. W. 25000 (SDS-polyacrylamide gel electrophoresis) I. P. 7.52.

Kinetic data

Calcium ions (10 mM) and ATP (2 mM) required for maximal activity (5.7 fold increase)

pH optimum 7.5

K_m 128.2×10^{-12}M

V_{max} 117.6 pgh^{-1} per 100 μg of the enzyme

p-chloromercuribenzoate and heating inactivate the enzyme.

The enzyme does not hydrolyse other proteins.

Carbohydrate contents (%)

total sugars (7.1); glucose (1.5), galactose (1.4) mannose (1.8), glucosamine (1.5), galactosamine (0.9) sialic acid (traces).

References

[1] Fisher, J. M., Taylor, K. E., The intracellular localization of Castle's intrinsic factor by an immunofluorescent technique using autoantibodies, Immunol. **16**, 779-784, 1969.

[2] Jacob, E., Glass, G. B. J., Localization of intrinsic factor and complement, fixing intrinsic factor- intrinsic factor antibody complex in parietal cell of man, Clin. Exptl. Immunol. **8**, 517-527, 1971.

[3] Inada, M., Glass, G. B. J., Effect of prolonged administration of homologous and heterologous intrinsic factor antibodies to rats, 14th Ann. Mtng. Am. Soc. Haematol., San Francisco, Calif., 1971 (Abstr.).

[4] Hoedemaeker, P. J., Abels, J., Watchers, J. J., Arends, A., Nieweg, H. O., Investigations about the site of production of Castle's gastric intrinsic factor, Lab. Invest. **13**, 1394-1399, 1964.

[5] Hoedemaeker, P. J., Investigations on site of production of Castle's gastric intrinsic factor, Thesis Univ. Groningen, 1965.

[6] Hoedemaeker, P. J., Abels, J., Watchers, J. J., Arends, H., Nieweg, H. O., Further investigations about the site of production of Castle's gastric intrinsic factor, Lab. Invest. **15**, 1163-1177, 1966.

[7] Sharpe, P. C., Horton, M. A., Evidence for the chief cell as the source of intrinsic factor secretion in the rat, in Vitamin B_{12}, Eds. B. Zagalak, W. Friedrich, W. de Gruyter, Berlin 1979, pp. 921-927.

[8] Finkler, A. E., Hall, C. A., Nature of the relationship between vitamin B_{12} binding and cell uptake, Arch. Biochem. Biophys. **120**, 79-85, 1967.

[9] Chanarin, J., The megaloblastic anaemias, Blackwell Scientific Publication, Oxford and Edinburgh 1969

[10] Ardeman, S., Chanarin, J., and Doyle, J. C., Studies on secretion of gastric intrinsic factor in man, Br. Med. J. **2**, 600-603, 1964.

[11] Weir, D. G., Temperley, I. J., and Collery, D., Intrinsic factor secretion in response to continuous histamine infusion, Gastroenterology **52**, 23, 1967.

[12] Dotevall, G., Walan, A., and Weinfeld, A., Gastric secretion of acid and intrinsic factor during histamine infusion in 14 cases of duodenal ulcer, Scand. J. Haemat. **4**, 164, 1967.

[13] Irvine, W. J., Cullen, D. R., Scarth, L., Simpson, J. D., Intrinsic-factor secretion assessed by direct radioimmunoassay and totalbody counting in patients with achlorhydria and in acid secretors, Lancet **2**, 184-188, 1968.

[14] Bitsch, V., Christiansen, P. M., Faber, V., Rodbro, P., Gastric secretory patterns before and after vagatomy, Lancet **1**, 1288-1291, 1966.

[15] Ardeman, S., Studies on gastric intrinsic factor, Proc. R. Soc. Med. **58**, 916, 1965.

[16] Irvine, W. J., Davies, S. H., Haynes, R. C., and Scarth, L., Secretion of intrinsic factor in response to histamine and to gastrin in the diagnosis of Addisomian pernicious anaemia, Lancet **2**, 397, 1965.

[17] Schipperijn, A. J. M., Een bepaling van Castle's intrinsic factor in menselijk maagsap in vitro, M. D. Thesis, Groningen 1965.

[18] Wangel, A. G., Callender, S. T., Effect of gastrin I and II on the secretion of intrinsic factor, Br. Med. J. **2**, 908, 1965.

[19] Røbro, P., Cristiansen, P. M., Schwartz, M., Intrinsic factor secretion in stomach diseases, Lancet **2**, 1200, 1965.

[20] Katz, J. H., Dimase, J., Donalson, R. M., Simultaneous administration of gastric juice bound and free radioactive cyanocobalamin, J. Lab. Clin. Med. **61**, 266, 1963.

[21 a] Gendre, J. P., Cosnes, J., Le Ouintrec, Y., Late radiation injury to the small intestine: 1. Malabsorption in relation to anatomical damage, Gastroenterol. Clin. Biol. **7**, 664-670, 1983.

[21 b] Kin, A. C., and Lant, B., Vitamin B_{12} deficiency after irradiation for bladder carcinoma, J. Urol. **131**, 888-890, 1984.

[21 c] Desai, H. G., and Kalro, R. H., Functional and histological abnormalities of small bowel mucosa in giardiasis, Indian J. Med. Res. **77**, 239-244, 1983.

[21 d] Morishita, R., Uchino, H., Vitamin B_{12} absorption studies with plastic whole counter in patients with gastroenterlogical operation, in Vitamin B_{12}, Eds. B. Zagalak, W. Friedrich, W. de Gruyter, Berlin 1979, pp. 945-948.

[21 e] Nillsson, L. O., Myrvold, H. E., Swolin, B., and Ojerskog, B., Vitamin B_{12} in plasma in patients with continent ileostomy and long observation time, Scand. J. Gastroenterol. **19**, 369-374, 1984.

[21 f] Bambach, Ch. P., and Hill, G. L., Long-term nutritional effects of extensive resection of the small intestine, Anst. N. Z., J. Surg. **52**, 500-506, 1982.

[22] Allen, R. H., Br. J. Haemat. **33**, 161-169, 1976.

[23] Chanarin, I., Muir, M., Hughes, A., Hoffbrand, A. V., Evidence for intestinal origin of transcobalamin II during vitamin B_{12} absorption, Brit. Med. J. **1**, 1453-1455, 1978.

[24] Burman, J. F., Malleson, P. N., Sourial, N. A., Mollin, D. L., TC II Deficiency: Observations with deoxyuridine suppression test, in Vitamin B_{12}, Eds. B. Zagalak, W. Friedrich, W. de Gruyter, Berlin 1979, pp. 889-893.

[25] Allen, R. H., Seetharam, B., Podell, E., Alpers, D. H., Effect of proteolytic enzymes on the binding of cobalamin to R protein and intrinsic factor, J. Clin. Invest. **61**, 47-54, 1978.

[26] Jacob, E., lnd O'Brien, H. A. W., J. Clin. Pathol. **25**, 320, 1972.

[27] Veeger, W., Abels, J., Hellemans, N., and Nieweg, H. O., Effect of sodium bicarbonate on the absorption of vitamin B_{12} and fat in pancreatic insufficiency, N. Engl. J. Med. **267**, 1341, 1962.

[28] LeBauer, E., Smith, K., and Greenberger, N. J., Pancreatic insufficiency and vitamin B_{12} malabsorption. Arch. Intern. Med. **122**, 423, 1965.

[29] Schwartz, M., Lous, P., and Meulengracht, E., Absorption of vitamin B_{12} in pernicious anaemia. Defective absorption induced by prolonged oral treatment, Lancet **2**, 1200, 1958.

[30] Abels, J., Intrinsic factor van Castle on resorptie van Vitamin B_{12}, Thesis, University of Groningen, 1959.

[31] Hippe, E., and Schwartz, M., Scand. J. Haematol. **8**, 276, 1971.

[32] Wilkinson, J. F., Megalocytic anaemias, Lancet **1**, 249, 1949.

[33] Wilkinson, J. F., Lancet **1**, 291, 1949.

[34] Wilkinson, J. F., Lancet **1**, 336, 1949.

[35] Glass, G. B. J., Gastric intrinsic factor and other vitamin B_{12} binders, Georg Thieme, Stuttgart 1974.

[36] Gottlieb, C. W., Retief, F. P., Herbert, V., Blockade of vitamin B_{12} binding sites in gastric juice, serum and saliva by analogues and derivatives of vitamin B_{12} and by antibody to intrinsic factor, Biochim. Biophys. Acta **141**, 560-572, 1967.

[37] Bunge, M. B., Schilling, R. F., Intrinsic factor studies. VI. competition for vitamin B_{12} binding sites offered by analogues of the vitamin, Proc. Soc. Exptl. Biol. Med. **96**, 587-592, 1971.

[38] Hippe, E., Haber, E., Olesen, H., Nature of vitamin B_{12} binding. II. Steric orientation of vitamin B_{12} on binding and number of combining sites of human intrinsic factor and the transcobalamins, Biochim. Biophys. Acta **243**, 75-82, 1971.

[39 a] Kolhouse, J. F., Allen, R. H., Absorption, plasma transport and cellular retention of cobalamin analogues in the rabbit, J. Clin. Invest. **60**, 1381-1392, 1977.

[39 b] Elsenhans, B., Rosenberg, H. I., Influence of metal substitution on vitamin B_{12} binding to human intrinsic factor, transcobalamin I and transcobalamin II, Biochemistry **23**, 805-808, 1984.

[40] Herbert, V., Landan, L., Bash, R., Grosberg, S., and Colman, N., Ability of megadoses of vitamin C to destroy vitamin B_{12} and cobinamide and to reduce absorption of vitamin B_{12}, in Vitamin B_{12}, Eds. B. Zagalak, W. Friedrich, W. de Gruyter, Berlin 1979, pp. 1069-1077.

[41 a] Kolhouse, J. F., Kondo, H., Allen, N. C., Podell, E., Allen, R. H., Cobalamin analogues are present in human plasma and can mask cobalamin deficiency because current radioisotope dilution assays are not specific for

true cobalamin, New Engl. J. Med. **299**, 785-792, 1978.

[41b] Kondo, H., Kolhouse, J.F., and Allen, R.H., Presence of cobalamin analogues in animal tissues, Proc. Natl. Acad. Sci. USA **77**, 817-821, 1980.

[41c] Muir, M., and Chanarin, I., Separation of cobalamin analogs in human sera binding to intrinsic factor and to R-type vitamin B_{12} binders, Br. J. Haematol. **54**, 613-622, 1983.

[41d] Kanazawa, S., and Herbert, V., Noncobalamin vitamin B_{12} analogs in human red cells, liver and brain, Am. J. Clin. Nutr. **37**, 774-777, 1983.

[42] Gräsbeck, R., Intrinsic factor and the transcobalamins with reflections on the general function and evolution of soluble transport proteins, Scand. J. Clin. Lab. Invest. Suppl. **95**, 7-18, 1967.

[43] Gräsbeck, R., Nature and action of intrinsic factor, XII. Congr. Intern. Soc. Hematol., Plenary Session Papers, New York, 124-132, 1968.

[44] Nexø, E., Olesen, H., Hansen, M.R., Spectral studies on hog intrinsic factor and hog man intrinsic factor, in Vitamin B_{12}, Eds. B. Zagalak, W. Friedrich, W. de Gruyter, Berlin 1979, pp. 895-903.

[45] Ellenbogen, L., and Highley, D.R., Hog intrinsic factor I isolation of vitamin B_{12} binding fractions from hog pylorus, J. Biol. Chem. **242**, 1004, 1967.

[46] Schade, S.G., and Schilling, R.F., Effect of pepsin on the absorption of food vitamin B_{12} and iron, Am. J. Clin. Nutr. **20**, 636, 1967.

[47] Reizenstein, R., Effect of digestive enzymes on bound vitamin B_{12}, Acta Med. Scand. **165**, 481, 1959.

[48] Mac Donald, R.M., Indelfinger, F.J., and Belding, H.W., Late effects of total gastrectomy in man, N. Engl. J. Med. **237**, 887, 1974.

[49] Castle, W.B., Current concepts of pernicious anaemia, Am. J. Med. **48**, 541, 1970.

[50] Te Velde, K., Hoedemaeker, P.J., Anders, G.J.P.A., Arends, A., and Nieweg, H.O., A comparative morphological and functional study of gastric mucosa with and without antibodies, Gastroenterology **51**, 138, 1966.

[51] Glass, G.B.J., Speer, F.D., Nieburgs, H.E., Ishimori, A., Jones, L.E., Baker, H., Schwartz, S., and Smith, R., Gastric atrophic gastritis and gastric secretory failure, Gastroenterology **39**, 429, 1960.

[52] Lowenstein, F., Absorption of cobalt^{60}-labeled vitamin B_{12} after subtotal gastrectomy, Blood **13**, 339, 1958.

[53] Stelsberg, H., Sigbjorn, T., Stomach cancer following gastric surgery for benign conditions. Lancet **2**, 1175, 1971.

[54] Deller, D.J., and Witte, L.J., Changes in the blood after partial gastrectomy with special reference to vitamin B_{12}, Q. J. Med. **31**, 71, 1962.

[55a] Hines, J.D., Hoffbrand, A.V., and Malin, D.L., The hematologic complications following partial gastrectomy, a study of 292 patients, Am. J. Med. **43**, 555, 1967.

[55b] Paimela, H., Ahonen, J., Hockerstedt, K., Scheinin, T.M., Serum vit. B_{12} concentration after proximal gastric vagotomy, Scand. J. Gastroenterol. **19**, 3, 445-448, 1984.

[56] Carmel, R., Rosenberg, A., Lan, K.S., Streiff, R.R., and Herbert, V., Vitamin B_{12} uptake in human small bowel homogenate and its enhancement by intrinsic factor, Gastroenterology **56**, 548, 1969.

[57] Hines, J.D., Rosenberg, A., and Harris, J., Intrinsic factor mediated radio B_{12} uptake in sequential incubation studies using everted sacs of guinea pig small intestine; evidence that IF is not absorbed into the intestinal cell, Proc. Soc. Exp. Biol. Med. **129**, 653, 1968.

[58] Schade, S.C., Feich, P., Muckerheide, M., and Schilling, R.F., Occurrence in gastric juice of antibody to a complex of intrinsic factor and vitamin B_{12}, N. Engl. J. Med. **275**, 528, 1966.

[59] Goldberg, L.G., Shuster, J., Stuckey, M., and Fudenberg, H.H., Secretony immunoglobulin. A. Autoantibody in gastric juice, Science **160**, 1240, 1968.

[60] Witts, W.J., The stomach and anaemia, Athlone Press. London 1966.

[61] Jeffries, G.H., Pernicious anaemia and atrophic gastritis, Immunological Diseases, Vol. II, Ed. Samter, M., Little Brown, Boston, Massachusetts 1971, p. 1228.

[62] McIntyre, O.R., Sullivan, L.W., Jeffries, G.H., and Silver, R.H., Pernicious anaemia in childhood, N. Engl. J. Med. **272**, 981, 1965.

[63] Spurting, C.L., Sacks, M.S., and Jiji, R.M., Juvenile pernicious anaemia, N. Engl. Med. **271**, 995, 1964.

[64] Katz, M., Lee, S.K., and Cooper, B.A., Vitamin B_{12} malabsorption due to a biologically inert intrinsic factor, N. Engl. J. Med. **287**, 425, 1972.

[65] McIntyre, P.A., Sachs, M.V., Krevans, J.R.,

and Conley, C.L., Pathogenesis and treatment of macrocytic anaemia. Information obtained with radioactive vitamin B_{12}, Arch. Intern. Med. **98**, 541, 1956.

[66] Toskes, P.P., Hansell, J., and Deren, J.J., Vitamin B_{12} malabsorption in chronic pancreatic insufficiency. Studies suggesting the presence of a pancreatic intrinsic factor, N. Engl. J. Med. **284**, 627, 1971.

[67] Herbert, V., Studies on the role of intrinsic factor in vitamin B_{12} absorption, transport and storage, Am. J. Clin. Nutr. **7**, 433, 1959.

[68a] Morishita, R., Fujii, M., Yamamoto, T., Wakisaka, G., Matsumoto, Y., and Ono, H., Effect of pancreatin on Vitamin B_{12} malabsorption in patients with total pancreatectomy, Digestion **11**, 240-248, 1974.

[68b] Lindemans, J., Neijens, H.J., Kerrebijn, K.F., and Abels, J., Vitamin B_{12} absorption in cystic fibrosis, Acta Paediatr. Scand. **73**, 537-540, 1984.

[69] Gräsbeck, R., and Nyberg, W., Inhibition of radiovitamin B_{12} absorption by ethylenediamine tetraacetate (EDTA) and its reversal by calcium ions, Scand. J. Clin. Lab. Invest. **10**, 448, 1958.

[70] Imerslund, O., Idiopathic chronic megaloblastic anaemia in children, Acta Pediatr. Suppl. **119**, 1960.

[71] Gräsbeck, R., Gordin, R., Kantero, I., and Kuhlback, B., Selective vitamin B_{12} malabsorption and proteinuria in young people, Acta Med. Scand. **167**, 289, 1960.

[72a] Schimoda, S., Saunders, D.R., and Rubin, C.E., The Zollinger-Ellison syndrome with steatorrhea. The mechanism of fat and vitamin B_{12} malabsorption, Gastroenterology **55**, 705, 1968.

[72b] Bandera, R., Rovetta, P., Impairment of cerebrospinal fluid circulation in alcoholism as studied by isotopic cisternography, Subst. Alcohol Actions Misuse **2**, 259-266, 1981.

[72c] Anderoen, K.J., Schjonsby, H., Skagen, D.W., Haga, J., Enzyme activities in jejunol biopsy samples from patients with adult celiac disease with and without steatorrhea, Scand. J. Gastroenterol. **18**, 365-368, 1983.

[73] Shum, H.Y., O'Neill, B.J., and Streeter, A.M., Vitamin B_{12} absorption and the Zollinger-Ellison syndrome, Lancet **1**, 1303, 1971.

[74] Heinivaara, O., and Pavla, I.P., Malabsorption and deficiency of vitamin B_{12} caused by treatment with para-aminosalicylic acid, Acta Med. Scand. **177**, 337, 1965.

[75a] Tomkin, G.H., Hadden, D.R., Weaver, J.A., and Montgomery, D.A.D., Vitamin B_{12} status of patients on long metformin therapy, Brit. Med. J. **1**, 685, 1971.

[75b] Adams, J.F., Clark, J.S., Ireland, J.T., Kesson, C.M., and Watson, W.S., Malabsorption of vitamin B_{12} and intrinsic factor secretion during biguanide therapy, Diabetologia **24**, 16-18, 1983.

[76] Webb, D.J., Chods, R.B., Mahar, C.Q., and Faloon, W.W., Mechanism of vitamin B_{12} malabsorption in patients receiving colchicine, N. Engl. J. Med. **279**, 845, 1968.

[77] Jacobson, E.D., Chodos, R.B., and Faloon, W.W., An experimental malapsorption syndrome induced by neomycin, Am. J. Med. **28**, 524, 1960.

[78] Ellenbogen, L., Burson, S.L., and Williams, W.L., Purification of intrisic factor, Proc. Soc. Exp. Biol. Med. **97**, 760, 1958.

[79] Ellenbogen, L., and Williams, W.L., Preparation properties and purification of intrinsic factor, Biochem. Biophys. Res. Commun. **2**, 340, 1960.

[80] Bromer, W.W., and Davisson, E.O., Preparation and characterization of a clinically effective porcine intrinsic factor vitamin B_{12} complex, in Vitamin B_{12} and Intrinsic Factor, 2. Europäisches Symposion, Hamburg 1961, Ed. H.C. Heinrich, Ferdinand Enke Verlag, Stuttgart 1962, pp. 457-462

[81] Gräsbeck, R., Simons, K., Sinkkonen, I., Isolation of intrinsic factor and its probable degration product, as their vitamin B_{12} complexes, from human gastric juice, Biochim. Biophys. Acta **127**, 47-58, 1966.

[82] Rapp, W., Taylor, K.B., Purification of human intrinsic factor by use of homologous and heterologeous immunoadsorbents, 2. Tagung Gesellsch. f. Immunologie, Wien 61 (Abstr.), 1970.

[83] Allen, R.H., Isolation of homogeneous human and hog intrinsic factor using affinity chromatography, 15th Ann. Meet. Am. Soc. Haematol., Hollywood, Fla. Dec. 3-6, 1972 (Abstr.), p. 175.

[84] Allen, R.H., Majerus, P.W., Isolation of vitamin B_{12}-binding proteins using affinity chromatography. I. Preparation and properties of vitamin B_{12}-Sepharose, J. Biol. Chem. **247**, 7695-7701, 1972.

[85] Allen, R.H., Majerus, P.W., Isolation of vitamin B_{12} binding proteins using affinity chromatography. III. Purification and properties of human plasma transcobalamin II, J. Biol. Chem. **247**, 7709-7717, 1972.

[86] Allen, R. H., Majerus, P. W., Isolation of vitamin B_{12} binding proteins using affinity chromatography. II. Purification and properties of human granulocyte vitamin B_{12}-binding protein, J. Biol. Chem. **247**, 7702-7708, 1972.

[87] Nexø, E., A new principle in biospecific affinity chromatography used for purification of cobalamin-binding proteins, Biochim. Biophys. Acta **379**, 189-192, 1975.

[88] Jacobsen, D., and Montejano, Y., Affinity photo-release purification of coabalamin-binding proteins, in Vitamin B_{12}, Eds. B. Zagalak, W. Friedrich, W. de Gruyter, Berlin 1979, pp. 949-954.

[89] Gräsbeck, R., Puutula, L., Aro, H., Vitamin B_{12} and analogues covalently coupled to cellulose for use in purification of B_{12} binding proteins, Scand. J. Clin. Lab. Invest. Suppl. **113**, 64 (Abstr.), 1970.

[90] Schilling, R. F., Intrinsic factor studies. II. The effect of gastric juice on the urinary excretion of radioactivity after the oral administration of radioactive Vitamin B_{12}. J. Lab. Clin. Med. **42**, 860-866, 1953.

[91] The National Formulary, 11th Ed., First Interim Announcement. Mack Printing Company, Easton, Pennsylvania 1960.

[92a] Lous, P., Schwartz, M., Acta Med. Scand. **164**, 407-417, 1959.

[92b] Doscherholmen, A., Silvis, S., and McMahon, J., Dual isotope Schilling test for measuring absorption of food-bound and free vitamin B_{12} simultaneously, Am. J. Clin. Pathol **80**, 490-495, 1983.

[92c] Cattan, D., Zittoun, J., and Yvart, J., Effect of folate deficiency on vitamin B_{12} absorption, Ann. Nutr. Metab. **26**, 367-373, 1982.

[92d] Fromm, H., Sarra, R. P., Ravitch, M. M., McJunkin, B., Farivar, S., and Amin, P., Effects of jejunoileal bypass on the enterohepatic circulation of bile acids, bacterial flora in the upper small intestine and absorption of vitamin B_{12}, Metab. Clin. Exp. **32**, 1133-1141, 1983.

[93a] Hippe, E., Gimsing, P., Holländer, N. H., A simplified method for quantitative determination of vitamin B_{12} absorption, in Vitamin B_{12}, Eds. B. Zagalak, W. Friedrich, W. de Gruyter, Berlin 1979, pp. 939-944.

[93b] Cardarelli, J. A., Burrows, B. A., Miller, A., and Slingerland, D. W., Correction of attenuation in whole-body determination of cobalt 57-labelled vitamin B_{12} absorption, J. Nucl. Med. **24**, 949-951, 1983.

[94] Strauss, E. W., and Wilson, T. H., Species differences in the intrinsic factor stimulation for B_{12} uptake by small intestine in vitro, (Abstr.), Proc. Fedn. Am. Socs. Exp. Biol. **18**, 333, 1959.

[95] Wolff, R., and Nabet, P., La recherche du facteur intrinsegue dans le suc gastrique chez les sujets normaux et chez les sujets affeinte d'anemie de biermer, Bull. Acad. Natn. Med. **145**, 150, 1961.

[96] Wolff, R., The determination of intrinsic factor activity by hog gastric mucosal extract by the use of guinea pig intestine in situ, in Vitamin B_{12} and Intrinsic Factor, 2. Europäisches Symposion, Hamburg 1961, Ed. H. C. Heinrich, Ferdinand Enke, Stuttgart 1962, pp. 512-514.

[97] Wolff, R., and Weis, J., Sur l'absorption de la vitamine B_{12} radioactive par l'intestin isole du rat, in Vitamin B_{12} und Intrinsic Factor, I. Europäisches Symposion, Hamburg 1956, Ed. H. C. Heinrich, Ferdinand Enke, Stuttgart 1957, p. 247.

[98] Lataner, A. L., and Reine, L., The uptake of radioactive vitamin B_{12} by rat intestine, in Vitamin B_{12} und Intrinsic Factor, I. Europäisches Symposion, Hamburg 1956, Ed. H. C. Heinrich, Ferdinand Enke, Stuttgart 1957, p. 243.

[99] Herbert, V., and Spaet, T. H., Distribution of intrinsic factor acitivity, Am. J. Physiol. **195**, 194, 1958.

[100] Boas, A., and Wilson, T. H., An assay for gastric intrinsic factor, Am. J. Physiol. **204**, 97, 1963.

[101] Wilson, T. H., and Wiseman, G., The use of sacs of everted small intestine for the study of the transference of substances from the musocal to the serosal surface, J. Physiol. Lond. **123**, 116, 1954.

[102] Sullivan, L. W., Herbert, V., and Castle, W. B., In vitro assay for human intrinsic factor, J. Clin. Invest. **42**, 1443, 1963.

[103] Castro-Curel, Z., and Glass, G. B. J., Assay of intrinsic factor activity on guinea pig intestinal mucosa homogenate, Proc. Soc. Exp. Biol. Med. **112**, 715, 1963.

[104] Herbert, V., Gottlieb, C., Lau, K. S., Wasserman, L. R., Intrinsic factor assay, Lancet **2**, 1017-1018, 1964.

[105] Ellenbogen, L., Highley, D. R., Baker, H. A., and Smyth, R. D., Inhibition of cobamide coenzyme activity by intrinsic factor, Biochem. Biophys. Res. Commun. **3**, 178, 1960.

[106] Baker, H. A., Smyth, R. D., Weissbach, H.,

Toohey, J. I., Ladd, J. N., and Volcani, B. E., Isolation and properties of crystalline cobamide coenzymes containing benzimidazole or 5,6-dimethylbenzimidazole, J. Biol. Chem. **235**, 480, 1960.
[107] Herbert, V., Gottlieb, C. W., and Ian, K. S., Hemoglobin-coated charcoal assay for serum vitamin B_{12} (Letter.), Blood **28**, 130, 1966.
[108] Begley, J. A., Trachtenberg, A., Hall, Ch. A., Cobinamide blocking assay for intrinsic factor, in Vitamin B_{12}, Eds. B. Zagalak, W. Friedrich, Walter de Gruyter, Berlin 1979, p. 917-918.
[109] Mathan, V., Babior, B., Donaldson, R., J. Lab. Clin. Invest. **54**, 598-608, 1974.
[110] Hansen, H. J., Miller, O. N., Gallo-Torres, H., and Goldsmith, G. A., Assay of intrinsic factor activity of human gastric juice with zirconyl phosphate gel, Anal. Biochem. **16**, 287, 1966.
[111] Wolff, R., and Nicolas, J. P., Gastric juice intrinsic factor assay with anti-saliva serum, Lancet **1**, 1008, 1967.
[112] Abels, J., Bouma, W., Nieweg, H. O., Assay of intrinsic factor with antiintrinsic factor serum in vitro, Bioch. Biophys. Acta **71**, 227, 1963.
[113] Ardeman, S., and Chanarin, I., Method for assay of human gastric intrinsic factor and for detection and titration of antibodies against intrinsic factor, Lancet **ii**, 1350, 1963.
[114] Irivine, W. J., Immunoassay of gastric intrinsic factor and the titration of antibody to intrinsic factor, Clin. Exp. Immunol. **1**, 99, 1966.
[115] Schwartz, M., Antibody to intrinsic factor, Scand. J. Clin. Lab. Invest. **19**, Suppl. **95**, 19, 1967.
[116] Schade, S. G., Abels, J., Schilling, R. F., Studies on antibody to intrinsic factor, J. Clin. Invest. **46**, 615-586, 1967.
[117] Schade, S. G., Feick, P. L., Imrie, M. H., Schilling, R. F., Clin. Exptl. Immunol. **2**, 399-413, 1967.
[118] Gottlieb, Ch., Lau, K., Wasserman, L., Herbert, V., Rapid charcoal assay for intrinsic factor (IF), gastric juice unsaturated B_{12}-binding capacity, antibody to IF and serum unsaturated B_{12}-binding capacity, Blood **25**, 875-884, 1965.
[119] Wide, L., Killander, A., A radiosorbent technique for the assay of serum vitamin B_{12}, Scand. J. Clin. Lab. Invest. **27**, 151, 1971.
[120] Garrido-Pinson, G. G., Turner, M., Miller, L., Segal, H., Gel filtration of vitamin B_{12}-binding materials from human gastric juice and gastric mucosa, Biochem. Biophys. Acta **127**, 478-487, 1966.
[121] Glass, G. B. J., Castro, Z., Uchino, H., Schwartz, G., Ishimori, A., and Morkin, P., In vitro test on gastric juice for pernicious anaemia and its precursor states, in Vitamin B_{12} und Intrinsic Factor, 2. Europäisches Symposion, Hamburg 1961, Ed. H. C. Heinrich, Ferdinand Enke, Stuttgart 1962, pp. 520-534.
[122] Gullberg, R., Gastric macromolecules with special reference to intrinsic factor, a methodology of experimental and clinical investigation, Acta Med. Scand. Suppl. **463**, 1966.
[123] Uchino, H., Schwartz, G., Glass, G. B. J., Paper electrophoretic analysis of B_{12} binding in human gastric juice, Clin. Chim. Acta **9**, 461-473, 1964.
[124] Heinrich, H. C., Lahann, H., Physiologie, Pathologie und biochemischer Wirkungsmechanismus der B_{12}-Vitamine, Vitamin-Hormon-Fermentforsch. **6**, 126-200, 1954.
[125] Hippe, E., Olesen, H., Nature of vitamin B_{12} binding. III. Thermodynamics of binding to human intrinsic factor and transcobalamins, Biochim. Biophys. Acta **243**, 83-89, 1971.
[126] McGuigan, J. E., Measurement of the affinity of human gastric intrinsic factor for cyanocobalamin, J. Lab. Clin. Med. **70**, 666-672, 1967.
[127] Nexø, E., Strength of binding of methyl-5′-deoxyadenosyl-, cyano- and hydroxocobalamin, to human transcobalamin I and II and intrinsic factor, in Vitamin B_{12}, Eds. B. Zagalak, and W. Friedrich, Walter de Gruyter, Berlin 1979, pp. 851-854.
[128] Gräsbeck, R., Salonen, E. M., Prog. Food Nutr. Sci. **2**, 193-231, 1976.
[129] Gordon, M. Y., Blackett, N. M., Douglas, I. D. C., Colony formation by hemopoietic precursor cells cultured in semi-solid agar in diffusion chambers, Brit. J. Haematol. **31**, 103-110, 1975.
[130] Marcoullis, G., Merivuori, H., Gräsbeck, R., Comparative studies on intrinsic factor and cobalophilin in different parts of the gastrointestinal tract of the pig. Biochem. J. **173**, 705-712, 1978.
[131] Olesen, H., Nexø, E., Lous, P., Thomsen, J., Bucher, D., Scand. J. Clin. Lab. Invest. **36**, 527-529, 1976.
[132] Gräsbeck, R., and Marcoullis, G., Studies on a fraction of human gastric mucosa contain-

ing intrinsic factor isoproteins typical of musoca, Lab. Invest. **35**, 13-18, 1975.

[133] Holdsworth, E.S., The isolation and properties if intrinsic factor and vitamin B_{12}-binding substances from pig pylorus, Biochem. Biophys. Acta **51**, 295-308, 1961.

[134] Yamaguchi, N., Rosenthal, W.S., Weisberg, H., Glass, G.B.J., Labelled intrinsic factor: Study of its intestinal absorption, Fed. Proc. (abstr.), **29**, 695, 1967.

[135] Mackenzie, I.L., Donaldson, jr., R.M., Schilling, R.F., Radioiodination of human intrinsic factor. J. Clin. Invest. **48**, 516-524, 1969.

[136] Aro, H., Gräsbeck, R., Electrofocusing of purified intrinsic factor (IF) and a non-IF-active vitamin B_{12}-binding protein from hog pyloric mucosa, Scand. J. Clin. Lab. Invest. Suppl. **113**, 1970.

[137] Allen, R.H., Mehlman, C.S., Isolation of gastric vitamin B_{12} binding proteins using affinity chromatography. Purification and properties of hog intrinsic factor and hog nonintrinsic factor, J. Biol. Chem. **248**, 3670-3680, 1973.

[138] Allen, R.H., and Mehlman, C.S., J. Biol. Chem. **248**, 3660, 1973.

[139] Visuri, K., and Gräsbeck, R., Biochem. Biophys. Acta **310**, 508, 1973.

[140] Highley, D.R., Davies, M.C., Ellenbogen, L., Hog intrinsic factor. Some physico-chemical properties of vitamin B_{12}-binding fractions from hog pylorus, J. Biol. Chem. **242**, 1010-1015, 1967.

[141] Abels, J., and Schilling, R.F., Protection of intrinsic factor by vitamin B_{12}, J. Lab. Clin. Med. **64**, 375, 1964.

[142] Prusoff, W.H., Welch, A.D., Heinle, R.W., and Meacham, G.C., Concentration of intrinsic factor and vitamin B_{12}-binding activities of fractions of desiccated hog stomach, Blood **8**, 491, 1953.

[143] Gräsbeck, R., Stenman, U.H., Puutula, L., Visuri, K., A procedure for detaching bound vitamin B_{12} from its transport proteins, Biochem. Biophys. Acta **158**, 292-295, 1968.

[144] Spray, G.H., The effect of heat on the microbiological and antianaemia properties of human gastric juice mixed with vitamin B_{12}, Biochem. J. **50**, 587-590, 1952.

[145] Campell, D.C., Hall, B.E., and Morgan, Oral administration of vitamin B_{12} in pernicious anaemia, Studies on nature and source of intrinsic factor, J. Lab. Clin. Med. **34**, 1590, 1949.

[146] Holdsworth, E.S., The isolation and properties of intrinsic factor and vitamin B_{12}-binding substances from pig pylorus, Biochim. Biophys. Acta **51**, 295, 1961.

[147] Hall, B.E., Studies on the nature of the intrinsic factor, Br. Med. J. **2**, 585, 1950.

[148] Glass, G.B.J., Gastric intrinsic factor and its function in the metabolism of vitamin B_{12}, Physiol. Rev. **43**, 529, 1963.

[149] Gräsbeck, R., Physiology and pathology of vitamin B_{12} absorption, distribution and excretion, Adv. Clin. Chem. **3**, 299, 1960.

[150] Havez, R., Carlier, A., Boulanger, C., Biserte, G., Ètude de deux transporteurs de vitamine B_{12} isoles de la muqueuse gasatrique de porc, in Protides of The Biological Fluids, Ed. H.Peeters, Pergamon Press, Oxford 1969, pp. 409-421.

[151] Gräsbeck, R., Action of group blocking agents and enzymes on intrinsic factor preparations, Acta Chem. Scand. **12**, 142-144, 1958.

[152] Gräsbeck, R., Influence of specific group inhibitors and enzymes in intrinsic factor activity, Acta Physiol. Scand. **45**, 88-95, 1959.

[153] Gräsbeck, R., Influence of specific group inhibitors on rat intrinsic factor, Acta Physiol. Scand. **45**, 116-121, 1959.

[154] Glass, G.B.J., Castro-Curel, Z., Assay of intrinsic factor activity on guinea pig intestinal musoca homogenates, Proc. IX.Congr. Europ. Soc. Haematol., Lisboa 1963, S.Karger, New York 1963, pp. 1470-1482.

[155] Gräsbeck, R., Studies on the vitamin B_{12}-binding principle and other biocolloids of human gastric juice, Acta Med. Scand. 154, Suppl. **314**, 1, 1956.

[156] Faillard, H., Pribilla, W., Untersuchungen zur physiologischen Bedeutung der Neuraminasäure-haltigen Koltigen Kohlehydratgruppe von Intrinsic Factor-Mucoiden menschlicher Magenschleimhaut, Klin. Wschr. **42**, 686-693, 1964.

[157] Kapadin, C.R., Mathan, V.J., Backer, S.J., Gastroenterology **70**, 704-706, 1976.

[158] Gräsbeck, R., Soluble and membrane bound vitamin B_{12} transport proteins, in Vitamin B_{12}, Eds. B.Zagalak, W.Friedrich, Walter de Gruyter, Berlin 1979, pp. 743-763.

[159] Faillard, H., Probilla, W., Posth, H.E., Die Einwirkung von Neuraminidase und Papain auf Intrinsic Factor aktive Mucoide des Magens von Mensch und Schwein, Z. Physiol. Chem. **327**, 100-108, 1962.

[160] Faillard, H., Probilla, W., Wolff, R., Nabet,

P., Posth, H. E., Investigations of highly purified human and animal intrinsic factor preparations with special consideration of neuraminic acid, Proc. IX. Intern. Congr. Hematol., Mexico City 1962, (Abstr.), p. 129.
[161] Latner, A. L., Merrills, R. J., Raine, L. C. D. P., Preparation of highly potent intrinsic factor mucoprotein, Biochem. J. **63**, 501-507, 1956.
[162] Glass, G. B. J., Jacob, T. A., Williams, D. E., Howe, E. E., Correlation of intrinsic factor activity of hog stomach preparations with their paper electrophoretic partners sedimentation constants and B_{12}-binding capacity, Tohoku. J. Exptl. Med. **71**, 1-17, 1959.
[163] Nexø, E., Olesen, H., Hansen, M. R., Bucher, D., Thomsen, J., Scand. J. Clin. Lab. Invest. **38**, 649-653, 1978.
[164] Bucher, D., Thomsen, J., Nexø, E., Comp. Biochem. Physiol. **6213**, 175-177, 1979.
[165] Nexø, E., Transcobalamin I and other human R-binders. Purification, structural, spectral and physiological studies, Scand. J. Haematol. **20**, 221-236, 1978.
[166] Bucher, D., Thomsen, J., Nexø, E., Olesen, H., Amino terminal sequence of hog nonintrinsic factor and hog intrinsic factor, in Vitamin B_{12}, Eds. B. Zagalak, W. Friedrich, Walter de Gruyter, Berlin 1979, pp. 905-908.
[167] Inada, M., Studies on hog intrinsic factor. I. Purification of hog intrinsic factor and its physicochemical properties, J. Vitaminology **16**, 10-16, 1970.
[168] Marcoullis, G., and Gräsbeck, R., Vitamin B_{12}-binding proteins in human gastric mucosa, Scand. J. Clin. Lab. Invest. **35**, 5-11, 1975.
[169] Marcoullis, G., Gräsbeck, R., and Salonen, E. M., Biochim. Biophys. Acta **491**, 663-672, 1977.
[170] Löwenstein, L., Cooper, B. A., Burton, L., Gartha, S., Kerner, K., An immunological basis for acquired resistance to oral administration of hog intrinsic factor and vitamin B_{12} in pernicious anaemia. J. Clin. Invest. **40**, 1956-1962, 1961.
[171] Gullberg, R., Kistner, S., A study of precipitating antibodies against human intrinsic factor, Acta Med. Scand. **172**, 385-388, 1962.
[172] Nabet, P., Wolff, R., Role antigenique a l'égard du lapin du facteur intrinsique contenu dans l'extrait de muquouse gastrigue du porc, Bull. Soc. Chim. Biol. **45**, 623-629, 1963.
[173] Gullberg, R., Separation and immunologic characterization of hog intrinsic factor in comparison with human intrinsic factor, Acta Med. Scand. **180**, 317-329, 1966.
[174] Samloff, I. M., Turner, M. O., Rabbit blocking and binding antibodies to human intrinsic factor and intrinsic factor-B_{12} complex. J. Immunol. **101**, 578-586, 1968.
[175] Bennet, S. J., Hansen, H. J., Antigenicity of vitamin B_{12} binders of porcine intrinsic factor concentrate (P, FE), Clin. Res. **17**, 297, (Abstr.), 1969.
[176] Goldberg, L. S., Samloff, I. M., Barnett, E. V., Studies on the antigenic structure of human and hog intrinsic factor, Clin. Exp. Immunol. **5**, 371-379, 1969.
[177] Hurliman, J., Zuber, C., Vitamin B_{12}-binders in human body fluids. I. Antigenic and physicochemical characteristics, Clin. Exptl. Immunol. **4**, 125-140, 1969.
[178] Golberg, W. S., Fudenberg, R. H., Effect of pH on the vitamin B_{12}-binding capacity of intrinsic factor, J. Lab. Clin. Med. **73**, 469-475, 1969.
[179] Morishita, R., Studies on human gastric intrinsic factor. III. Immunological observations of vitamin B_{12} binders in human gastric juice and other body fluids, J. Vitaminology (Kyoto) **15**, 104-115, 1969.
[180] Inada, M., Glass, G. B. J., Effects of prolonged administration of intrinsic factor antibodies on gastric morphology and secretion in rats, Federation Proc. **31**, 299, 1972.
[181] Taylor, K. B., Clin. Haematol. **5**, 497-519, 1976.
[182] Neuman, E., Honetz, N., Antikörper gegen Intrinsic Factor bei oraler Behandlung der perniziösen Anämie mit hochgereinigtem Schweine-Intrinsic-Factor-Vitamin-B_{12}-Komplex, Blut **24**, 15-21, 1972.
[183] Nicolas, J. P., Heterogeneity of antibodies against B_{12} binders in pernicious anaemia, in Vitamin B_{12}, Eds. B. Zagalak, W. Friedrich, Walter de Gruyter, Berlin 1979, pp. 791-802.
[184] Taylor, K. B., Inhibition of intrinsic factor by pernicious anaemia sera, Lancet **2**, 106, 1959.
[185] Schwartz, M., Intrinsic factor antibody in serum of patients with pernicious anemia. Lancet **2**, 1263, 1960.
[186] Bernier, G. M., Hines, J. D., Immunologic heterogenity of autoantibodies in patients with pernicious anaemia, N. Engl. J. Med. **277**, 1386-1391, 1967.
[187] Wolff, R., Nicolas, J. P., Kaise, W., Réactions croisées avec le facteur intrinséque du Poro d'autoanticorps anti-facteur intrinséque hu-

mains du type II présents dans quelques, C. R. Soc. Biol. **164**, 1384-1387, 1970.

[188] Jacob, E., Schilling, R. F., An in vitro test for the detection of serum antibody to intrinsic factor-vitamin B_{12} complex, J. Lab. Clin. Med. **67**, 510-515, 1966.

[189] Strickland, R. G., Baur, S., Ashworth, L. A. E., Taylor, K. B., Clin. Exp. Immunol. **8**, 25-36, 1971.

[190] Baur, S., Koo, N., Taylor, K. B., The immunoglobulin class of autoantibody containing cells in the gastric musoca. Immunology **19**, 891-894, 1970.

[191] Abels, J., Bouna, W., Janz, A., Woldring, M. G., Bakker, A., Nieweg, H. O., Experiments on the intrinsic factor antibody in serum from patients with pernicious anaemia. J. Lab. Clin. Med. **61**, 893-906, 1963.

[192] Goldberg, L. S., Shuster, J., Fudenberg, H. H., Gastric autoimmunity in pernicious anaemia, J. Lab. Chim. Med. **273**, 249-257, 1969.

[193] Ramsay, G., Herbert, V., Dialysis assay for intrinsic factor and its antibody: Demonstration of species specificity of antibodies to human and hog intrinsic factor, J. Lab. Clin. Med. **65**, 143-152, 1965.

[194a] Gullberg, R., Reaction of human and hog intrinsic factor with Type I antibody to intrinsic factor. Clin. Exp. Immunol. **7**, 453-462, 1970.

[194b] Hansen, H. J., Jacob, E., and Frentz, G. D., Assay of antiintrinsic factor antibodies. Methods in Enzymology **18C**, 130-133, 1971.

[195] Kauvonen, I., Gräsbeck, R., Subunit structure of the pig ileal intrinsic factor receptor, in Vitamin B_{12}, Eds. B. Zagalak, W. Friedrich, Walter de Gruyter, Berlin 1979, pp. 909-916.

[196] Yamada, S., Itaya, H., Nazakawa, O., Fukuda, M., Purification of rat intestinal receptor for intrinsic factor vitamin B_{12} complex by affinity chromatography, Biochim. Biophys. Acta **496**, 571-575, 1977.

[197] Cotter, R., Rothenberg, S. P., Weiss, I. P., Purification of the intestinal receptor for intrinsic factor by affinity chromatography, Biochim. Biophys. Acta **490**, 19-26, 1977.

[198] Marcoullis, G., Gräsbeck, R., Isolation of the porcine ileal intrinsic factor receptor by sequential affinity chromatography, Biochim. Biophys. Acta **499**, 309-314, 1977.

[199] Marcoullis, G., Gräsbeck, R., Solubilised intrinsic factor receptor from pig ileum; its characteristics, Biochim. Biophys. Acta **496**, 36-51, 1977.

[200] Katz, M., Stopa, E., Effect of pentagastrin in vitamin B_{12} absorption in the guinea pig, in Vitamin B_{12}, Eds. B. Zagalak, W. Friedrich, Walter de Gruyter, Berlin 1979, pp. 955-960.

[201] Katz, M., Cooper, B. A., Brit. J. Haematol. **26**, 569-579, 1977.

[202] Katz, M., Cooper, B. A., Solubilized receptor for vitamin B_{12} intrinsic factor complex from guinea pig intestinal mucosa, J. Clin. Invest. **54**, 733-739, 1974.

[203] Stopa, E. G., O'Brien, R., Katz, M., Effect of colchicine on guinea pig intrinsic factor vitamin B_{12} receptor, Gastroenterology **76**, 309-314, 1979.

[204] Okuda, K., and Sasayama, P., Intestinal distribution of intrinsic factor vitamin B_{12} absorption, Am. J. Physiol. **208**, 14, 1965.

[205a] Herbert, V., and Castle, W. B., Divalent cation and pH dependence of net intrinsic factor action in everted sacs and mucosal homogenates of net small intestine. J. Clin. Invest. **40**, 1978, 1961.

[205b] Kouvonen, I., and Gräsbeck, R., The role of sialic acid in the binding of calcium ions to intrinsic factor its intestinal receptor, Biochim. Biophys. Acta **797**, 163-170, 1984.

[206] Donaldson, R. M., Mac.Kenzie, jr., I. L., and Trier, J. S., Intrinsic factor-mediated attachment of vitamin B_{12} to brush borders and microvillons membranes of hamster intestine, J. Clin. Invest. **46**, 1215, 1967.

[207] Marhan, V. I., Babior, B. M., Donaldson, R. M., Kinetics of the attachment of intrinsic factor-bound cobamides to ileal receptor, J. Clin. Invest. **54**, 598-603, 1974.

[208] Hellegers, A., Okuda, K., Nesbett, R. E. L. Jr., Vitamin B_{12} absorption in pregnancy and in the newborn, Am. J. Clin. Nutr. **5**, 327-331, 1957.

[209] Katz, M., Stopa, E., de Leon, A., O'Brien, R., Hormonal control of B_{12} absorption in pregnancy, Clin. Res. **25**, 694 A, 1977.

[210] Brown, J., Robertson, J., Gallagher, N., Hormonal regulation of vitamin B_{12} absorption by pregnant mouse small intestine, Gastroenterology **72**, 881-885, 1977.

[211] Heyssel, R. M., Bozian, R. C., Darby, W. J., Bell, M. C., Vitamin B_{12} turnover in man. The assimilation of vitamin B_{12} from natural foodstuff by man and estimates of minimal daily dietary requirements, Am. J. Clin. Natur **18**, 176, 1966.

[212] Donaldson, R. M., Katz, J. M., Exchange be-

tween free and gastric juice-bound cyanocobalamin, J. Clin. Invest. **42**, 534, 1963.

[213] Khadapkar, S.V., and Vakil, U.K., Release of vitamin B_{12} in vitro from intrinsic factor-vitamin B_{12} complex. Purification and characterization of a releasing factor from rat ileal brush border, Biochim. Biophys. Acta **719**, 619-632, 1982.

[214] Kouvonen, I., and Gräsbeck, R., A simplified technique to isolate the porcine and human ileal intrinsic factor receptors and studies on their subunit structures. Biochem. Biophys. Res. Commun. **86**, 358-364, 1979.

[215] Marcoullis, G., Merivuori, H., and Gräsbeck, R., Comparative studies on intrinsic factor and cobalophilin in different parts of the gastrointestinal tract of the pig, Biochem. J. **173**, 705-712, 1978.

[216] Stopa, E.G., O'Brien, R., and Katz, M., Effect of colchicine on guinea pig intrinsic factor-vitamin B_{12} receptor, Gastroenterology **76**, 309-314, 1979.

[217] Parmentier, Y., Marcoullis, G., and Nicolas, J.P., The intraluminal transport of vitamin B_{12} and exocrine pancreatic insufficiency, Proc. Soc. Exp. Biol. Med. **160**, 396-400, 1979.

[218] Rothenberg, S.P., Weiss, J.P., and Cotter, Formation of transcobalamin. II. vitamin B_{12} complex by guinea-pig ileal mucosa in organ culture after in vitro incubation with intrinsic factor-vitamin B_{12}. Br. J. Haematol. **43**, 401-414, 1978.

[219] Hirota, A., Clinical study on output of gastric hydrochloric acid and intrinsic factor in juvenile diabetics, Tokyo Jikeikai Med. J. **97** (2), 208-224, 1982.

[220] Beesley, R., Solubilization of the ileal receptor for intrinsic factor-vitamin B_{12} complex by digestion with papain, Biochem. J. **200**, 225-230, 1981.

[221] Batt, R.M., Needham, J.R., and Carter, M.W., Bacterial overgrowth associated with a naturally occurring enteropathy in the German shepherd dog, Res. Vet. Sci. **35**, 42-46, 1983.

[222] Dekker, B.G., Van Dongen, C.J., Schipper, N.P.M., and De Ligny, C.L., The influence of experimental conditions on the association constant for binding of vitamin B_{12} by human intrinsic factor coupled to Sepharose 4B, Int. J. Appl. Radiat. Isot. **34**, 1351-1356, 1983.

[223] Dulloo, R.M., Majumdar, S., Chakravarti, R.N., and Mahmood, A., Intestinal brush border membrane structure and function: Effect of chronic vitamin C deficiency in guinea pigs, Biochem. Med. **27**, 325-333, 1982.

8.2 Transcobalamin (TC)

8.2.1 Synonyms and abbreviations

Transcobalamin II (TC II)
Betaglobulin B_{12} binder,
Component B, Small binder

8.2.2 Source

Serum and plasma, blood cells and tissues of human and animals.

Two forms of TC are distinguished:
apo-TC or free TC
holo-TC or TC-Cbl complex

Man unlike monkeys and other animal species have the bulk of the plasma Cbl bound to *Cobalophilin (CP),* another Cbl binder (see section 8.3). Thus in the greater part of human plasma TC appears in the apo-form. Free Cbl, antibiotics and diseases affect TC concentration (Tables 8-13, 8-14, 8-15 and Figs. 8-14, 8-15). The quantity

Table 8-13 Concentration of apo-TC in serum of man and various other species. Data are expressed by concentration of exogeneous vitamin B_{12} bound to TC assuming a conversion rate: 136 pg B_{12}/ml = 100 p mole B_{12}/l (data from [6, 7]).

Species	apo-TC	
	pg B_{12}/ml	pmol B_{12}/l
Human	1150	846
Rhesus monkey	1970	1448
African green monkey	4370	3213
Horse	780	570
Bovine	1880	1382
Pig	1510	1110
Sheep	5510	4051
Dog	2250	1650
Cat	31040	22820
Rabbit	20000	14705
Rat	2280	1676
Guinea pig	3400	2500
Mouse	42880	31530
Bat *(Rosuettus aegyptiacus)*	1559	1146

Table 8-14 Concentration of holo-TC in plasma from man and various other species (from England and Linnell [8]).

Species	holo-TC	
	pg/ml*	pmol/l
Healthy human	29±27	21±20
Puma *(Felis concolor)*	128	94
Capuchin *(Cebus apella)*	196	144
Baboon *(Papio papio)*	292	215
Dog *(canis familaris)*	354	260
Rat	359	264
Sheep	551	405
Cat *(Felis catus)*	635	467
Guinea-pig	5081	3736
Rabbit	8353	6142
Monkey *(Erythrocebus patas)*	8448	6212

* expressed in pg of B_{12} bound to TC per ml (see section 8.2.2)

of the binding proteins is expressed throughout the literature by the amount of Cbl bound to the protein (e.g. pg B_{12}/ml serum) assayed mainly in a radiodilution assay. The molar concentration of the binding proteins is calculated from the concentration of bound Cbl as well. It is assumed in these calculations that one protein molecule binds one Cbl molecule. If CN-Cbl is used for quantitation (M.W. 1360) the following conversion rate can be applied: 136 pg B_{12}/ml = 100 pmol B_{12}/l

Table 8-15 Concentration of apo-TC and holo-TC in serum of man affected by various diseases. Data are expressed by concentration of exogeneous vitamin B_{12} bound to apo-TC or endogeneous vitamin B_{12} bound to holo-TC. A conversion rate: 136 pg B_{12}/ml = 100 pmol B_{12}/l, is assumed (data from Rachmilewitz et al. [1] and Wierusz-Wysocka [120-121]).

Disease	No	apo-TC, mean ± S.D. or range		holo-TC, mean ± S.D.	
		pg B_{12}/ml	pmol B_{12}/l	pg B_{12}/ml	pmol B_{12}/l
Normals	25	750 – 1500	551 – 1103		
	40	980 ± 220	721 ± 162		
Chronic myeloid leukemia (CML)	20	900 – 1600	662 – 1176		
	32	1250 ± 340	919 ± 250		
Acute myeloid leukemia (AML)	10	750 – 1600	551 – 1176		
Acute promyelocytic leukemia (APL)	6	750 – 1750	551 – 1287		
	6	1650 ± 480	1213 ± 353		
Polycytemia vera	30	1380 ± 320	1015 ± 235		
Acute leukemia	20	3048 ± 1419	2241 ± 1043		
Gauchers disease	9	3195 ± 209	2349 ± 154		
Acute immunological diseases	13	3244 ± 688	2385 ± 506		
Chronic monocytic leukemia	4	3160 – 15646	2324 – 11504		
Liver diseases:					
healthy controls	50	881 ± 246	648 ± 181	90 ± 42	66 ± 31
virus hepatitis: 1 st week	40	453 ± 107	333 ± 79	1000 ± 374	735 ± 275
3 rd week	40	605 ± 115	445 ± 85	400 ± 57	294 ± 42
6 th week	40	697 ± 133	513 ± 98	266 ± 43	196 ± 32
		apo-TC + apo-CP			
Chronic active hepatitis	19	1268 ± 637	932 ± 468	175 ± 100	129 ± 81
Liver cirrhosis	40	1103 ± 340	811 ± 250	696 ± 344	512 ± 253
Hepatic coma	5	0 – 426	0 – 313	3230 ± 1422	2375 ± 1046

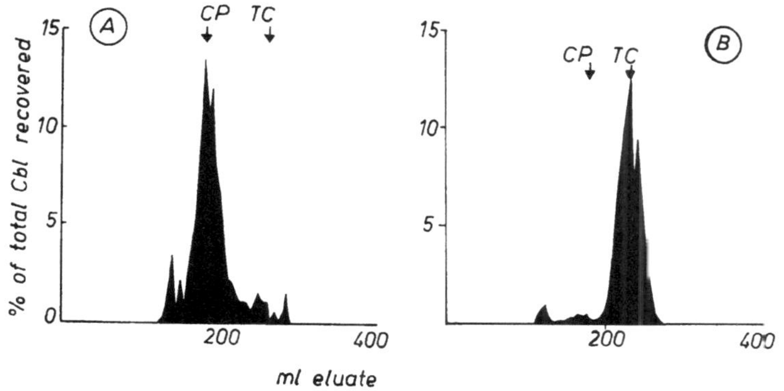

Fig. 8-14 Distribution of endogenous serum Cbl between TC and CP in man (A) and Patas monkey (B). Elution pattern obtained by gel filtration of serum on Sephadex G-200 column and assay of Cbl contents of the eluted fractions (from England and Linnel [8]). The Figure shows that mans serum cobalamin is almost entirely bound to CP, while that of monkey to TC.

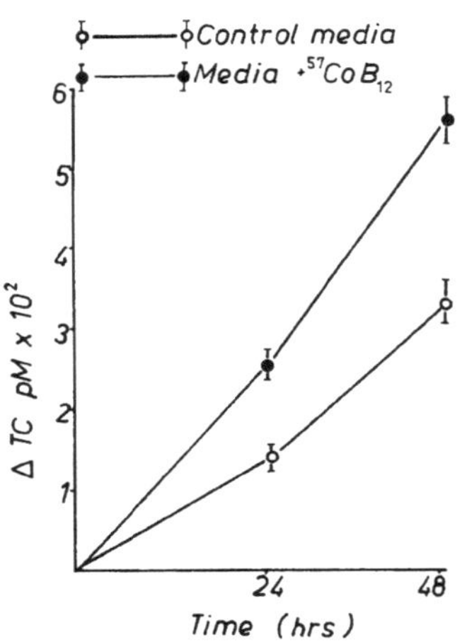

Fig. 8-15 Effect of free Cbl on TC synthesis in rabbit liver cell culture (from Tan and Brockelman [6])

Transcobalamin in tissues
murine bone marrow tissue and spleen [9], murine fibroblasts [10], rat liver [11-13], rat hepatocytes [10, 14], mononuclear phagocytes of mouse [9], canine liver, spleen and heart [15-16], ileum [17], brain gray matter [18], cerobrospinal fluid [19], enterocytes [20].

8.2.3 Site of synthesis

liver [6, 21], monocytes and macrophages [9, 22], spleen cells and peritoneal exudate cells [9, 15, 16], fibroblasts [10], enterocytes [20] and myeloblasts [2].

8.2.4 Biological function (see Fig. 8-23)

TC in circulating blood binds Cbl from the gut epithelium and carries it to the cells of organs and to the fetus [122]. The transport through the cell membrane is mediated by a specific protein receptor which binds the TC-Cbl complex (see section 8.2.10.). Inside the cell cobalamin is slipped off its carrier and becomes functional as coenzyme, in particular, for methionine synthetase and for methylmalonyl-CoA mutase (see section 7.1.3.1. and 7.1.2.2.). Once in the cell, the TC appears to be degraded [23-24]. Vitamin B_{12} transported by TC into cells appears partly in the free form [25-28]. The promotion of uptake of Cbl by TC has been demonstrated by the following cells or tissues:

human reticulocytes [29-30],
human bone marrow [31-32],
human lymphocytes [33-35],
human and murine fibroblasts [10, 24],
human placenta [36-37],
HeLa cells [38-39],

murine tumor cells [38, 40-41],
rat liver [42],
rat liver mitochondria [43-45],
canine liver [15].
TC is not needed for Cbl uptake in all tissues.

8.2.4.1 TC deficiency in man

The cause of TC deficiency can be congenital [28, 47-50] or due to autoimmunological response [3, 10, 19, 47, 51-56]. Because of intracellular B_{12} depletion, many symptoms of TC deficiency are similar to those of Cbl deficiency (see Chapter 9):

megaloblastic anaemia [48, 49, 52],
methylmalonic aciduria [52],
neurological disorders [48, 49, 57],
deficiency to mediate the uptake
of Cbl by bone marrow cells [48, 57],
oral ulcers, glossitis, intestinal
atrophy, skin changes and failure
of synthesis of immunoglobulins [58].

Abnormal high levels
various autoimmune disorders [127]
sickle cell disease [128]
malnutrition [132]
hepatocellular carcinoma [130]
renal adenocarcinoma [133]
various forms of leukemia (Table 8-15)

8.2.4.2 Half time and turnover

man: 5 minutes [59]
6 minutes [60]
30-90 minutes [61, 62]
rabbit: 90 minutes [23]
turnover in man: 10030 pmol/day [8, 60]
fractional turnover rate: 1500-2000 per cent/day [63]

8.2.4.3 Species specificity of TC

Cells of one species responsive to their own TC-Cbl vary in uptake of Cbl bound to TC from other species (Table 8-16).

Table 8-16 Species specificity of TC mediated Cbl uptake (from Haus et al. [64]).

Source of TC-Cbl	TC-Cbl acceptor cells (2.5–3.10^6 cells, 3 h at 37 °C)		
	HeLa Cells pg Cbl/10^6 cells	RPMI 6410 Human Lymphoblasts pg Cbl/10^6 cells	L-929 Murine Fibroblasts pg Cbl/mg protein
Free Cbl	0.7	0.4	2.8
Man	35.3	11.1	32.5
Baboon	24.5	8.7	18.4
Dog	14.7	7.5	35.3
Pig	11.9	2.4	11.4
Goat	11.4	4.7	28.1
Rabbit	7.5	2.1	6.0
Mouse	6.8	Not done	15.1
Horse	2.8	1.6	13.7
Guinea-pig	0.4	0	1.2

8.2.5 Methods of TC assay

The quantities of two forms of TC (holo-TC and apo-TC) circulating in blood or present in tissues are termed by the amount of endogenous Cbl in holo-TC or by the capacity to bind exogenous Cbl to apo-TC. Serum and tissues also contain other Cbl binders and the common term "unsaturated B_{12}-binding capacity" (UBBC or $UB_{12}BC$) refers to the sum of all B_{12} binders in the apo-form. Besides TC, the cobalophilin (CP) contributes mainly to the Cbl binding capacity in serum. In tissues, in addition to TC and CP, the apoenzymes of methionine synthetase and methylmalonyl-CoA mutase bind Cbl strongly as well. The concentration of the TC in serum and tissues is of the order 10^{-10} M. Such low concentration requires very sensitive methods.

Radiolabelling techniques

The most convenient approach is the labelling of the B_{12} binders with [^{57}Co]CN-Cbl of high specific radioactivity (180–300 μCi/μg, Amersham). The following analytical techniques based on radiolabelling of Cbl allow the assessment of the concentration of TC specifically or the sum of TC and other Cbl binders:

Ammonium sulfate precipitation – separation of holo-TC from holo-CP [65],
Adsorption of both, holo-TC and holo-CP, on zirconyl phosphate gel [66],
Radioimmunosorbent technique (RIST) – adsorption to specific anti-TC antibodies against TC immobilized on Sepharose beads or cellulose [69–72],
Adsorption of holo-TC to cellulose nitrate membrane filters [73],
Different affinity of TC and CP to Cbl at pH 2 [74–75],
TC mediated transport of labelled Cbl into cells [48],
Complex formation of labelled TC-Cbl with specific acceptor protein derived from membranes of responsive cells [76],
Adsorption of the excess of free [^{57}Co]Cbl on hemoglobin coated charcoal: [77–78],
Deoxyuridine suppression test [79]

Microbiological assays

The Cbl bound to TC can be assayed by microbiological methods described in Chapter 5.

8.2.6 Methods of purification

ammonium sulfate fractionation [65, 80], filtrations through Sephadex G-100, G-200 [18, 118] and Sephacryl S-200 [7, 76], DEAE-Sephadex, CM-Sephadex [81], CM-cellulose [59], affinity chromatography [82, 83], affinity photo-release chromatography [84] (Table 8-22) and by affinity thermo-release chromatography [123].

8.2.7 Physical and chemical properties

Thermodynamic parameters of TC-Cbl complex formation [85]

$\Delta G°$ (26 °C)	−15.777 kcal/mol	(− 66.06 kJ/mol)
$\Delta H°$	− 1.01 kcal/mol	(− 4.23 kJ/mol)
$\Delta S°$	+49.36 cal/K/mol	(+206.67 J/K/mol)

The binding strength in terms of $\Delta G°$ is pH dependent (Fig. 8-6).

Association constants (human TC)

K_{ass}	cobalamin	conditions	reference
1.85×10^{11} M^{-1}	CN-Cbl	38 °C, 0.1 M sodium phosphate	[87]
2.81×10^{11} M^{-1}	CN-Cbl	26 °C, pH 7.4, 1 M NaCl	[85]
3.03×10^{11} M^{-1}	CN-Cbl	22 °C, −″−	[85]
2.00×10^{11} M^{-1}	CN-Cbl	16 °C, −″−	[85]
1.0×10^{11} M^{-1}	CN-Cbl	20 °C, pH 7.4	[86]
0.5×10^{11} M^{-1}	OH-Cbl	20 °C, pH 7.4	[86]
1.1×10^{11} M^{-1}	AdoCbl	20 °C, pH 7.4	[86]
2×10^{11} M^{-1}	MeCbl	20 °C, pH 7.4	[86]

TC-Cbl complex dissociates reversibly at pH 2.

The strength of Cbl binding in TC-Cbl is weakened in 8 M urea and 7 M guanidin [82].

Molecular weight

human TC [82, 83]:

53900 (ultra-centrifugation)

60000 (gel filtration)

38000 (SDS polyacrylamide gel electrophoresis)

rabbit holo-TC [84]: 47000 (SDS polyacrylamide gel electrophoresis)

bat holo-TC [7]: 42000 (gel filtration)

TC aggregates at low ionic strength with itself or other proteins [67, 88-92].

Stokes radius (human TC) [92, 93]

apo-TC 2.67 nm

holo-TC 2.53 nm

Spectral data of TC-Cbl

The absorbance bands of aqCbl in the region 350–560 nm change after binding to TC. The shift of absorbance bands of Cbl occurs generally due to substitutions of the α and β ligands at the cobalt atom e.g. aqueous solution of aqCbl remains in equilibrium with OH-Cbl:

$$\text{aqCbl} \underset{\text{pH}<7}{\overset{\text{pH}>7}{\rightleftharpoons}} \text{OH-Cbl}$$

The equilibrium seems to be shifted towards OH-Cbl in TC-aqCbl complex as concluded from the absorbance bands in Table 8-17.

The contribution of the TC protein to the spectrum of TC-aqCbl is insignificant in this wavelength region (see section 8.1.6.).

Table 8-17 Absorbance bands of TC-aqCbl. (From Jacobsen and Montejano [84, 94]).

cobalamin	nm		
	α	β	γ
free aqCbl	526	500	351
free OH-Cbl	538	516	359
TC-aqCbl human	535	508	358
TC-aqCbl rabbit	538	510	359

B_{12}-binding specificity

TC binds cobalamin stronger than does IF as assessed by association constants of these two binders. It demonstrates lower specificity toward corrinoids than IF. Except for various forms of cobalamin, the association constants for other corrinoids have not been estimated so far (see section 8.2.7). Binding studies of Cbl analogs were based mainly on competition with [^{57}Co]CN-Cbl for the binding site on TC [67, 85, 95–99]. From these studies the following can be concluded:

1. The β ligands of cobalamin do not interfere with the strength of binding to TC.
2. The nucleotide moiety of the corrinoid is necessary for effective binding. The nucleotide deprived corrinoid, cobinamide (Cbi), is bound weakly. TC is less selective than IF with regard to the nucleotide since it effectively binds pseudovitamin B_{12} (AdeCba), a corrinoid containing adenine in the nucleotide part instead of dimethylbenzimidazole. Studies with metal free-, Zn-, Cu- and Rh analogues of vitamin B_{12} show that the coordination of benzimidazole moiety to the central metal ion is required for strong binding to the protein [99b] (see also section 8.1.3).
3. The pyrrol rings A and B of the corrin core and the side chains extending from them interact with TC in complex formation.
4. The phosphate group and the aminopropanol moiety do not interact in complex formation.

Isoelectric point (I. P.)
human serum TC I. P. 6.0-7.0 [100-102]
I. P. heterogeneity is observed in isoelectric focusing and electrophoresis of human serum [101-105].
I. P. patterns differ depending on individual [100] (see section 8.2.8).

Electrophoretic mobility [61]
- β-globulin mobility on paper electrophoresis in veronal buffer pH 8.6.
- moves towards the cathode on starch block electrophoresis in acetate buffer pH 4.5.

Chemical composition
carbohydrate content – none [106]
active SH groups – none
aminoacid composition [82, 83, 106]
N-terminal aminoacid sequence [107-108].

8.2.8 Polymorphic variants of TC

TC in human serum consists of genetically determined isoproteins which may be resolved by polyacrylamide electrophoresis [101-102, 110-112] (Fig. 8-16 and Table 8-18) and by isoelectric focusing [100, 105]. Five alleles of TC can be distinguished: Tc^1, Tc^2, Tc^3, Tc^4, and Tc^5 represented by two isoproteins (two bands) each. The capacity for Cbl of TC variants in serum is differentiated. This suggests a genetic

Table 8-18 Frequencies of TC alleles in European and African populations (from Porck et al. [113] and Frater-Schröder et al. [110]).

Population	number studied	Tc alleles				
		Tc^1	Tc^2	Tc^3	Tc^4	Tc^5
European (Swiss)	510	0.380	0.017	0.590	0.013	0.002
West-African, Bozo	90	0.022	0.300	0.500	0.178	–
East-African, Bantu	46	0.076	0.109	0.608	0.206	–

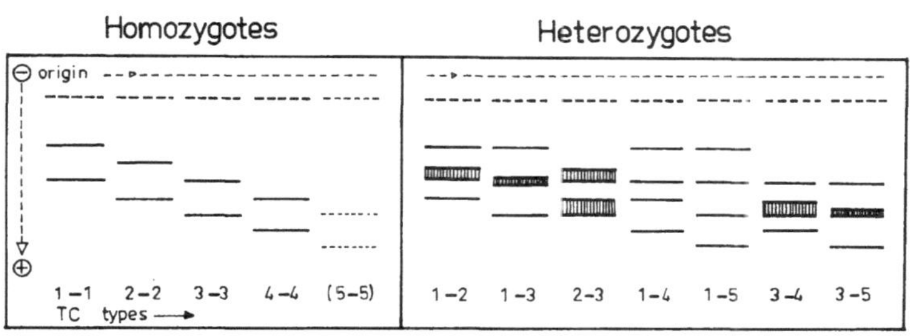

Fig. 8-16 TC isoprotein pattern of human serum. The isoproteins refer to 5 alleles of TC, each represented by two bands. The fifth allele however present in heterozygotes has not been observed in the homozygous form. The pattern was obtained by polyacrylamide gel electrophoresis of the neuramidase treated serum. (from Frater-Schröder et al. [115])

basis for the observed differences of UBBC of individual sera [113–115]. The frequencies of TC alleles are different in various ethnic groups [113, 115].

8.2.9 Antibodies against TC (TC-Abs)

Auto anti TC antibodies appear in pernicious anaemia patients overloaded with Cbl in course of intensive treatments [3, 10, 51, 52].

Two types of TC-Abs complexes [67, 116]

type I: TC-Abs able to bind Cbl.

type II: TC-Abs unable to bind Cbl presumably because of blocked Cbl binding site by the antibody.

Molecular weight of TC-Cbl-Abs complexes [19, 53]

240000–350000

Electrophoretic mobility [19, 53]

$\beta-\gamma$ mobility on paper electrophoresis in veronal buffer pH 8.6

Methods of assay

Gel filtration on Sephadex G-200 of [^{57}Co]TC-Cbl-Abs [116], immunodiffusion [116], inhibition of TC mediated Cbl uptake by cells [39, 67].

8.2.10 TC acceptor

The transport of Cbl bound to TC accross the cell membrane is mediated by an acceptor (receptor) protein which forms a complex with TC-Cbl in presence of calcium ions (see section 8.2.4). Among the two forms of Cbl: CNCbl and OHCbl complexed with TC, the latter is faster internalized and converted to active coenzyme forms [131]. The transport of Cbl to the fetus is affected by Cd- and Hg poisening [129].

Source

Human placenta, fibroblasts, rat liver parenchymal cells [24, 36, 50, 76, 117, 118].

Assay

By binding labelled TC-Cbl followed centrifugation to remove excess TC-Cbl [75].

Isolation and purification

The acceptor was solubilized from the membranes by treatment with the detergent Triton X-100 and purified by affinity chromatography with rabbit TC-Cbl immobilized on Sepharose [117] or wheat germ agglutinin (WGA) immobilized on Sepharose.

Molecular weight [117]

Two subunits, the larger subunit (50000), binds TC-Cbl.

Stokes radius [76]

human TC-Cbl [92, 93]	2.53 nm
human acceptor	5.1 nm
human acceptor-TC-Cbl complex	6.7 nm
rabbit liver acceptor-TC-Cbl complex	5.7 nm

Carbohydrate contents

ca. 33% [117].

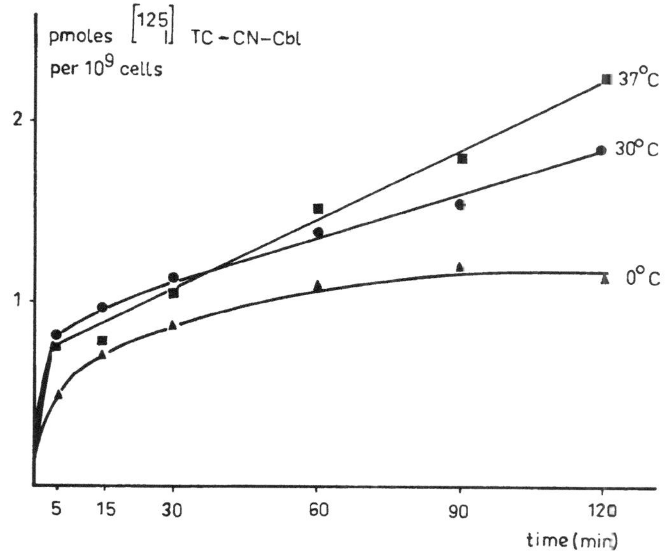

Fig. 8-17 Time course of the binding and uptake of human [^{125}I] TC-CN-Cbl by rat liver parenchymal cells at three temperatures. The concentration of the labelled holo-TC was 320 pM (from Lindemans et al. [107])

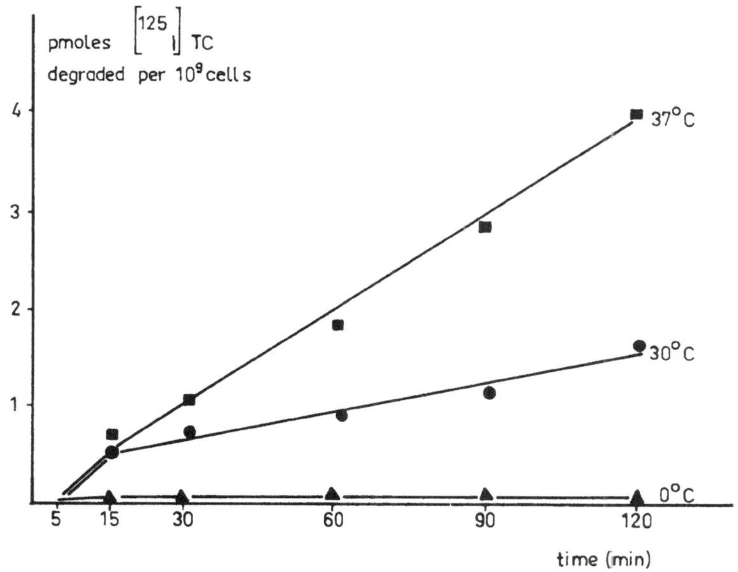

Fig. 8-18 Time course of the degradation of [^{125}I] TC-CN-Cbl by isolated liver parenchymal cells at three temperatures. The concentration of the labelled holo-TC was 320 pM (from Lindemans et al. [107])

Table 8-19 Association constants for binding human- and rabbit TC-Cbl with membrane bound human- and rabbit TC-receptors.

source		K_{ass} (M^{-1})	reference
receptor	TC-Cbl		
human placenta	human (Cbl free)	3×10^8	[117]
human placenta	human	6×10^8	[117]
human placenta	human	2.3×10^9	[76]
human placenta	rabbit	1.1×10^9	[76]
rabbit liver	human	6.7×10^9	[76]
rabbit liver	rabbit	9×10^9	[76]

Association constant

The association constants have been estimated with the use of [^{57}Co]Cbl labelled holo-TC obtained from serum and the membrane bound receptor of the appropriate tissue (Table 8-19). The binding of TC-Cbl to cell membranes proceeds also at 0 °C (Fig. 8-17). The cell uptake and degradation of TC-Cbl in the cell is strongly inhibited, however, at low temperature (Fig. 8-18). Calcium ions (5 mM) are required for binding TC-Cbl [24, 101, 119].

References

[1] Rachmilewitz, B., Schlesinger, M., Rabinowitz, R., and Rachmilewitz, M., The origin and clinical implications of vitamin B_{12} binders – the transcobalamins, in Vitamin B_{12}, Eds. B. Zagalak, W. Friedrich, Walter de Gruyter, Berlin 1979, pp. 765-775.

[2] Zittoun, J., Zittoun, R., Marquet, J., Britt, J. Haematol. **31**, 287-298, 1975.

[3] Rachmilewitz, B., Rachmilewitz, M., Israel J. Med. Sci. **12**, 583-586, 1976.

[4] Wierusz-Wysocka, B., Vitamin B_{12} binding by serum proteins in virus hepatitis, Pol. Arch. Med. Wewn. **56**, 445-456, 1976.

[5] Tan, C., Balisdell, S., Comp. Biochem. Physiol. **50 B**, 541-548, 1975.

[6] Tan, C. H., and Brockelman, T. J., The effect of antibiotics on transcobalamin II synthesis in rabbit liver cell culture, in Vitamin B_{12}, Eds. B. Zagalak, W. Friedrich, Walter de Gruyter, Berlin 1979, pp. 837-842.

[7] Green, R., Myers, P., and Jacobsen, D., Transcobalamins during induction of nutritional cobalamin deficiency in the fruit bat, in Vitamin B_{12}, Eds. B. Zagalak, W. Friedrich, Walter de Gruyter, Berlin 1979, pp. 867-872.

[8] England, J. M., and Linnell, J. C., Haematological aspects of cobalamin deficiency, in Vitamin B_{12}, Eds. B. Zagalak, W. Friedrich, Walter de Gruyter, Berlin 1979, pp. 991-1000.

[9] Rachmilewitz, B., Rachmilewitz, M., Chaouat, M., Schlesinger, M., Blood **52**, 1089-1098, 1978.

[10] Green, P. D., Savage, C. R., Hall, C. A., Mouse transcobalamin II biosynthesis and uptake by L-929 cells, Arch. Biochem. Biophys. **176**, 683-689, 1976.

[11] Haught, J. E., Miller, O. N., Smith, J. L., Hansen, H. J., Fed. Proc. **29**, 633, 1970.

[12] England, J. M., Tavill, A. S., Chanarin, I., Hepatic synthesis of transcobalamin II in the rat, Clin. Sci. Mol. Med. **45**, 479-483, 1973.

[13] Cooksley, W. G. E., England, J. M., Louis, L., Down, M. C., Tavill, A. S., Hepatic vitamin B_{12} release and transcobalamin II synthesis in the rat, Clin. Sci. Mol. Med. **47**, 531-545, 1974.

[14] Savage, C. R., Green, P. D., Biosynthesis of transcobalamin II by adult rat liver parenchymal cells in culture, Arch. Biochem. Biophys. **173**, 691-702, 1976.

[15] Rappazzo, M. E., Hall, Ch. A., Transport function of transcobalamin II, J. Clin. Invest. **51**, 1915-1918, 1972.

[16] Hall, C. A., Rappazzo, M. E., Green, P. D., Proc. Soc. Exp. Bio. Med. **148**, 1202-1205, 1975.

[17] Chanarin, I., Muir, M., Hughes, A., Hoffbrand, A. V., Evidence for intestinal origin of transcobalamin II during vitamin B_{12} absorption, Br. M. Journal **1**, 1453–1455, 1978.

[18] Inada, M., Kameyama, M., and Toyoshima, M., The significance of vitamin B_{12} binders in the central nervous system, in Vitamin B_{12}, Eds. B. Zagalak, W. Friedrich, Walter de Gruyter, Berlin 1979, pp. 821–826.

[19] Carmel, R., Large vitamin B_{12} binding proteins and complex in human serum, in Vitamin B_{12}, Eds. B. Zagalak, W. Friedrich, Walter de Gruyter, Berlin 1979, pp. 777–790.

[20] Rothenberg, S. P., Weiss, J. P., Cotter, R., Formation of transcobalamin II-vitamin B_{12} complex by guinea pig ileal mucosa in organ culture after in vivo incubation with intrinsic factor-vitamin B_{12}, Brit. J. Haematol. **40**, 401–414, 1978.

[21] Tan, C. H., Hansen, H. J., Studies on the site of the synthesis of Transcobalamin II, Proc. Soc. Exp. Biol. Med. **127**, 740–744, 1968.

[22] Tan, C. H., Blaisdell, S. I., Mammalian transcobalamin II metabolism. The immunological and the biological cross-reactivity of mammalian transcobalamin II, Biochim. Biophys. Acta **444**, 416–427, 1976.

[23] Schneider, R. J., Burger, R. L., Mehlman, C. S., Allen, R. H., The role and fate of rabbit and human transcobalamin II in the plasma transport of vitamin B_{12} in the rabbit, J. Clin. Invest. **57**, 27–38, 1976.

[24] Youngdahl-Turner, P., Rosenberg, L. E., Allen, R. H., Binding and uptake of transcobalamin II by human fibroblasts, J. Clin. Invest. **61**, 133–141, 1978.

[25] Pletsch, Q. A., Coffey, J. W., Properties of the proteins that bind vitamin B_{12} in subcellular fractions of rat liver, Arch. Biochem. Biophys. **151**, 157–167, 1972.

[26] Kolhouse, J. F., Allen, R. H., Recognition of the two intracellular cobalamin binding proteins and their identification as methylmalonyl-CoA mutase and methionine synthetase, Proc. Natl. Acad. Sci. **74**, 921–925, 1977.

[27] Rosenberg, L. E., Patel, L., Lilljequist, A. C., Proc. Nat. Acad. Sci. **72**, 4617–4621, 1975.

[28 a] Hall, C. A., Hitzig, W. H., Begley, J. A., Blood **53**, 251–263, 1979.

[28 b] Thomas, P. K., Hoffbrand, A. V., and Smith, J. S., Neurological involvement in hereditary transcobalamin II deficiency, J. Neurol. Neurosurg. Psychiatry **45**, 74–77, 1982.

[29] Retief, F. P., Gottlieb, C. W., Herbert, V., Mechanism of vitamin B_{12} uptake by erythrocytes, J. Clin. Invest. **45**, 1907–1915, 1966.

[30] Retief, F. P., Gottlieb, C. W., Herbert, V., Delivery of $[Co^{57}]B_{12}$ to erythrocytes from α and β globulin of normal B_{12}-deficient and chronic myeloid leukemia serum, Blood **29**, 837–851, 1967.

[31] Wickramasinghe, S. N., Carmel, R., Br. J. Haematol. **23**, 307–312, 1972.

[32] Wickramasinghe, S. N., England, J. M., Saunders, J. E., Down, M. C., Role of transcobalamins I, II and III in the transfer of vitamin B_{12} to human bone marrow cells in vitro, Acta Haem. **54**, 89–94, 1975.

[33] Hoffbrand, A. V., Tripp, E., Das, K. C., Uptake of vitamin B_{12} by phytohemagglutinin-transformed lymphocytes, Br. J. Haematol. **24**, 147–156, 1973.

[34] Meyer, L., Garms, R. A., Ryel, E., Miller, E., Kumar, S., Proc. Soc. Exp. Biol. Med. **147**, 679–680, 1974.

[35] Hall, C. A., Green, P. D., Proc. Soc. Exp. Biol. Med. **158**, 206–209, 1978.

[36] Friedman, P. A., Shia, M. A., Wallace, J. K., A saturable high affinity binding site for transcobalamin II-vitamin B_{12} complexes in human placental membrane preparations, J. Clin. Invest. **59**, 51–58, 1977.

[37] Seligman, P. A., Allen, R. H., Blood **50**, Suppl. 1, 86, 1977.

[38] Cooper, B. A., Paranchych, W., Physiology selective uptake of specifically bound cobalt-58 vitamin B_{12} by human and mouse tumour cells, Nature **191**, 393–395, 1961.

[39] Finkler, A. E., Hall, C. A., Nature of the relationship between vitamin B_{12}-binding and cell uptake, Arch. Biochem. Biophys. **120**, 79–85, 1967.

[40] Paranchych, W., Cooper, B. A., Factors influencing the uptake of cyanocobalamin (vitamin B_{12}) by Ehrlich ascites carcinoma cells, Biochim. Biophys. Acta **60**, 393–403, 1962.

[41] Coffey, J. W., Hansen, H. J., Miller, O. N., Studies on the interaction of vitamin B_{12} intrinsic factor and receptors. III. Chemical and biological properties of vitamin B_{12} combining substances from the stomach, serum and ascites fluid of the mouse, Arch. Biochim. Biophys. **110**, 117–123, 1965.

[42] England, J. M., Down, M. C., Wise, I. J., Linell, J. C., The transport of endogenous vitamin B_{12} in normal human serum, Clin. Sci. Mol. Med. **51**, 47, 1976.

[43] Beck, W. S., Cahen, R., Jorgensen, J., Mitochondrial cobalamins: types, sources and functions with evidence of their noninvolvement in mitochondrial DNA synthesis, in Vi-

tamin B_{12}, Eds. B. Zagalak, W. Friedrich, Walter de Gruyter, Berlin 1979, pp. 975-977.

[44] Fiedler-Nagy, C., Rowley, G.R., Coffey, J.W., Miller, O.N., Br. J. Haematol. **31**, 311-321, 1975.

[45] Gams, R.A., Ryel, E.M., Ostroy, F., Blood **47**, 923-930, 1976.

[46] Hall, C.A., Rappazzo, M.E., Proc. Soc. Exp. Biol. Med. **146**, 898-900, 1974.

[47] Hitzig, W.H., Fráter-Schröder, M., Seger, R., Clinical diseases related to deficiencies of vitamin B_{12} transport proteins, in Vitamin B_{12}, Eds. B. Zagalak, W. Friedrich, Walter de Gruyter, Berlin 1979, pp. 1009-1016.

[48] Burman, J.F., Mollin, D.L., Sladden, R.A., Sourial, N.A., Greany, M., Inherited deficiency of transcobalamin II causing megaloblastic anaemia, Brit. J. Haem. **35**, 676-677, 1977.

[49] Hakami, N., Neiman, P.E., Canellos, G.P., Lazerson, J., Neonatal megaloblastic anaemia due to inherited transcobalamin II deficiency two siblings, New Engl. J. Med. **285**, 1163-1170, 1971.

[50] Hitzig, W.H., Dohmann, V., Pluss, H.J., Vischer, P., Hereditary transcobalamin II deficiency: clinical findings in a new family, J. Pediat. **85**, 622-628, 1974.

[51] Skouby, A., Hippe, E., Olesen, H., Antibody to transcobalamin II and B_{12} binding capacity in patients treated with hydroxocobalamin, Blood **38**, 769-774, 1971.

[52] Burman, J.F., Malleson, P.N., Sourial, N.A., and Mollin, D.L., London, U.K., TC II deficiency: observations with deoxyuridine suppression test, in Vitamin B_{12}, Eds. B. Zagalak, W. Friedrich, Walter de Gruyter, Berlin 1979, pp. 889-893.

[53] Carmel, R., Tatsis, B., Baril, L., Blood **49**, 987-1000, 1977.

[54] Carmel, R., Baril, L., Circulating immunoglobulin-transcobalamin I (R binder) complexes, J. Lab. Clin. Med. **91**, 769-779, 1978.

[55] Nicolas, J.P., Perrin, M.O., Parmentier, Y., Ann. Immunol. (Inst. Pasteur) **128c**, 287-290, 1977.

[56] Nicolas, J.P., Perrin, M.O., Parmentier, Y., Gastric intrinsic factor (IF) autoantibodies of types I and II, Scand. J. Haematol. **20**, 479, 1978.

[57] Burman, J.F., Mollin, D.L., Sourial, N.A., Sladden, R.A., Brit. J. Haem. **42**, 1979.

[58] Hitzig, W.H., Kenny, A.B., Clin. Exp. Immunol. **20**, 105-111, 1975.

[59] Hall, C.A., Finkler, A.E., The dynamics of transcobalamin II a vitamin B_{12}-binding substance in plasma, J. Lab. Clin. Med., 459-468, 1965.

[60] Allen, R.H., Br. J. Haematol. **33**, 161-169, 1976.

[61] Glass, G.B.J., Gastric intrinsic factor and other vitamin B_{12} binders, Georg Thieme, Stuttgart 1974.

[62] Ellenbogen, L., Absorption and transport of cobalamin, in Cobalamin-Biochemistry and Pathophysiology, Ed. B.M. Babior, John Wiley and Sons, New York 1974, pp. 216-286.

[63] Hom, B.L., Olesen, H.A., Plasma clearance of ^{57}cobalt-labelled vitamin B_{12} bound in vivo and in vitro to transcobalamin I and II, Scand. J. Clin. Invest. **23**, 201-211, 1969.

[64] Haus, M., Green, P.D., and Hall, Ch.A., Species specificity between TC II, TC II-Cbl uptake and anti TC II, in Vitamin B_{12}, Eds. B. Zagalak, W. Friedrich, Walter de Gruyter, Berlin 1979, p. 861.

[65] Morelli, T.A., Savage, C.R. Jr., Begley, J.A., Hall, C.A., Radioimmunoassay for serum transcobalamin II, J. Lab. Clin. Med. **89**, 645-652, 1977.

[66] Hansen, H., Tan, C., Clin. Res. **15**, 279, 1967.

[67] Finkler, A.E., Green, P.D., Hall, C.A., Immunological properties of human vitamin B_{12} binders, Biochim. Biophys. Acta **200**, 151-159, 1970.

[68] Hom, B.A., Olesen, H., Molecular weights of vitamin B_{12} binding proteins in human serum determined by Sephadex G-200 gel filtration, Scand. J. Clin. Lab. Invest. **19**, 269-274, 1967.

[69] Frater-Schröder, M., Vitins, P., Hitzig, W.H., Radioimmunosorbent determinations of unsaturated and total transcobalamin II in human serum, in Vitamin B_{12}, Eds. B. Zagalak, W. Friedrich, Walter de Gruyter, Berlin 1979, pp. 878-886.

[70] Wilde, L., Acta Endocrinol. Suppl. **152**, 207-218, 1969.

[71] Frater-Schröder, M., Hitzig, W.H., Grob, P.J., Kenny, A.B., Lancet **II**, 238-239, 1978.

[72] Axen, R., Vretblad, P., Acta Chem. Scand. **25**, 2711-2716, 1971.

[73] Selhub, J., Rachmilewitz, B., Grossowicz, N., Fractionation of Serum transcobalamins on charged cellulose filters, Proc. Soc. Exp. Biol. Med. **152**, 204-209, 1976.

[74] Gilbert, H., In vitro acid dissociation of cobalamin-transcobalamin II complexes formed in vivo: A probe in the study of TC II kinetics, in Vitamin B_{12}, Eds. B. Zagalak,

W. Friedrich, Walter de Gruyter, Berlin 1979, pp. 827-832.

[75] Gilbert, H.S., J. Lab. Clin. Med. **89**, 13-24, 1977.

[76] Nexø, E., Hollenberg, M.D., Olesen, H., Solubilization and characterization of the transcobalamin II acceptor from human placenta and rabbit liver, in Vitamin B_{12}, Eds. B. Zagalak, W. Friedrich, Walter de Gruyter, Berlin 1979, pp. 843-850.

[77] Herbert, V., Gottlieb, C.W., Lau, K.S., Hemoglobin-coated charcoal assay for serum vitamin B_{12}, (Letter), Blood **28**, 130, 1966.

[78] Røbro, P., Cristiansen, P.M., Schwartz, M., Intrinsic factor secretion in stomach diseases, Lancet **2**, 1200, 1965.

[79] Wickramasinghe, S.N., England, J.E., Assessment of deoxyuridine suppression test in diagnosis of vitamin B_{12} or folate deficiency, Brit. Med. J. **3**, 148-152, 1974.

[80] Hom, B.L., Olesen, H.A., Lous, P., Separation of transcobalamin I and transcobalamin II in human serum, in Protides of the Biological Fluids, Ed. H. Peeters, vol. 14, Elsevier, Amsterdam 1966, pp. 667-669.

[81] Hom, B.L., Ahluwalia, B.K., The vitamin B_{12}-binding capacity of transcobalamin I and II of normal human serum, Scand. J. Haematol. **5**, 64-74, 1968.

[82] Allen, R.H., Majerus, P.W., Isolation of vitamin B_{12}-binding proteins using affinity chromatography, III. Purification and properties of human plasma transcobalamin, J. Biol. Chem. **247**, 7709-7717, 1972.

[83] Allen, R.H., Majerus, P.W., Isolation of vitamin B_{12}-binding proteins using affinity chromatography. II. Purification and properties of human granulocyte vitamin B_{12}-binding protein. J. Biol. Chem. **247**, 7702-7707, 1972.

[84] Jacobsen, D., Montejano, Y., Affinity photorelease purification of cobalamin-binding proteins, in Vitamin B_{12}, Eds. B. Zagalak, W. Friedrich, Walter de Gruyter, Berlin 1979, pp. 949-954.

[85] Hippe, E., Olesen, H., Nature of vitamin B_{12} binding. III. Thermodynamics of binding to human intrinsic factor and transcobalamins, Biochim, Biophys. Acta **243**, 83-89, 1971.

[86] Nexø, E., Olesen, H., Hansen, M.R., Strength of binding of methyl-5-deoxyadenosyl-cyano- and hydroxocobalamin to human transcobalamin I and II and intrinsic factor, in Vitamin B_{12}, Eds. B. Zagalak, W. Friedrich, Walter de Gruyter, Berlin 1979, pp. 851-854.

[87] Kolhouse, J.F., Allen, R.H., Absorption, plasma transport and cellular retention of cobalamin analogues in the rabbit, J. Clin. Invest. **60**, 1381-1392, 1977.

[88] Stenman, U.H. Intrinsic factor and the vitamin B_{12}-binding proteins, Clin. Haematol. **5**, 473-495, 1976.

[89] Hom, B.L., Demonstration of transcobalamin II complex formation and binding to Sephadex G-200 at low ionic strength, Clin. Chim. Acta **18**, 315-317, 1967.

[90] Green, R., Van Tonder, S.V., Cettle, G.J., Cole, G., Metz, J., Evidence for the B_{12}-dependent enzyme ethanolamine deaminase in *Salmonella*, Nature **254**, 148-150, 1975.

[91] Heller, P., Epstein, R., Cunningham, B., Henderson, W., Yakulis, V., Vitamin B_{12} binding capacity in B_{12} deficiency, Proc. Soc. Exp. Biol. Med. **115**, 342-346, 1964.

[92] Olesen, H., Rehfeld, J., Hom, B.L., Hippe, E., Stokes radius of ^{57}Co-labelled vitamin B_{12} transcobalamin I and II and ^{125}I-labelled insulin estimated by Sephadex G-200 gel filtration in human plasma at 37 °C, Biochim. Biophys. Acta **194**, 67-70, 1969.

[93] Hippe, E., Changes in Stokes radius on binding of vitamin B_{12} to human intrinsic factor and transcobalamins II, Biochim. Biophys. Acta **208**, 337-339, 1970.

[94] Nexø, E., Transcobalamin I and other human R-binders, purification, structural, spectral and physiological studies, Scand. J. Haematol. **20**, 221-236, 1978.

[95] Bunge, M.M., and Schilling, R.F., Intrinsic factor studies. VI. Competition for vitamin B_{12} binding sites offered by analogues of the vitamin, Proc. Soc. Exptl. Biol. Med. **96**, 587-592, 1957.

[96] Gottlieb, C.W., Retief, F.P., and Herbert, V., Blockade of vitamin B_{12}-binding sites in gastric juice, serum and saliva by analogues and derivatives of vitamin B_{12} and by antibody to intrinsic factor, Biochim. Biophys. Acta **141**, 560-572, 1967.

[97] Hippe, E., Haber, E., and Olesen, H., Nature of vitamin B_{12}-binding. II. steric orientation of vitamin B_{12} on binding and number of combining sites of human intrinsic factor and the transcobalamins, Biochim. Biophys. Acta **243**, 75-82, 1971.

[98] Gregory, M.E., and Holdsworth, E.S., The binding of cyanocobalamin and naturally occurring analogues by certain body fluids

and tissue extracts, Biochim. Biophys. Acta **42**, 462, 1960.

[99a] Meyer, L.M., Reizenstein, P.G., Cronkite, E.P., Miller, I.F., and Mulzac, C.W., Serum binding of vitamin B_{12} analogues: Identification of binding groups in the B_{12} molecule, Brit. J. Haematol. **9**, 158, 1963.

[99b] Elsenhans, B., Rosenberg, H.I., Influence of metal substitution on vitamin B_{12} binding to human intrinsic factor, transcobalamin I and transcobalamin II, Biochemistry, **23**, 805-808, 1984.

[100] Stenman, V.H., Salonen, E.M., Heterogeneity of transcobalamin II demonstrated by isoelectric focusing in urea, in Vitamin B_{12}, Eds. B. Zagalak, and W. Friedrich, Walter de Gruyter, Berlin 1979, pp. 873-876.

[101] Daiger, S.P., Labowe, M.L., Paeaons, M., Hang, L., Cavalli-Storza, L.L., Detection of genetic variation with radiactive ligands. III. Genetic polymorphism of transcobalamin II in human plasma, Am. J. Hum. Genet. **30**, 202-214, 1978.

[102] Frater-Schröder, M., Hitzig, W.H., The transcobalamin II – its protein pattern, Experientia **33**, 791, 1977.

[103] Puutula, L., Gräsbeck, R., One millionfold purification of transcobalamin II from human plasma, Biochim. Biophys. Acta **263**, 734-746, 1972.

[104] Natori, H., Physicochemical analysis of vitamin B_{12}-binding proteins in sera, gastric juice and leucocytes, Tumor Res. **6**, 1-79, 1971.

[105] Stenman, U.H., Gräsbeck, R., Gradients for isoelectric focusing at low pH, Biochim. Biophys. Acta **286**, 243-251, 1972.

[106a] Allen, R.H., and Mehlman, G.S., Isolation of gastric vitamin B_{12} binding proteins using affinity chromatography. Purification and properties of hog intrinsic factor and hog nonintrinsic factor, J. Biol. Chem. **248**, 3660, 1973.

[106b] Lindemans, J., van Kapel, J., Abels, J., Adsorptive endocytosis of transcobalamin II-vitamin B_{12} by isolated rat liver parenchymal cells, in Vitamin B_{12}, Eds. B. Zagalak and W. Friedrich, Walter de Gruyter, Berlin 1979, pp. 855-861.

[107] Bucher, D., Thomsen, J., Nexø, E., Olesen, H., Amino terminal sequence of hog nonintrinsic factor and hog intrinsic factor, in Vitamin B_{12}, Eds. B. Zagalak, W. Friedrich, Walter de Gruyter, Berlin 1979, pp. 905-908.

[108] Thomsen, J., Bucher, D., Brunfeldt, K., Nexø, E., Olesen, H., An improved procedure for automated Edman degradation used for determination of the N-terminal amino acid sequence of human transcobalamin I and human intrinsic factor, Eur. J. Biochem. **69**, 87-96, 1976.

[109] Nexø, E., A new principle in biospecific affinity chromatography used for purification of cobalamin-binding proteins, Biochem. Biophys. Acta **379**, 189-192, 1975.

[110] Frater-Schröder, M., Hitzig, W.H., Butler, R., Hakkinen, A.K., Experientia **34**, 909, 1978.

[111] Frater-Schröder, M., Hitzig, W.H., Butler, R., Blood **53**, 193-203, 1979.

[112] Frater-Schröder, M., Haller, O., Gmur, R., Kierat, L., Anastasi, S., Allelic forms of mouse transcobalamin II, Biochem. Genet. **20** (9/10), 1000-1014, 1982.

[113] Porck, H.J., Frater-Schröder, M., Hakkinen, A.K., Transcobalamin II polymorphism in african populations, in Vitamin B_{12}, Eds. B. Zagalak, W. Friedrich, Walter de Gruyter, Berlin 1979, pp. 863-866.

[114] Gilbert, H.S., J. Lab. Clin. Med. **89**, 13-24, 1977.

[115] Frater-Schröder, M., Osusky, R., Hitzig, W.H., Butler, R., and Hasler-Hakkinen, A., Polymorphic variants of transcobalamin II rare alleles in family studies, in Vitamin B_{12}, Eds. B. Zagalak, W. Friedrich, Walter de Gruyter, Berlin 1979, pp. 807-812.

[116] Nicolas, J.P., Heterogenecity of antibodies against B_{12}-binders in pernicious anaemia, in Vitamin B_{12}, Eds. B. Zagalak, W. Friedrich, W. de Gruyter, Berlin 1979, pp. 791-801.

[117] Seligman, P.A., Allen, R.H., Characterization of the receptor for transcobalamin II isolated from human placenta, J. Biol. Chem. **253**, 1766-1772, 1978.

[118] Lindemans, J., van Kapel, J., Abels, J., Abstracts, XVII. Internat. Congr. Haematol, Paris 1978, p. 749.

[119] Becker, C.M., and Beck, W.S., Calcium dependencies in the binding of transcobalamins to subcellular particles of liver cells, in Vitamin B_{12}, Eds. B. Zagalak, W. Friedrich, W. de Gruyter, Berlin 1979, 833-836.

[120] Wierusz-Wysocka, B., Vitamin B_{12} binding by serum proteins in chronic liver disease, Pol. Arch. Med. Wewn. **58**, 139-147, 1977.

[121] Wierusz-Wysocka, B., Vitamin B_{12}-binding by serum proteins in virus hepatitis, Pol. Arch. Med. Wew. **56**, 445-456, 1976.

[122a] Fernades-Costa, F., and Metz, J., Transplacental transport in the rabbit of vitamin B_{12}

bound to human transcobalamin I, transcobalamin II and transcobalamin III, Br. J. Haematol. **43**, 625-630, 1979.

[122b] Dencker, L., Placental accumulation of cobalt-57 vitamin B_{12} in mice studied by light-microscopic and electron-microscopic autoradiography, Placenta **4**, 207-216, 1983.

[123] Van Kapel, J., Loef, B. G., Lindemans, J., and Abels, J., An improved method for large scale purification of human holo-transcobalamin II, Biochim. Biophys. Acta **676**, 307-313, 1981.

[124] Rachmilewitz, B., Rachmilewitz, M., Chaonat, M., and Schlesinger, M., Production of TC II (Vitamin B_{12} transport protein) by mouse mononuclear phagocytes, Blood **52**, 1089-1098, 1978.

[125] Masina, P., Ramunno, L., and Iannelli, D., Polymorphism of vitamin B_{12} [^{57}Co] binding proteins in rabbit serum. Anim. Blood. Groups. Biochem. Genet. **9**, 51-54, 1978.

[126] Polak, D. M., Elliot, J. M., and Haluska, M., Vitamin B_{12}-binding proteins in bovine serum, J. Dair. Sci. **62**, 697-701, 1979.

[127] Christensen, P. A., Brynskow, J., Gimsing, P., Petersen, J., Vitamin B_{12} binding proteins transcobalamin and haptocorrin in serum and synovial fluid of patients with rheumatoid arthritis and traumatic synovits, Scand. J. Rheumatol. **12**, 268-272, 1983.

[128] Osifo, B. O. A., Adeyokunnu, A., Parmentier, Y., Gerard, P., Nicolas, J. P., Abnormalities of serum transcobalamin in sickle cell disease hemoglobin S in Black Africa, Scand. J. Haematol. **30**, 135-140, 1983.

[129] Danielsson, B. R. G., Dencker, L., Khayat, A., and Orsen, L, Fetotoxicity of inorganic in the mouse: Distribution and effects on nutrient uptake by placenta and fetus, Biol. Res. Pregnancy Perinatol **5**, 102-109, 1984. Danielsson, B. R. G., and Dencker, L., Effects of cadmium on the placental uptake and transport to the fetus of nutrients, Biol. Res. Pregnancy Perinatol. **5**, 93-101, 1984.

[130] Paradinas, F. J., Melia, W. M., Wilkinson, M. L., Portmann, B., Johnson, P J., Murroy-Lyon, I. M., Williams, R., High serum vitamin B_{12} binding capacity as a marker of the fibrolamellar variant of hapatocellular carcinoma, Br. Med. J. **285**, 840-842, 1982.

[131] Hall, Ch. A., Begley, J. A., Green-Colligan, P. D., The availability of therapeutic hydroxocobalamin to cells, Blood **63**, 2, 335-341, 1984.

[132] Nava, H., Transcobalamin II in protein-energy malnutrition among residents of Kivu area Zaire, Clin. Chim. Acta **124**, 157-162, 1982.

[133] Jensen, H. S., Gimsing, P., Pedersen, F., Hippe, E., Transcobalamin II as an indicator of activity in metastatic renal adenocarcinoma, Cancer (Phila.) **52**, 1700-1704, 1983.

8.3 Cobalophilin (CP)

A group of immunologically identical glycoproteins from man and animals which firmly bind vitamin B_{12} has been classified under this term. These proteins found in various locations are functionally and immunologically distinct from two other B_{12} binders, transcobalamin and intrinsic factor.

8.3.1 Synonyms and abbreviations

The term *Cobalophilin* (CP) has been used since 1979 when it became apparent that a number of B_{12}-binding proteins from different sources had several similar features. The CP's obtained from various sources have been referred to in the past under several names: Those isolated from the gastric juice were named *fast binder* [1, 2], *tertiary B_{12} binder* [3, 4], *rapid binder, binder R* (*R* refers to the movement of the B_{12}-binder ahead of IF in the course of paper electrophoresis in borate buffer pH 9.0) or *Non Intrinsic Factor* (NIF). The CP's from serum have been named *alpha-globulin B_{12} binder, plasma R-binder, component A, native serum B_{12} binder* or *Transcobalamin I* (TC I) [5].

Another serum CP which differed slightly from TC I by charge was named *Transcobalamin III* (TC III) or *PV binder* [6, 7]. The usage of synonyms TC I and TC III for these B_{12} binders is justified since they are distinct isoproteins of CP which are synthetized by granulocytes at various stages of maturation [7, 8].

8.3.2 Physiological significance of cobalophilin

This vitamin B_{12} binding protein, widely occurring in tissues and body fluids does not seem to be indispensable for man. Congenital lack of cobalophilin in man does not produce any noticable defect or disease [9, 10]. It has been shown to act in transport of cobalamin out of the cell [11-15]. It does not promote the uptake of Cbl by isolated cells or organs except by the liver cells which contain a specific membrane-bound cobalophilin acceptor [12, 16-20]. Human serum of healthy subjects contains the cobalophilin in the holo-form, mainly complexed with MeCbl and aqCbl [21-23] (see Fig. 8-14). Cobalophilin is the dominant binding protein excreted in bile, milk, saliva and urine. It seems to be the primary binder of Cbl in the digestion system before it is bound to IF. Cobalophilin binds preferentially dietary vitamin B_{12} at low pH of the upper gastrointestinal tract [24]. On the other hand appreciable quantities of cobalophilin-bound Cbl is secreted with the bile. The amount of the biliary Cbl may exceed several-fold of that ingested with the food. The function of CP in the digestion system in unknown. Presumably it protects the Cbl from being chemically deamidated at acid pH in the stomach and also prevents Cbl from being taken up by microorganisms in the intestine. CP does not promote the absorption of Cbl by the intestine. Cobalamin, in order to be absorbed, must be liberated from CP and bound to intrinsic factor. Cobalophilin is one of the strongest among the Cbl-binders. The liberation of Cbl from CP-Cbl complex results through digestive action of the pancreatic enzymes. Faulty function of the pancreas as in the *exocrine pancreatic insufficiency* (EPI) may lead to severe B_{12} malabsorption [25-29, 143] (see also Fig. 8-23).

Clearance from plasma

CP is cleared from plasma by the liver unspecific asialoglycoprotein clearance mechanism [30]. CP-Cbl is taken up by the liver [150]. Subsequently Cbl is transferred to the TC or secreted as CP-Cbl with the bile [14, 31-32].

Turnover

54 pmol/day [33]
Biological half time: 10 days [34].
Fractional catabolic rate: 0.15/day [35].
Fractional 50% clearance of Cbl bound to CP lasts one day [35].
Peak binding of Cbl, after oral intake, to plasma CP occurs within 1-3 days [15].

8.3.3 Site of synthesis

Salivary and gastric glands [36],
granulocytes and their precursors [37-42].
Clotting process may activate the granulocytes to release the sialic poor isoprotein TC III [46-47].

Phagocytosis effects the release of CP from normal polymorphonuclear (PMN) granulocytes [48].
The CP isoproteins are considered to be coded by one gene [43-45].
Separate sites of synthesis for the protein- and carbohydrate moieties are postulated [49].

8.3.4 Occurrence

Cobalophilin occurs in two forms:

apo-CP or free CP
holo-CP or CP-Cbl complex

The quantity of these two forms presented in tables and figures is expressed by the capacity to bind exogenous Cbl in *apo-CP* or the amount of endogenous Cbl in *holo-CP*. Serum and tissues contain other Cbl binders as well, and the common term unsaturated B_{12} binding capacity (UBBC or $UB_{12}BC$) refers to the sum of all B_{12} binders in the apo-form. The amount of CP is designated below in terms of pg B_{12} or pmol B_{12} bound to this protein (see section 8.2.2):

1. Human plasma and serum (see Tables 8-20 and 8-21), high plasma levels in malignant conditions [50-52a, b].
2. Umbilical cord serum [53-55] and cord blood [151].
3. Leukocytes [8j]
   Granulocytes [8a]
   TC I in young differentiated cells
   TC III in mature cells
   The level of CP isoproteins TC I and TC III have medical diagnostic values since it permits the differentiation between various leukemias (Table 8-21).
   Neutrophils [8c-j]

Table 8-20 Concentration of serum holo-CP in 8 normal human subjects (100 pmol B_{12}/l = 136 pg B_{12}/ml, see section 8.3.4) (from England and Linnel [33]).

Subject	Holo-CP	
	pmol/l	pg B_{12}/ml
A	349.4	475.1
B	243.8	331.5
C	266.1	361.8
D	229.2	311.7
E	424.1	576.7
F	248.4	337.8
G	171.6	233.3
H	148.2	201.5
mean	260.0	353.6

Table 8-21 Levels of serum CP in normal subjects and in various diseases (100 pmol B_{12}/l = 136 pg B_{12}/ml) (data from Rachmilewitz et al. [8], Wierusz-Wysocka [137-138] and Itaya [147]).

Disease	No. of cases	CP isoprotein	apo-CP Mean ± S. D. or range		Method of assay	holo-CP Mean ± S. D.		Reference
			pg/ml	pmol/l		pg/ml	pmol/ml	
Normal subjects	40	TC I	217 ± 64	160 ± 47	charged cellulose filter			[8]
	40	TC III	229 ± 115	168 ± 85				[8]
	25	TC I	150 – 300	110 – 221	column chromatography			[8]
	25	TC III	100 – 350	74 – 257	column chromatography			[8]
Chronic myeloid leukemia (CML)	32	TC I	3950 ± 1630	2904 ± 1199	charged cellulose filter			[8]
	32	TC III	1340 ± 670	985 ± 493	charged cellulose filter			[8]
	20	TC I	850 – 6000	625 – 4412	column chromatography			[8]
	20	TC III	500 – 3000	368 – 2206	column chromatography			[8]
		TC I	3629 ± 1169					[147]
Acute myeloid leukemia (AML)	10	TC I	100 – 350	74 – 257	column chromatography			[8]
	10	TC III	50 – 250	37 – 184	column chromatography			[8]
Acute promyelocytic leukemia (APL)	6	TC I	3000 – 30000	2206 – 22059	column chromatography			[8]
	6	TC III	0 – 1000	0 – 735	column chromatography			[8]
	6	TC I	8540 ± 1430	6279 ± 1051	charged cellulose filter			[8]
	6	TC III	650 ± 420	478 ± 309	charged cellulose filter			[8]
Polycythemia vera	30	TC I	570 ± 230	419 ± 169	charged cellulose filter			[8]
	30	TC III	2940 ± 1240	2162 ± 912	charged cellulose filter			[8]
		TC III	1271 ± 541					[8]
Acute + malignant lymphoma	14 + 15	TC I	230 ± 49	169 ± 36	charged cellulose filter			[8]
	14 + 15	TC III	380 ± 80	279 ± 59	charged cellulose filter			[8]
Virus hepatitis: healthy controlls	50	TC I	135 ± 52	99 ± 38		393 ± 31	289 ± 23	[137] [138]
1 st week	40	TC I	334 ± 40	246 ± 29		155 ± 69	114 ± 51	[137] [138]
3 rd week	40	TC I	379 ± 30	279 ± 22		125 ± 46	92 ± 34	[137] [138]
6 th week	40	TC I	371 ± 36	273 ± 26		136 ± 56	100 ± 41	[137] [138]
Chronic active hepatitis	19	TC I	–	–		260 ± 141	191 ± 104	[137] [138]
Liver cirrhosis	40	TC I	–	–		643 ± 182	473 ± 134	[137] [138]
Hepatic coma	5	TC I	–	–		1332 ± 399	979 ± 293	[137] [138]
Breast cancer				up to 50000				[148]

4. Erythrocytes: apo-CP 1540 pmol/l, holo-CP 40 pmol/l [23]
5. Liver as source of TC I [16, 35, 44, 56]
6. Seminal plasma [57]
7. Cerebrospinal fluid [57]
8. Brain: [58]
   gray matter 4183 ± 1816 pg B_{12}/g tissue
   white matter 4428 ± 2686 pg B_{12}/g tissue
9. Human gastric juice (NIF) [59],
   Hog pyloro-duodenal mucosa [60]
10. Bile:
    human (1 ng B_{12}/ml) [31-32]
    rat (0.1 ng B_{12}/g dry weight) [59]
11. Milk contains a B_{12}-binder which shares the physicochemical properties of the CPs [61-62]: women (72 ng B_{12}/ml, colostrum 27 ng B_{12}/ml) [145], sow (245 ng B_{12}/ml), rat (126 ng B_{12}/ml), goat (2,7 ng B_{12}/ml), cow (0.5 ng B_{12}/ml) [63-65].
12. Urine [66-68] and tears (250 ng B_{12}/ml) [69] contain B_{12} binding proteins which seem to be of the CP type.
13. Egg yolk and egg white [150].

8.3.5 Methods of cobalophilin assay

Several methods applied for assay of the two other animal vitamin B_{12} binders, intrinsic factor (IF) and transcobalamin (TC), may be employed to cobalophilin as well. The most common two techniques:

- adsorption of the [^{57}Co]Cbl-CP on zirconyl phosphate gel (the free [^{57}Co]Cbl is not being adsorbed)
- charcoal adsorption of the excess of free [^{57}Co]Cbl (see assay of Cbl section 5.3.3),

do not distinguish the cobalophilin from TC or IF (see assays of IF and TC section 8.1.5 and 8.2.5 respectively). There are, however, a number of methods which permit the assay of CP if accompanied by other binders:

Column chromatography on DEAE-cellulose [70-71, 88], Sephadex G-100 or G-200 [48, 72] and Cbl-Sepharose.

Thin layer chromatography [21].

Paper electrophoresis of holo-CP at pH 8.6 [73-77] or pH 4.5 [78] combined with autoradiography or with microbiological assay of the eluates of the paper segments.

Cobinamide blocking assay. The method is based on the stronger affinity of cobalophilin for cobinamide (Cbi) in contrast to the other two B_{12}-binders [79].

Acid dissociation technique – makes use of the stronger affinity of cobalophilin for Cbl at acidic pH as compared with the other binders [80-81].

Charged cellulose filter technique. This technique permits a rapid separation of holo-TC from holo-CP by retention of the latter on charged cellulose filters [8, 82].

Radioimmunosorbent technique (RIST) employs specific anti-CP antibodies immobilized on Sepharose 6 B or cellulose [23, 83].

Radioactive ^{57}Co-labelled CN-Cbl of high specific activity (180-300 μCi/μg, Amersham) is used mainly in these techniques. The quantity of cobalophilin is expressed by its capacity to bind Cbl. The quantity of the protein bound Cbl may be estimated by microbiological techniques as well. The microbiological methods, however laborious, are still in use due to high sensitivity and specificity toward Cbl (see microbiological methods of Cbl assay in Chapter 5).

8.3.6 Purification

The cobalophilin is present in a variety of biological material and the choice of isolation and purification procedure is determined by the accompanying proteins. The purification techniques may not work universally since the properties of CP's are not identical. This is mainly due to variable carbohydrate content of this glycoprotein (up to 40%) which greatly affects the charge and the size of the CP molecule. For this reason it seems that even the purest preparations obtained by very elegant and unique techniques will not meed the criteria of homogeneity. Since the discovery of B_{12}-binding proteins, a number of methods for separation of CP from TC in serum and CP from IF in gastric juice have been developed with much effort. Some of these techniques have been untilized for analytical purposes (see section 8.3.5). Besides conventional approaches based on salting out procedures, ion exchange chromatography or gel filtration, known for almost two decades, unique and tricky techniques like the affinity photorelease method have been recently developed.

Methods of purification

1. ***Salt precipitation*** [84-85]: The CP from serum (TC I) is precipitated by ammonium sulfate in the concentration range 2.5-3.5 M.
2. ***Ion exchange column chromatography*** [2, 60, 70, 86-94]:
   Separation of gastric CP (NIF) from IF [2] and serum CP (TC I) from TC [90]
   CM-cellulose-separation of gastric CP from IF [91] and serum CP from TC
   CM-Sephadex and DEAE-Sephadex-purification of gastric CP [2, 60, 92]
   Amberlite IRC-50-purification of gastric CP [94]
3. ***Gel filtration*** on Sephadex G-100 and G-200 columns [85, 94-97]
4. ***Polyacrylamide gel electrophoresis*** [98-99]: This technique has been applied so far on analytical scale only.
5. ***Immuno chromatography*** [100]
6. ***Affinity chromatography*** [100, 113, 141] (Table 8-22)
7. ***Affinity photo-release chromatography*** [101]
8. ***Lectin — sepharose affinity chromatography*** [140]
9. ***Affinity thermo-release chromatography*** [149]: This method successfully applied for TC purification may also be used for CP.

Table 8-22 Purification of vitamin B_{12} binding proteins by affinity chromatography [141] (Reprinted from Allen et al., in: Methods in Enzymology **34**, 305, 1974).

Vitamin B_{12} binding protein	Starting material	Steps before affinity chromatography	Affinity chromatography on vitamin B_{12}-Sepharose							
			Applied			Volume	Protein	Recovered		
			Volume	Protein	B_{12}-binding ability			B_{12}-binding ability	Purification fold	Yield
			l	g	µg	ml	mg	µg		%
Granulocyte binder (cobalophilin)	2.7×10^{11} granulocytes	Sonication Centrifugation filtration	0.63	6.25	22.1	40	0.58	20.4	9860	92
Milk binder (cobalophilin)	0.6 liters milk	Centrifugation filtration	0.52	7.75	117	32	4.76	97.9	1360	84
Saliva binder (cobalophilin)	4.8 liters saliva	Centrifugation filtration	4.44	7.76	101	33	3.90	80.0	1510	79
Intrinsic factor (IF)	3.3 liters gastric juice	Filtration	3.75	7.54	290	44	8.25	248	780	86
Transcobalamin (TC)	1400 liters plasma	Cohn fractionation, CM-Sephadex, filtration	19.70	760	447	62	24.7	353	24300	79

8.3.7 Physical and chemical properties

Thermodynamic parameters of human serum CP-Cbl complex formation [102]

$\Delta G°$ (26 °C)	−15.79	kcal/mol	(−66.11 kJ mol)
$\Delta H°$	+ 4.20	kcal/mol	(+17.59 kJ mol)
$\Delta S°$	+66.86	cal/K/mol	(+279.94 J/K/mol)

The binding strength (in terms of $\Delta G°$) is pH-dependent (see Fig. 8-6).

Binding of corrinoids to CP

Cobalophilin is the strongest but the least specific binder [31, 89, 103-104] (see association constants in Table 8-23). The specificity of three animal B_{12} binders decrease in the following order IF ≫ TC > CP [13, 31, 105]. CP also binds effectively corrinoids substituted in the nucleotide part by other than benzimidazole bases such as adenine (pseudovitamin B_{12}) or corrinoids deprived of the nucleotide part (cobinamides) [29]. In contrast to IF and TC it binds Cbl at low pH [34, 102]. The contact between the corrinoid and CP is in the area of the edge of the B ring in the corrin moiety. This was concluded from binding studies of various B_{12} derivatives. There is one combining site on the CP molecule for the corrinoid (see binding capacity of CP).

Table 8-23 Association constants of CP from human serum and saliva.

K_{ass}	cobalamin	source of CP	conditions	reference
1.3×10^{11} M^{-1}	MeCbl	human serum	pH = 7.4, 20 °C	[106]
0.8×10^{11} M^{-1}	AdoCbl	human serum	pH = 7.4, 20 °C	[106]
0.6×10^{11} M^{-1}	CN-Cbl	human serum	pH = 7.4, 20 °C	[106]
0.4×10^{11} M^{-1}	aqCbl	human serum	pH = 7.4, 20 °C	[106]
15×10^{9} M^{-1}	CN-Cbl	human saliva	pH = 7.4, 20 °C	[34]
3.03×10^{11} M^{-1}	CN-Cbl	human serum	pH = 7.4, 38 °C	[102]
2.92×10^{11} M^{-1}	CN-Cbl	human serum	pH = 7.4, 26 °C	[102]
1.77×10^{11} M^{-1}	CN-Cbl	human serum	pH = 7.4, 16 °C	[102]

Binding capacity of pure CP

CP's from hog stomach:

24.3-27.7 μg B_{12}/mg [107]
18 μg B_{12}/mg [91]
19.8 μg B_{12}/mg [108]
18-25 μg B_{12}/mg [108]

CP from sows milk: [61, 63]
23.6 μg B_{12}/mg

Molecular weight of CP

There is a significant discrepancy in molecular weight data for CP obtained by gel filtration, gel electrophoresis and other methods. While the molecular weight estimated by gel filtration amounts about 120000, the values obtained by other methods are in the range of 55000-75000 (Table 8-24). The elevated values obtained by gel electrophoresis and gel filtration are due to high carbohydrate content.

Table 8-24 Molecular weight of CP's.

MW (daltons)	Source	Method of estimation	Reference
100000-115000	human stomach	gel filtration	[52, 112]
120000	human serum	gel filtration	[72]
120000-122000	human serum	gel filtration	[72]
56000	human granulocytes	ultracentrifugation	[113]
58200	human granulocytes	aminoacid and carbohydrate analysis	[113]
121000	human granulocytes	gel electrophoresis	[113]
138000	human granulocytes	gel filtration	[113]
73000	human intestinal juice		[25]
66000	human saliva	ultracentrifugation	[114-116]
70300	human saliva	gel electrophoresis	[101]
120000	umbilical cord serum	gel filtration	[53, 54]
102000	human milk	gel filtration	[52]
75000	hog stomach mucosa	ultracentrifugation	[107]
70500	hog stomach	gel electrophoresis	[101]
75000	hog gastric mucosa	gel electrophoresis	[111]
340000	hog intestine (Triton X-100 extract)	gel filtration	[117]
55000	sows milk	calculated from Cbl content	[61, 63]

Stokes radius

CP from human serum [109-110] 5.12 nm apo-CP
5.15 nm holo-CP

Diffusion coefficient ($D^{\circ}_{20,w}$)

4.5×10^{-7} cm^2/s for CP from hog stomach [107].

Absorbance bands of holo-CP (see absorbance bands of Cbl in Fig. 3-8)

CP-aqCbl complex from human saliva [101]:

			Absorbance ratio	
α	β	γ	γ/α	β/α
530 nm	512 nm	353 nm	2.73	1.00

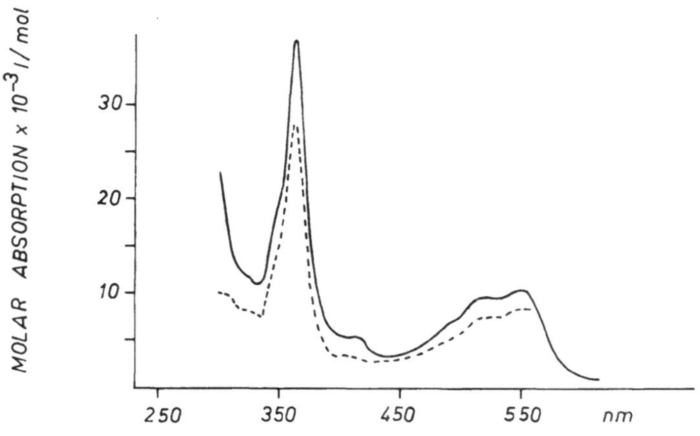

Fig. 8-19 UV-VIS spectra of CP-CN-Cbl complex (——) and free CN-Cbl (- - -). The CP was isolated from hog gastric mucosa (from Nexø et al. [111])

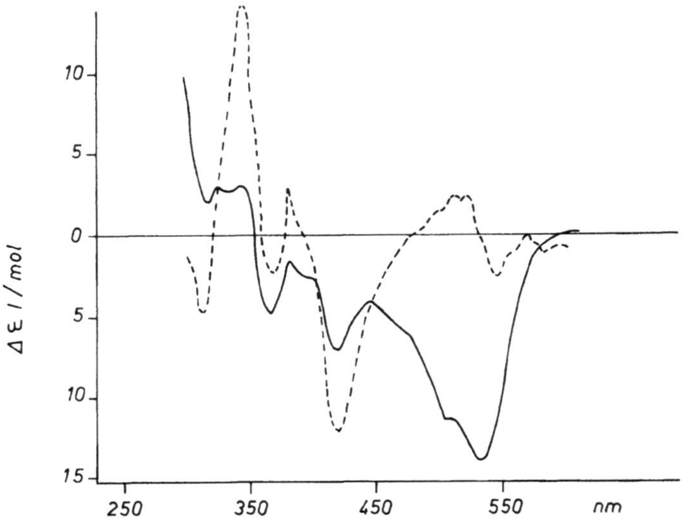

Fig. 8-20 Circular dichroism spectra of CP-aqCbl complex (——) and free aqCbl (- - -). The CP was isolated from hog gastric mucosa (from Nexø et al. [111])

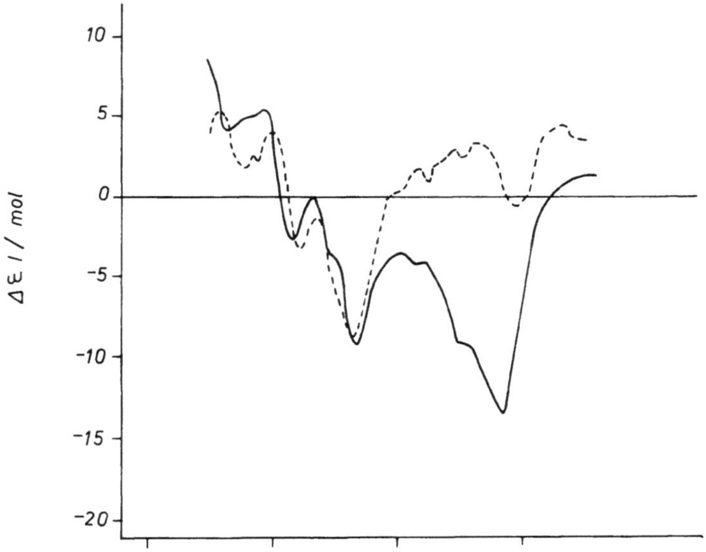

Fig. 8-21 Circular dichroism spectra of CP-OH-Cbl complex (——) and free OH-Cbl (- - -). The CP was isolated from hog gastric mucosa (from Nexø et al. [111])

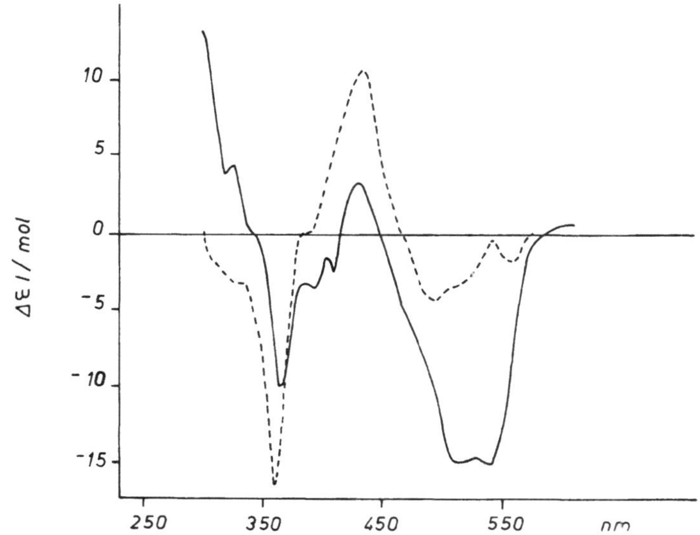

Fig. 8-22 Circular dichroism spectra of CP-CN-Cbl complex (——) and free CN-Cbl (- - -). The CP was isolated from hog gastric mucosa (from Nexø et al. [111])

Spectra of holo-CP [111]

The UV-VIS spectra of various forms of cobalamin differ slightly only from corresponding spectra obtained from cobalamins bound to CP (Fig. 8-19). Significant differences are observed, however, in circular dichroism spectra of free and CP-bound cobalamins (see Figs. 8-20, 8-21, 8-22).

Isoelectric point (I. P.)

Variable sialic acid content is the cause of microheterogeneity of CP as demonstrated by isoelectric focusing [118-123, 144a-b]:

4.55-5.28	from human saliva [118]
3.2 -3.9	from human serum [118]
4.5 -5.0	from human serum after sialidase treatment [119]
3.0 -5.0	holo-CP from human stomach [120]
3.75-4.45	holo-CP from hog stomach [121]
3.9 -4.36	holo-CP from hog stomach [122]

Electrophoretic mobility

3.0×10^{-5} $cm^2/V/s$ for CP of hog stomach at pH 7.8 [91].

Mobility of the CP's in paper electrophoresis at pH 8.6 related to the globulin mobility of human serum:

α-β globulin mobility,	from hog stomach [112]
$\beta_2$globulin mobility,	from human saliva [116]
$\alpha_1$globulin mobility,	from human serum [116]
$\alpha_2$globulin mobility,	from human leukocytes [116]
β globulin mobility,	from human serum sialidase treated [119]

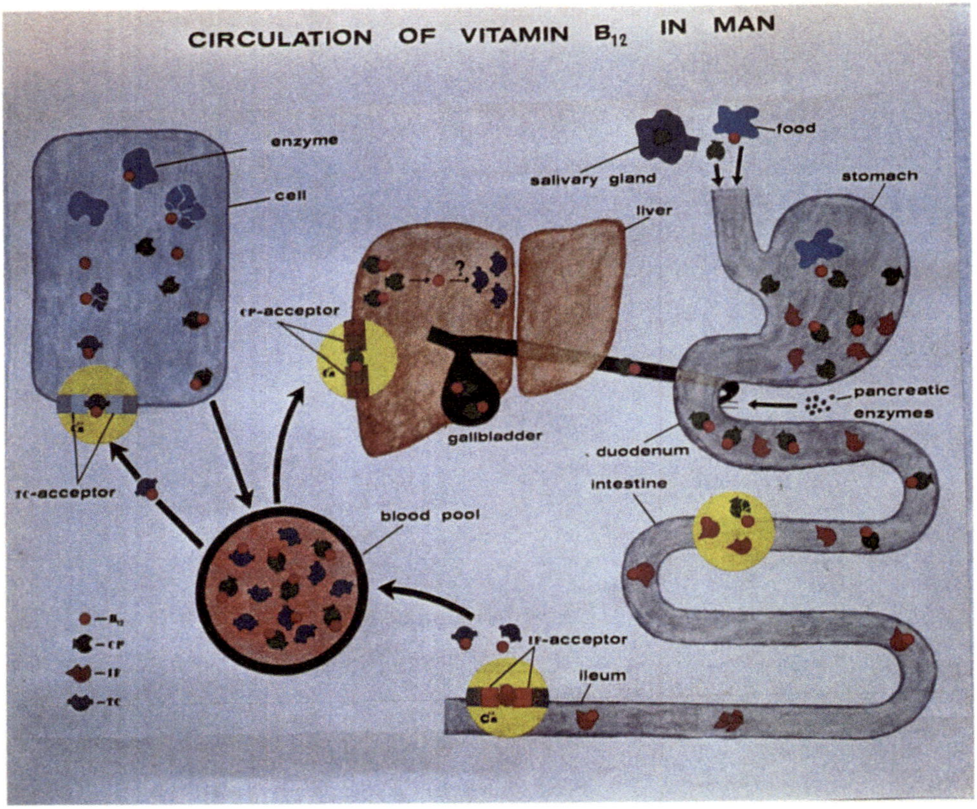

Fig. 8-23 *Circulation of vitamin B_{12} in man.* Vitamin B_{12} ingested with food (3–5 µg/day) is released in the stomach from the food protein due to acidic- and digestive action of the stomach juice. The freed cobalamin is bound to cobalophilin (CP) produced by salivary glands as well as by the stomach mucosa. Another B_{12}-binding protein secreted by the stomach mucosa (parietal cells) intrinsic factor (IF) has little binding affinity in the acidic stomach juice. Both CP and IF are glycoproteins resistant to digestion by the stomach juice. Appreciable quantities of cobalamin (5-8 µg daily) bound to cobalophilin is secreted with the bile into the duodenum (Table 6-15). As the food passes along the intestine the vitamin B_{12} is released from the CP-complex due to action of pancreatic enzymes. The free vitamin B_{12} is bound immediately by IF. The IF-Cbl complex is resistant to digestion and absorption by intestinal microorganisms. The IF-Cbl complex binds specifically to a receptor protein in the cell membrane of the terminal ileum. The binding requires calcium ions and $pH > 5$. Subsequently the cobalamin is transported within 2–4 hours by an energy dependent process across the intestinal epithelium. Within the ileal cell it becomes attached to transcobalamin (TC), the third B_{12}-binding protein. It reaches the blood of the portal vein in form of this complex. TC-Cbl reaches via the blood stream the cells of all tissues. The crossing of the cell membrane is mediated by TC-acceptor. Once in the cell the Cbl is slipped off its carrier and undergoes enzymatic conversion to coenzymatic forms AdoCbl and MeCbl. The coenzymatic forms then become integrated as cofactors of the two enzymes so far known: methylmalonyl-CoA-mutase and homocysteine methyltransferase (see Chapter 7). In the course of metabolic degradation of the enzyme systems the cobalamin is released and rebound to new apoenzymes or trapped presumably by cobalophilin. The CP-Cbl complex is removed from the cell into the blood. It is being taken up by the liver cells which are exclusively equipped with an CP-acceptor. Cells of other tissues do not seem to be capable of taking up Cbl from CP-Cbl complex. The transfer of Cbl from CP to TC (CP-Cbl→TC-Cbl) within the liver is postulated. Part of Cbl reaches the bile in the form of CP-Cbl. It is then secreted with the bile into the digestive system. The gastric absorption system, although very effective, is functioning with limited efficiency since part of the cobalamin (2–5 µg) is excreted with faeces (not shown in this figure).

Chemical composition

The number of carbohydrate residues in CP of granulocytes are following (in brackets the number of residues) [119]:
fucose (10), galactose (24), mannose (14), N-acetylgalactosamine (16), N-acetylglucosamine (34), sialic acid (8). Carbohydrate composition of hog stomach CP is indicated in Table 8-25. Lectin-sepharose affinity chromatography permits the separation of isoproteins species which differ by carbohydrate contents [140]. The

Table 8-25 Composition of the carbohydrates and nitrogen content in CP from hog stomach (in % wt/wt) obtained by various authors (compiled by Glass [59]).

Components	Cresseri (1957) [107] CP	Holdsworth (1961) [91] CP	Highley et al. (1967) [108]	
			CP	CP-Cbl
Galactose		10.2	5.7	6.0
Mannose		4.2	4.3	4.9
Fucose	2.6-4.0	5.2	3.2	3.3
Galactosamine		3.4	4.1	4.6
Glucosamine		6.0	10.2	10.1
Sialic acid			4.0	4.0
Reducing substances	17.6-18.2	29.0	13.2	14.0
Nitrogen	7.8-8.0	8.9		

Table 8-26 Amino Acid composition of CP from the hog stomach (from Cresseri [107] and Highley et al. [108]).

Components	CP	CP-Cbl
	In % protein of dry wt.	
Aspartic acid	7.0	6.8
Threonine	4.0	3.1
Serine	4.0	3.1
Proline	2.8	2.1
Glutamic acid	6.8	6.8
Glycine	1.7	1.5
Alanine	2.8	2.3
Valine	3.2	3.3
Cystine	small amount	
Methionine	1.1	0.7
Isoleucine	3.4	3.7
Leucine	5.9	5.4
Tyrosine	2.8	2.7
Phenylalanine	3.3	2.5
Lysine	3.5	2.9
Histidine	1.7	1.3
Arginine	2.4	2.2
Tryptophan	0.9	1.2
Total	57.3	51.6

N-terminal sequence of 28 aminoacids of human serum CP is known [124, 125]. Aminoacid composition of CP is specified in Table 8-26.

8.3.8 Cobalophilin receptor

The biological function of the two other animal B_{12} binding proteins, IF and TC, is to promote transport of cobalamin into the cell. The cell membranes of the target cell contain specific receptors which strongly bind the IF-Cbl or TC-Cbl complex in the first stage of the transport across the cell membrane. While a great variety of cells, if not all, from different organs are equipped with the TC receptor, only the cells of the terminal part of the ileum (human) contain the IF receptor. CP does not promote the absorption of Cbl in the digestion system, although it is present there in appreciable quantities and does not seem to have CP-receptors in the ileum. CP does not promote the uptake of Cbl by other organs except by the liver [150]. The liver receptor of CP has been isolated and partly characterized [126, 127].

Source: rabbit liver
binding [128]
association constant $K_{ass} = 15 \times 10^{11}\ M^{-1}$ (5 mM Ca^{2+})

Molecular weight
CP receptor from rabbit liver [126]: 500 000 Da
subunits: 40 000 Da and 48 000 Da

Table 8-27 Immunological properties of cobalamin binders (data from Hall and Finkler [71], and Ellenbogen [137]).

Antigen		Antiserum			
		Saliva	TC I	TC II	Intrinsic Factor
Transcobalamins:	Normal TC II of serum	–	–	+	–
	Leukemic TC II	–	–	+	–
	CSF TC II	–	–	+	–
	Seminal fluid TC II	–	–	+	–
Cobalophilins:	Saliva	+	+	–	–
	Normal TC I of serum	+	+	–	–
	CSF TC I	+	+	–	–
	Seminal fluid TC I	+	+	–	–
	Leucemic TC I	+	+	–	–
	Leukocyte binder	+	+	–	–
	Nonintrinsic factor	+	+	–	–
Intrinsic factor		–	–	–	+

8.3.9 Antibodies against CP

Antibodies against CP do not bind to IF or TC (see Table 8-27). They are produced spontaneously in pernicious anaemia patients [7, 99, 129-130] or are induced by injection of foreign CP (isolated e.g. from human saliva, hog leucocytes or hog pyloric mucosa) to rabbit [89, 133].

The CP-Abs complexes are present exclusively in the serum. Neither secretions such as saliva and gastric juice nor leukocyte extracts contained CP bound to antibodies [7].

Two types of Abs [13]:
blocking Abs – antibodies which block the binding of Cbl to CP (IgG-type)
binding Abs – antibodies which do not prevent the CP-Cbl complex formation.

Molecular weight of CP-Cbl-Abs complexes from human serum [7, 130, 135-136]: 450000-600000 Da (gel filtration)

Electrophoretic mobility of CP-Cbl-Abs: like α_1, β serum globulin fractions.

Assay is similar to the assay of TC antibodies (see section 8.2.9)
Abs immobilized on cellulose were used for CP assay [23].

There is a potential to make use of the adsorptive property of the zirconyl phosphate gel for the assay [142].

References

[1] Gullberg, R., Electrophoretic fraction of B_{12}-binders in gastric juices from patients with pernicious anaemia and from controls, Proc. Soc. Exptl. Biol. Med. **105,** 62-66, 1960.

[2] Gräsbeck, R., Simons, K., Sinkkonen, I., Purification of intrinsic factor and vitamin B_{12}-binders from human gastric juice, Ann. Med. Exptl. Biol. Fenniae. (Helsinki) **40,** Suppl. 6, 1962.

[3] Glass, G. B. J., Castro-Curel, Z., Uchino, H., Schwartz, G., Ishimori, A., Morkim, P., In vitro test on gastric juice for pernicious anaemia and its precursor states, in Vitamin B_{12} and Intrinsic Factor, 2. Europ. Symposium Hamburg, 1961, Ed. H. C. Heinrich. Ferdinand Enke, Stuttgart 1962, pp. 520-534.

[4] Uchino, H., Schwartz, G., Glass, G. J., Paper electrophoretic analysis of B_{12} binding in human gastric juice, Clin. Chim. Acta **9,** 461-473, 1964.

[5] Hall, C. A., Finkler, A. E., Function of transcobalamin II. A B_{12}-binding protein in human plasma, Proc. Soc. Exp. Med. **123,** 55-58, 1966.

[6] Bloomfield, F. J., Scott, I. M., Identification of a new vitamin B_{12} binder transcobalamin III in normal human serum, Brit. J. Haemat. **22,** 33-42, 1972.

[7] Carmel, R., Tatsis, B., Barill, L., Blood **49,** 987-1000, 1977.

[8 a] Rachmilewitz, B., Schlesinger, M., Rabinowitz, R., Rachmilewitz, M., The origin and clinical implications of vitamin B_{12}-binders. The trans-cobalamins, in Vitamin B_{12}, Eds. B. Zagalak, W. Friedrich, Walter de Gruyter, Berlin 1979, pp. 765-776.

[8 b] Naef, A., Damerau, B., Keller, H. U., Relationship between the transient cyclic AMP increase, exocytosis from specific and asurophil granules and chemotaxis in neutrophil granulocytes, Agents Actions. **14,** 63-71, 1984.

[8 c] Maallem, H., Sheppard, K., and Fletcher, J., The discharge of primary and secondary granules during immune phagocytosis by normal and chronic granulocytic leukemia polymorphonuclear neutrophils, Br. J. Haematol. **51,** 201-208, 1982.

[8 d] Borregaard, N., Heiple, J. M., Simons, E. R. and Clark, R. A., Subcellular localization of the b-cytochrome component of the human neutrophil microbicidal oxidase translocation during activation, J. Cell. Biol. **97,** 52-61, 1983.

[8 e] Ringel, E. W., Soter, N. A. and Austen, K. F., Localization of histaminase to the specific granule of the human neutrophil, Immunology **52**, 649-658, 1984.

[8 f] Bloomfield, F. J., Young, M. M., Enhanced release of inflammatory mediators from lithium-stimulated neutrophil in psoriasis, Br. J. Dermatol. **109**, 9-14, 1983. Bloomfield, F. J., Young, M. M., Influence of lithium and fluoride on degranulation from human neutrophils in vitro, Inflammation **6**, 257-268, 1982.

[8 g] Katka, K., Seger, R. A., Matsunaga, T., Toivauem, A. and Hitzig, W. H., Granulocyte function in untreated pernicious anaemia, Br. J. Haematol **53**, 23-30, 1983.

[8 h] Davies, J. M., Fletcher, K. S., The effects of surgery on the activity of neutrophil granule proteins, Br. J. Haematol. **53**, 5-14, 1983.

[8 i] O Morain, C., Smethurst, P., Levi, A. J. and Peters, T. J., Biochemical analysis of enzymic markers of inflammation in rectal biopsies from patients with ulcerative colitis and Crohns disease, J. Clin. Pathol. Lond. **36**, 1312-1316, 1983.

[8 j] Smith, G. P. and Peters, T. J., Release of granule components from human polymorphonuclear leukocytes in response to both phagocytic and chemical stimuli, Biochim. Biophys. Acta **719**, 304-308, 1982.

[9] Carmel, R., Herbert, V., Deficiency of vitamin B_{12}-binding alpha globulin in two brothers, Blood, **33**, 1-12, 1969.

[10] Hall, C. A., Begley, J. A., Congenital deficiency of human R-type binding proteins of cobalamin, Am. J. Human Genetics **29**, 619-629, 1977.

[11] Finkler, A. E., Hall, C. A., Nature of the relationship between vitamin B_{12} binding and cell uptake, Arch. Biochem. Biophys. **120**, 79-85, 1967.

[12] Finkler, A. E., Green, P. D., Hall, C. A., Immunological properties of human vitamin B_{12}-binders, Biochim. Biophys. Acta, **200**, 151-159, 1970.

[13] Nicolas, J. P., Heterogeneity of antibodies against B_{12}-binders in pernicious anaemia, in Vitamin B_{12}, Eds. B. Zagalak, W. Friedrich, Walter de Gruyter, Berlin 1979, pp. 791-801.

[14] Hall, Ch. A., The plasma transport of cobalamin, in Vitamin B_{12} Eds. B. Zagalak, W. Friedrich, Walter de Gruyter, Berlin 1979, pp. 725-739.

[15] Hall, C. A., Transcobalamins I and II as natural transport proteins of vitamin B_{12}, J. Clin. Invest. **56**, 1125-1131, 1975.

[16] Rappazzo, M. E., Hall, C. A., Transport function of transcobalamin, J. Clin. Invest. **51**, 1915-1918, 1972.

[17] Rosenthal, W. S., Weisberg, H., Glass, G. B. J., Tissue distribution of radio-B_{12} after intravenous injection of gastric juice, saliva and plasmabound ^{57}Co-B_{12}. I. In man. Blood **30**, 198-209, 1967.

[18] Rosenthal, W. S., Weisberg, H., Glass, G. B. J., Tissue distribution of radio-B_{12} after intravenous injection of gastric juice, saliva and plasmabound ^{57}Co-B_{12}. II. In rats. Blood **30**, 210-218, 1967.

[19] Friedman, P. A., Shia, M. A., Wallace, J. K., J. Clin. Invest. **59**, 51-58, 1977.

[20] Hall, C. A., Rappazzo, M., Proc. Soc. Exp. Biol. Med. **146**, 898-900, 1974.

[21] Nexø, E., Scand. J. Haematol. **18**, 358-360, 1977.

[22] MacDonald, C. M. L. A., Farquharsan, J., Bessent, R. G., Adams, J. F., Clin. Sci. Mol. Med. **52**, 215-218, 1977.

[23] Nexø, E., Gimsing, P., The pattern of cobalamins in human erythrocytes, in Vitamin B_{12}, Eds. B. Zagalak, W. Friedrich, Walter de Gruyter, Berlin 1979, pp. 967-970.

[24] Allen, R. H., Seetharam, B., Podell, E., and Alpers, D. H., Effect of proteolytic enzymes on the binding of cobalamin to R-protein and intrinsic factor, J. Clin. Invest., **61**, 47-54, 1978.

[25] Paramentier, Y., Marcoullis, G., Nicolas, J. P., and Perrin, M. O., The intraluminal phase of vitamin B_{12} transport in humans, in Vitamin B_{12}, Eds. B. Zagalak, W. Friedrich, Walter de Gruyter, Berlin 1979, pp. 803-806.

[26] Parmentier, Y., Marcoullis, G., and Nicolas, J. P., The intraluminal transport of vitamin B_{12} and the exocrine pancreatic insufficiency, Proc. Soc. Exp. Biol. Med., 1979.

[27] Marcoullis, G., Merivuori, H., and Gräsbeck, R., Comparative studies on intrinsic factor and cobalophilin in different parts of the gastrointestinal tract of the pig, Biochem. J. **173**, 705-712, 1978.

[28] Allen, R. H., Seetharam, B., Podell, E., Alpers, D. H., J. Clin. Invest. **61**, 47-54, 1978.

[29] Steinberg, W., Curington, C., Toskes, P., Clin. Res. **26**, 794, 1978.

[30] Ashwell, G., Morell, A. G., The role of surface carbohydrates in the hepatic recognition and transport of circulating glycoproteins, Adv. Enzym. **41**, 99-128, 1974.

[31] Kolhouse, J. F., Allen, R. H., Absorption, plasma transport, and cellular retention of

cobalamin analogues in the rabbit, J. Clin. Invest. **60**, 1381-1392, 1977.

[32] Burger, R.L., Mehlman, C.C., Schneider, R.L., Allen, R.H., Human plasma R-type vitamin B_{12}-binding proteins, J. Biol. Chem. **250**, 7707-7713, 1975.

[33] England, J.M., Linnell, J.C., Haematological aspects of cobalamin deficiency, in Vitamin B_{12}, Eds. B. Zagalak, W. Friedrich, Walter de Gruyter Berlin 1979, pp. 991-1000.

[34] Olssen, H., Scand. J. Gastroent. **9**, 13-16, 1974.

[35] Nexø, E., Gimsing, P., Scand. J. Clin. Lab. Invest. **35**, 391-397, 1975.

[36] Hurlimann, J., Zuber, C., Vitamin B_{12}-binders in human body fluids. II. Synthesis in vitro, Clin. Exptl. Immunol. **4**, 141-148, 1969.

[37] Simons, K., Weber, T., The vitamin B_{12}-binding protein in human leukocytes, Biochim. Biophys. Acta **117**, 201-208, 1966.

[38] Rachmilewitz, B., Rachmilewitz, M., Gross, J., Brit. J. Haematol. **26**, 557-567, 1974.

[39] Gilbert, H.S., Demonstration of the PV B_{12}-binding protein in leukocytes of normals and patients with myeloproliferative disorders (MPD), Blood **38**, 805, 1971.

[40] Carmel, R., Br. J. Haematol. **22**, 43-51, 53-62, 1972.

[41] Stenman, U.H., Scand. J. Haematol. **13**, 129-134, 1974.

[42] Scott, J.M., Bloomfield, F.J., Stebbins, R., Herbert, V., Studies on derivation of transcobalamin III from granulocytes. Enhancement by lithium and elimination by fluoride of in vitro increments in vitamin B_{12}-binding capacity, J. Clin. Invest. **53**, 228-239, 1974.

[43] Hippe, E., Haber, E., and Olesen, H., Nature of vitamin B_{12}-binding. II. steric orientation of vitamin B_{12} on binding and number of combining sites of human intrinsic factor and the transcobalamins, Biochim. Biophys. Acta **243**, 75-82, 1971.

[44] Hall, C.A., Rappazzo, M.E., Green, P.D., Proc. Soc. Exp. Bio. Med. **148**, 1202-1205, 1975.

[45] Thomsen, J., Bucher, D., Brunfeldt, K., Nexø, E., Olesen, H., An Improved procedure for automated Edman degradation used for determination of the N-terminal amino-acid sequence of human transcobalamin I and human intrinsic factor, Eur. J. Biochem. **69**, 87-96, 1976.

[46] Gullberg, R., Scand. J. Gastroenterol. **9**, (Suppl. 29), 23-26, 1974.

[47] Carmel, R., Am. J. Clin. Path. **69**, 319-325, 1978.

[48] Fenrych, W., Hansz, J., Boduch, K., Pytlak, B., Relationship between cobalophilin releasing and functional state of polymorphonuclear granulocytes, in Vitamin B_{12}, Eds. B. Zagalak and W. Friedrich, Walter de Gruyter, Berlin 1979, pp. 813-819.

[49] Stenman, U.H., Simons, K., Gäsbeck, R., Vitamin B_{12}-binding proteins in normal and leukemic human leukocytes and sera, Scand. J. Clin. Lab. Invest. **21**, 202-210, 1968.

[50] Nexø, E., Scand. J. Haematol. **20**, 221-236, 1978.

[51] Burger, R.L., Waxman, S., Gilbert, H.S., Mahlman, C.S., Allen, R.H., J. Clin. Invest. **56**, 1262-1270, 1975.

[52a] Gimsing, P., Hippe, E., Scand. J. Haematol. **21**, 243-249, 1978.

[52b] Zittoun, J., Farcet, J.P., Marquet, J., Sultan, C., Zittoun, R., Cobalamin vitamin B_{12} and B_{12} binding proteins in hypereosinophilic syndromes and secondary eosinophilia, Blood **63**, 779-783, 1984.

[53] Kumento, A., Lopez, R., Luhby, A.L., Hall, C.A., B_{12}-binders in human cord serum, Clin. Res. (abstr.) **15**, 283, 1967.

[54] Kumento, A., Studies on the serum binding of vitamin B_{12} in the newborn human infant, Acta Paediatrica Scand. (Suppl.) **194**, 1-55, 1969.

[55] Stenman, U.H., Nyberg, W., Vitamin B_{12}-binding proteins in fetal blood and amniotic fluid, Scand. J. Clin. Lab. Invest. (Suppl.) **108**, 33, 1969.

[56] Nexø, E., Olesen, H., Chrristensen, J.M., Thomsen, J., Kristiansen, K., Scand. J. Clin. Lab. Invest. **6**, 683-690, 1976.

[57] Simons, K., Vitamin B_{12}-binders in human body fluids and blood cells. Societas Scientiarum Fennica Comment. Biol. **27**, **5**, 1-94, 1964.

[58] Inada, M., Kameyama, M., Toyoshima, M., The significance of vitamin B_{12}-binders in the central nerveous system, in Vitamin B_{12}, Eds. B. Zagalak, W. Friedrich, Walter de Gruyter, Berlin 1979, pp. 821-826.

[59] Glass, G.B.J., Gastric intrinsic factor and other vitamin B_{12}-binders, Georg Thieme Publishers, Stuttgart 1974, p. 96.

[60] Ellenbogen, L., Highley, D.R., Hog intrinsic factor. Isolation of vitamin B_{12}-binding fractions from hog pylorus, J. Biol. Chem. **242**, 1004-1009, 1967.

[61] Gregory, M.E., Holdsworth, E.S., The oc-

currence of a cyanocobalaminbinding protein in milk and the isolation of a cyanocobalamin protein complex from sow's milk. Biochem. J. **59,** 329-334, 1955.

[62] Finkler, A. E., Rappazzo, M., Hall, C. A., Cyanocobalamin binding protein in human milk. Fed. Proc. (abstr.), **26,** 695, 1967.

[63] Gregory, M., Holdsworth, E. S., Some properties of the cyanocobalaminprotein complex from sow's milk and mode of linkage of cyanocobalamin with protein, Biochem. J. **59,** 335-340, 1955.

[64] Gizis, E., Kim, Y. P., Brunner, J. R., Schweigert, B. S., Vitamin B_{12} content and binding capacity of cow's milk proteins. J. Nutrition **87,** 349-352, 1965.

[65] Kim, Y. P., Gizis, E., Brunner, I. R., Schweigert, B. S., Vitamin B_{12} distribution in cow milk. J. Nutr. **86,** 394, 1965.

[66] Kallee, E., Debiasi, S., Karypidis, C., Heide, K., Schwick, H. G., ^{60}Co-Vitamin-B_{12}-bindende Clycoproteine, I. Mitteilung: Anreicherung und Charakterisierung ^{60}Co-Cyanocobalamin-bindende Glykoproteine aus Leukämikerharn und Normalharn, Acta Isotop. **4,** 103-136, 1964.

[67] Kallee, E., Castrillon-Oberndorfer, W. L., Feine, U., Fine, I., Vitamin-B_{12}-bindende Glycoproteine. IV. ^{57}Co-Cyanocobalamin-Bindungskapazität des Harnes bei einigen Blutkrankheiten und bei multiplen Sklerosen, Klin. Wschr. **44,** 845, 1966.

[68] Karypidis, C., Kallee, E., Vitamin-B_{12}-bindende Glykoproteine. V. Immunologische Charakterisierung ^{60}Co-Cyanocobalamin-bindender Uroglykoproteine von normalen Personen und Leukämiepatienten, Acta Isotop. **5,** 407-413, 1965.

[69] Gräsbeck, R., Takki-Luukkainen, I. T., Vitamin B_{12}-binding substance in human tear fluid, Acta Ophthalmol. **36,** 860-864, 1958.

[70] Silverstein, E., Herbert, V., Rapid determination of vitamin B_{12}-binding α and β globulin in serum, Blood **31,** 519-521, 1968.

[71] Hall, C. A., Finkler, A. E., Isolation and evaluation of various B_{12}-binding proteins in human plasma, Methods in Enzymology **18,** 108-126, 1971.

[72] Hom, B. A., Olesen, H., Molecular weight of vitamin B_{12}-binding proteins in human serum determined by Sephadex G-200 gel filtration, Scand. J. Clin. Lab. Intrest. **19,** 269-274, 1967.

[73] Latner, A. L., Ungley, C. C., Cox, E. V., McEvoy-Bowe, E., Raine, L., Electrophoresis of human gastric juice in relation to Castle's intrinsic factor, Brit. Med. J. **1,** 467-473, 1953.

[74] Ostrowski, W., Skarżynski, B., Żak, Z., Vitamin B_{12} in the protein fraction of human blood serum, Bull. Acad. Pol. Sci. Ci. II Ser. Sci. Biol. **2,** 9, 1954.

[75] Pitney, W. R., Beard, M. D., van Loon, E. J., Observations on the bound form of vitamin B_{12} in human serum, J. Biol. Chem. **207,** 143, 1954.

[76] Heinrich, H. C., Erdmann-Oehlecker, S., Sommer, L., Radel, G., Der Vitamin-B_{12}-Stoffwechsel bei Hämoblastosen. II. Die intravitale Bindung (Transport) der B_{12}-Vitamine an die Serumproteinfraktionen bei Hämoblastosen, Clinica Chim. Acta **1,** 311-325, 1956.

[77] Mendelsohn, R., Watkin, D. M., Horbett, A. P., Fahey, J. L., Identification of the vitamin B_{12}-binding protein in the serum of normal and of patients with chronic myelocytic leukemia, Blood **13,** 740, 1958.

[78] Miller, A., Sullivan, J. E., Electrophoretic studies of the vitamin B_{12}-binding protein of normal and chronic myelogenous leukemia serum, J. Clin. Invest. **38,** 2135-2143, 1959.

[79] Bergely, J. A., Trachtenberg, A., Hall, Ch. A., Cobinamide blocking assay for intrinsic factor, in Vitamin B_{12}, Eds. B. Zagalak, W. Friedrich, Walter de Gruyter, Berlin 1979, p. 917.

[80] Gillbert, H. S., J. Lab. Clin. Med. **89,** 13-24, 1977.

[81] Gilbert, In Vitro acid dissociation of cobalamin-transcobalamin II complexes formed in vivo: A probe in the study of TC II kinetics, in Vitamin B_{12}, Eds. B. Zagalak, W. Friedrich, Walter de Gruyter, Berlin 1979, pp. 827-832.

[82] Selhub, J., Rachmilewitz, B., Grossowicz, N., Fractionation of serum transcobalamins on charged cellulose filters, Proc. Soc. Exp. Biol. Med. **152,** 204-209, 1976.

[83] Nexø, E., Andersen, J., Scand. J. Clin. Lab. Invest. **37,** 723-728, 1977.

[84] Nexø, E., Biochim. Biophys. Acta, **379,** 189-192, 1975.

[85] Hom, B. L., Olesen, H., Lous, P., Separation of transcobalamin I and transcobalamin II in human serum, in Protides of the Biological Fluids, Ed. H. Peeters, vol. 14, Elsevier, Amsterdam 1966, pp. 567-669.

[86] Gizis, E., Meyer, I. M., A DEAE-cellulose chromatographic separation of [^{57}Co] vitamin B_{12} binders in normal human serum,

Proc. Soc. Exptl. Biol. Med. **126,** 577-579, 1967.

[87] Gizis, E., Meyer, L. M., Separation of transcobalamins on small DEAE-cellulose columns, Proc. Soc. Exp. Biol. Med. **140,** 326-328, 1972.

[88] Retief, F., Gottlieb, C. W., Kochwa, S., Pratt, P. W., Herbert, V., Separation of vitamin B_{12}-binding proteins of serum, gastric juice and saliva by rapid DEAE-cellulose chromatography, Blood **29,** 501-516, 1967.

[89] Gottlieb, C. W., Retief, F. P., Herbert, V., Blockade of vitamin B_{12} binding sites in gastric juice, serum and saliva by analogues and derivatives of vitamin B_{12} and antibody to intrinsic factor, Biochim. Biophys. Acta **141,** 560-572, 1967.

[90] Hardwicke, J., Jones, J. H., The nature of vitamin B_{12}-binding protein in human serum, Brit. J. Haematol. **12,** 529-535, 1966.

[91] Holdsworth, E. S., The isolation and properties of intrinsic factor and vitamin B_{12}-binding substances from pig pylorus, Biochem. Biophys. Acta **51,** 295-308, 1961.

[92] Gräsbeck, R., Simons, K., Sinkkonen, J., Isolation of intrinsic factor and its probable degradation product, as their vitamin B_{12} complexes, from human gastric juice, Biochim. Biophys. Acta **127,** 47-58, 1966.

[93] Gizis, E., and Meyer, L. M., A DEAE-cellulose chromatographic separation of [^{57}Co] vitamin B_{12} binders in normal serum, Methods in Enzymology **18,** 127-129, 1971.

[94] Garrido-Pinson, G. C., Turner, M. D., Miller, L. L., Segal, H. L., Gel filtration of vitamin B_{12}-binding materials from human gastric juice and gastric mucosa, Biochim. Biophys. Acta **127,** 478-487, 1966.

[95] Parmentier, Y., Marcoullis, G., Nicolas, J. P., and Perrin, M. O., The intraluminal phase of vitamin B_{12} transport in human, in Vitamin B_{12}, Eds. B. Zagalak, W. Friedrich, Walter de Gruyter, Berlin 1979, pp. 791-801.

[96] Hom, B. A., Olesen, H., Molecular weights of vitamin B_{12}-binding proteins in human serum determined by Sephadex G-200 gel filtration, Scand. J. Clin. Lab. Invest. **19,** 269-274, 1967.

[97] Hom, B. L., Demonstration of transcobalamin II complex formation and binding to Sephadex G-200 at low ionic strength, Clin. Chim. Acta, **18,** 315-317, 1967.

[98] Frater-Schröder, M., Osusky, R., Hitzig, W. H., Butler, R., Hakkinen, A. H., Polymorphic variants of transcobalamin II rare alleles in family studies, in Vitamin B_{12}, Eds. B. Zagalak and W. Friedrich, Walter de Gruyter, Berlin 1979, pp. 807-812.

[99] Nicolas, J. P., Perrin, M. O., Parmentier, Y., Ann. Immunol. (Inst. Pasteur) **128c,** 287-290, 1977.

[100] Marcoullis, G., Salonen, E. M., and Gräsbeck, R., Isolation of vitamin B_{12}-binding proteins by combined immuno- and affinity chromatography. Comparative studies on the isolated and unisolated proteins, Biochim. Biophys. Acta, **495,** 336-348, 1977.

[101] Jacobsen, D., Montejano, Y., Affinity photorelease purification of cobalamin-binding proteins, in Vitamin B_{12}, Eds. B. Zagalak, W. Friedrich, Walter de Gruyter, Berlin 1979, pp. 949-954.

[102] Hippe, E., Olesen, H., Nature of vitamin B_{12} binding. III. Thermodynamics of binding to human intrinsic factor and transcobalamins, Biochem. Biophys. Acta, **243,** 83-89, 1971.

[103] Bunge, M. B., Schilling, R. F., Proc. Soc. Exptl. Biol. Med. **96,** 587-592, 1971.

[104] Hippe, E., Haber, E., Olesen, H., Nature of vitamin B_{12}-binding, Biochim. Biophys. Acta, **243,** 5-82, 1971.

[105] Hippe, E., Scand. J. Clin. Lab. Invest. **29,** 59-68, 1972.

[106] Nexø, E., Strength of binding of methyl-,5'-deoxyadenosyl-,cyano- and hydroxocobalamin to human transcobalamin I and II and intrinsic factor, in Vitamin B_{12}, Eds. B. Zagalak, Walter de Gruyter, Berlin 1979, pp. 851-854.

[107] Cresseri, A., Isolation and physical and chemical properties of a vitamin B_{12}-binding factor and other mucopolysaccharides from hog gastric mucosa, in Vitamin B_{12} and intrinsic factor, I. Europ. Symposion, Hamburg, 1956, Ed. H. D. Heinrich, Ferdinand Enke, Stuttgart 1957, pp. 268-283.

[108] Highley, D. R., Davies, M. C., Ellenbogen, L. L., Hog intrinsic factor. Some physicochemical properties of vitamin B_{12}-binding fractions from hog pylorus, J. Biol. Chem. **242,** 1010-1015, 1967.

[109] Hippe, E., Changes in Stoke's radius on binding of vitamin B_{12} to human intrinsic factor and transcobalamins, Biochim. Biophys. Acta. **208,** 337-339, 1970.

[110] Hom, B. L., Stokes radius of ^{57}Co-labelled vitamin B_{12}-transcobalamin I and II, Biochim. Biophys. Acta **175,** 20-23, 1969.

[111] Nexø, E., Olesen, H., Hansen, M. R., Spectral studies on hog intrinsic factor and hog

non-intrinsic factor, in Vitamin B_{12}, Eds. B. Zagalak, W. Friedrich, Walter de Gruyter, Berlin 1979, pp. 895-903.

[112] Gräsbeck, R., Intrinsic factor and the transcobalamins with reflections on the general function and evolution of soluble transport proteins, Scand. J. Clin. Lab. Invest. Suppl. **95,** 7-18, 1967.

[113] Allen, E. H., Majerus, P. W., Isolation of vitamin B_{12} binding proteins using affinity chromatography. II. Purification and properties of a human granulocyte vitamin B_{12}-binding protein, J. Biol. Chem. **247,** 7702-7708, 1972.

[114] Gräsbeck, R., Visuri, K., Isolation of the vitamin B_{12}-binding protein of human saliva (abstr.), Scand. J. Clin. Lab. Invest. **21,** Suppl. 101, 13, 1968.

[115] Gräsbeck, R., Visuri, K., Stenman, U. H., The vitamin B_{12}-binding protein in human saliva. Isolation, characterization and comparison with the unpurified protein, Biochim. Biophys. Acta **263,** 721, 1972.

[116] Glass, G. B. J., Gastric Intrinsic factor and other vitamin B_{12} binders, Georg Thieme Publishers, Stuttgart 1974, p. 101.

[117] Marcoullis, G., Biochim. Biophys. Acta **499,** 373-381, 1977.

[118] Glass, G. B. J., Gastric intrinsic factor and other vitamin B_{12} binders, Georg Thieme Publishers, Stuttgart 1974, p. 104.

[119] Stenman, U. H., Simons, K., Gräsbeck, R., The effect of neuraminidase on human transcobalamin I and leukocyte vitamin B_{12}-binding protein, Scand. J. Clin. Invest. **21,** Suppl. 101, 13, 1968.

[120] Gräsbeck, R., Intrinsic factor and other vitamin B_{12} transport proteins, in Progress in Hematology, Eds. Grune and Stratton, vol. 6, Brown and Moore, New York, 1969, pp. 233-260.

[121] Aro, H., Gräsbeck, R., Electrofocusing of purified intrinsic factor (IF) and a non-IF-active vitamin B_{12}-binding protein from hog pyloric mucosa (abstr.), Scand. J. Clin. Lab. Invest. Suppl. **113,** 65, 1970.

[122] Aro, H., Gräsbeck, R., Visuri, K., Purification of intrinsic factor (IF) and a non-IF-active vitamin B_{12} binding protein from hog pyloric mucosa (abstr.), Scand. J. Clin. Lab. Invest. **23,** Suppl. 108, 28, 1969.

[123] Stenman, U. H., Clinics in Hematol. **5,** 473-494, 1976.

[124] Nexø, E., Olesen, H., Bucher, D., Thomsen, J., Purification and characterization of rabbit transcobalamin II, Biochim. Biophys. Acta **494,** 395-402, 1977.

[125] Bucher, D., Thomsen, J., Nexø, E., Olesen, H., Amino terminal sequence of hog non intrinsic factor and hog intrinsic factor, in Vitamin B_{12}, Eds. B. Zagalak, W. Friedrich, Walter de Gruyter, Berlin 1979, pp. 905-908.

[126] Hudgin, R. L., Pricer, W. E., Jr., Ashwell, G., Stockert, R. J., Morell, A. G., The isolation and properties of a rabbit liver binding protein specific for asialoglucoproteins, J. Biol. Chem. **249,** 5536-5543, 1974.

[127] Kawasaki, T., Ashwell, G., Chemical and physical properties of a hepatic membrane protein that specifically binds asialoglycoprotein, J. Biol. Chem. **251,** 1296-1302, 1976.

[128] Becker, C. M., Beck, W. S., Calcium dependencies in the binding of transcobalamins to subcellular particles of liver cells, in Vitamin B_{12}, Eds. B. Zagalak, W. Friedrich, Walter de Gruyter, Berlin 1979, pp. 833-836.

[129] Nicolas, J. P., Perrin, M. O., Parmentier, Y., Gastric intrinsic factor (IF) autoantibodies of types I and II, Scand. J. Haematol. **20,** 479, 1978.

[130] Carmel, R., Baril, L., J. Lab. Clin. Med. **91,** 769-799, 1978.

[131] Wolff, R., Nicolas, J. P., Lancet **1,** 1008-1009, 1967.

[132] Gräsbeck, R., Aro, H., Biochem. Biophys. Acta. **252,** 217-220, 1971.

[133] Marcoullis, G., Gräsbeck, R., Biochem. Biophys. Acta, **476,** 36-51, 1977.

[134] Hurliman, J., Zuber, C., Vitamin B_{12}-binders in human body fluids. I. Antigenic and physicochemical charakteristics, Clin. Exptl. Immunol. **4,** 125-140, 1969.

[135] Carmel, R., Large vitamin B_{12}-binders proteins and complexes in human serum, in Vitamin B_{12}, Eds. B. Zagalak, W. Friedrich, Walter de Gruyter, Berlin 1979, pp. 777-790.

[136] Schwartz, M., Bastrup-Madsen, P., A new Vitamin B_{12}-binding protein in serum causing excessively high serum vitamin B_{12} values, Scand. J. Haematol. **5,** 35-40, 1968.

[137] Ellenbogen, L., Absorption and transport of cobalamin, in Cobalamin, Ed. B. Babior, John Willey & Sons, New York, 1975, pp. 215-286.

[138] Wierusz-Wysocka, B., Vitamin B_{12}-binding by serum proteins in virus hepatitis, Pol. Arch. Med. Wewn. **56,** 445-456, 1976.

[139] Wierusz-Wysocka, B., Vitamin B_{12}-binding by serum proteins in chronic liver disease, Pol. Arch. Med. Wewn. **58,** 139-147, 1977.

[140] Nochumson, S. and Meyer, L. M., Behaviour of human serum R-type vitamin B_{12}-binding proteins (cobalophilins) on lectin-Sepharose affinity resins (41023), Proc. Soc. Exp. Biol. Med. **166,** 51-56, 1981.

[141] Allen, R. H., Burger, R. L., Mehlman, C. S., Majerus, P. W., Affinity Techniques, Vitamin B_{12}, in Methods in Enzymology **34,** 305-314, 1974.

[142] Hansen, H. J., Jacob, E. and Frentz, G. D., Assay of antintrinsic factor antibodies, Methods in Enzymology **18,** 130-133, 1971.

[143] Francis, G. L., Intrinsic factor mediated cobalamin absorption, Ann. Clin. Lab. Sci. **10,** 334-337, 1980.

[144a] Azen, E. A., and Denniston, C., Genetic polymorphism of vitamin B_{12}-binding R-proteins of human saliva detected by isoelectric focusing. Biochem. Genet. **17,** 909-920, 1979.

[144b] Yang, S. Y., Coleman, P. S. and Dupont, B., The biochemical and genetic basis for the microheterogeneity of human R-type vitamin B_{12}-binding proteins, Blood **59,** 747-755, 1982.

[145] Samson, R. R., and McClelland, D. B. L., Vitamin B_{12} in human colostrum and milk: Quantitation of the vitamin and its binder and the uptake of bound vitamin B_{12} by intestinal bacteria, Acta. Pediatr. Scand. **69,** 93-100, 1980.

[146] Andersen, K. J., and Von der Lippe, G., The effect of proteolytic enzymes on the vitamin B_{12}-binding proteins of human gastric juice and saliva, Scand. J. Gastroenterol. **14,** 833-838, 1979.

[147] Itaya, H., and Fukuda, M., Studies on serum vitamin B_{12}-binding proteins in chronic myeloproliferative disorders: Especially chronic myelogenous leukemia and polycythemia vera, Sappor. Med. J. **49,** 81-98, 1980.

[148] Gimsing, P., and Hippe, E., Increased concentration of transcobalamin I in a patient with metastatic carcinoma of the breast. Scand. J. Haematol. **21,** 243-249, 1978.

[149] Van Kapel, J., Loef, B. G., Lindemans, J. and Abels, J., An improved method for large scale purification of human holo-transcobalamin II. Biochim. Biophys. Acta **676,** 307-313, 1981.

[150] Lindemans, J., De Jongh, E. J. M., Brand, F. C. M., Schoester, M., van Kapel, J. and Abels, J., The uptake of R-type cobalamin-binding protein by isolated rat liver cells, Biochim. Biophys. Acta **720,** 203-210, 1982.

[151] Areekul, S. and Churdehu, K., Vitamin B_{12} and vitamin B_{12}-binding proteins in cord blood, J. Med. Assoc. Thail. **64,** 604-610, 1981.

9 Medical Aspects of Vitamin B_{12}

A. Stroinski

9.1 Introduction

Vitamin B_{12} was applied in medicine in the early twienties when two American physicians Minot and Murphy obtained positive results in curing pernicious anaemia with liver diet. This disease was reported as early as 1824. It remained fatal, however, until this discovery. The discovery and application of the antipernicious factor is one of the most fascinating events in the history of medicine. Minot, Murphy and Whipple received in 1934 the Nobel Prize for their life saving contribution (see references in Chapter 1). Liver contains appreciable concentration of vitamin B_{12} and it served for the next two decades as the main source of the unknown, at that times, curing factor, called also *extrinsic factor*. In 1929 Castle demonstrated that the curing effect of the liver diet could be greatly enhanced by a factor present in stomach juice of healthy humans. This substance appearing to be a protein derived from the gastric secretion was called by him *intrinsic factor*. It was required for the absorption of the hemopoietic substance from liver and other animal derived food. Vitamin B_{12} was obtained in pure cristalline state in 1948 and its structure was elucidated in the following years (see Table 1-1). The production of vitamin B_{12} on industrial scale in the early fifties enabled its worldwide application in medicine to treat pernicious anaemia. It was observed already in the 19th century that some neuropathies are associated with pernicious anaemia (for references see Chapter 1). The neurological disturbances appearing at moderate vitamin B_{12} deficiencies are treated also with vitamin B_{12}, however, its curing effect appears not so spectacular as it is in pernicious anaemia cases. Parallel to medical application basic studies on absorption and transport of vitamin B_{12} into the tissues has begun in many laboratories and they are continued to this time. Besides *intrinsic factor* (IF) two other proteins: *transcobalamin* (TC) and *cobalophilin* (CP) were isolated and characterized (for details see Chapter 8). The mechanism of action of vitamin B_{12} in human which revealed to be a cofactor for enzymes, was studied as well. Although more than ten B_{12}-dependent enzymes or enzyme systems were isolated from bacteria, only two of them have been found in man (for details see Chapter 7). These two enzymes require for activity two different forms of vitamin B_{12}: 5'deoxyadenosyl cobalamin (AdoCbl) [1] for methylmalonyl CoA-mutase and methylcobalamin (MeCbl) [2] for homocysteine methyltransferase (see Chapter 7). These forms are interconvertible within the organism (see section 6.6.2). Besides the two coenzymatic forms three other cobalamins: hydroxocobalamin (OH-Cbl), sulphitocobalamin (SO_3-Cbl) [4] and cyanocobalamin (CN-Cbl) [2, 3] have been found in human. The latter may originate from medical treatment (CN-Cbl is used for injection), smoking, or special diet (for occurrence of various forms of Cbl in man see Chapter 6, their structure and properties are specified in Table 3-1). The cobalamins are usually determined by thin-layer chromatography with bioautography [8, 10] (see Chapter 5), but SO_3-Cbl is not easily assessed with this method. The one-dimensional thin-layer chromatographic method described recently by Nexø et al. [4, 5, 7, 9] is capable of distinguishing all five cobalamins. Figure 9-1 shows the cobalamin pattern in erythrocytes from healthy donor obtained by means of this method. The result seems to show more B_{12} forms than previously presented by Linnell et al. [10] (see also section 6.3). The spot moving between AdoCbl and CN-Cbl suggests existance of another form of cobalamin. For the same purpose cation exchange chromatography with SP-Sepha-

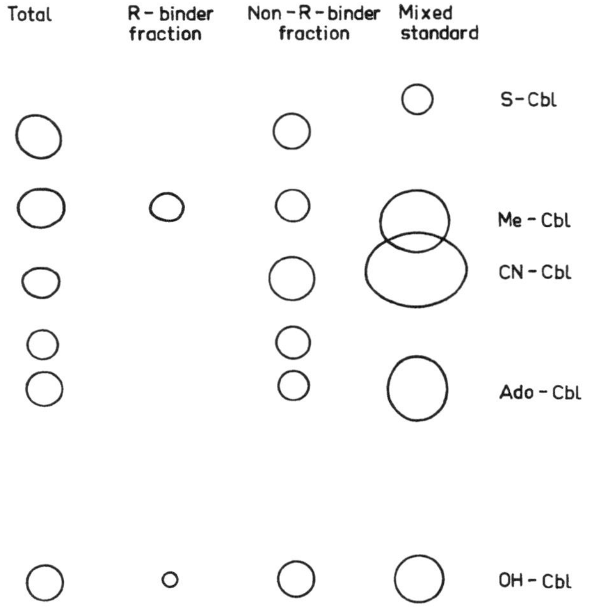

Fig. 9-1 A typical example of the cobalamin pattern in erythrocyte and the distribution on R-binder fraction and non-R-binder fraction (reprinted from Nexø and Gimsing [5])

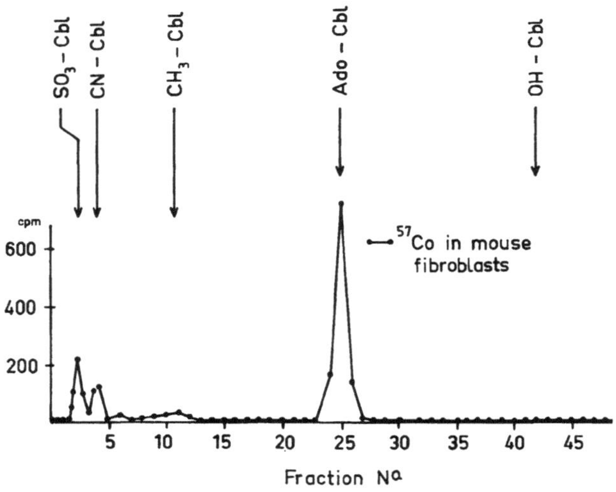

Fig. 9-2 Reprinted from Begley and Hall [6]

dex C-25 may be used [6]. Fig. 9-2 presents the cobalamins, estimated in this way, in mouse fibroblasts after uptake of transcobalamin-CN [^{57}Co]Cbl complex for 24 hours [6]. The absorption and circulation of Cbl in human are presented in Fig. 8-23. Phato-physiological and biochemical aspects of vitamin B_{12} disorders in man are reviewed in this chapter only. This concerns in particular to such clinical features, as: megaloblastic anaemia, neurological disorders, malignant neoplasms and the mechanism of these disorders related to the absorption, transport and utilization within tissues and cells.

9.2 Clinical feature of the cobalamin abnormalities syndrom

9.2.1 Megaloblastic anaemia and related changes

The "megaloblastic anaemia" concerns morphological and functional disorders of the blood and the bone morrow cells and is caused by cobalamin deficiency in the system.

This syndrom may have the following clinical features:

- ineffective erythropoiesis and moderate hemolysis [11-18], high death rate of arrested E_3 and E_4 cells [19-18],
- inefficient leukopoiesis and thrombopoiesis [20, 21 a],
- abnormal morphology of blood elements [21 b]: erythrocytes are normochromic and macrocytic, reticulocyte count is lower than normal, nucleis of neutrophils often have more than four segments,
- abnormal morphology of bone marrow cells [21 c]; megaloblastic changes in the erythrocyte, granulocyte and platelet precursors, larger proerythroblast with higher than normal ratio of cytoplasmic area to nuclear area, during maturation the chromatin maintains its granular structure and coarsness, characteristic nuclear-cytoplasmic asynchronism for precursors of the erythrocyte and the granulocyte,
- the features of morphological abnormalities of the chromosomes in the bone morrow are presented in the quotation and the figures from Menzies et al. [22]:

1. The centromeric constrictions were of exaggerated size (Figs. 9-3 and 9-4). Similarly, the normal sites of secondary constriction showed marked negative heteropycnosis and appeared as gaps (Figs. 9-3 and 9-4). The satellites of the acrocentric chromosomes (numbers 13-15 and 21-22) were prominent in most cells, mainly because the heteropycnotic secondary constrictions which separate them from the body of the chromosome were attenuated (Fig. 9-4).

2. Chromatid breakage was the most common of all abnormalities, being present in 5-33 per cent of the scorable cells in nine of the ten cases, and individual cells often showed several breaks (Fig. 9-5). Significantly, the breaks involved one chromatid only, and there was no evidence of reunion or interchanges. A considerable number of metaphases from some patients showed more extensive chromosome fragmentation, with multiple constrictions and breaks affecting so many chromosomes that counting became impossible. In some cells fracture with separation of the broken chromatids occurred, particularly near the centromeric region (Fig. 9-6).

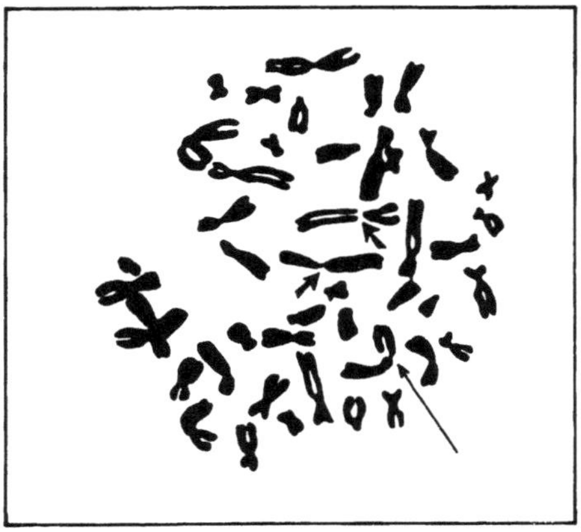

Fig. 9-3 Methaphase from case 1 showing a prominent secondary constriction in an A group chromosome (long arrow) and prominent centromeric constrictions (short arrows) (Reprinted from Menzies et al. [22] Blood **28**, 584, 1966, with permission)

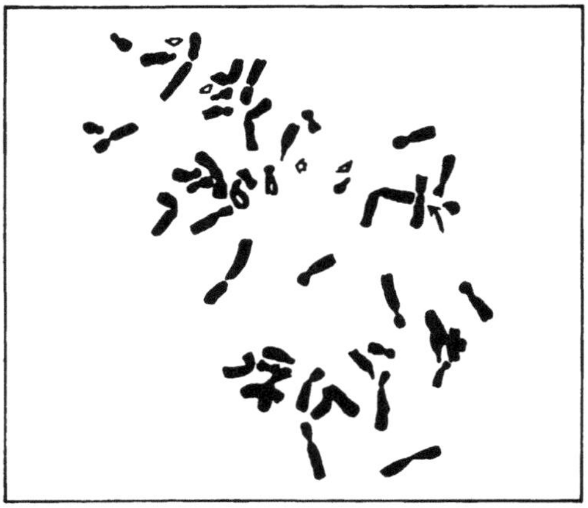

Fig. 9-4 Methaphase from case 2 showing prominent satellites on B and G group chromosomes (light arrows) and a secondary constriction in a C group chromosome (dark arrow) (Reprinted from Menzies et al. [22], Blood **28**, 585, 1966, with permission)

3. A number of cells showed thin, elongated chromosomes which probably resulted either from reduced contraction or despiralization. In these cells the centromeric and secondary constrictions were characteristically attenuated.

4. In patients showing the greatest number of abnormalities, some metaphases had abnormally small and poorly defined chromosomes whose chromatids had separated; others showed more pronounced degeneration of the chromosomes (Fig. 9-7).

5. The colcemid-treated megaloblastic marrows showed sizable numbers of mitoses in prophase (up to 52 per cent as against about 15 per cent in a normoblastic marrow) [11].

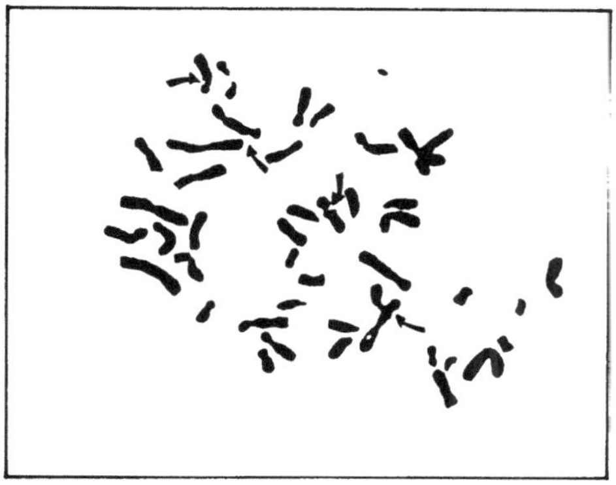

Fig. 9-5 Methaphase from case 2 showing chromatid breaks (arrows) (Reprinted from Menzies et al. [22], Blood **28**, 586, 1966, with permission)

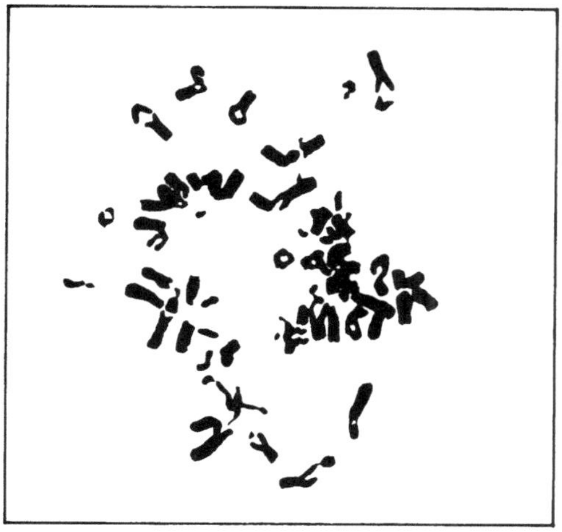

Fig. 9-6 Methaphase from case 1 showing multiple breakage of chromosomes, especially near the centromeres and separation of chromatids (Reprinted from Menzies et al. [22], Blood **28**, 587, 1966, with permission)

- megaloblastic cells are in a state of unbalanced growth with impairment of DNA synthesis, RNA/DNA ratio increases [23-46],
- sharp increase in the number of unlabeled S-phase cells in bone morrow employing ^{3}H-thymidine [22, 26, 27, 49, 50],
- a correlation between the severity of the anaemia and the macrocytosis [47],
- the principal effects of AdoCbl and MeCbl deficiencies in man, according England and Linnell [47], are presented in Fig. 9-8.
- metabolic changes:
  - decreased level of serum total vitamin B_{12}, MeCbl is reduced disproportionately [10, 51-55 a-c] (Figs. 9-9 and 9-10, and Table 9-1).

Fig. 9-7 Methaphase from case 1 showing advanced degeneration of chromosomes (Reprinted from Menzies et al. [22], Blood **28**, 588, 1966, with permission)

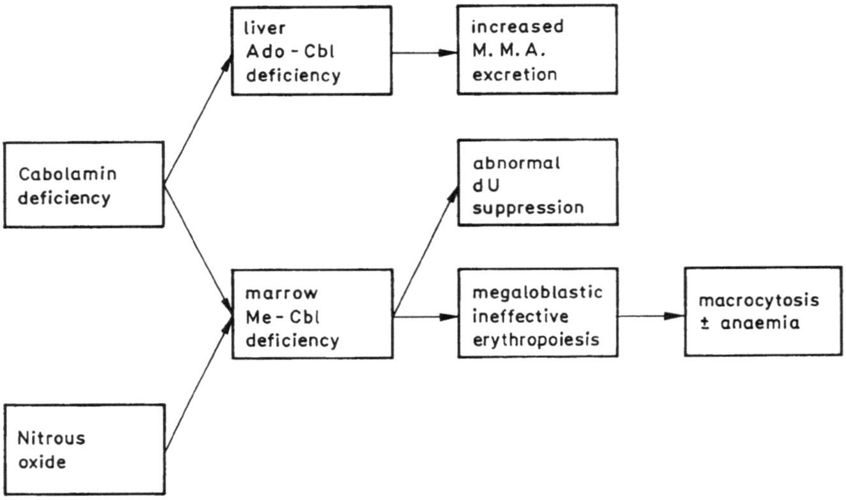

Fig. 9-8 Consequence of AdoCbl and MeCbl deficiency in man (Reprinted from England and Linnell [121])

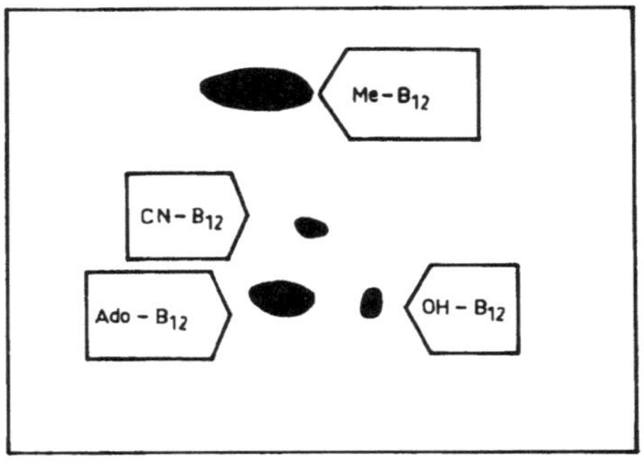

Fig. 9-9 Plasma cobalamins from a normal subject two-dimensional chromatography and bioautography. A trace of CN-Cbl_{12} is present in this sample [10] (Reprinted by permission from Clin. Sci. Mol. Med. **46**, 1963, 1974, copyright (c) 1974 The Biochemical Society, London)

Table 9-1 Blood and bone marrow cobalamins in control subjects and patients with pernicious anaemia [10] (Reprinted by permission from Clin. Sci. Mol. Med **46**, 163, 1974, copyright (c) 1974 The Biochemical Society, London)

	No. of samples	Total vitamin B_{12}	Me-Cbl		CN-Cbl		Ado-Cbl	
		(pg/ml)	(pg/ml)	(%)	(pg/ml)	(%)	(pg/ml)	(%)
Plasma								
Controls	22	385 ± 31[1]	251 ± 22	65	7.8 ± 1.2	2.0	126 ± 12	33
Pernicious anaemia	12	88 ± 11	22 ± 3.5	25	8.2 ± 2.2	9.3	58 ± 7.6	66
Erythrocytes								
Controls	12	202 ± 25	31 ± 4.2	15	11 ± 2.3	5.5	159 ± 24	79
Pernicious anaemia	9	168 ± 39	23 ± 3.2	14	16 ± 6.1	9.5	129 ± 29	77
		(pg/g)	(pg/g)	(%)	(pg/g)	(%)	(pg/g)	(%)
Leucocytes								
Controls	18	3830 ± 437	657 ± 90	20	139 ± 40	4.1	2978 ± 458	76
Pernicious anaemia	3	2575 ± 1467	518 ± 243	20	243 ± 202	9.4	1813 ± 1364	70
Bone marrow								
Controls	13	12960 ± 1440	1700 ± 316	13	274 ± 108	2.1	10940 ± 1275	85
Pernicious anaemia	10	6450 ± 1404	984 ± 462	15	375 ± 139	5.8	5091 ± 1101	79

[1] Mean ± SEM.

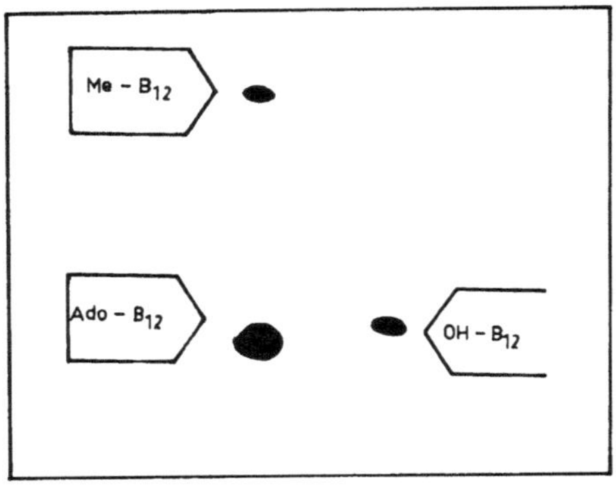

Fig. 9-10 Plasma cobalamins from a patient with untreated pernicious anaemia (plasma total vitamin B_{12} 80 pg/ml). MeCbl is disproportionately reduced. No CN-Cbl is detectable in this sample [10] (Reprinted by permission from Clin. Sci. Mol. Med. **46**, 163, 1974, copyright (c) 1974 The Biochemical Society, London)

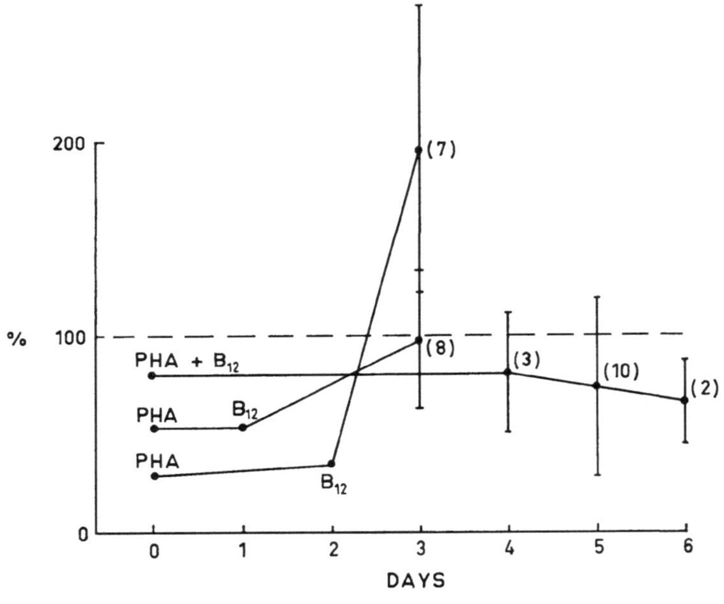

Fig. 9-11 The effect of vitamin B_{12} on thymidylate activity when added to cultures of stimulated lymphocytes (reprinted from Haurani [81], Science **182**, 78, 1973. Copyright 1973 by the AAAS)

- in erythrocytes, leukocytes, and bone morrow cells total cobalamin levels are reduced without change of proportion between cobalamins, only CN-Cbl increased [10, 54, 55] (Figs. from 9-11 to 9-14 and Table 9-1)
- reduced level of AdoCbl in liver is responsible for methylmalonicaciduria [56],
- depression of leukocyte methylmalonyl-CoA mutase activity [51, 57],
- urinary excretion of methylmalonic acid is increased from a few mg to sometimes more than 300 mg/24 h [10, 58-60 a-b]
- urine propionic acid is increased [60 a],

Table 9-2 Thymidylate synthetase activity in phytohemagglutinin-stimulated lymphocytes. The enzyme activity is expressed as the uptake of dUMP by 10^7 cells. Two of the six patients with pernicious anaemia were studied both in relapse and in remission; the others were studied only in remission (Reprinted from Haurani [80])

Condition	Number of individuals	Enzyme activity (100 pmol/h)	Average activity
Vitamin B_{12} deficiency	8	0.0-2.9	1.1*
Pernicious anaemia			
Relapse	2	0.0	
Remission	2	6.8, 6.1	
Relapse			
Remission	4	3.0-12.9	7.1+
Folic acid deficiency	4	7.4-10.5	8.9
Normal	15	4.7-12.8	8.2

* The enzyme activity is the average of the activities of the enzymes from patients with vitamin B_{12} deficiency including the activities of the two patients with pernicious anaemia in relapse
+ The enzyme activity is the average for the activities of the enzymes from all six patients with pernicious anaemia in remission

- excretion of formiminoglutamic acid is elevated after an oral loading of histidine [61, 61 a, 61 b],
- N^5-methyltetrahydrofolate accumulates in the serum [62-67],
- serum folate level is increased but erythrocyte folate is depressed [66 a],
- bone morrow MeCbl deficiency is a reason why the deoxyuridine (dU) suppression test results abnormal values [46, 48, 68-76], and this test can be corrected by addition of vitamin B_{12} or folate [70, 71, 73, 76-78, 78 a],
- in vitro methionine increases and homocysteine decreases the incorporation of ^{125}I UdR to normal and megaloblastic bone morrow cells [46, 71],
- uptake of N^5-methyltetrahydrofolate by bone morrow cells is decreased [79],
- the lack of thymidylate synthetase activity [80, 81] (Table 9-2),
- vitamin B_{12} stimulates the activity of thymidylate synthetase in lymphocytes [81] (Fig. 9-11),
- the methionine synthetase activity in bone morrow cells is significantly lower than in normal (Figs. 9-12 and 9-13),
- increase of enzyme activity in serum: lactate dehydrogenase [14, 84-86], muramidase [20], 6-phosphogluconic dehydrogenase [87], malic dehydrogenase [87],
- decrease of activity of serum alkaline phosphatase [88],
- nitrogen balance is negative [89],
- in the serum and in erythrocytes the lactic dehydrogenase isoenzyme 1 (LDH-1) is more active than LDH-2 in normal and in nonmegaloblastic anaemias subjects reversal dependence [90],
- intrinsic factor (IF) deficiency in gastric juice [91-106],
- serum contains various populations of anti-IF antibodies [107-109],
- exocrine pancreatic insufficiency [110, 111],
- elevated level of serum deoxytymidine kinase [111 a]

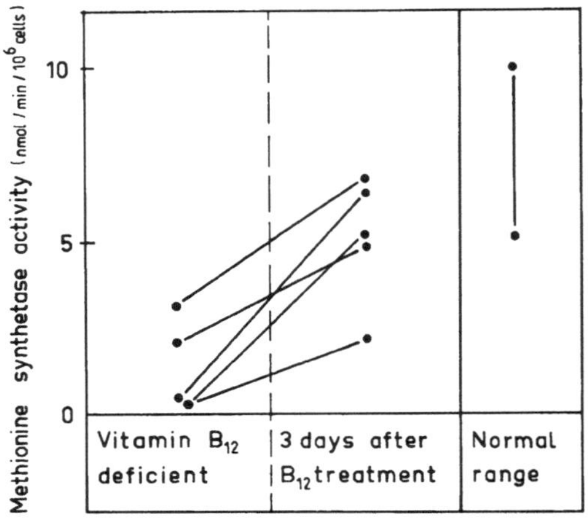

Fig. 9-12 Methionine synthetase activity in bone morrow cells from patients with vitamin B_{12} deficiency before and after B_{12} therapy (Reprinted from Sauer and Wilmanns [66])

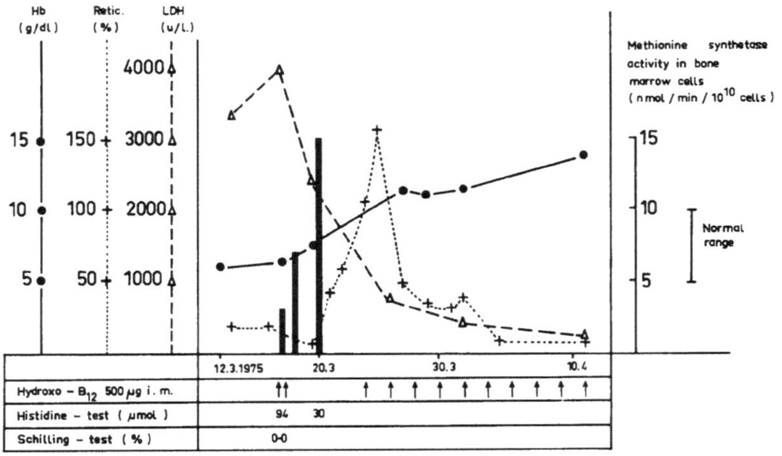

Fig. 9-13 The effect of vitamin B_{12} treatment on the activity of methionine synthetase in a patient with pernicious anaemia (Reprinted from Sauer and Wilmanns [66])

In some cases megaloblastic anaemia is due to inborn metabolic error [77, 112, 113, 113a, b]. In Tables 9-3 and 9-4, and Fig. 9-14 are shown hematological results, dU suppression test and other investigations for two patients with inherited deficiency of transcobalamin (TC II), (see section 8.2)

Clinical response to cobalamin therapy:

- Addison's pernicious anaemia was associated with paralysis of upward gaze [113b]
- Imerslund - Gräsbeck anaemia - congenital, hereditary selective malabsorption of vitamin B_{12} [113c]
- highly elevated transcobalamin concentrations were found in serum of patients with rheumatoid arthritis and traumatic synovits [113d]

Table 9-3 Haematology data of two patients with inherited TC-deficiency (from Burman et al. [77]).

	Patient 1	Patient 2
Hemoglobin (g/dl)	14.9	9.7
Packed Cell Volume (l/l)	0.44	0.28
Red Cells (10^9/l)	4.6	3.17
MCV (fl)	98	89
MCH (pg)	33	31
MCHC (g/dl)	34	35
Reticulocytes (%)	1.0	0.3
White cells (10^9/l)	6.3	8.6
Neutrophils (10^9/l) 70%	4.4	6.9
Eosinophils 1%		
Lymphocytes(10^9/l)23%	1.4	1.4
Monocytes 6%		
Platelets (10^9/l)	241	282
Blood film: Macrocytic red cells		Macrocytic red cells
Bone marrow: Megaloblastic		Megaloblastic
Serum Vitamin B_{12} (ng/l)	800	505
Serum folate (μg/l)	33	18
Red cell folate (μg/l)	1439	590
Methylmalonic acid excretion	+	+++

Table 9-4 Serum levels of transcobalamin (TC), cobalophilin (CP) and unsaturated B_{12} binding capacity (UBBC) in two patients with inheriled TC-deficiency (from Burman et al [77]). The level is expressed in picograms of the vitamin B_{12} bound to these proteins present in 1 ml of serum (see section 8.2.2)

Measurement	Method	Control	Patient 1	Patient 2
UBBC	charcoal	1225	438	235
CP (TC I)		345	372	204
TC		880	66	30
UBBC	gel filtration	1515	304	153
CP (TC I)		182	269	131
TC		1311	0	0[c]

[a] Patient 1 was receiving folic acid and had reccived B_{12} four weeks previously.
[b] Patient 2 had received folic acid and blood transfusion
[c] No endogenous B_{12} was found by microbiological assay of the gel filtration fraction corresponding to TC.

- increased circulating levels of transcobalamin II was found in Gaucher's disease [121, 123], multiple myeloma [122, 123] and others diseases (Table 9-5 A)
- decrease in plasma bilirubin, iron, lactic dehydrogenase activity, iron turnover, and faecal urobilinogen [114] (Fig. 9-15),
- reticulocyte crisis [114, 115],
- increase in serum cobalamin [116-118, 118 a, b, c]
- MeCbl and AdoCbl levels are elevated in bone morrow cells [55],
- decrease in serum folate [89],

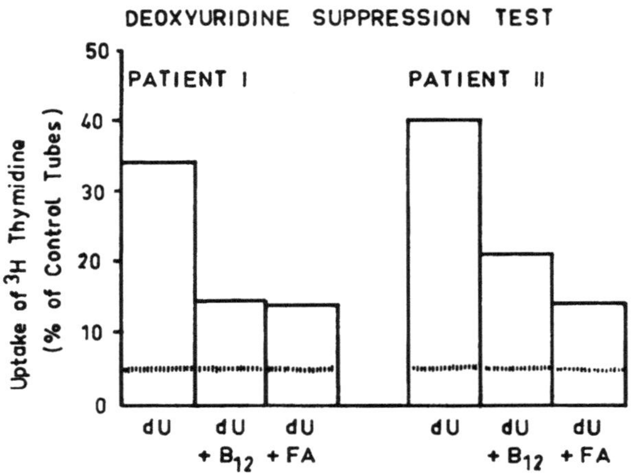

Fig. 9-14 Reprinted from Burman et al. [77].

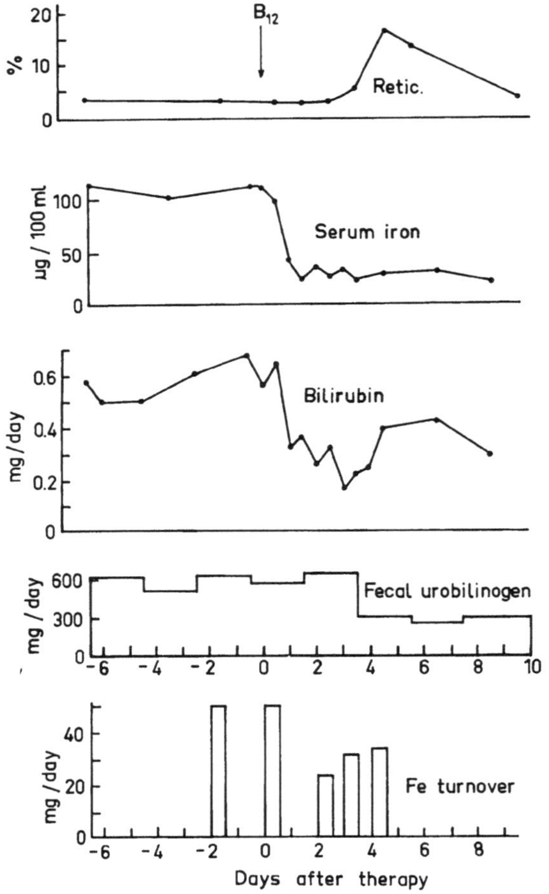

Fig. 9-15 Effect of therapy with cyanocobalamin on reticulocyte count, plasma iron, serum bilirubin, fecal urobilinogen, and plasma iron turnover, in a patient with the cobalamin deficiency of pernicious anaemia (Reprinted from Finch et al. [114], Blood **11,** 807, 1956, with permission)

- increase in uptake of ^{3}H-N^{5}-methyltetrahydrofolate by bone marrow cells [79],
- high activity of thymidylate synthetase (Table 9-2 and Fig. 9-11),
- increase in activity of methionine synthetase [80, 81] (Figs. 9-12 and 9-13) and serum alkaline phosphatase [88],
- sharp decrease in serum potassium, which should be monitored, and in the same time red blood cell potassium rise [119, 120],
- rapid increase in serum and urine uric acid [89],
- decrease in urine phosphorus after administration, increase of phosphorus in course of reticulocytosis, then return to normal [89],
- excretion of formiminoglutamic acid is reduced [61],
- rapid improvement in the frame of mind [88].

Table 9-5 A Distribution of endogenous cobalamin in sera carried on CP and TC from patients with various disorders (reprinted from Carmel, Am. J. Clin. Nutr. **41,** 713-719, 1985)

	Serum cobalamin (pg/ml)	Percent of recovered endogenous cobalamin carried on	
		CP	TC
Myeloproliferative disease			
CML 1	9,574	99	0
CML 2	2,065	98	2
CML 3	1,790	60	40
Eosinophilic leukemia	12,720	100	0
Myelofibrosis	651	84	16
Polycythemia vera	1,952	96	4
Liver disease			
Alcoholic hepatitis	978	83	15
Cirrhosis 1	1,477	99	1
Cirrhosis 2	1,866	56	30
Cirrhosis 3	1,149	88	12
Cirrhosis 4	3,520	22	78
Hemochromatosis	552	37	63
Miscellaneous			
Pernicious anaemia	105	87	2
Pregnancy	82	66	28
Plasmacytoma	41	77	13
Multiple myeloma 1	136	67	33
Multiple myeloma 2	233	48	28
Multiple myeloma 3	324	82	18
Multiple myeloma 4	499	64	28
Multiple myeloma 5	581	58	34
Multiple myeloma 6	999	10	90
Aplastic anaemia	716	32	68
Gaucher's disease	1,018	27	69
Felty's syndrome	555	90	6

References

[1] Toohey, J. I., Barker, H. A., Isolation of coenzyme B_{12} from liver, J. Biol. Chem. **236,** 560-563.

[2] Lindstrand, K., Stahlberg, K. G., On vitamin B_{12} forms in human plasma, Acta Med. Scand., **174,** 665-669, 1963.

[3] Matthews, D. M., Distribution of cobalamins in the animal body, in Vitamin B_{12}, Eds. B. Zagalak and W. Friedrich, Walter de Gruyter, Berlin 1979, pp. 681-694.

[4] Gimsing, P., Hippe, E., Nexø, E., Determination of the plasma cobalamins by one-dimensional thin-layer chromatography, in Vitamin B_{12}, Eds. B. Zagalak and W. Friedrich, Walter de Gruyter, Berlin 1979, pp. 665-669.

[5] Nexø, E., Gimsing, P., The pattern of cobalamin in human erythrocytes, in Vitamin B_{12}, Eds. B. Zagalak and W. Friedrich, Walter de Gruyter, Berlin 1979, pp. 967-970.

[6] Begley, J. A., and Hall, C. A., Presence of sulfitocobalamin in cell extracts. Resolution and identification by SP-Sephadex C-25. Cation exchange chromatography, in Vitamin B_{12}, Eds. B. Zagalak, W. Friedrich, Walter de Gruyter, Berlin 1979, pp. 971-974.

[7] Nexø, E., Characterization of the cobalamin attached to transcobalamin I and transcobalamin II in human plasma, Scan. J. Haematol. **18,** 358-360, 1977.

[8] Linnell, J. C., Mackenzie, H. M., Wilson, J., Matthews, D. M., Patterns of plasma cobalamins in control subjects and in cases of vitamin B_{12} deficiency, J. Clin. Path. **22,** 545-550, 1969.

[9] Nexø, Ebba, Andersen, J., Unsaturated and cobalamin saturated transcobalamin I and II in normal human plasma, Scand. J. clin. Lab. Invest. **37,** 723-728, 1977.

[10] Linnell, J. C., Hoffbrand, A. V., Hussein, H. A. A., Wise, I. J., Matthews, D. M., Tissue distribution of coenzyme and other forms of vitamin B_{12} in control subjects and patients with pernicious anaemia, Clin. Sci. Mol. Med. **46,** 163-172, 1974.

[10 a] Zimran, A., Hershke, C., The changing pattern of megaloblastic anemia. Megaloblastic anemia in Israel, Am. J. Clin. Nutr. **37,** 855-861, 1983.

[11] Finch, C. A., Coleman, D. H., Motulsky, A. G., Donohue, D. M., Reiff, R. H., Erythrokinetics in pernicious anaemia. Blood **11,** 80, 1956.

[12] Myhre, E., Studies on erythrokinetics in pernicious anaemia, Scand. J. Clin. Lab. Invest. **16,** 391, 1964.

[13] Heller, P., Weinstein, H. G., West, M., Zimmerman, H. J., Enzymes in anaemia: a study of abnormalities of several enzymes of carbohydrate metabolism in the plasma and erythrocytes in patients with anaemia, with preliminary observations of bone marrow enzymes, Ann. Inter. Med. **53,** 898-913, 1960.

[14] Elliot, B. A., Fleming, A. F., Source of elevated serum enzyme activities in patients with megaloblastic erythropoiesis secondary to folic-acid deficiency, Brit. Med. J. **1,** 626-628, 1965.

[15] Libnoch, H. A., Yakulis, V. J., Heller, P., Lactate dehydrogenase in megaloblastic bone marrow, Am. J. Clin. Pathol. **45,** 302-305, 1966.

[16] London, J. M., West, R., The formation of bile pigment in pernicious anaemia, J. Biol. Chem. **184,** 359-364, 1950.

[17] White, P., Coburn, R. F., Williams, W. J., Goldwein, M. I., Rother, M. L., Shafer, B. C., Carbon monoxide production associated with ineffective erythropoiesis, J. Clin. Invest. **46,** 1986, 1967.

[18] Lynch, E. C., Alfrey, C. P., Jr., Studies in vitro on the autohemolysis of normoblastic and megaloblastic morrow cells, Texas Rep. Biol. Med. **24,** 180, 1966.

[19] Wickeramasinghe, S. N., Kinetics and morphology of homopoiesis in pernicious anaemia, Brit. J. Haemat. **22,** 111-115, 1972.

[20] Perillie, P. E., Kaplan, S. S., Finch, S. C., Significance of changes in serum muramidase activity in megaloblastic anaemia, New Engl. J. Med. **277,** 10-12, 1967.

[21 a] Harker, L. A., Finch, C. A., Thrombokinetics in man, J. Clin. Invest. **48,** 963-974, 1969.

[21 b] Magnus, E. M., Red cell changes in top, middle and bottom layer of packed red cells in vitamin B_{12} deficiency, Scand. J. Haematol. **24,** 299-306, 1980.

[21 c] Ingeberg, S., Stoffersen, E., Platelet disfunction in patients with vitamin B_{12} deficiency, Acta Haematol. **61,** 75-79, 1979.

[22] Menzies, R. C., Crossen, P. E., Fitzgerald, P. H., Gunz, F. W., Cytogenetic and cytochemical studies in marrow cells in B_{12} and folate deficiency, Blood **28,** 581-594, 1966.

[23] Glazer, H. S., Mueller, J. F., Jarrold, T., Sakurai, K., Will, J. J., Vilter, R. R., Effect of vi-

tamin B_{12} and folic acid on nucleic acid composition of the bone marrow of patients with megaloblastic anaemia, J. Lab. Clin. Med., **43,** 905, 1954.
[24] White, J.C., Leslie, I., Davidson, J.N., Nucleic acids of bone marrow cells with special references to pernicious anaemia, J. Pathol. Bacteriol. **66,** 291-306, 1953.
[25] Lessner, H.E., Friedkin, M., In vitro incorporation of deoxyuridine and thymidine into deoxyribonucleic acid of human megaloblastic and normoblastic bone marrow, Clin. Res., **7,** 207, 1959.
[26] Wichramasinghe, S.N., Cooper, E.H., Chalmers, D.E., A study of erythropoiesis by combined morphologic, quantitative cytochemical and autoradiographic methods. Normal human bone marrow, vitamin B_{12} deficiency and iron deficiency anaemia, Blood **31,** 304, 1968.
[27] Yoshida, Y., Todo, A., Shirakawa, S., Wakisaka, G., Uchino, H., Proliferation of megaloblasts in pernicious anaemia as observed from nucleic acid metabolism, Blood **31,** 292, 1968.
[28] Beck, W.S., The metabolic functions of vitamin B_{12}, N. Engl. J. Med. **266,** 708-714, 1962.
[29] Beck, W.S., The metabolic functions of vitamin B_{12}, N. Engl. J. Med. **266,** 814-817, 1962.
[30] Beck, W.S., Medicine (Balt.) **43,** 715, 1964.
[31] Cohen, S.S., Barner, H.E., Studies on unbalanced growth in *Escherichia coli,* Proc. Natl. Acad. Sci. USA **40,** 885-893, 1954.
[32] Beck, W.S., Hook, S., Barnett, B.H., The metabolic functions of vitamin B_{12}. I. Distinctive modes of unbalanced growth behavior in *Lactobacillus Leichmanii,* Biochem. Biophys. Acta, **55,** 455-469, 1962.
[33] Freifelder, D., Maaløe, O., Energy requirement from thymineless death in cells of *Escherichia coli,* J. Bacteriol. **88,** 987-990, 1964.
[34] Smith, B.J., Burton, K., The integrity of deoxyribonucleic acid extracted from *Escherichia coli* 15 T after thymine-less death, Biochem. J. **97,** 240-246, 1965.
[35] Freifelder, D., Single-strand breaks in bacterial DNA associated with thymine starvation, J. Mol. Biol. **45,** 1-7, 1969.
[36] Ruekert, R.R., Mueller, G.C., Studies on unbalanced growth in tissue culture. I. Induction and consequence of thymidine deficiency. Cancer Res. **20,** 1584-1591, 1960.
[37] Kim, J.H., Perez, A.G., Djordjevic, B., Studies on unbalanced growth in synchronized HeLa cells, Cancer Res. **28,** 2443-2447, 1968.
[38] Myhre, E., Studies on megaloblasts in vitro. II. Maturation of nucleated red cells in pernicious anaemia before and during treatment with vitamin B_{12}, Scand. J. Clin. Lab. Invest. **16,** 320, 1964.
[39] Nathan, D.G., Gardner, F.H., Erythroid cell maturation and hemoglobin synthesis in megaloblastic anaemia, J. Clin. Invest. **41,** 1086, 1962.
[40] Rondanelli, E.G., Gorini, P., Magkulo, E., Fiori, G.P., Differences in proliferative activity between normoblasts and pernicious anaemia megaloblasts, Blood **24,** 542-552, 1964.
[41] Wichramasinghe, S.N., Chalmers, D.G., Cooper, E.H., Distributed proliferation of erythropoietis cells in pernicious anaemia, Nature **215,** 189, 1967.
[42] Williams, A.M., Chosy, J.J., Schilling, R.F., Effect of vitamin B_{12} in vitro on incorporation of nucleic acid precursors by pernicious anaemia bone marrow, J. Clin. Invest., **42,** 670-674, 1963.
[43] Lajtha, L.G., Kumatori, T., Nucleic acid metabolism in megaloblastic Morrows in vitro, Nature **180,** 991-992, 1957.
[44] Bock, H.F., Hartje, J., Müller, D., Williams, W., Thyminnucleotid-Synthese und Proliferation von Knochenmarkzellen bei megaloblastären Anämien unter Einwirkung von Vitamin B_{12}, Klin. Wochenschr. **45,** 176, 1967.
[45] Rotherham, J., Price, F.M., Otani, T.T., Evans, V.J., J. Natl. Cancer Inst. **47,** 277, 1971.
[46] Waxman, S., Metz, J., Herbert, V., Defective DNA synthesis in human megaloblastic bone marrow: Effects of homocysteine and methionine, J. Clin. Invest. **48,** 284-289, 1969.
[46a] Iwata, N., Omine, M., Yamauchi, H., Maekawa, T., Characteristic abnormality of deoxyribonucleoside triphosphate metabolism in megaloblastic anemia, Blood **60,** 918-923, 1982.
[47] England, J.M., Linnell, J.C., Hematological aspects of cobalamin deficiency, in Vitamin B_{12}, Eds. B. Zagalak and W. Friedrich, Walter de Gruyter, Berlin 1979, pp. 991-1000.
[48] Wichramasinghe, S.N., Longland, J.E., Rate of incorporation of tritiated thymidine into DNA of normoblastic and megaloblastic bone marrow cells in vitro, Acta Hemat. **52,** 14-23, 1974.

[49] Schmid, J. K., Moeschlin, S., Haegi, V., Pernicious anaemia: erythrokinetic and autoradiographic study, using ^{3}H-thymine, ^{3}H-uridine, and ^{3}H-cystidine. Acta Hemat., **32**, 65, 1964.

[50] Messner, H., Fliender, T. M., Cronkite, E. P., Kinetics of erythropoietic cell proliferation in pernicious anaemia. Ser. Haemat. **2**, 44, 1969.

[51] Linnell, J. C., Mackenzie, H. M., Matthews, D. M., Normal values for individual plasma cobalamins, J. Clin. Pathol. **22**, 506, 1969.

[52] Linnell, J. C., Mackenzie, H. M., Wilson, J., Matthews, D. M., Patterns of plasma cobalamins in control subjects and in cases of vitamin B_{12} deficiency, J. Clin. Pathol. **22**, 545-550, 1969.

[53] Linnell, J. C., Hoffbrand, A. V., Peters, T. J., Matthews, D. M., Chromatographic and bioautographic estimation of plasma cobalamins in various disturbances of vitamin B_{12} metabolism, Clin. Sci. **40**, 1-16, 1971.

[54] Linnell, J. C., Identification and quantitation of cobalamins in plasma and tissues, Scand. J. Clin. Lab. Invest. **29** (Suppl. 126), 66, 1972.

[55a] Abe, T., Gibbs, B., Cooper, B. A., Forms of vitamin B_{12} in blood and bone marrow in patients with pernicious anaemia, Brit. J. Haemat. **31**, 493-499, 1975.

[55b] Shojania, A. M., Physicians management of suspected vitamin B_{12} deficiency, CMA Journal, **123**, 127-1130, 1980.

[55c] Berlin, R., Berlin, H., Brante, G., Pilbrant, A., Vitamin B_{12} body during oral and parenteral treatment of pernicious anaemia, Acta Med. Scand. **204**, 81-84, 1978.

[56] Chanarin, I., England, J. M., Mollin, C., Perry, J., Methylmalonic acid excretion studies, Brit. J. Haematol. **25**, 45-53, 1973.

[56a] Baumgartner, R., Giardini, O., Cantani, A., Sabetta, G., Castro, M., Methylmalonic acidemia due to mutase apoenzyme defect: Responsive to vit. B_{12} in intact fibroblastis but not in vivo, J. Inherited Metab. Dis. **5**, 137-141, 1982.

[56b] Robb, R. M., Dowton, S. B., Fulton, A. B., Levy, H. L., Retinal degeneration in vitamin B_{12} disorder associated with methylmalonic aciduria and sulfur amino acid abnormalities, Am. J. Ophthalmol., **97**, 691-696, 1984.

[57] Contreras, E., Giorgio, A. J., Leukocyte methylmalonyl CoA mutase activity in human vitamin B_{12} deficiency, Fed. Proc. **30**, 267, 1971. Am. J. Clin. Nutr. **25**, 695, 1972.

[58] Cox, E. V., White, A. M., Methylmalonic acid excretion: an index of vitamin B_{12} deficiency, Lancet **2**, 853-856, 1962.

[59] Kahn, S. B., Williams, W. J., Barness, L. A., Young, D., Shafer, B., Vicacqua, R. J., Beaupre, E. M., Methylmalonic acid excretion, a sensitive indicator of vitamin B_{12} deficiency in man, J. Lab. Clin. Med. **66**, 75-83, 1965.

[60] Bashir, H. V., Hinterberger, H., Jones, B. P., Brit. J. Haematol. **12**, 704, 1966.

[60a] White, A. M., The effect of vitamin B_{12} deficiency on the excretion of propionic acid by the human, Biochem. J. **95**, 17 p, 1965.

[60b] Sarnaik, A. P., Ravindranath, Y., Krieger, I., Reversal of bone marrow hypoplasia in methylmalonic acidemia with vitamin B_{12}, Dev. Pharmacology and Ther. **1**, 372-377, 1980.

[61] Luhby, A. L., Cooperman, J. M., Aminoimidazdecarboxamine excretion in vitamin B_{12} and folic acid deficiencies, Lancet **2**, 1381-1382, 1962.

[61a] Stokstad, E. L. R., Webb, R. S., Shah, E., Proc. Soc. Exp. Biol. Med. **123**, 752-754, 1966.

[61b] Batra, K. K., Buehring, K. U., Stokstad, E. L. R., Proc. Soc. Exp. Biol. Med. **147**, 72-79, 1974.

[62] Herbert, V., Zalusky, R., Interrelations of vitamin B_{12} and folic acid metabolism: folic acid clearance studies, J. Clin. Invest. **41**, 1263-1276, 1962.

[63] Herbert, V., Larrabee, A. R., Buchanan, J. M., Studies on the identifications of folate compound of human serum. J. Clin. Invest. **41**, 1134, 1962.

[64] Waters, A. H., Mollin, D. L., Observations on the metabolism of folic acid in pernicious anaemia. Brit. J. Haematol. **9**, 319-327, 1963.

[65] Cooper, B. A., Abe, T., Variable response of bone marrow to feeding DL-5-formyltetrahydrofolate in pernicious anaemia, Brit. J. Haematol. **32**, 387-394, 1976.

[66] Sauer, H., Wilmanns, W., Cobalamin dependent methionine synthesis and methyl folate trap in human vitamin B_{12} deficiency, Brit. J. Haematol. **36**, 189-198, 1977.

[66a] Beck, W. S., Metabolic features of cobalamin deficiency in man, in Cobalamin, Ed. B. M. Babior, J. Wiley and Sons, New York 1975, pp. 403-450.

[66b] Cattan, D., Zittoun, J., Yvart, J., Effect of folate deficiency on vitamin B_{12} absorption, Ann. Nutr. Metab. **26**, 367-373, 1982.

[67] Fujii, K., Nagasaki, T., Huennekens, F. M., Vitamin B_{12} requirement for replication of mouse leukemia L 1210 cells: functional rel-

evance to folate metabolism, in Vitamin B_{12}, Eds. B. Zagalak and W. Friedrich, Walter de Gruyter, Berlin 1979, pp. 1127-1130.

[68] Wickramasinghe, S. N., Saunders, J., Correlation between the deoxyuridine-suppressed value and some conventional haematological parameters in patients with folate or vitamin B_{12} deficiency. Scand. J. Haematol. **16,** 121-127, 1976.

[69] Killmann, S. A., Effect of deoxyuridine on incorporation of tritiated thymidine: difference between normoblast and megaloblasts, Acta Med. Scand. **175,** 483, 1964.

[70] Metz, J., Kelly, A., Swett, C., Waxman, S., Herbert, V., Deranged DNA synthesis by bone marrow from vitamin B_{12}-deficient humans, Brit. J. Haematol. **14,** 575, 1968.

[71] Stebbins, R., Scott, J., Herbert, V., Therapeutic trial in the test tube: the "dU supression test" using "physiologic" doses of B_{12} and folic acid to replace therapeutic trial in vivo for diagnosis of B_{12} and folate deficiency, Blood **40,** 927, 1972.

[72] Van der Weyden, M. B., Cooper, M., Firkin, B. G., Defective DNA synthesis in human Megaloblastic bone marrow effects of hydroxy-B_{12}, 5-deoxyadenosyl-B_{12} and methyl-B_{12}, Blood **41,** 299-308, 1973.

[73] Van der Weyden, M. B., Rother, M., Firkin, B. G., The metabolic significance of reduced serum B_{12} in folate deficiency, Blood **40,** 23, 1973.

[74] Maley, F., Maley, G. F., Nucleotide interconversions. II. Elevation of deoxycytidylate deaminase and thymidylate synthetase in regenerating rat liver, J. Biol. Chem. **235,** 2968-2970, 1960.

[75] Gelbard, A. S., Kim, J. H., Perez, A. G., Fluctuations in deoxycytidine monophosphate deaminase activity during the cell cycle in synchronous populations of HeLa cells. Biochim. Biophys. Acta **182,** 564-566, 1969.

[76] Das, K. C., Herbert, V., The "dU supression test" and "thymidine supression test": evidence for reciprocal relationship between the "de novo" and "salvage" pathways of DNA synthesis, in, Vitamin B_{12}, Eds. B. Zagalak, and W. Friedrich, Walter de Gruyter, Berlin 1979, pp. 881-887.

[77] Burman, J. F., Malleson, P. N., Sourial, N. A., Mollin, D. L., TC II deficiency: observations with deoxyuridine supression test, in, Vitamin B_{12}, Eds. B. Zagalak, W. Friedrich, Walter de Gruyter, Berlin 1979, pp. 889-893.

[78] Zittoun, I., Marquet, J., Zittoun, R., Effect of folate and cobalamin compounds on the deoxyuridine suppression test in vitamin B_{12} and folate deficiency, Blood **51,** 119-128, 1978.

[78 a] Das, K. C., Manusselis, C., Herbert, V., In vitro DNA synthesis by bone marrow cells and PHA-stimulated lymphocytes. Suppression by nonradioactive thymidine of the incorporation of ^{3}H-deoxyuridine into DNA: enhancement of incorporation when inadequate vitamin B_{12} or folate is corrected, British J. Haemat. **44,** 51-63, 1980.

[79] Tisman, G., Herbert, V., B_{12} dependence of cell uptake of serum folate: an explanation for high serum folate and cell folate depletion in vitamin B_{12} deficiency. Blood **41,** 465-471, 1973.

[80] Haurani, F. I., Vitamin B_{12} and the megaloblastic development, Science **182,** 78-79, 1973.

[81] Haurani, F. I., Vitamin B_{12} and folic acid interrelations: a new look, in, Vitamin B_{12}, Eds. B. Zagalak and W. Friedrich, Walter de Gruyter, Berlin 1979, pp. 1131-1138.

[82] Sauer, H., Wilmanns, W., Cobalamin dependent methionine synthesis and methyl-folate-trap in human vitamin B_{12} deficiency, Brit. J. Haematol. **36,** 189-198, 1977.

[83] Sauer, H., Cobalamin-dependent methionine synthesis and the regulation of the metabolism of activated folic acid derivatives in cultured human lymphoblasts, in, Vitamin B_{12}, Eds. B. Zagalak and W. Friedrich, Walter de Gruyter, Berlin 1979, pp. 1145-1148.

[84] Hess, B., Gehm, E., Über die Milchsäuredehydrogenase im menschlichen Serum, Klin. Wochenschr. **33,** 91-93, 1955.

[85] Anderssen, N., Scand. J. Haematol. **1,** 212, 1964.

[86] Emerson, P. M., Wilkinson, J. H., Brit. J. Haematol. **12,** 678, 1966.

[87] Heller, P., Weinstein, H. G., West, M., Zimmerman, H. J., Glycolytic, cytric acid cycle, and hexosemonophosphate shunt enzymes of plasma and erythrocytes in megaloblastic anaemia. J. Lab Clin. Med. **55,** 425-434, 1960.

[88] van Dommelen, C. K. V., Klaassen, C. H. L., N. Engl. J. Med. **271,** 541, 1964.

[89] Amos, R. J., Amess, J. A. L., Nancekievill, D. G., Ress, G. M., Prevention of nitrous oxide - induced megaloblastic changes in bone marrow using folinic acid, Br. J. Anaesth. **56,** 103-108, 1984.

[90] Winston, R. M., Warburton, F. G., Stott, A., Brit. J. Haematol. **19,** 587, 1970.

[91] McIntyre, O.R., Sullivan, L.W., Jeffries, G.H., Silver, R.H., Pernicious anaemia in childhood, N. Engl. J. Med. **272,** 981, 1965.

[92] Jeffries, G.H., Pernicious anaemia and atrophic gastritis, in Immunological Disease, Vol. II, Ed. M. Samter, Little Brown, Boston, Massachusetts, 1971, p. 1228.

[93] Schwartz, M., Intrinsic-factor-inhibiting substance in serum of orally treated patients with pernicious anaemia, Lancet **2,** 61, 1958.

[94] Taylor, K.B., Inhibition of intrinsic factor by pernicious anaemia sera, Lancet, **2,** 106, 1959.

[95] Schade, S.G., Abels, J., Schilling, R.F., Studies on antibody to intrinsic factor, J. Clin. Invest. **46,** 615, 1967.

[96] Ardeman, S., Chanarin, I., A method for the assay of human gastric intrinsic factor and for the detection and titration of antibodies against intrinsic factor. Lancet **2,** 1350, 1963.

[97] Abels, J., Bouma, W., Jansz, A., Woldring, M.G., Bakker, A., Nieweg, H.O., Experiments on the intrinsic factor antibody in serum from patients with pernicious anaemia. J. Lab. Clin. Med., **61,** 893, 1963.

[98] Jeffries, G.H., Sleisenger, M.H., Studies of parietal cell antibody in pernicious anaemia. J. Clin. Invest. **44,** 2021, 1965.

[99] Carmel, R., Rosenberg, A., Lau, K.S., Streiff, R.R., Herbert, V., Vitamin B_{12} uptake in human small bowel homogenate and its enhancement by intrinsic factor. Gastroenterology **56,** 548, 1969.

[100] Hines, J.D., Rosenberg, A., Harris, J.W., Intrinsic factor-mediated radio-B_{12} uptake in sequential incubation studies using everted sacs of guinea pig small intestine: Evidence that IF is not absorbed into the intestinal cell. Proc. Soc. Exp. Biol. Med. **129,** 653, 1968.

[101] Spurling, C.L., Sacks, M.S., Jiji, R.M., Juvenile pernicious anaemia. N. Engl. J. Med. **271,** 995, 1964.

[102] Katz, M., Lee, S.K., Cooper, B.A., Vitamin B_{12} malabsorption due to a biologically inert intrinsic factor, N. Engl. J. Med. **287,** 425, 1972.

[103] Katz, M., Mehlman, C.S., Allen, R.H., Isolation and characterization of abnormal human intrinsic factor, J. Clin. Invest. **53,** 1274, 1974.

[104] Begley, J.A., Trachtenberg, A., Hall, C.A., Cobinamide blocking assay for intrinsic factor, in, Vitamin B_{12}, Eds. B. Zagalak and W. Friedrich, Walter de Gruyter, Berlin 1979, pp. 1131-1138.

[105] Hauptman, E., Long-term treatment of pernicious anaemia patients with intrinsic factor, in, Vitamin B_{12}, Eds. B. Zagalak, W. Friedrich, Walter de Gruyter, Berlin 1979, pp. 929-932.

[106] Hippe, E., Gimsing, P., Holländer, N.H., A simplified method for quantitative determination of vitamin B_{12} absorption, in, Vitamin B_{12}, Eds. B. Zagalak and N. Friedrich, Walter de Gruyter, Berlin 1979, pp. 939-944.

[107] Taylor, K.B., Inhibition of intrinsic factor by pernicious anaemia sera, Lancet **2,** 196-108, 1959.

[108] Schwartz, M., Intrinsic factor antibody in serum from patients with pernicious anaemia, Lancet **2,** 1263-1267, 1960.

[109] Nicolas, J.P., Heterogeneity of antibodies against B_{12} binders in pernicious anaemia, in, Vitamin B_{12}, Eds. B. Zagalak and W. Friedrich, Walter de Gruyter, Berlin 1979, pp. 791-802.

[110] Parmentier, Y., Marcoullis, Nicolas, J.P., Perrin, M.O., The intraluminal phase of vitamin B_{12} transport in humans, in Vitamin B_{12}, Eds. B. Zagalak, and W. Friedrich, Walter de Gruyter, Berlin 1979, pp. 803-806.

[111] Allen, R.H., Seetharam, B., Podell, E., Alpers, D.H., Effects of proteolytic enzymes on the binding of cobalamin to R-protein and intrinsic factor, J. Clin. Invest. **61,** 47-54, 1978.

[111a] Hagberg, H., Gronowitz, S., Killander, A., Kallander, C., Serum thymidine kinase in vitamin B_{12} deficiency, Scand. J. Haematol. **32,** 41-45, 1984.

[112] Rosenberg, L.E., Lilljequist, A.C., Hsia, Y.E., Methylmalonic aciduria, N. Engl. J. Med. **278,** 1319, 1969.

[113] Mudel, S.H., Levy, H.L., Abeles, R.H., A derangement in B_{12} metabolism leading to homocystinemia, cystathioninemia and methylmalonic aciduria, Biochem. Biophys. Res. Commun. **35,** 121-126, 1969.

[113a] Schuh, S., Rosenblatt, D.S., Cooper, B.A., Schroeder, M.L., Bishop, A.J., Seargeant, L.E., Haworth, J.C., Homocystinuria and megaloblastic anemia responsive to vit B_{12} therapy: An inborn error of metabolism due to a defect in cobalamin metabolism, N. Engl. J. Med. **310,** 686-690, 1984.

[113b] Sandyk, R., Paralysis of upward gaze as a presenting symptom of vitamin B_{12} deficiency, Eur. Neurol. **23,** 198-200, 1984.

[113c] Broch, H., Inerslund, O., Monn, E., Hovig, T., Seip, M., Imerslund-Grasbeck anemia: A

long-term follow-up study, Acta Paediatr. Scand. **73**, 248-253, 1984.

[113 d] Christensen, P. A., Brynskov, J., Gimsing, P., Petersen, J., Vitamin B_{12} binding proteins (transcobalamin and haptocorrin) in serum and synovial fluid of patients with rheumatoid arthritis and traumatic synovits, Scand. J. Rheumatol. **12**, 268-272, 1983.

[114] Finch, C. A., Coleman, D. H., Motulsky, A. G., Donohue, D. M., Reiff, R. H., Erythrokinetics in pernicious anaemia, Blood **11**, 807, 1956.

[115] Hillman, R. S., Adamson, J., Burba, E., Characteristics of vitamin B_{12} correction of the abnormal erythropoiesis of pernicious anaemia, Blood **31**, 419-432, 1968.

[116] Adams, J. F., Hume, R., Kennedy, E. H., Pirrie, T. G., Whitelaw, J. W., Brit. J. Nutr. **22**, 575, 1968.

[117] Berlin, H., Berlin, R., Brante, G., Oral treatment of pernicious anaemia with high doses of vitamin B_{12} without intrinsic factor, Acta Med. Scand. **184**, 247-258, 1968.

[118] Adams, J. F., Therapeutic effect, utilization, and fate of injected vitamin B_{12} in man, Brit. Med. J. **1**, 1735-1737, 1961.

[118 a] Lindemans, J., Neijens, H. J., Kerrebijn, K. F., Abels, J., Vitamin B_{12} absorption in cystic fibrosis, Acta Paediatr. Scand. **73**, 537-540, 1984.

[119] Lawson, D. H., Murray, R. M., Parker, J. L. W., Hay, G., Hypokalaemia in megaloblastic anaemias, Lancet **2**, 588-590, 1970.

[120] Lawson, D. H., Murray, R. M., Parker, J. L. W., Early mortality in the megaloblastic anaemias, Q. J. Med. **41**, 1-14, 1972.

[121] England, J. M., and Linell, J. C., in Vitamin B_{12}, Eds. B. Zagalak, W. Friedrich, Walter de Gruyter, Berlin 1979, p. 992.

[122] Gilbert, H. S., Weinreb, N., Increased circulating levels of transcobalamin II in Gaucher's disease, N. Engl. J. Med. **295**, 1096-1101, 1976.

[123] Carmel, R., Hollander, D., Extreme elevation of transcobalamin II levels in multiple myeloma and other disorders, Blood **51**, 1057-1063, 1978.

[124] Carmel, R., The distribution of endogenous cobalamin among cobalamin-binding proteins in the blood in normal and abnormal states[1-3], Am. J. Clin. Nutr. **41**, 713-719, 1985.

9.2.2 Neurological abnormalities

Clinical features of neurological syndrome:

- cerebral manifestations: irritability, somnolence, "megaloblastic madness", perversion of taste, smell, and vision including optic atrophy [1-4, 4 a],
- pathological changes of the spinal cord or peripheral nerve are due to a demyelination of nerve fiber [5-8, 8 a, b], which swell in multifocal, asymetric and uneven fashion [1],
- N_2O alters the pattern of myelin proteins and induces vitamin B_{12} deficiency [49],
- There is close association between folate-responsive neuropsychiatric symptoms, vitamin B_{12} and folate deficiencies, and low level of 5-hydroxyindoleacetic acid in human cerebrospinal fluid [50-56]
- a decrease in conduction velocity in the most distal portion of sensory nerve fibers [9],
- neurological disorders may be corrected by cobalamin therapy [1, 10, 10 a], in patients with megaloblastic anaemia [11-13], and in non-anaemic patients [14, 15],
- serum vitamin B_{12} levels:
  - many investigators have found a high frequency of subnormal serum cobalamin levels among psychiatric patients [8, 13, 16-20, 20 a] (Table 9-5),
  - in many results the serum levels of vitamin B_{12} are low in only up to 10 per cent of patients [10, 21-29], and up to 22 per cent of patients in one study [30],
- these results did not confirm a simple relationship between the total vitamin B_{12} level in serum and mental disease,

Table 9-5 B Neuropsychiatric findings with megaloblastosis (reprinted from Reynolds [8])

	Vitamin B_{12} deficiency (N = 50)	Folic acid deficiency (N = 34)
Normal	32%	35%
Organic Mental Change	26%	27%
Affective Disorder	20%	56%
Subacute Combined Degeneration of Cord	16%	0%
Peripheral Neuropathy	40%	18%
Ataxia	8%	3%
Optic Atrophy	2%	0%

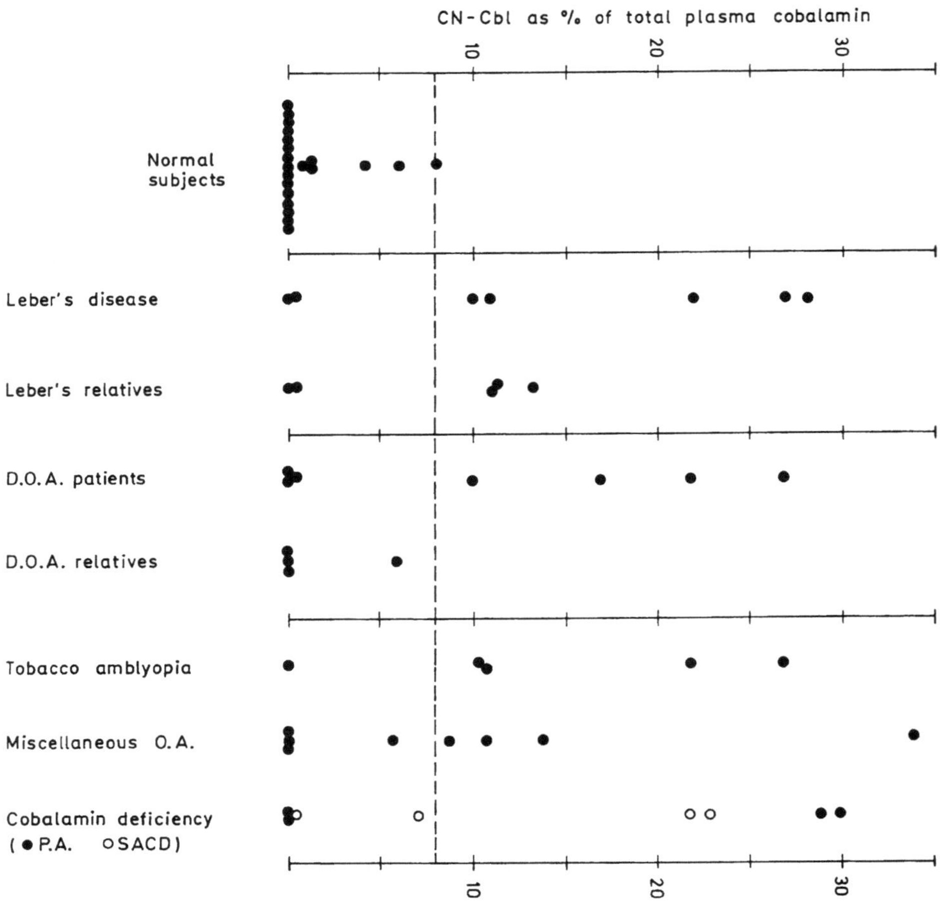

Fig. 9-16 Plasma CN-Cbl in neuroophtalmological diseases. D.O.A. = dominantly inherited optic atrophy; SACD = subacute combined degeneration (reprinted from Wilson et al. [42])

- in pernicious anaemia may be found psychiatric disturbances [8, 13, 31–36] and electroencephalogramm abnormalities [37],
- subnormal levels of cerebrospinal fluid vitamin B_{12} in patients with anticonvulsant megaloblastic anaemia (mean 1.2 pg/ml) and in non-anaemic, anticonvulsant-treated epileptics (9.7 pg/ml) in comparison with normal subjects (17.8 pg/ml), there was not a difference in serum levels of vitamin B_{12} [11],
- increased levels of the 15- and 17-carbon acids in red cells and plasma lipids from patients with pernicious anaemia [2],
- large amounts of C_{15} and C_{17} odd-chain fatty acids and the presence of significant amounts of branched-chain fatty acid in the nervous cells from patient with homocystinuria and methylmalonic aciduria [38],
- nerve biopsy slices from patients with pernicious anaemia incorporate [^{14}C-]propionate into branched-chain fatty acids [39, 40],
- in neuroophtalmological diseases: decrease in plasma MeCbl and increase in erythrocyte MeCbl, elevation of CN-Cbl in plasma [41, 42] (Fig. 9-16),
- patients with tropical ataxic neuropathy suffer from a cyanide intoxication, which is due to the eating of cassava root [43–46]:
  - optic atrophy, deafness, sensory changes
  - rise in plasma thiocyanate,
  - reduced levels of cysteine and methionine in plasma,

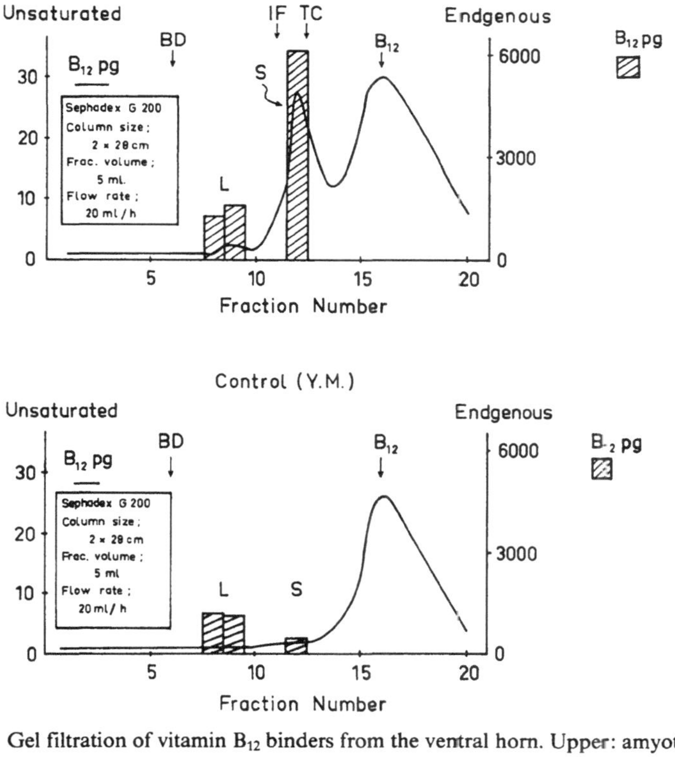

Fig. 9-17 Gel filtration of vitamin B_{12} binders from the ventral horn. Upper: amyotrophic lateral sclerosis, lower: control (reprinted from Inada et al. [47])

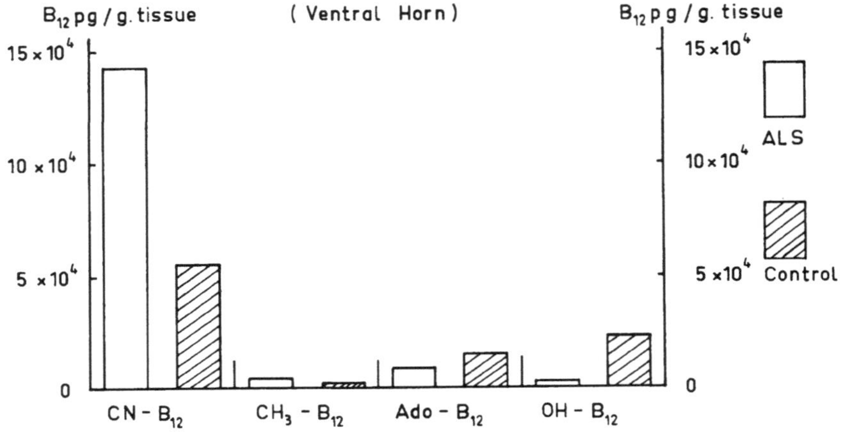

Fig. 9-18 B_{12} derivatives in spinal cords from ALS and control cases (reprinted from Inada et al. [47])

  - total plasma cobalamin level remains unchanged, or is decreased, but plasma CN-Cbl increases six times,
- in the gray matter of the brain with arteriosclerotic dementia [47]:
  - B_{12} contents are decreased in frontal (1428 ± 321 pg/g tissue) and temporal lobes (1665 ± 356 pg/g tissue)
  - B_{12} binders are decreased, especially large binder (CP) in temporal gray matter (325 ± 106 pg of B_{12} bound to binders per gram tissue) and small binder (TC) in the frontal white matter (163 ± 9 pgBn/g tissue),
- in the brains with severe abnormalities were found substantial decrease of B_{12} and TC contents [47],
- in cases of amyotrophic lateral sclerosis was found [47, 48]:
  - in ventral horn higher levels of CN-Cbl and TC (Fig. 9-17 and 9-18),
  - increase in serum and urine CN-Cbl,

References

[1] Beck, W. S., Metabolic features of cobalamin deficiency in man, in Cobalamin, Ed. B. M. Babior, J. Wiley and Sons, New York 1975, pp. 403-450.

[2] Mumenthaler, M., in Neurologie, Georg Thieme Verlag, Stuttgart 1976.

[3] Smith, A. D. M., Megaloblastic madness, Brit. J. Med. **2,** 1840-1843, 1960.

[4] Smith, A. D. M., Duckett, S., Cyanide, vitamin B_{12}, experimental demyelination and tobacco amblyopia, Brit. J. Exp. Pathol. **46,** 615-622, 1965.

[4a] Spatz, R., Perniziöse Anämie - Funikuläre Spinalerkrankung - Perniciosa-Psychose, Der Nervenarzt **40,** 475-481, 1969.

[5] Pant, S. S., Asbury, A. K., Richardson, E. P., The myelopathy of pernicious anaemia. A neuropathological reappraisal, Acta Neurol. Scand. (Suppl. 35), **44,** 7-36, 1968.

[6] Adams, R. D., Kubik, C. S., Subacute degeneration of the brain in pernicious anaemia, N. Engl. J. Med., **231,** 1-9, 1944.

[7] Grienfield, J. G., Carmichael, E. A., The peripheral nerves in cases of subacute combined degeneration of the cord, Brain **58,** 483-491, 1935.

[8] Reynolds, E. H., The neurology of vitamin B_{12} deficiency, in, Vitamin B_{12}, Eds. B. Zagalak and W. Friedrich, Walter de Gruyter, Berlin 1979, pp. 1001-1008.

[8a] Bandera, R., Rovetta, P., Impairment of cerebrospinal fluid circulation in alcoholism as

studied by isotopic cisternography, Subst. Alcohol Actions Misuse **2**, 259-266.

[8b] Van der Westhuyzen, J., Cantrill, R.C., Femandes, Costa, F., Metz, J., Effect of a vitamin B_{12}-deficient diet on lipid and fatty acid composition of spinal cord myelin in the fruit bat *(Rousettus aegyptiacus)*, J. Nutr. **113**, 531-537, 1983.

[9] Mayer, R.F., Peripheral nerve function in vitamin B_{12} deficiency, Arch. Neurol. **13**, 355-362, 1965.

[10] Reynolds, E.H., Neurological aspects of folate and vitamin B_{12} metabolism, Clin. Haemat. **5**, 661-696, 1976.

[10a] Havelius, U., Hindfelt, B., Rosen, I., Reversibility of neurological deficits in vitamin B_{12} deficiency, Arch. Psychiatr. Nervenkr. **232**, 473-478, 1982.

[11] Frenkel, E.P., Mc Call, M.S., Sheehan, R.G., Cerebrospinal fluid folate and vitamin B_{12} in anticonvulsant-induced megaloblastosis, J. Lab. Clin. Med. **81**, 105-115, 1973.

[12] Reynolds, E.H., Schizophrenia-like psychoses of epilepsy and disturbances of folate and vitamin B_{12} metabolism induced by anticonvulsant drugs, Brit. J. Psychiatry **113**, 911-913, 1967.

[13] Roose, D., Neurological symptoms and signs in a selected group of partially gastroectomized patients with particular reference to B_{12} deficiency, Acta Neurol. Scand. **50**, 719-752, 1974.

[14] Reynolds, E.H., Effects of folic acid on the mental state and fit frequency of drug-treated epileptic patients. Lancet **1**, 1086-1088, 1967.

[15] Neubauer, C., Mental detorioration in epilepsy due to folate deficiency, Brit. Med. J. **2**, 759-761, 1970.

[16] Droller, H., Dossett, J.A., Vitamin B_{12} levels in senile dementia and confusional states, Geront. Clin. **1**, 96-106, 1959.

[17] Edwin, E., Holten, K., Norum, K.R., Schrumph, A., Skaug, O.E., Vitamin B_{12} hypovitaminosis in mental diseases. Acta med. scand. **117**, 689-699, 1965.

[18] Shulman, R., The present status of vitamin B_{12} and folic acid deficiency in psychiatric illness, Canad. psychiat. Ass. J. **17**, 205-216, 1972.

[19] Isaksson, A., Nyrstener, A., Ottosson, J.O., Screeming for vitamin B_{12} deficiency in psychiatric patients, Acta psychiat. scand. (Suppl.), **221**, 133-142, 1971.

[20] Carney, M.W.P., Shefield, B.F., Serum folic acid and B_{12} in 272 psychiatric patients, Physiol. Med. **8**, 139-144, 1978.

[20a] Roos, D., Neurological complications in patients with impaired vitamin B_{12} absorption following partial gastrectomy, Acta Neurol. Scand. (Suppl.) **59**, 1-78, 1978.

[21] Elsborg, L., Hansen, T., Kafaelscn, O.J., Vitamin B_{12} concentrations in psychiatric patients, Acta psychiat. scand. **59**, 145-152, 1979.

[22] Klipstein, F.A., Subnormal serum folate and macrocytosis associated with anticonvulsant drug therapy, Blood **23**, 68-86, 1964.

[23] Malpas, J.S., Spray, G.H., Witts, L.J., Serum folic acid and vitamin B_{12} levels in anticonvulsant therapy, Brit. Med. J. **1**, 955-957, 1966.

[24] Reynolds, E.H., Milner, G., Matthews, D.M., Charnarin, I., Anticonvulsant therapy, megaloblastic haemopoiesis and folic acid metabolism, Quart. J. Med. **35**, 521-537, 1966.

[25] Ibbotson, R.N., Dilena, B.A., Horwood, J.M., Studies on deficiency and absorption of folates in patients on anticonvulsant drugs. Austral. Ann. Med. **16**, 144-150, 1967.

[26] Gordon, N., Folic acid deficiency from anticonvulsant therapy, Develop. Med. Child Neurol. **10**, 497-504, 1968.

[27] Horowitz, S.J., Klipstein, F.A., Lovelaca, R.E., Relation of abnormal folate metabolism to neuropathy developing during anticonvulsant drug therapy, Lancet **1**, 563-565, 1968.

[28] Snaith, K.P., Mehta, S., Raby, A.H., Serum folate and vitamin B_{12} in epileptics with and without mental illness. Brit. J. Psych. **116**, 179-183, 1970.

[29] Preece, J., Reynolds, E.H., Johnson, A.L., Relationship of serum to red cell folate concentrations in drug treated epileptic patients, Epilepsia **2**, 335-340, 1971.

[30] Spaans, F., No effect of folic acid supplement on CSF folate and serum vitamin B_{12} in patients on anticonvulsants, Epilepsia **11**, 403-411, 1970.

[31] Warburg, E.J., Jargensen, S., Psychoses and neurasthenia associated with achylia gastrica and megalocytosis and the relation between this syndrome and pernicious anaemia, Acta Med. Scand. **69**, 537-592, 1928.

[32] Warburg, E.S., Jørgensen, S., Psychoses and neurasthenia associated with achylia gastrica and megalocytosis and the relation between this syndrome and pernicious anaemia, Acta Med. Scand. **70**, 193-215, 1929.

[33] Holmes, J.M., Cerebral manifestations of vitamin B_{12} deficiency. Brit. med. J. **2**, 1394-1398, 1956.

[34] Cabot, R.C., in A system of medicine, Vol. 4, Eds. Osler, W., Mc Crea, T., Hodder and Stoughton, London, Chapter 10.

[35] Woltman, H.W., The nervous symptoms in pernicious anaemia: an analysis of one hundred and fifty cases, Amer. J. Med. Sci. **157,** 400-409, 1919.

[36] Cox, E.V., The clinical manifestations of vitamin B_{12} deficiency in Addisonian pernicious anaemia, in, Vitamin B_{12} and Intrinsic Factor, Ed. H.C. Heinrich, Enke, Stuttgart 1961, pp.590-602.

[37] Wallace, P.W., Westmoreland, B.F., The electroencephalogram in pernicious anaemia, Mayo Clin. Proc. **51,** 281-285, 1976.

[38] Kishimoto, Y., Williams, M., Moser, H.W., Hignite, C., Biemann, K., Branched-chain and odd-numbered fatty acids and aldehydes in the nervous system of a patient with deranged vitamin B_{12} metabolism, J. Lipid Res. **14,** 69, 1973.

[39] Frenkel, E.P., Serum level of erythropoietin in anaemias associated with chronic infection, malignancy, and primary hematopoietic disease, J. Clin. Invest. **50,** 332-355, 1971.

[40] Frenkel, E.P., Abnormal fatty acid metabolism in peripheral nerves of patients with pernicious anaemia, J. Clin. Invest. 1237-1245, 1971.

[41] Linnell, J.C., The fate of cobalamins in vivo, in Cobalamin, Ed. B.M. Babior, J. Wiley and Sons, New York 1975, pp.287-333.

[42] Wilson, J., Linnell, J.C., Matthews, D.M., Plasma cobalamins in neuro-ophtalmological diseases, Lancet **1,** 259-261, 1971.

[43] Montgomery, R.D., in toxic constituents of plant foodstuffs, Academic Press, New York, p.144.

[44] Monekosso, G.L., Wilson, J., Plasma thiocyanate and vitamin B_{12} in Nigerian patients with degenerative neurological disease, Lancet **1,** 1062-1064, 1966.

[45] Osuntokun, B.O., Durowoju, J.E., Mc Farlane, H., Wilson, J., Brit. Med. J. **2,** 647, 1968.

[46] Osuntokun, B.O., Matthews, D.M., Hussein, H.A.A., Wise, I.J., Linnell, J.C., Plasma and hepatic cobalamins in tropical ataxic neuropathy, Clin. Sci. Mol. Med., **46,** 563-567, 1974.

[47] Inada, M., Kameyama, M., Toyoshima, M., The significance of vitamin B_{12} binders in the central nervous system, in, Vitamin B_{12}, Eds. B. Zagalak and W. Friedrich, Walter de Gruyter, Berlin 1979, pp.851-856.

[47 a] Baker, H., Frank, O., Chen, T., Feingold, S., Deangelis, B., Baker, E., Vitamin content of some normal human brain segments, J. Neurosci. Res. **11,** 419-436, 1984.

[48] Kameyama, M., Inada, M., Toyoshima, M., Ogawa, M., Neurol. Med. **8,** 570, 1978.

[49] Maloughlin, J.L., Cantrill, R.C., Nitrous oxide alters the pattern of myelin proteins in the nervous system of fruit bat, Neurosci. Lett. **42,** 99-104, 1983.

[50] Carney, M.W.P., Sheffield, B.F., Serum folate acid and B_{12} in 272 psychiatric patients, Psychol. Med. **8,** 139-144, 1978.

[51] Botez, M.I., Peyronnard, J.M., Bachevalier, J., Charron, L., Polyneuropathy and folate deficiency, Arch. Neurol. **35,** 581-584, 1978.

[52] Botez, M.I., Peyronnard, J.M., Charron, L., Polyneuropathies responsive to folic acid therapy. In: Botez, M.I., Reynolds, E.H. (eds), Folic Acid in Neurology, Psychiatry, and Internal Medicine, pp.401-412, Raven, New York 1979.

[53] Botez, M.I., Botez, T., Leveille, J., Bielman, P., Cadotte M., Neuropsychological correlates of folic acid deficiency: facts and hypotheses. In: Botez, M.I., Reynolds, E.H. (eds), Folic Acid in Neurology, Psychiatry, and Internal Medicine, pp.435-461, Raven, New York 1979.

[54] Shorvon, S.D., Carney, M.W.P., Chanarin, I., Reynolds, E.H., The neuropsychiatry of megaloblastic anaemia, Br. Med. J. **281,** 1036-1038, 1980.

[55] Lowe, T.L., Cohen, D.J., Miller, S., Young, J.G., Folic acid and B_{12} in autism and neuropsychiatric disturbances of childhood, J. Am. Acad. Child Psychiatry **20,** 104-111, 1981.

[56] Botez, M.I., Young, S.N., Bachevalier, J., Gauthier, S., Effect of Folic Acid and Vitamin B_{12} Deficiencies on 5-Hydroxyindoleacetic Acid in Human Cerebrospinal Fluid, Ann. Neurol. **12,** 479-484, 1982.

9.2.3 Malignant neoplasms and other clinical conditions

Some malignancies occurring in bone marrow and in liver are associated with well-marked changes in serum vitamin B_{12} binding proteins (TC and CP) levels, which increase serum B_{12} concentration, or serum unsaturated B_{12}-binding capacity (UBBC), or increase of both parameters [1-9]. The following features of vitamin B_{12}-binding proteins may have diagnostic values in clinical treatment [8, 9, 12]:

CP isoproteins TC I and TC III:

TC I and TC III are glycoproteins of similar molecular weight (120000) but different charge due to slight difference in the sialic acid content (see section 8.3)

TC I is the carrier of the endogenous B_{12} in the blood. The serum B_{12} concentration is a function of TC I in the blood. TC I is synthesized in young differentiated granulocytes. TC III does not bind B_{12} in vivo, but can bind it in vitro. It derives from mature granulocytes (see section 8.3.3)

TC protein:

TC is a protein of 38000 molecular weight, the UBBC of the serum is dependent on the serum TC. It is synthetized by the cells of monocytes and macrophages (see sec-

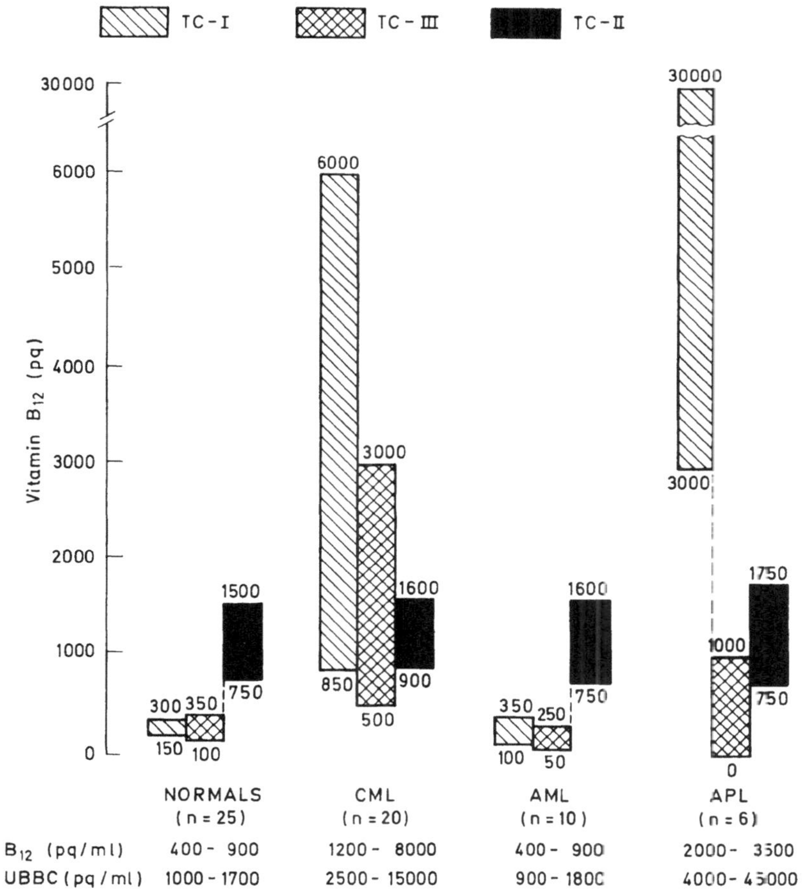

Fig. 9-19 Ranges of serum vitamin B_{12}-binding proteins TC and CP isoproteins (TC I and TC III) levels in normal subjects and in patients with chronic myeloid leukemia (CML) acute myeloid leukemia (AML) and acute promyelocytic leukemia (APL). The levels of the binding proteins are expressed in terms of pg B_{12} bound to these proteins in course of assay per ml of serum (see also the legend to Table 9-7 and section 8.2.2) (from Rachmilewitz et al. [8])

tion 8.2.3). TC transports the cobalamin across the cell membrane [19, 20]. There seems to be a possibility that the different cell lines synthetize heterogenous molecules of TC [8, 25, 27, 28] (for details of TC occurrence and biological properties see section 8.2).

In Fig. 9-19 are presented serum TC- and CP levels, serum B_{12} levels and serum UBBC in normal subjects and in patients with chronic myeloid leukemia (CML), acute myeloid leukemia (AML), and acute promyelocytic leukemia (APL):

CML: increase in serum B_{12} concentration, and serum UBBC, the majority of the proliferating cells of immature granulocytes, increase in TC I of serum, TC III is elevated as well. TC level is unchanged.

AML: parameters of vitamin B_{12} binding proteins are unchanged. The majority of the proliferating cell is the young undifferentiated myeloblast [8].

APL: Is characterized by high UBBC and serum B_{12} levels and enormous increase of TC I. The predominant proliferating myeloid cell is in the stage of the promyelocyte, lack of mature granulocytes [8]. TC level remains unchanged (see also Table 6-8).

In peripheral mononuclear cells from patients with acute myelogenous leukemia, the rate of receptor-mediated uptake and degradation of low-density lipoprotein is higher as compared with cells from healthy patients and subjects with acute and chronic lymphocytic leukemias, infectious mononucleosis, and nonhematologic malignancies [29].

In Table 9-6 the results estimated by means of the rapid charged cellulose filter technique are summarized (for assay by this method see section 5.3.3)

Table 9-6 *Serum B_{12} concentration, UBBC and serum B_{12}-binding proteins (apo-form) in different clinical conditions determined by the charged cellulose filter technique.* The unsaturated B_{12} binding capacity (UBBC) of the vitamin B_{12}-binding proteins, the levels of transcobalamin (TC) and cobalophilin isoproteins (TC I and TC III) is expressed in terms of picograms B_{12} being bound to the proteins, in course of assay, per ml of serum (for vitamin B_{12} assay see Chapter 5 and B_{12}-binding proteins assay Chapter 8). Serum contains vitamin B_{12} bound – as well as free of vitamin B_{12} binding proteins. The values in the Table refer to free TC, TC I and TC III in serum. (Reprinted from Rachmilewitz et al. [8])

	Serum B_{12}	Serum UBBC	TC	CP isoproteins	
				TC I	TC III
		pg/ml Mean ± S.D.			
Chronic myeloid leukemia (n = 32)	3210 ± 1105	6890 ± 1935	1250 ± 340	3950 ± 1630	1340 ± 670
Acute promyelocytic l. (n = 6)	4890 ± 1940	9860 ± 2150	1650 ± 480	8540 ± 1430	650 ± 420
Acute l. + Malignant lymphoma (n = 14) (n = 15)	690 ± 170	4100 ± 800	3350 ± 760	230 ± 49	380 ± 80
Polycythemia vera (n = 30)	860 ± 290	3900 ± 1220	1380 ± 320	570 ± 230	2940 ± 1240
Normals (n = 40)	600 ± 120	1200 ± 207	980 ± 220	217 ± 94	229 ± 115

CML patients: elevated serum CP isoproteins (TC I and TC III), vitamin B_{12} and UBBC. The TC III derives from mature granulocytes.

APL patients: elevated CP isoprotein TC I, B_{12} and UBBC. Granulocytes are mainly at the immature stage. The levels of TC and CP isoprotein, TC III, are also elevated, however, to a lesser extent.

Acute leukemia and malignant lymphoma: higher levels of TC and UBBC.

PV patients: levels of UBBC and TC III are increased. After chemotherapy transcobalamins reach normal levels (Table 9-7).

In Table 9-8 is shown a distribution of cobalamins of tissues in a case of human primary hepatoma [6]. All tissues, except the spleen, contain high levels of MeCbl.

Table 9-7 Changes in serum B_{12}-binding protein (TC, TC I, TC III) levels in different clinical conditions following chemotherapy (from Rachmilewitz and Rachmilewitz [26])

Diagnosis	TC*	CP isoproteins		Remarks
		TC I*	TC III*	
CML	1,468	2,294	1,016	
	1,326	379	400	FC
PV 1	979	295	1,289	
	902	229	339	FC
PV 2	939	372	2,118	
	1,110	248	222	FC
PV 3	1,200	280	1,450	
	3,250	320	250	AMC
AML	2,545	121	364	Active stage
	984	136	320	FC remission
Normal range	217 ± 64	1,148 ± 241	329 ± 115	

* Values are expressed by the quantity of vitamin B_{12} bound to the proteins in course of assay (pg B_{12}/ml of serum) (see remarks in the legend to Table 9-7) CML = chronic myeloid leukemia, PV = polycythemia vera, AML = acute myeloblastic leukemia, FC = following chemotherapy, AMC = acute myeloblastic crisis.

Table 9-8 Tissue cobalamins in a case of human primary hepatoma (reprinted from Linnell and Matthews [6])

Tissue	Total Cbl	MeCbl	AdoCbl	OH-Cbl	CN-Cbl
			picograms per ml		
Plasma	104	77 (74%)	27	0	0
Erythrocytes	180	136 (76%)	23	21	0
Ascitic fluid	223	102 (46%)	96	21	4
			picograms per mg		
Liver	344	73 (21%)	225	46	0
Tumour	86	21 (24%)	55	10	0
Kidney	54	14 (26%)	25	15	0
Spleen	24	2 (8%)	20	2	0
Brain	3	1.1 (37%)	0.9	1	0

Table 9-9 Serum vitamin B_{12} and vitamin B_{12}-binding protein patterns of adolescent hepatoma compared to chronic myeloid leukemia (see also the remarks in the legend to Table 9-7) (from Waxman and Gilbert [3]. Reprinted by permission of the New England Journal of Medicine **289**, 1053, 1973)

Specimen	White Cell Count	Average Serum Vitamin B_{12} level	Unsaturated Vitamin B_{12} Binding Capacity	α-Globulin Vitamin B_{12} Binding Protein in Unsaturated Vitamin B_{12} Binding Capacity[1]	β-Globulin Vitamin B_{12} Binding Protein in Unsaturated Vitamin B_{12} Binding Capacity[2]
		pg/ml		%	
Normal[3]		386	850- 1400	12.1-29.1	70.9-87.9
Chronic myeloid leukemia	40000	4480	7500-22960	68.6-97.3	2.7-31.4
Hepatoma (adolescent):					
Case 1[4]	5500	18185	243000	90.0; 92.4	10.0; 7.6
Case 1		24800	180000	91.8	8.2
Case 2	7100	53100	480000	99.5; 95.5	0.5; 4.5
Case 3	4200	5166	4570[5]	94.0	6.0

1 – Transcobalamin I
2 – Transcobalamin II-III
3 – Values obtained from Herbert V. (Blood, 32, 305-312, 1968)
4 – Pleural fluid
5 – Not diluted

Table 9-10 Increased unsaturated B_{12}-binding capacity (UBBC) of the serum due to increased serum TC in heterogenous clinical conditions (from Rachmilewitz et al. [8])

	No of cases	Serum TC	Serum UBBC
		pg/ml (Mean ± SD)**	
Normal values	40	886 ± 214	1200 ± 207
Acute leukemia	20	3048 ± 1419	3857 ± 1483
Hodgkin and Non-Hodgkin lymphoma	18	3346 ± 769	4100 ± 808
Gaucher's disease	9	3195 ± 209	3797 ± 2280
Acute immunological diseases*	13	3244 ± 688	4090 ± 638
		Range	
Chronic monocytic leukemia	4	3160-15646	3590-16130

* Rheumatoid Arthritis – 5 cases
Systemic lupus erythematosus – 6 cases
Dermatomyositis – 1 case
Scleroderma – 1 case

** The serum TC level refers to the apo-TC form. The level is expressed in terms of $pgB_{12}n$ bound to TC in course of assay per ml of serum.

Adolescent hepatoma cases are characterized by extraordinary increase of serum vitamin B_{12} levels and serum vitamin B_{12}-binding proteins levels [3] (Table 9-9).

Elevated level of serum TC is characteristic for other clinical conditions such as Gaucher's disease [22, 23], autoimmuno diseases [23, 24], chronic monocytic leukemia [23] and inflammatory bowel disease [8] (Table 9-10).

The elevation of a vitamin B_{12} uptake to stimulated lymphocytes may be an explanation of the rise of TC during active phases of autoimmunological disease [19, 24]. TC was significantly elevated with activity of metastatic renal carcinoma [30].

References

[1] Ståhlberg, K.G., Scand. J. Haematol. **1**, 220-222, 1964.

[2] Myasishcheva, N.V., Areshkina, L., Kutzeva, L.S., Skorobogatova, E.P., Cobalamin compounds and folic acid derivatives in human blood serum, Voprosy Meditsinskoi Khimii **14**, 273-277, 1968.

[3] Waxman, S., Gilbert, H.S., A tumor-related vitamin B_{12}-binding protein in adolescent hepatoma, N. Engl. J. Med. **289**, 1053-1056, 1973.

[4] Hall, C.A., Finkler, A.E., Measurement of the amounts of the individual vitamin B_{12}-binding proteins in plasma. I. Studies of normal plasma. II. Abnormalities in leukemia and pernicious anaemia. Blood **27**, 611-622, 1966.

[5] Linnell, J.C., The fate of cobalamins in vivo, in Cobalamin, Ed. B.M. Babior, J. Wiley and Sons, New York 1975, pp. 287-333.

[6] Linnell, J.C., Matthews, D.M., Recent advances in cobalamin metabolism: abnormalities in coenzyme distribution in tumour development and in inherited metabolic disease, in Vitamin B_{12}, Eds. B. Zagalak and W. Friedrich, Walter de Gruyter, Berlin 1979, pp. 1101-1111.

[7] Hansen, O.P., Drivsholm, A., Hippe, E., Quadros, E.V., Linnell, J.C., Interrelationships between vitamin B_{12} and folic acid in myelomatosis: Cobalamin coenzyme and tetrahydrofolic acid function, Scand. J. Haemat. **20**, 360-370, 1978.

[8] Rachmilewitz, B., Schlesinger, M., Rabinowitz, R., and Rachmilewitz, M., The origin and clinical implications of vitamin B_{12} binders – the transcobalamins, in Vitamin B_{12}, Eds. B. Zagalak and W. Friedrich, Walter de Gruyter, Berlin 1979, pp. 765-776.

[9] Burger, R.L., Mehlman, C.S., Allen, R.H., Human plasma R-type vitamin B_{12}-binding proteins. I. Isolation and characterization of transcobalamin I, transcobalamin III, and the normal granulocyte vitamin B_{12}-binding protein, J. Biol. Chem. **25**, 7700-7706, 1975.

[10] Beard, M.F., Pitney, W.R., Sanneman, E.H., Serum concentrations of vitamin B_{12} in patients suffering from leukemia, Blood **9**, 789, 1954.

[11] Rachmilewitz, B., Rachmilewitz, M., Moshkowitz, B., Gross, J., Serum transcobalamin in myeloid leukemia. J. Lab. Clin. Med. **78**, 275-288, 1971.

[12] Rachmilewitz, B., Rachmilewitz, M., Gross, J., A Vitamin B_{12} binder with transcobalamin I. Characteristics synthesized and released by human granulocytes in vitro, Brit. J. Haematol. **26**, 557-567, 1974.

[13] Selhub, J., Rachmilewitz, B., Grossowicz, N., Fractionation of serum transcobalamins on charged cellulose filters, Proc. Soc. Exp. Biol. Med. **152**, 204-209, 1976.

[14] Hall, C.A., Finkler, A.E., Vitamin B_{12}-binding protein in polycythemia vera plasma, J. Lab. Clin. Med. **73**, 60-69, 1969.

[15] Rachmilewitz, B., Manny, N., Rachmilewitz, M., The transcobalamins in polycythaemia vera, Scand. J. Haematol. **19**, 453-462, 1977.

[16] Zittoun, J., Zittoun, R., Marquet, J., Sultan, C., The three transcobalamins in myeloproliferative disorders and acute leukemia. Brit. J. Haemat. **31**, 287-298, 1975.

[17] Rachmilewitz, B., Rachmilewitz, M., Serum transcobalamin II levels in acute leukemia and lymphoma. Israel. J. Med. Sci. **12**, 583-586, 1976.

[18] Rachmilewitz, B., Rachmilewitz, M., Chemotherapy-induced changes in serum vitamin B_{12}-binding proteins in myeloid leukemia, Israel. J. Med. Sci. **7**, 1140-1145, 1971.

[19] Hoffbrand, A.V., Tripp, E., Das, K.C., Uptake of vitamin B_{12} by phytohemagglutinin transformed lymphocytes, Brit. J. Haemat. **24**, 147-151, 1973.

[20] Rosenberg, L. E., Lilljequist, A., Allen, R. H., Transcobalamin II facilitated uptake of vitamin B_{12} by cultured fibroblasts; studies in methylmalonic aciduria (Abstr.), J. Clin. Invest. **52**, 69 a, 1973.
[21] Hitzig, W. H., Dohmann, U., Pluss, H. J., Vischer, D., Hereditary transcobalamin II deficiency: clinical findings in a new family, J. Pediatr. **85**, 622-628, 1974.
[22] Gilbert, H. S., Weinreb, N., Increased circulating levels of transcobalamin II in Gaucher's disease, N. Engl. J. Med. **295**, 1036-1101, 1976.
[23] Rachmilewitz, B., Rachmilewitz, M., Serum transcobalamin II concentration (Letter), N. Engl. J. Med. **295**, 1174, 1977.
[24] Träter-Schröder, M., Hitzig, H., Grob, P. J., Kenny, A. B., Increased unsaturated transcobalamin II in active autoimmune disease, Lancet **2**, 238-239, 1978.
[25] Tan, C. H., Blaisdell, S. J., Mammalian transcobalamin II metabolism. The immunological and biological cross reactivity of mammalian transcobalamin II, Biochim. Biophys. Acta, **444**, 416-427, 1976.
[26] Rachmilewitz, B., Rachmilewitz, M., The clinical significance of serum transcobalamins in myeloid leukemia and other myeloproliferative disorders, Israel. J. Med. Sci. **13**, 710-715, 1977.
[27] Rachmilewitz, B., Rachmilewitz, M., Chaouat, M., Schlesinger, M., Production of TC II (Vitamin B_{12} transport protein) by Mouse Mononuclear Phagocytes, Blood **52**, 1089-1098, 1978.
[28] Green, P. D., Savage, R. C., Hall, C. A., Mouse transcobalamin. II. biosynthesis and uptake by L-929 cells, Arch. Biochem. Biophys. **176**, 683-689, 1976.
[29] Ho, Y. K., Smith, R. G., Brown, M. S., Goldstein, J. L., Low density lipoprotein (LDL) receptor activity in human acute myelogenous leukemia cells. Blood **52**, 1099-1114, 1978.
[30] Jensen, H. S., Gimsing, P., Pedersen, F., Hippe, E., Transcobalamin II as an indicator of activity in metastatic renal adenocarcinoma, Cancer (Phila) **52**, 1700-1704, 1983.

9.3 Mechanisms of cobalamin metabolism disorders

9.3.1 Cobalamin malabsorption (see also section 8.1.3)

Phase of Absorption	Nature of abnormal events
A. Intragastric events	1. Nutritional cobalamin deficiency 2. Defective cobalamin release from food 3. Intrinsic factor (IF) deficiency – defect of secretion of IF, – secretion of an abnormal IF – autoimmunological reactions – surgical resection
B. Small bowel events	1. Competitive parasites: bacterial overgrowth, fish tapeworm 2. Pancreatic insufficiency
C. Intestinal absorption	1. Ileal resection 2. Ileal disease 3. Lack of divalent cations 4. Unfunctional transport system through membrane 5. Drugs 6. Deficiency of intracellular binders

A.1: – The nutritional defect is associated with low serum levels of vitamin B_{12} and may be corrected by a change of diet [1–4, 4a] (see vitamin B_{12} occurrence in food Table 6–7a–b).

A.2: – The release of vitamin from food requires the presence of acid and proteolytic enzymes [5–10].

A.3: – In most cases, IF deficiency has autoimmunological character [11–23].
- The pernicious anaemia sera contain various populations of IF antibodies, which bind to different antigenic determinants on the IF molecule [19, 23] or on molecules of isoproteins [20–22, 24] (see also section 8.1.8).
- Two distinct antibodies inhibit the IF activity [25–30, 30a].
  - blocking antibody (antibody I),
  - binding antibody (antibody II) can bind the IF-Cbl complex and prevents its uptake by the intestinal mucosa (Fig. 9–20)
- The sera of 70% of patients with pernicious anaemia have antibody I and of 30% of patients have antibody II [11, 31].
- The antibodies I and II were found in gastric juice and saliva of PA patients [32–34].
- Antiparietal cells antibodies were found in the sera of PA patients [29, 35, 36], but these organ-specific immunoglobulins were found in healthy subjects and in other different diseases [11, 30].

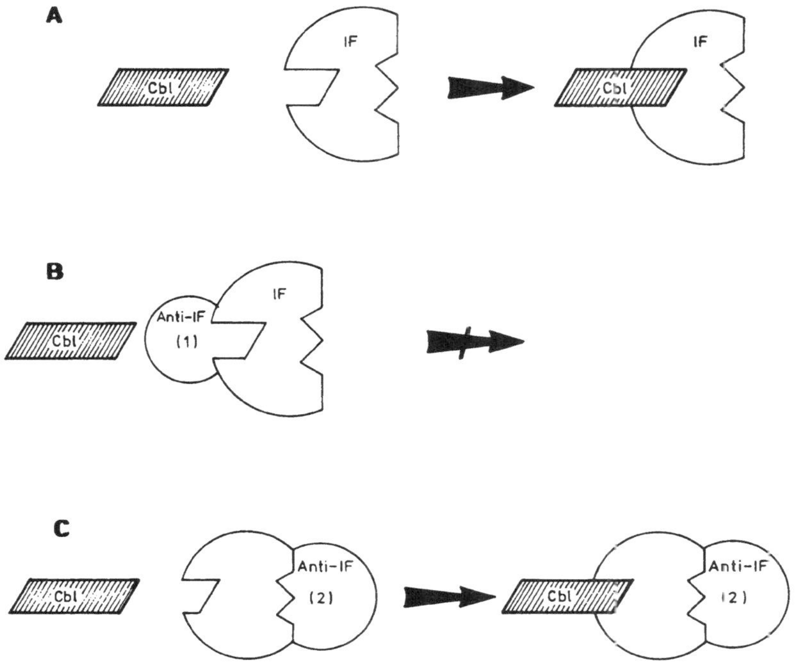

Fig. 9-20 Hypothetical illustration of intrinsic factor antibodies according to Schade et al. [26]. A. Normal absorption, B. Blocking type antibodies. C. Binding type antibodies (from Ellenbogen [30] et al., in: Cobalamin, Babior (ed.), 1975, p. 236. Copyright 1975 John Wiley & Sons, Inc. Reprinted by permission of John Wiley & Sons, Inc.)

- In some cases of PA anti IF antibodies were not detected; on the other hand sometimes these antibodies were present in patients without PA [30-40].
- Congenital IF deficiency is due to the defect of gastric parietal cells [41-44] or to the production of a biologically inert IF [45, 46].
- drug therapy can decrease IF secretion [46 a].

B.1: – The bacteria and fish tapeworm can take up the vitamin B_{12} and prevent its absorption by intestine [47-60].

B.2: – Some cases of exocrine pancreatic insufficiency (EPI) are associated with a reduction of the cobalamin absorption [63-66], and this defect is not corrected by exogeneous IF but is eliminated by the pancreatic proteolytic enzymes [67] (see Fig. 8-23).
In healthy subjects the vitamin is transported in the intestine by IF, in EPI by the cobalophilins, which are not degraded [68, 69] (see section 8.1.3).

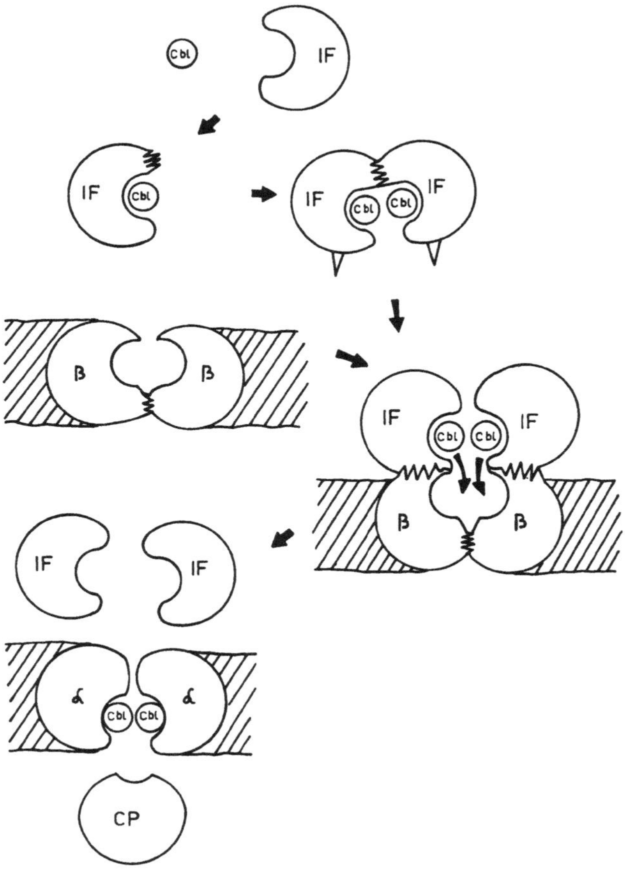

Fig. 9-21 A very hypothetical model for the working of the "B_{12} pump". The Cbl-IF complex has a tendency to aggregate. It forms a pseudo-oligomer with its "relative", the IF receptor. It is probably attached to the Cbl-free subunit beta. Following binding of Cbl beta becomes alpha. Intracellular binders accept Cbl from alpha subunit (for IF-Cbl receptor see section 8.1.9) (reprinted from Gräsbeck [22])

C.1: – The resection of the terminal ileum is associated with cobalamin malabsorption and the absorption is not corrected by IF [70–76] (see Fig. 8–13).

C.2: – Mucosal diseases [77–85 c]:
celiac sprue – a damage of the microvillous membrane of the epithelial absorptive cells,
tropical sprue – treatment with antibiotics such as tetracycline, causes a restoration of the mucosa and an increase in vitamin B_{12} absorption.
– radiation injuring [85 d]

C.3: – The presence of divalent cations is necessary for the binding of IF-Cbl complex to specific receptor on the microvillous membrane in the terminal part of the ileum (see Fig. 8–13) [86, 87 a].

C.4: – the defect concerns the membrane process (-es) (Fig. 9–21)
- it can not be corrected by IF, pancreatic juice or antibiotic treatment [88, 89].
- ileal mucosa is normal when examined by light- and electron microscopy [88],
- the complex of IF-Cbl is bound by ileal mucosal homogenates
- the level of transcobalamin (TC) in serum is normal [88, 91].
- intestinal transport was depressed in scurvy [91 a]

C.5: – Cobalamin malabsorption is induced by some medicine [92–96, 96 a–e] (see section 8.1.3).

C.6: – Congenital deficiency of TC can stop the "B_{12} pump".
B_{12}-binding proteins are faulty [96 e]

References

[1] Droller, H., Dossett, J. A., Vitamin B_{12} levels in senile dementia and confusional state, Geront. Clin. **1**, 96–106, 1959.

[2] Elsborg, L., Lund, V., Bastrup-Madsen, P., Serum vitamin B_{12} levels in the aged, Acta med. scand. **200**, 309–314, 1976.

[3] Hessov, I., Elsborg, L., Nutritional studies on long-term surgical patients with special reference to the intakes of vitamin B_{12} and folic acid, Int. J. Vitamin Nutr. Res. **46**, 427–432, 1976.

[4] Linnell, J. C., Mackenzie, H. M., Wilson, J., Matthews, D. M., Patterns of plasma cobalamins in control subjects and in cases of vitamin-B_{12} deficiency. J. Clin. Path. **22**, 545–550, 1969.

[4 a] Carmel, R., Subtle cobalamin malabsorption in a vegan patient, Arch. Intern. Med. **142**, 2206–2207, 1982.

[4 b] Johnson, A., Latham, M. C., Roe, D. A., Nutritional anemias among preschool children in Guyana, Ecol. Food Nutr. **13**, 9–14, 1983.

[4 c] Shinwell, E. D., Gorodischer, R., Totally vegetarian diets and infant nutrition, Pediatrics **70**, 582–586, 1982.

[4 d] Levine, A. S., Doscherholmen, Vitamin B_{12} bioavailability from egg yolk and egg white: Relationship to binding proteins, Am. J. Clin. Nutr. **38**, 436–439, 1983.

[5] Schade, S. G., Schilling, R. F., Effect of pepsin on the absorption of food vitamin B_{12} and iron, Am. J. Clin. Nutr. **20**, 636, 1967.

[6] Reizenstein, R., Effect of digestive enzymes on bound vitamin B_{12}, Acta Med. Scand. **165**, 481, 1959.

[7] Mahmud, K., Ripley, D., Doscherholmen, A., Vitamin B_{12} absorption tests. Their unreliability in post gastrectomy states, JAMA **216**, 1167, 1971.

[8] Chang, S., Ripley, D., Swain, W., Doscherholmen, A., Gastric achlorhydria and hypochlorhydria with low serum B_{12} levels and normal B_{12} absorption, Clin. Res. **19**, 655, 1971.

[9] Doscherholmen, A., Swain, W. R., Impaired

assimilation of egg ^{57}Co-vitamin B_{12} in patients with hypochlorhydria and achlorhydria and after gastric resection, Gastroenterology **64,** 913, 1973.

[10] Heyssell, R.M., Bozian, R.C., Darby, W.J., Vitamin B_{12} turnover in man. The assimilation of vitamin B_{12} from natural food stuff by man and estimates of minimal daily dietary requirements, Am. J. Clin. Nutr. **18,** 176, 1966.

[10 a] Silingerland, D.W., Cardarelli, J.A., Burrows, B.A., Miller, A., The utility of serum gastrin levels in assessing the significance of low serum vit. B_{12} levels, Arch. Intern. Med. **144,** 1167-1168, 1984.

[11] Jeffries, G.H., Pernicious anaemia and atrophic gastritis, Immunological Disease, Vol. II, ed. Samter, M., Little Brown, Boston, 1971, p. 1228.

[12] Donaldson, R.M., Jr., Mechanisms of malabsorption of cobalamin, in Cobalamin, Ed. B.M. Babior, J. Wiley and Sons, New York 1975, pp. 335-368.

[13] Castl, W.B., Current concepts of pernicious anaemia. Am. J. Med. **48,** 541, 1970.

[14] Schwartz, M., Intrinsic-factor-inhibiting substances in serum of orally treated patients with pernicious anaemia, Lancet **2,** 61, 1958.

[15] Taylor, K.B., Inhibition of intrinsic factor by pernicious anaemia sera, Lancet, **2,** 106, 1959.

[16] Glass, G.B.J., Physiol. Rev. **43,** 529-849, 1963.

[17] Glass, G.B.J., In Gastric intrinsic factor and other vitamin B_{12} binders, Georg Thieme, Stuttgart 1974, p. 165.

[18] Taylor, K.B., Clin. Haemat. **5,** 497-519, 1976.

[19] Jacob, E., O'Brien, H.A.W., Mollin, D.L., Scand. J. Haemat. **19,** 201-206, 1977.

[20] Marcoullis, G., Gräsbeck, R., Scand, J. Clin. Lab. Invest. **35,** 5-11, 1975.

[21] Gräsbeck, R., Marcoullis, G., Scand, J. Clin. Lab. Invest. **35,** 13-18, 1975.

[22] Gräsbeck, R., Soluble and membrane-bound vitamin B_{12} transport proteins, in Vitamin B_{12}, Eds. B. Zagalak and W. Friedrich, Walter de Gruyter, Berlin, 1979, pp. 743-763.

[23] Nicolas, J.P., Heterogeneity of antibodies against vitamin B_{12} binders in pernicious anaemia, in Vitamin B_{12}, Eds. B. Zagalak and H. Friedrich, Walter de Gruyter, Berlin 1979, pp. 791-802.

[23 a] Borch, K., Liedberg, G., Prevalence and incidence of pernicious anemia: An evaluation for gastric screening, Scand. J. Gastroenterol **19,** 154-160, 1984.

[24] Marcoullis, G., Gräsbeck, R., Biochim. Biophys. Acta **491,** 663-672, 1977.

[25] Schade, S.G., Abels, J., Schilling, R.F., Studies on antibody to intrinsic factor, J. Clin. Invest. **46,** 615, 1967.

[26] Schade, S.G., Feick, P.L., Imrie, M.H., Schilling, R.F., In vitro studies on antibodies to intrinsic factor, Clin. Exp. Immunol. **2,** 399, 1967.

[27] Ardeman, S., Chanarin, I., A method for the assay of human gastric intrinsic factor and for the detection and titration of antibodies against intrinsic factor, Lancet **2,** 1350, 1963.

[28] Abels, J., Bouma, W., Jansz, A., Woldring, M.G., Bakker, A., Nieweg, H.O., Experiments on the intrinsic factor antibody in serum from patients with pernicious anaemia, J. Lab. Clin. Med. **61,** 893, 1963.

[29] Jeffries, G.H., Sleisenger, M.H., Studies of parietal cell antibody in pernicious anaemia, J. Clin. Invest. **44,** 2021, 1965.

[30] Ellenbogen, L., Absorption and transport of cobalamin. Intrinsic factor and the transcobalamins, in Cobalamin, Ed. B.M. Babior, J. Wiley and Sons, New York 1975, p. 236.

[30 a] Zittoun, J., Tuliez, M., Estournet, B., Goulon, M., Humoral and cellular immunity to intrinsic factor in myasthenia gravis, Scand. J. Haematol. **23,** 442-448, 1979.

[31] Samloff, I.M., Kleinman, M.S., and Turner, M.D., Blocking and binding antibodies to intrinsic factor and parietal cell antibody in pernicious anaemia, Gastroenterology **55,** 575, 1968.

[32] Schade, S.C., Feick, P., Muckerheide, M., and Schilling, R.F., Occurrence in gastric juice of antibody to a complex of intrinsic factor and vitamin B_{12}, N. Engl. J. Med. **275,** 528, 1966.

[33] Goldberg, L.G., Shuster, J., Stuckey, M., and Fudenberg, H.H., Secretory immunoglobulin. A. Autoantibody in gastric juice, Science **160,** 1240, 1968.

[34] Carmel, R., and Herbert V., Intrinsic-factor antibody in the saliva of a patient with pernicious anaemia, Lancet **1,** 80, 1967.

[35] Hoedemacker, P.J., Abels, J., Wachters, J.J., Arends, A. and Bieweg, H.O., Investigations about the site of production of Castle's gastric intrinsic factor, Lab. Invest. **13,** 1394, 1964.

[36] Markson, J.I., Moore, J.M., Autoimmunity in pernicious anaemia and iron-deficiency anaemia: A complement-fixation test using human gastric mucosa, Lancet **2,** 1240, 1962.

[37] Rose, M.S., Chanarin, I., Doniach, D., Brostoff, J., Arodeman, S., Intrinsic factor antibodies in absence of pernicious anaemia. Lancet **2,** 9, 1970.

[38] Samloff, I. M., Kleinman, M. S., Turner, M. D., Sobel, M. V., Jeffries, G. H., Blocking and binding antibody to intrinsic factor and parietal cell antibody in pernicious anaemia. Gastroenterology **55,** 575, 1968.

[39] Wangel, A. G., Schiller, K. F. R., Diagnostic significance of antibody to intrinsic factor, Brit. Med. J. **1,** 1274, 1966.

[40] Doniach, D., Roitt, I. M., Taylor, K. B., Autoimmune phenomena in pernicious anaemia: Serologic overlap with thyroiditis thyrotoxicosis and systemic lupus crythematosus, Brit. Med. J. **1,** 1374, 1963.

[41] McIntyre, O. R., Sullivan, L. W., Jeffries, G. H., Silver, R. H., Pernicious anaemia in childhood, N. Engl. J. Med. **272,** 981, 1965.

[42] Spurling, C. L., Sacks, M. S., Jiji, R. M., Juvenile pernicious anaemia, N. Engl. J. Med. **271,** 995, 1964.

[43] Lillibridge, S. B., Brandborg, L. L., Rubin, C. E., Childhood pernicious anaemia: Gastrointestinal secretory, histological and electron microscopic aspects, Gastroenterology **52,** 792, 1967.

[44] Miller, D. R., Bloom, G. E., Streiff, R. R., LoBuglio, A. F., Diamond, L. K., Juvenile "congenital" pernicious anaemia, Clinical and immunologic studies, N. Engl. J. Med. **275,** 978, 1966.

[44a] Hirota, A., Clinical study on output of hydrochloric acid and intrinsic factor in juvenile diabetics, Tokyo, Jikeika, Med. J. **97,** 208–224, 1982.

[45] Katz, M., Lee, S. K., Cooper, B. A., Vitamin B_{12} malabsorption due to a biologically inert intrinsic factor, N. Engl. J. Med. **287,** 425, 1972.

[46] Katz, M., Mehlman, C. S., Allen, R. H., Isolation and characterization of abnormal human intrinsic factor, J. Clin. Invest. **53,** 1274.

[46a] Adams, J. F., Clark, J. S., Ireland, J. T., Kesson, C. M., Watson, W. S., Melabsorpthion of vitamin B_{12} and intrinsic factor secretion during biguanide therapy, Diabetologia **24,** 16–18, 1983.

[47] Donaldson, R. M., Small bowel bacterial overgrowth, Adv. Intern. Med. **16,** 191, 1970.

[48] Tabaqchali, S., The pathophysiological role of small intestinal bacterial flora, Scand. J. Gastroenterol. (Suppl. 6), 139, 1970.

[49] Donaldson, R. M., McConnel, C., Defner, N., Bacteriological studies in clinical and experimental blind loop syndromes, Gastroenterology **52,** 1082, 1967.

[50] Donaldson, R. M., Malabsorption of Co-labelled cyanocobalamin in rats with intestinal diverticula. I. Evaluation of possible mechanisms, Gastroenterology **43,** 271, 1962.

[51] Halstead, J. A., Lewis, P. M., Gasster, M., Absorption of radioactive vitamin B_{12} in the syndrome of megaloblastic anemia associated with intestinal structure or anastomosis, Am. J. Med. **20,** 52, 1956.

[52] Ament, M. E., Shimoda, S. S., Saunders, D. R., Rubin, C. F., Pathogenesis of steatorrhea in three cases of small intestinal stasis syndrome, Gastroenterology **63,** 728, 1972.

[53] Booth, C. C., Heath, J., The effect of *E. coli* on the absorption of vitamin B_{12}, Gut **3,** 70, 1962.

[54] Dack, G. M., Retran, E., Bacterial activity in different levels of the intestine and in isolated segments of small and large bowel in monkeys and in dogs. J. Infect. Dis. **54,** 204, 1934.

[55] Dixon, J. M. S., The fate of bacteria in the small intestine. J. Pathol. Bacteriol. **79,** 131, 1960.

[56] von Bonsdorff, B., Pathogenesis of vitamin B_{12} deficiency, with special reference to tapeworm pernicious anaemia, in Vitamin B_{12} and Intrinsic Factor, Ed. Heinrich, H. C., Enke, Stuttgart 1967.

[57] Brante, G., Ernberg, T., The in vitro uptake of vitamin B_{12} by *Diphyllobothrium latum* and its blockage by intrinsic factor, Scand. J. Clin. Lab. Invest. **9,** 313, 1957.

[58] Nyberg, W., The influence of *Diphyllobothrium latum* on the vitamin B_{12}-intrinsic factor complex. I. In vivo studies with Schilling test technique, Acta Med. Scand. **167,** 185, 1960.

[59] Nyberg, W., The influence of Diphyllobothrium latum on the vitamin B_{12}-intrinsic factor complex. II. In vitro studies, Acta Med. Scand. **167,** 189, 1960.

[60] Salokannel, J., Intrinsic factor in tapeworm anaemia. Acta Med. Scand. **188** (Suppl. 517), 1970.

[61] Giannella, R. A., Broitman, S. A., Zamcheek, N., Vitamin B_{12} uptake by intestinal microorganisms: Mechanism and relevance to syndromes of intestinal bacterial overgrowth, J. Clin. Invest. **50,** 1100, 1971.

[62] Schjonsby, H., The effect of bacteria on intestinal uptake of vitamin B_{12}. I. Effect of cultures of blind-loop contents, Scand. J. Gastroenterol. **7,** 119, 1972.

[63] McIntyre, P. A., Sachs, M. V., Krevans, J. R., Conley, C. L., Pathogenesis and treatment of macrocytic anaemia: Information obtained with radioactive vitamin B_{12}, Arch. Intern. Med., **98,** 541, 1956.

[64] Veeger, W., Abels, J., Hellemans, N., Nieweg, H.O., Effect of sodium bocarbonate and pancreatin on the absorption of vitamin B_{12} and fat in pancreatic insufficiency, N. Engl. J. Med. **267**, 1341, 1962.
[65] Toskes, P.O., Deren, J.J., Vitamin B_{12} absorption and malabsorption, Gastroenterology **65**, 662, 1973.
[66] Toskes, P.P., Hansell, J., Cerda, J., Deren, J.J., Vitamin B_{12} malabsorption in chronic pancreatic insufficiency. Studies suggesting the presence of a pancreatic intrinsic factor, N. Engl. J. Med. **284**, 627, 1971.
[67] Toskes, P.P., Ginsberg, A., Conrad, M.E., Deren, J.J., The physical chemical characteristics and mode of action of pancreatic intrinsic factor, Blood **38**, 609, 1971.
[68] Allen, R.H., Seetharam, B., Podell, E., Alpers, D.H., Effect of proteolytic enzymes on the binding of cobalamin to R-protein and intrinsic factor, J. Clin. Invest. **61**, 47-54, 1978.
[69] Parmentier, Y., Marcoullis, G., Nicolas, J.P., Perrin, M.O., The intraluminal phase of vitamin B_{12} transport in the human, in Vitamin B_{12}, Eds. B. Zagalak and W. Friedrich, Walter de Gruyter, Berlin 1979, pp. 803-806.
[70] Booth, C.C., Mollin, D.L., The site of absorption of vitamin B_{12} in man. Lancet **1**, 18, 1959.
[71] Smith, A.N., Falconer, C.W.A., Small, W.P., Malabsorption in Crohn's disease, in Malabsorption, Eds. Girdwood, R.H., Smith, A.N., Edinburgh Press, Edinburgh 1969.
[72] Fromm, H., Thomas, P.J., Hofman, A.F., Sensitivity and specificity in tests of distal ileal function, Gastroenterology **64**, 1077, 1973.
[73] Katz, J.H., DiMase, J., Donaldson, R.M., Simultaneous administration of gastric juice bound and free radioactive cyanocobalamin, J. Lab. Clin. Med. **61**, 266, 1963.
[74] Booth, C.C., Metabolic effects of intestinal resection in man, Postgrad. Med. J. **37**, 725, 1961.
[75] Donaldson, R.M., Absorption of vitamin B_{12}, Lancet **1**, 790, 1959.
[76] Booth, C.C., Effect of location along the small intestine on absorption of nutrients, in Handbook of Physiology. Sect. 6, Vol. III, Eds. C.F. Code and W. Weidel, Williams and Wilkins, Baltimore 1968, p. 1513.
[76a] Bambach, Ch.P., Hill, G.L., Long-term nutritional effects of extensive resection of the small intestine, Anst. N.Z. J. Surg. **52**, 500-506, 1982.
[76b] Kennedy, H.J., Callender, S.T., Trufelove, S.C., Warner, G.T., Hematological aspects of life with an ileostomy, Br. J. Haematol. **52**, 445-454, 1982.
[77] Donaldson, R.M., Mackenzie, I.L., Trier, J.S., Intrinsic factor-mediated attachment of vitamin B_{12} to brush borders and microvillous membranes of hamster intestine, J. Clin. Invest. **46**, 1215, 1967.
[78] Rubin, C.E., Brandborg, L.L., Phelps, P.C., Taylor, H.C., Studies of celiac disease, Gastroenterology **38**, 28, 1960.
[79] MacKenzie, I.L., Donaldson, R.M., Kopp, W.L., Trier, J.S., Antibodies to intestinal microvillous membrane. I. Characterization and morphologic localization, J. Exp. Med. **128**, 357, 1968.
[80] Stewart, J.J., Pollock, D.J., Hoffbrand, A.V., Mollin, D.L., Booth, C.C., A study of proximal and distal intestinal structure and absorptive function in intestinal steatorrhea, Q. J. Med. **36**, 425, 1967.
[81] Mathan, V.I., Tropical sprue in Gastrointestinal Disease, Eds. Sleisenger, M., and Fordtran, J., Saunders, Philadelphia, Pennsylvania, 1973, p. 978.
[82] Brunser, O., Eidelman, S., Klipstein, F., Intestinal morphology of rural Haitians. A comparison between overt tropical sprue and asymptomatic subjects. Gastroenterology **58**, 655, 1970.
[83] Bayless, T.M., Wheby, M.S., Swanson, V.L., Tropical sprue in Puerto Rico, Am. J. Clin. Nutr. **21**, 1030, 1968.
[84] Gorbach, S.L., Banwell, J.G., Jacobs, B., Tropical sprue and malnutrition in West Bengal. I. Intestinal microflora and absorption, Am. J. Clin. Nutr. **23**, 1545, 1970.
[85] Klipstein, F.A., Tropical sprue, Gastroenterology **54**, 275, 1969.
[85a] Andersen, K.J., Schjonsby, H., Skagen, D.W., Haga, J., Enzyme activities in jejunal biopsy simples from patients with adult celiac disease with and without steatorrhea, Scand. J. Gastroenterol. **18**, 365-368, 1983.
[85b] Batt, R.M., Bush, B.M., Subcellular biochemical studies of a naturally occurring enteropathy in the dry resembling chronic tropical sprue in humans, Am. J. Vet. Res. **44**, 1492-1496, 1983.
[85c] Fromm, H., Sarva, R.P., Ravitch, M.M., Mejunkin, B., Farivar, S., Amin, P., Effects of jejunoileal bypass on the enterohepatic circulation of bile acids bacterial flora in the upper small intestine and absorption of vitamin B_{12}, Metab. Clin. Exp. **32**, 1133-1141, 1983.
[85d] Gendre, J.P., Cosnes, J., Le Quintrec, Y., Late

radiation injury to the small intestine: 1. Malabsorption in relation to anatomical damage, Gastroenterol. Clin. Biol. **7**, 664-670, 1983.
[86] Gräsbeck, R., Nyberg, W., Inhibition of radiovitamin B_{12} absorption by ehtylenediamine tetraacetate (EDTA) and its reversal by calcium ions, Scand, J. Clin. Lab. Invest. **10**, 448, 1958.
[87] MacKenzie, I.L., Donaldson, R.M., Effect of divalent cations and pH on intrinsic factor-mediated attachment of vitamin B_{12} to intestinal microvillous membranes, J. Clin. Invest. **51**, 2465, 1972.
[87a] Kouvonen, I., Grasbect, R., The role of sialic acid in the binding of calcium ions to intrinsic factor its intestinal receptor, Biochim. Biophys. Acta **797**, 163-170, 1984.
[88] MacKenzie, I.L., Donaldson, R.M., Trier, J.S., Mathan, V.I., Ileal mucosa in familial selective vitamin B_{12} malabsorption, N. Engl. J. Med. **286**, 1021, 1972.
[89] Imerslund, O., Idiopathic chronic megaloblastic anaemia in children, Acta Pediatr. (Suppl.) **119**, 1960.
[90] Gräsbeck, R., Gordin, R., Knatero, I., Kuhlback, B., Selective vitamin B_{12} malabsorption and proteinuria in young people, Acta Med. Scand. **167**, 289, 1960.
[91] Gräsbeck, R., Familial selective vitamin B_{12} malabsorption, N. Engl. J. Med. **287**, 358, 1972.
[91a] Dulloo, R.M., Majumdar, S., Chakravarti, R.N., Mahmood, A., Intestinal brush border membrane structure and function: Effect of chronic vitamin C deficiency in guinea pigs, Biochem. Med. **27**, 325-333, 1982.
[92] Heinivaara, O., Pavla, I.P., Malabsorption and deficiency of vitamin B_{12} caused by treatment with para-aminosalicylic acid, Acta Med. Scand. **177**, 337, 1965.
[93] Tomkin, G.H., Hadden, D.R., Weaver, J.A., Montgomery, D.A.D., Vitamin B_{12} status of patients on long term Metformin therapy, Brit. Med. J. **1**, 685, 1971.
[94] Webb, D.L., Chodos, R.B., Mahar, C.Q., Falcon, W.W., Mechanism of vitamin B_{12} malabsorption in patients receiving colchicine, N. Engl. J. Med. **279**, 845, 1968.
[95] Jacobson, E.D., Chodos, R.B., Falcon, W.W., An experimental malabsorption syndrome induced by Neomycin, Am. J. Med. **28**, 524, 1960.
[96] Heinivaara, O., Palva, I., Siorala, M., Pelkonen, R., Selectivity of the PAS-induced malabsorption of vitamin B_{12}. Ann. Med. Intern. Fenn. **53**, 75, 1964.
[96a] Steinberg, W.M., King, C.E., Toskes, P.P., Malabsorption of protein-bound cobalamin but not unbound cobalamin during cimetidine administration, Dig. Dis. Sci. **25**, 188-192, 1980.
[96b] Binder, H.J., Donaldson, R.M., Effect of cimetidine on intrinsic factor and pepsin secretion in man, Gastroenterology **74**, 371-375, 1978.
[96c] Fielding, L.P., Chalmers, D.M., Chanarin, I., Levi, A.J., Inhibition of intrinsic factor secretion by cimetidine, Br. Med. J. **1**, 818-819, 1978.
[96d] King, C.E., Leibach, J.R., Toskes, P.P., Clinically significant vitamin B_{12} deficiency secondary to malabsorption of protein-bound vitamin B_{12}, Dig. Dis. Sci. **24**, 397-402, 1979.
[96e] Nakazawa, O., Fukuda, M., Study on vitamin B_{12}-binding protein in hepatoma: evidence and analyses of anamalous B_{12}-binding proteins, Sapporo Med. J. **49**, 99-116, 1980.
[97] Hakami, N., Neiman, P.E., Canellos, G.P., Lazerson, J., Neonatal megaloblastic anaemia due to inherited transcobalamin. II. deficiency in two siblings. N. Engl. J. Med. **285**, 1163, 1971.

9.3.2 Defective cobalamin transport

CP deficiency:

Possible mechanisms: congenital error of synthesis [1, 2, 2a] or autoimmunological effect [3a]. CP-Cbl complex contributes mainly to cobalamin reserve of plasma [4-8] (Fig. 8-14). Its deficiency in plasma, and absence in saliva, and peripheral leukocytes, however, is not vital for cobalamin metabolism [1, 2] (see section 8.3.2).

TC deficiency:

Possible mechanisms: inborn error [9-16, 16a, b] or effect of autoimmunological reaction [3, 3a]. Transcobalamin has a dynamic function in the cobalamin transport.

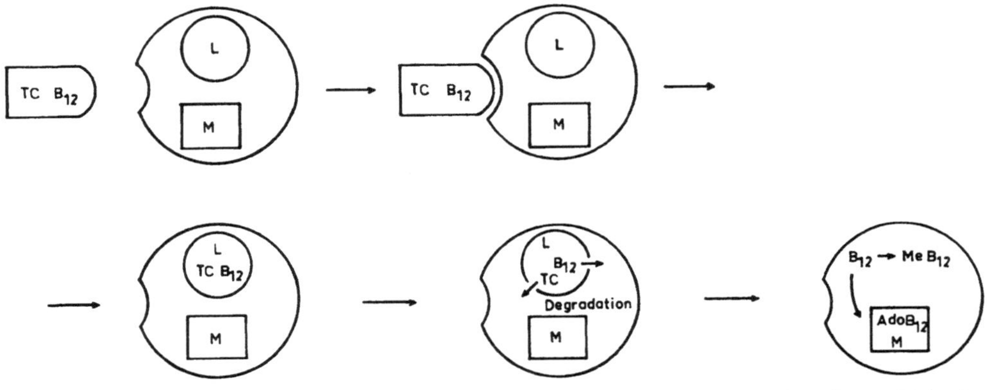

Fig. 9-22 Incorporation of cobalamins into mammalian cells. The receptor site on the cell has a high affinity for the transcobalamin-cobalamin complex but not for free transcobalamin. The complex then enters the cell in the lysosomal fraction (L). The transcobalamin-cobalamin complex is degraded in the lysosome. Some of the cobalamin enters the mitochondrion (M) and is used to synthesize adenosylcobalamin (Ado B_{12}): some of the cobalamin is also used to synthetize methylcobalamin (MoB_{12}) in the cytosol (reprinted from Ellenbogen [7]). (see also Fig. 8-23).

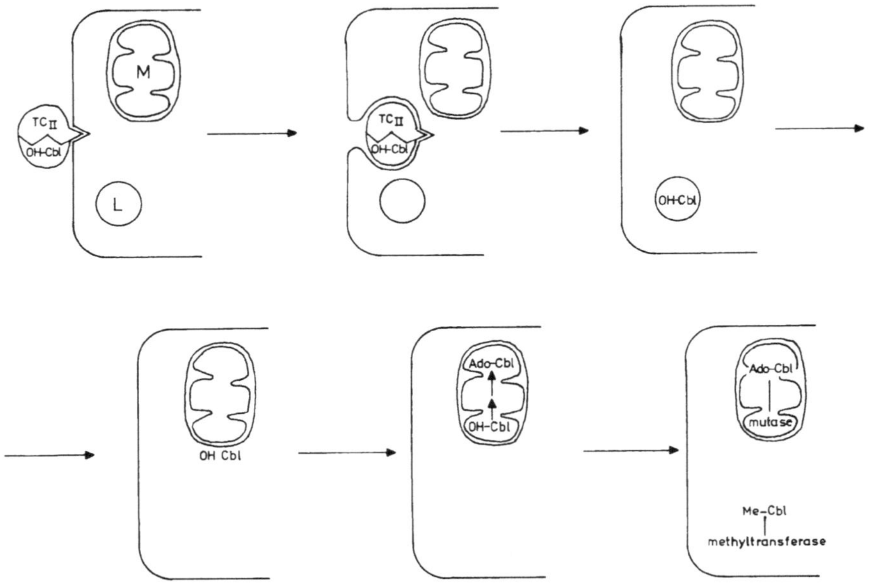

Fig. 9-23 Suggested intracellular processing of cobalamins. The circulating complex of TC and cobalamin (OH-Cbl) binds to recognition sites on the cell membrane and is taken into the cell by a mediated process that suggests pinocytosis. The pinocytotic vacuole fuses with a primary lysosome (L) and OH-Cbl is split from TC in the resultant secondary lysosome. From that organelle, some of the OH-Cbl moves to the mitochondrion (M), is transported across the mitochondrial membrane, and the converted to AdoCbl to function as coenzyme for MM-CoA mutase. Also, OH-Cbl is converted to MeCbl, at an unspecified site, and serves as coenzyme for the methyltetrahydrofolate homocysteine methyltransferase reaction in the cytosol. (from Mahoney and Rosenberg, in: Cobalamin, Babior (ed.), 1975, p. 236. Copyright 1975 John Wiley & Sons, Inc. Reprinted by permission of John Wiley & Sons, Inc.)

The presence of TC is the last essential element of the "B_{12} pump" in the intestinal epithelial cells [7, 17, 18] and is imperative for the transport of cobalamin through the membrane (Figs. 9-22, 9-23) [6, 18-37].

Abnormalities of transcobalamin: Increase in the level of the binding capacities of TC-I and TC-II with concomitant increase of UBBC, and a significant reduction in the level of endogenous B_{12} [38a].

9.3.3 Faults in tissue utilisation

The process of degradation of the OH-Cbl-TC complex releases OH-Cbl, which can be utilized in cells of destination. OH-Cbl is converted to coenzyme forms: AdoCbl and MeCbl (Figs. 9-23, 9-24).

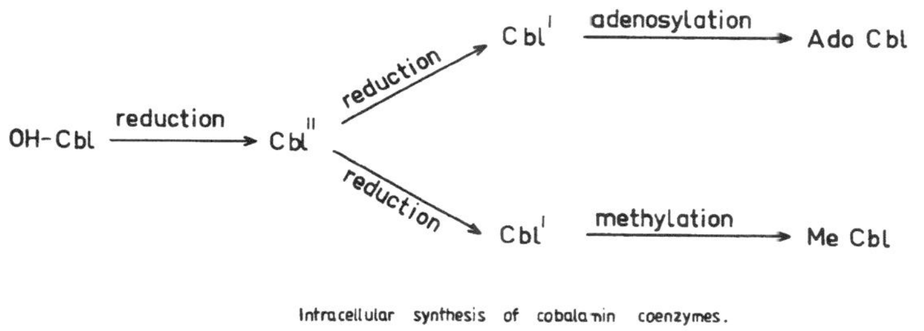

Fig. 9-24 *Scheme of enzymatic synthesis of coenzymatic forms of Cbl in the cell: AdoCbl and MeCbl from the common precursor OH-Cbl.* The paths run through a two step reduction stage catalized by two enzymes (for details see Chapter 4). Metabolic errors pertaining to faulty enzymes engaged in the synthesis of coenzymatic forms may occur at the stage of reduction or attachment of the methyl or adenosyl residue. Poor synthesis of AdoCbl in the cell is manifested by excretion of methylmalonic acid (methyl malonic aciduria) while excretion of homocysteine (homocystinuria) is an indication for poor MeCbl synthesis. In patients who excrete both methylmalonic acid and homocysteine a faulty reduction of Cbl is more likely.

The levels of AdoCbl and MeCbl may be disturbed by:
- congenital defects of coenzyme synthesis [38-61]
- presence of cyanide [62-69].

The utilisation of B_{12} coenzymes in enzymatic reactions can be disordered by some abnormalities of apoenzymes. The lack or impairment of methylmalonyl-CoA mutase activity is caused by the block of enzyme synthesis or the synthesis of an abnormal enzyme protein [43, 61, 70-72] (The B_{12}-dependent enzymes are described in details in Chapter 7).

References

[1] Carmel, R., Herbert, V., Deficiency of vitamin B_{12}-binding alpha globulin in two brothers, Blood **33**, 1, 1969.

[2] Linnell, J. C., Matthews, D. M., Mudd, S. H., Uhlendorf, B. W., Wise, I. J., Pediat. Res. **10**, 179-183, 1976.

[2a] Carmel, R., A new case of deficiency of the R binder for cobalamin, with observations on minor cobalamin-binding proteins in serum and saliva, Blood. **59**, 152-156, 1982.

[3] Nicolas, J.P., Heterogeneity of antibodies against B_{12} binders in pernicious anaemia, in, Vitamin B_{12}, Eds. B. Zagalak and W. Friedrich, Walter de Gruyter, Berlin 1979, pp. 791-802.

[3a] Carmel, R., An unusual case of autoimmune agranulocytosis with total absence of myeloid precursors: demonstration of diverse sources of R binder for cobalamin in plasma and secretions, Am. J. Clin. Pathol. **76**, 611-615, 1983.

[4] Linnell, J.C., Hoffbrand, A.V., Peters, T.J., Matthews, D.M., Chromatographic and bioautographic estimation of plasma cobalamins in various disturbances of vitamin B_{12} metabolism, Clin. Sci. **40**, 1-16, 1971.

[5] Linnell, J.C., Hoffbrand, A.V., Hussein, H.A.A., Wise, I.J., Matthews, D.M., Tissue distribution of coenzyme and other forms of vitamin B_{12} in control subjects and patients with pernicious anaemia, Clin. Sci. Mol. Med. **46**, 163-172, 1974.

[6] Ellenbogen, L., Absorption and transport of cobalamin. Intrinsic factor and the transcobalamins, in Cobalamin, Ed. B.M., Bavior, John Wiley and Sons, New York 1975, pp. 215-286.

[7] Ellenbogen, L., Uptake and transport of cobalamins, Biochem. Nutr. **27**, 45-96, 1979.

[8] Mahoney, M.J., Rosenberg, L.E., Inborn errors of cobalamin metabolism, in, Cobalamin, Ed. B.M., Bavior, John Wiley and Sons, New York 1975, pp. 369-402.

[9] Hakami, N., Neiman, P.E., Canellos, G.P., Lazeron, J., Neonatal megaloblastic anaemia due to inherited transcobalamin. II deficiency in two siblings, New Engl. J. Med. **285**, 1163, 1971.

[10] Scott, C.R., Hakami, N., Teng, C.C., Sagerson, R.N., Hereditary transcobalamin II deficiency: The role of transcobalamin II in vitamin B_{12}-mediated reactions, J. Pediatr. **81**, 1106, 1971.

[11] Hitzig, W.H., Dohmann, U., Pluss, H.J., Vischer, D., Hereditary transcobalamin II deficiency: Clinical findings in a new family, J. Pediatr. **85**, 622-628, 1974.

[12] Burman, J.F., Mollin, D.L., Sladden, R.A., Sourial, N., Greany, M., Inherited deficiency of transcobalamin II causing megaloblastic anaemia, Brit. J. Haemat. **35**, 676-677, 1977.

[13] Hitzig, W.H., Kenny, A.B., The role of vitamin B_{12} and its transport globulins in the production of antibodies, Clin. Exp. Immunol. **20**, 105-111, 1975.

[14] Hitzig, W.H., Frater-Schröder, M., Seger, R., Clinical diseases related to deficiencies of vitamin B_{12} transport proteins, in, Vitamin B_{12}, Eds. B. Zagalak and W. Friedrich, Walter de Gruyter, Berlin 1979, pp. 1009-1016.

[15] Linnell, J.C., Matthews, D.M., Recent advances in cobalamin metabolism: abnormalities in coenzyme distribution in tumor development and in inherited metabolic disease, in, Vitamin, B_{12}, Eds. B. Lagalak and W. Friedrich, Walter de Gruyter, Berlin 1979, pp. 1101-1111.

[16] Burman, J.F., Malleson, P.N., Sourial, N.A., Mollin, D.L., TC II deficiency: observations with deoxyuridine suppression test, In, Vitamin B_{12}, Eds. B. Zagalak and W. Friedrich, Walter de Gruyter, Berlin 1979, pp. 889-893.

[16a] Hall, C.A., Begley, J.A., A typical cobalamin binding in the serum of congenital deficiency of transcobalamin II, Br. J. Haematol. **51**, 65-71, 1982.

[16b] Carmel, R., Ravindranath, Y., Congenital transcobalamin II deficiency presenting a typically with a low serum cobalamin level: studies demonstrating the coexistence of a circulating transcobalamin I (R binder) complex., Blood. **63**, 598-605, 1984.

[17] Hall, C.A., Finkler, A.E., The dynamics of transcobalamin II. A vitamin B_{12}-binding substance in plasma, J. Lab. Clin. Med. **65**, 459, 1965.

[18] Hall, C.A., Transport of vitamin B_{12} in man, Br. J. Haematol. **16**, 429-433, 1969.

[19] Allen, R.H., Human vitamin B_{12} transport proteins, Prog. Haematol. **9**, 57-84, 1975.

[20] Finkler, A.E., Hall, C.A., Nature of the relationship between vitamin B_{12} binding and cell uptake, Arch. Biochem. Biophys. **120**, 79, 1967.

[21] Retief, F.P., Gottlieb, C.W., Herbert, V., Mechanism of vitamin B_{12} uptake by erythrocytes, J. Clin. Invest. **415**, 1907, 1966.

[22] Retief, F.P., Gottlieb, C.W., Herbert, V., Delivery of ^{57}Co-B_{12} to erythrocyte from α- and β-globulin of normal, B_{12}-deficient, and chronic myeloid leukemia serum. Blood, **29**, 837, 1967.

[23] Paranchych, W., Cooper, B.A., Factors influencing the uptake of cyanobobalamin (vitamin B_{12}) by Ehrlich ascites carcinoma cells, Biochem. Biophys. Acta **60**, 393, 1962.

[24] Ryel, E.M., Meyer, L.M., Gams, R.A., Uptake and subcellular distribution of vitamin

B_{12} in mouse L 1210 leukemic lymphoblasts, Blood **44**, 427-433, 1974.

[25] Meyer, L. M., Gams, R. A., Ryel, E. M., Miller, I. E., Kumar, S., Delivery of CoB_{12} to lymphoblasts derived from mice with transplanted 1210 ascites tumor cells by transcobalamins I, II and III, Proc. Soc. Exp. Biol. Med. **147**, 679-680, 1974.

[26] Gams, R. A., Ryel, E. M., Meyer, L. M., Metabolism of cyanocobalamin by L 1210 leukemic lymphoblasts, Proc. Soc. Exp. Biol. Med. **149**, 384, 1975.

[27] Chello, P. J., Bertino, J. R., Dependence of 5-methyl-tetrahydrofolate utilization by L 5178 Y murine leukemia cells in vitro on the presence of hydroxycobalamin and TC II, Cancer Res. **33**, 1898-1904, 1973.

[28] DiGirolami, P. M., Huennekens, F. M., Arch. Biochem. Biophys. **168**, 386, 1975.

[29] Wickramasinghe, S. N., Carmel, R., Uptake of Covitamin B_{12} by human haemopoietic cells in vitro, Br. J. Haematol. **23**, 307, 1972.

[30] Wickramasinghe, S. N., Moffatt, B., Factors influencing the uptake of vitamin B_{12} by normoblastic and vitamin B_{12}-deficient bone morrow cells, Acta Haemat. **49**, 65-73, 1973.

[31] Rosenberg, L. E., Lillhequist, A., Allen, R. H., Transcobalamin II-facilitated uptake of vitamin B_{12} by cultured fibroblasts: Studies in methylmalonic aciduria, J. Clin. Invest. **52**, 69A, 1973.

[32] Youngdahl-Turner, P., Allen, R. H., Rosenberg, L. E., Binding and uptake of transcobalamin. II by human fibroblasts, J. Clin. Invest. **61**, 133, 1978.

[33] Gams, R. A., Ryel, E. M., Ostroy, F., Protein-mediated uptake of vitamin B_{12} by isolated mitochondria, Blood **47**, 923-930, 1976.

[34] Newmark, P., Newman, G. E., O'Brien, J. R. P., Vitamin B_{12} in the rat kidney. Evidence for an association with lysosomes, Arch. Biochem. Biophys. **141**, 121, 1970.

[35] Pletsch, Q. A., Coffey, J. W., Intracellular distribution of radioactive vitamin B_{12} in rat liver, J. Biol. Chem. **246**, 4619, 1971.

[36] Pletsch, Q. A., Coffey, J. W., Properties of the proteins that bind vitamin B_{12} in subcellular fractions of rat liver, Arch. Biochem. Biophys. **151**, 157, 1972.

[37] Ostroy, F., Gams, R. A., Cellular fluxes of vitamin B_{12}, Blood **50**, 877-888, 1977.

[38] Osifo, B. O. A., Adeyokunnu, A., Parmentier, Y., Gerard, P., Nicolas, J. P., Abnormalitiés of serum transcobalamin in sickle cell disease (hemoglobin SS) in Black Africa, Scand. J. Haematol. **30**, 135-140, 1983.

[39] Stokke, O., Eldjarn, J., Norum, K. R., Steen-Johnson, J., Halvorsen, S., Methylmalonic acidemia. A new inborn error of metabolism which may cause fatal acidosis in the neonatal period, Scand. J. Clin. Lab. Invest. **20**, 313, 1967.

[40] Hsia, Y. E., Scully, K. J., Rosenberg, L. E., Defective propionate carboxylation in ketotic hyperglycinaemia, Lancet **1**, 757, 1969.

[41] Rosenberg, L. E., Lilljequist, A. Ch., Hsia, Y. E., Methylmalonic aciduria: Metabolic block localization and vitamin B_{12} dependency. Science **162**, 805, 1968.

[42] Lindblad, B., Lindstrand, K., Svanberg, B., Zetterström, R., The effect of cobamide coenzyme in methylmalonic acidemia, Acta Paediatr. Scand. **58**, 178, 1969.

[43] Morrow, G. III, Barness, L. A., Cardinale, G. J., Abeles, R. H., Flaks, J. G., Congenital methylmalonic acidemia: Enzymatic evidence for two forms of the disease, Proc. Natl. Acad. Sci. USA **63**, 191, 1969.

[44] Cox, E. V., White, A. M., Methylmalonic acid excretion: An index of vitamin B_{12} deficiency, Lançet **2**, 853, 1962.

[45] Cox, E. V., Robertson-Smith, D., Small, M., White, A. M., The excretion of propionate and acetate in vitamin B_{12} deficiency, Clin. Sci. **35**, 123, 1968.

[46] Rosenberg, L. E., Lilljequist, A. Ch., Hsia, Y. E., Rosenbloom, F. M., Vitamin B_{12}-dependent methylmalonicaciduria: Defective B_{12} metabolism in cultured fibroblasts, Biochem. Biophys. Res. Commun. **37**, 507, 1969.

[47] Mahoney, M. J., Rosenberg, L. E., Mudd, S. H., Uhlendorf, B. W., Defective metabolism of vitamin B_{12} in fibroblasts from children with methylmalonic aciduria. Biochem. Biophys. Res. Commun. **44**, 375, 1971.

[48] Mahoney, M. J., Hart, A. C., Rosenberg, L. E., Defect in 5'-deoxyadenosylcobalamin synthesizing enzyme in methylmalonic acidemia. Pediatr. Res. **8**, 436, 1974.

[49] Gravel, R. A., Mahoney, M. J., Ruddle, F. H., Rosenberg, L. E., Genetic complementation in heterokaryons of human fibroblasts defective in cobalamin (vitamin B_{12}) metabolism (abstr.), Am. J. Hum. Genet. **26** (6), 36 a, 1974.

[50] Kaye, C. I., Morrow, G., Nadler, H. L., In vitro "responsive" methylmalonic acidemia: a new variant. J. Pediatr. **85**, 55, 1974.

[51] Mudd, S. H., Levy, H. L., Abeles, R. H., A derangement in B_{12} metabolism leading to homocystinemia, cystathioninemia and methylmalonic aciduria, Biochem. Biophys. Res. Commun. **35**, 121, 1969.

[52] Levy, H.L., Mudd, S.H., Schulman, J.D., Dreyfus, P.M., Abeles, R.H., A derangement in B_{12} metabolism associated with homocystinemia, cystathioninemia, hypomethioninemia, and methylmalonic aciduria. Am. J. Med. **48,** 390, 1970.

[53] Mudd, S.H., Uhlendorf, B.W., Hinds, K.R., Levy, H.L., Deranged B_{12} metabolism: Studies of fibroblasts grown in tissue culture, Biochem. Med. **4,** 215, 1970.

[54] Mudd, S.H., Levy, H.L., Morrow, G., Deranged B_{12} metabolism. Effects on sulfur amino acid metabolism, Biochem. Med. **4,** 193, 1970.

[55] Rosenberg, L.E., Patel, L., Absence of a cobalamin binding protein in cultured fibroblasts from patients with defective cobalamin coenzyme synthesis (abstr.), J. Clin. Invest. **55,** 6, in press.

[56] Goodman, S.I., Moe, P.G., Hammond, K.P., Mudd, S.H., Uhlendorf, B.W., Homocystinuria with methylmalonic aciduria: Two cases in a sibship, Biochem. Med. **4,** 500, 1970.

[57] Goodman, S.I., Keyser, A.J., Mudd, S.H., Schulman, J.D., Turse, H., Lewy, J., Responsiveness of congenital methylmalonic aciduria to derivatives of vitamin B_{12}, Pediatr. Res. **6,** 138, 1972.

[58] Dillon, M.J., England, J.M., Gompertz, D., Goodey, P.A., Grant, D.B., Hussein, H.A.A., Linnell, J.C., Matthews, D.M., Mudd, S.H., Newns, G.H., Seakins, J.W.T., Uhlendorf, B.W., Wise, I.J., Mental retardation, megaloblastic anaemia, methylmalonic aciduria and abnormal homocysteine metabolism due to an error in vitamin B_{12} metabolism, Clin. Sci. Mol. Med. **47,** 43, 1974.

[59] Hollowell, J.G., Jr., Hall, W.K., Coryell, M.E., McPherson, J., Jr., Hahn, D.A., Homocystinuria and organic aciduria in a patient with vitamin B_{12} deficiency, Lancet **2,** 1428, 1969.

[60] see Chapter 4

[61] Baumgartner, R., Wick, H., Inherited deficiences of the deoxyadenosylcobalamin (AdoCbl) dependent L-methylmalonyl-CoA mutase system: Clinical and biochemical aspects, in Vitamin B_{12}, Eds. B. Zagalak and W. Friedrich, Walter de Gruyter, Berlin 1979, pp. 1019-1024.

[62] Matthews, D.M., Wilson, J., In the Cobalamins Eds. Arnstein, H.R.V., Weighton, R.J., Churchill-Livingstone, London, 1971, p. 115.

[63] Wokes, F., Picard, G.W., Am. J. Clin. Nutr. **3,** 383, 1955.

[64] Mushett, C.W., Kelley, K.L., Boxer, G.E., Rickards, J.C., Antidotal efficacy of hydroxocobalamin in experimental cyanide poisoning, Proc. Soc. Exp. Biol. Med. **81,** 234-237, 1952.

[65] Basil, W., Brown, J.K., Matthews, D.M., J. Clin. Pathol. **18,** 317, 1965.

[66] Matthews, D.M., Wilson, J., Zilkha, K.J., J. Neurol. Neurosurg. Psychiat. **28,** 426, 1965.

[67] Wilson, J., Matthews, D.M., Clin. Sci. **31,** 1, 1966.

[68] Linnell, J.C., Wilson, J., Matthews, D.M., Clin. Sci. **37,** 878, 1969.

[69] Beck, W.S., Metabolic features of cobalamin deficiency in man, in Vitamin B_{12}, Eds. B. Zagalak and W. Friedrich, Walter de Gruyter, Berlin 1979, pp. 403-450.

[70] Mahoney, M.J., Hart, A.C., Steen, V.D., Rosenberg, L.E., PNAS, USA, **72,** 2799, 1975.

[71] Gravel, R.A., Mahoney, M.J., Ruddle, F.H., Rosenberg, L.E., Genetic complementation in heterokryons of human fibroblasts defective in cobalamin metabolism, PNAS USA **72,** 3181, 1975.

[72] Willard, H.F., Mellman, I.S., Rosenberg, L.E., Genetic complementation among inherited deficiencies of methylmalonyl CoA mutase activity: evidence for a new class of human cobalamin mutants, Am. J. Hum. Genet. **30,** 1-13, 1978.

[73] Hall, Ch.A., Begley, J.A., Green, Colligan, P.D., The availability of therapeutic hydroxocobalamin to cells, Blood **83,** 335-341, 1984.

[74] Baumgartner, R., Giardini, O., Cantani, A., Sabetta, G., Castro, M., Methylmalonic Acidaemia due to Mutase Apoenzyme Defect: Responsive to Vit. B_{12} in Intact Fibroblasts but not in vivo, J. Inher. Metab. Dis. **5,** 137-141, 1982.

[75] Baumgartner, E.R., Wick, H., Linnell, J.C., Gaull, G.E., Bachmann, C., Steinmann, B., Congenital defect in intracellular cobalamin metabolism resulting in homocystinuria and methylmalonic aciduria. II. Biochemical investigations, Helv. Paediatr. Acta. **34,** 483-496, 1979.

[76] Morrow, G., Barness, L.A., Cardinale, G.J., Abeles, R.H., Flaks, J.G., Congenital methylmalonic acidemia: Enzymatic evidence for two forms of the disease, Proc. Natl. Acad. Sci (USA). **63,** 191-197, 1969.

[77] Morrow, G., Revsin, B., Clark, R., Lebowitz, J., Whelan, D.T., A new variant of methylmalonic acidemia. Defective coenzyme-apoen-

zyme binding in cultured fibroblasts, Clin. Chim. Acta. **85,** 67-72, 1978.

[78] Rosenberg, L. E., Disorders of propionate, methylmalonate and cobalamin metabolism. In: Stanbury, J. B., Wyngaarden, J. B., Fredrickson, D. S., The Metabolic Basis of Inherited Disease McGraw-Hill, New York, 4th edn., pp. 411-429, 1978.

[79] Willard, H. F., Rosenberg, L. E., Inherited deficiencies of human methylmalonyl CoA mutase activity: reduced affinity of mutant apoenzyme for adenosyl-cobalamin, Biochem. Biophys. Res. Commun. **78,** 927-934, 1977.

[80] Willard, H. F., Rosenberg, L. E., Inherited methylmalonyl CoA mutase apoenzyme deficiency in human fibroblasts. Evidence for allelic heterogeneity, genetic compounds, and codominant expression, J. Clin. Invest. **65,** 690-698, 1980b.

9.3.4 Biochemical effects of cobalamin deficiency

- The block of methylmalonyl-CoA mutase activity limits the catabolism of some amino-acids, fatty acids, and thymine (Fig. 9-25).
- The intracellular methylmalonyl-CoA and propionyl-CoA levels are increased, and may have the influence on fatty acid synthesis. An increase in C_{15} and C_{17} odd-chain fatty acids and branched-chain fatty acids in glycerolipids of the nervous system was observed in some cases [1-3].
- The folate metabolism disorder:
  - the membrane transport of N^5-methyltetrahydrofolate is decreased [4-6],
  - the lack or decreasing of methionine synthetase causes accumulation of folate as N^5-methyltetrahydrofolate ("the methylfolate trap") and it creates insufficiences of thymidylate synthetase activity and DNA synthesis (Fig. 9-26) [4-23, 23a-d].

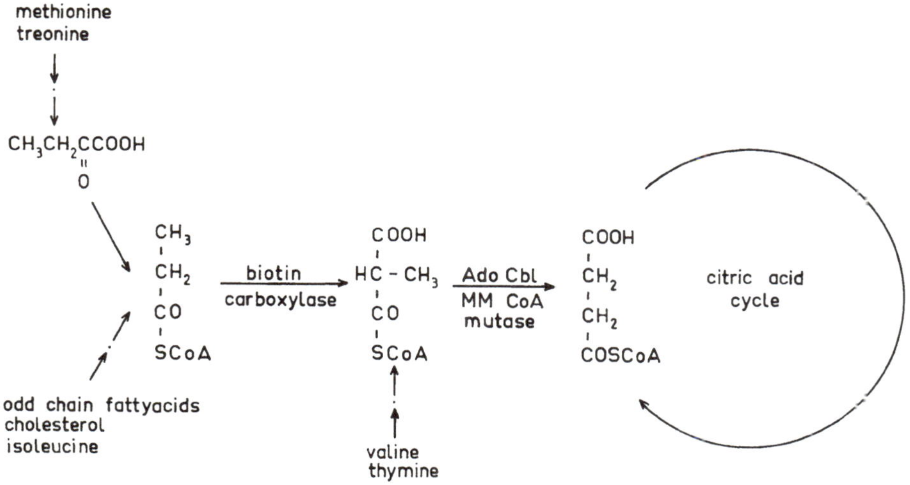

Fig. 9-25 The dependence of the catabolism of some aminoacids, fatty acids, and thymine upon methylmalonyl-CoA mutase

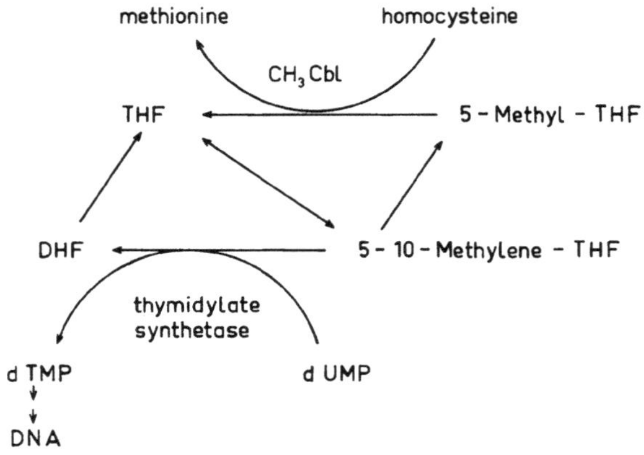

Fig. 9-26 The methylfolate trap

References

[1] Kishimoto, Y., Williams, M., Moser, H., Hignite, C., Biemann, K., Branched-chain and odd-numbered acids and aldehydes in the nervous system of a patient with deranged vitamin B_{12} metabolism, J. Lipid Res. **14**, 69, 1973.

[2] Barley, F.W., Sato, G.H., Abeles, R.H., An effect of vitamin B_{12} deficiency in tissue culture, J. Biol. Chem. **247**, 4270, 1972.

[3] Frenkel, E.P., Abnormal fatty acid metabolism in peripheral nerves of patients with pernicious anaemia, J. Clin. Invest. 1237-1245, 1971.

[4] Das, K.C., Hoffbrand, A.Y., Lymphocyte transformation in megaloblastic anaemias. Morphology and DNA synthesis, Brit. J. Haematol. **19**, 203, 1970.

[5] Tisman, G., Herbert, V., B_{12} dependence of cell uptake of serum folate: an explanation for high serum folate and cell folate depletion in vitamin B_{12} deficiency, Blood **41**, 464-471, 1973.

[6] Beck, W.S., Metabolic features of cobalamin deficiency in man, in Cobalamin, Ed. B.M., Babior, J. Wiley and Sons, New York, 1975, pp. 403-450.

[7] Noronha, J.M., Silverman, M., On folic acid, vitamin B_{12}, methionine and formiminoglutamic acid metabolism, in Vitamin B_{12} und Intrinsic Factor, Ed. H.C. Heinrich, Ferdinand Enke Verlag, Stuttgart 1962, pp. 728-736.

[8] Herbert, V., Zalusky, R., Interrelations of vitamin B_{12} and folic acid metabolism: folic acid clearance studies, J. Clin. Invest., **41**, 1263-1276, 1962.

[9] Killmann, S.-A., Effect of deoxyuridine on incorporation of tritiated thymidine: difference between normoblasts and megaloblasts, Acta Med. Scand. **175**, 483, 1964.

[10] Metz, J., Kelly, A., Swett, V.C., Waxman, S., Herbert, V., Deranged DNA synthesis by bone morrow from vitamin B_{12}-deficient humans, Brit. J. Haematol. **14**, 575, 1968.

[11] Waters, A.H., Mollin, D.L., Observations on the metabolism of folic acid in pernicious anaemia, Brit. J. Haematol., **9**, 319-327, 1963.

[12] Cooper, B.A., Abe, T., Variable response of bone morrow to feeding D,L-5-formyltetrahydrofolate in pernicious anaemia, Brit. J. Haematol., **32**, 387-394, 1976.

[13] Sauer, H., Wilmanns, W., Cobalamin dependent methionine synthesis and methyl folate trap in human vitamin B_{12} deficiency, Brit. J. Haematol. **36**, 189-198, 1977.

[14] Fuji, K., Nagasaki, T., Huennekens, F.M., Vitamin B_{12} requirement for replication of mouse leukemia L 1210 cells: functional relevance to folate metabolism, in, Vitamin B_{12}, Eds. B. Zagalak and W. Friedrich, Walter de Gruyter, Berlin 1979, pp. 1127-1130.

[15] Burman, J.F., Malleson, P.N., Sourial, N.A., Mollin, D.L., TC II deficiency: observations with deoxyuridine suppression test, in, Vitamin B_{12}, Eds. B. Zagalak and W. Friedrich, Walter de Gruyter, Berlin 1979, pp. 889-893.

[16] Das, K.C., Herbert, V., The "dU supression test" and "thymidine supression test" evidence for reciprocal relationship between the "de novo" and "salvage" pathways of DNA synthesis, in, Vitamin B_{12}, Eds. B. Zagalak and

W. Friedrich, Walter de Gruyter, Berlin 1979, pp. 881-887.

[16 a] Goulian, M., Bleile, B., Tseng, B. Y., Methotrexate-induced misincorporation of uracil into DNA, Proc. Natl. Acad. Sci. USA **77**, 1956-1960, 1980.

[17] Zittoun, J., Marquet, J., Zittoun, R., Effect of folate and cobalamin compounds on the deoxyuridine suppression test in vitamin B_{12} and folate deficiency, Blood **51**, 119-128, 1978.

[18] Haurani, F. I., Vitamin B_{12} and the megaloblastic development, Science **182**, 78-79, 1973.

[19] Haurani, F. I., Vitamin B_{12} and folic acid interrelations: a new look, in Vitamin B_{12}, Eds. B. Zagalak and W. Friedrich, Walter de Gruyter, Berlin 1979, pp. 1131-1138.

[20] Sauer, H., Wilmanns, W., Cobalamin dependent methionine synthesis and methyl-folate-trap in human vitamin B_{12} deficiency, Brit. J. Haemat. **36**, 189-198, 1977.

[21] Sauer, H., Cobalamin-dependent methionine synthesis and the regulation of the metabolism of activated folic acid derivatives in cultures human lymphoblasts, in Vitamin B_{12}, Eds. B. Zagalak and W. Friedrich, Walter de Gruyter, Berlin 1979, pp. 1145-1148.

[22] Kass, L., Detection of methionine in pernicious anaemia megaloblasts and other types of erythroid precursors. Am. J. Clin. Path. **65**, 504, 1976.

[23] Krebs, H. A., Hems, R., Tyler, B., The regulation of folate and methionine metabolism, Biochem. J. **158**, 341, 1976.

[23 a] Lumb, M., Deacon, R., Perry, J., Chanarin, I., Minty, B., Halsey, M. J., Nunn, J. F., The effect of nitrous oxide inactivation of vitamin B_{12} on rat hepatic folate: implications for the methylfolate-trap hypothesis, Biochem. J. **186**, 933-936, 1980.

[23 b] Deacon, R., Lumb, M., Perry, J., Chanarin, I., Minty, B., Halsey, M., Nunn, J., Inactivation of methionine synthase (EC 2.1.1.13) by nitrous oxide, Eur. J. Biochem. **104**, 419-422, 1980.

[23 c] Wickramasinghe, S. N., Saunders, J. E., Incorporation of radioactive precursors into the nucleic acids of bone marrow cells from patients with vitamin B_{12} or folate deficiency, Acta Haematol. **63**, 196-203, 1980.

[23 d] Wickremasinge, R., Hoffbrand, A. V., Reduced rate of DNA replication fork movement in megaloblastic anaemia., J. Clin. Invest. **65**, 23-36, 1980.

[24] Montgomery, J. A., Maner, O. A., Scriver, C. R., Metabolism of methylmalonic acid in rats: Is methylmalonyl coenzyme A racemase deficiency symptomatic in man, J. Clin. Invest. **72**, 1937-1947, 1983.

[25] Magnus, E. M., Bache-Wiig, J. E., Aanderson, F. R., Melbostad, E., Folate and vitamin B_{12} (cobalamin) blood levels in elderly persons in geriatric homes, Scand. J. Haematol. **28**, 360-366, 1982.

[26] Iordan, E. P., Novozhilova, Yu., Vorobeva, L. J., The correlation between DNA synthesis and vitamin B_{12} content in *Propionibacterium shermanii* cells, Mikrobiologiya **52**, 591-596, 1983.

9.4 Some methods of disorder diagnosis

Determination of the cobalamin absorption [1-5] using:
- Urinary excretion test (Schilling's test) [6],
- Faecal excretion test [7],
- Blood plasma radioactivity [8, 9],
- Hepatic uptake method [10, 11],
- Whole body counter [12],
- Stool sample test [12 a].

Assay of intrinsic factor by immunological techniques [13-15],
Assay of unsaturated B_{12}-binding capacity [15].
Determinations of transcobalamin and cobalophilin using:
- rapid DEAE-cellulose chromatography [16],
- charged cellulose filter [17],
- insolubilized antibody [17 a].

Estimation of vitamin B_{12} of body fluids using:
- microbiological methods [18-20],
- radioisotope method [21].

Determinations of cobalamins of body tissues and fluids by means of
- thin-layer chromatography and bioautography [22, 23],
- thin layer chromatography [17a, 24, 25],
- cation exchange chromatography [26, 26a]
- bioaffinity chromatography [26a]

Detections of antibodies against:
- gastric parietal cells [27],
- intrinsic factor [13],
- transcobalamins [28].

simultaneous assays of serum vitamin B_{12} and folic acid by semiantomated system / ARIA II, Becton Dickinson Immunodiagnostics [29]

prenatal and postnatal detection of inborn errors of propionate, methylmalonate and cobalamin metabolism [45]

genetic and biochemical analysis of human cobalamin mutants in cell culture [46]

detection of methylmalonic acid in urine [47, 48].

References

[1] Ellenbogen, L., in Newer Methods of Nutritional Biochemistry, Ed. A. Albanese, vol. I, Academic Press, New York 1963, pp. 235-287.

[2] Ellenbogen, L., Highley, D. R., Vitam. Horm. **21,** 1, 1963.

[3] Chanarin, I., The megaloblastic anaemias. Blackwell, Oxford 1968.

[4] Pawelski, S., Konopka, L., Rechowicz, K., Diagnostyka izotopowa w hematologii, PZWL, Warszawa, 1972, p. 177-215.

[5] Ellenbogen, L., Uptake and transport of cobalamins. Biochem. Nutr. **27,** 45-96, 1979.

[6] Schilling, R. F., Intrinsic factor studies. II. The effect of gastric juice on the urinary excretion of radioactivity after the oral administration of radioactive vitamin B_{12}, J. Lab. Clin. Med. **42,** 860, 1953.

[7] Heinle, R. W., Welch, A. D., Schart, V., Macham, G. C., Prusoff, W. H., Studies of excretion (and absorption) of ^{60}Co-labelled vitamin B_{12} in pernicious anaemia, Trans. Ass. Amer. Physic. **65,** 214, 1952.

[8] Booth, C. C., Mollin, D. L., Brit. J. Haemat. **2,** 223, 1956.

[9] Doscheholmen, A., Hagen, P. S., Liu, M., Blood **12,** 336, 1957.

[10] Glass, G. B. J., Boyd, L. J., Gellin, G. A., Stephanson, L., Uptake of radioactive vitamin B_{12} by the liver in humans; test for measurement of intestinal absorption of vitamin B_{12} and intrinsic factor activity, Arch. Biochem. Biophys. **51,** 251, 1954.

[11] Weisberg, H., Glass, G. B. J., A rapid quantitative method for measuring intestinal absorption of vitamin B_{12} in man using a double label hepatic uptake test, J. Lab. Clin. Med. **68,** 163, 1966.

[12] Heinrich, H. C., in Künstliche Radioaktive Isotope in Physiologie, Diagnostik und Therapie, Eds. H. Schwieg, F. Turba, Springer, Berlin 1961, p. 660.

[12a] Hippe, E., Gimsing, P., Holländer, N. H., A simplified method for quantitative determination of vitamin B_{12} absorption, in Vitamin B_{12}, Eds. B. Lagalak and W. Friedrich, Walter de Gruyter, Berlin 1979, pp. 939-944.

[13] Irvine, W. J., Immunoassay of gastric intrinsic factor and the titration of antibody to intrinsic factor, Clin. Exp. Immunolog. **1,** 99, 1966.

[14] Callender, S. T., J. Roy. Coll. Phys. Lond. **4,** 114, 1970.

[15] Gottlieb, C. W., Lau, K. S., Wasserman, L. R., Rapid charcoal assay for intrinsic factor (IF), gastric juice unsaturated B_{12}-binding capacity, Blood **25,** 875-884, 1965.

[16] Retief, F. P., Gottlieb, C. W., Kochwa, S., Pratt, P. W., Herbert, V., Separation of vitamin B_{12}-binding proteins of serum, gastric juice

and saliva by rapid DEAE-cellulose chromatography. Blood **29,** 501-516, 1967.

[17] Seldhub, J., Rachmilewitz, B., Grossowicz, N., Fractionation of serum transcobalamins on charged cellulose filters. Proc. Soc. Exp. Biol. Med. **152,** 204, 1976.

[17a] Nex, E., Characterization of the cobalamins attached to transcobalamin I and transcobalamin II in human plasma, Scand. J. Haemat. **18,** 358-360, 1977.

[18] Hutner, S.H., Bach, M.K., Ross, G.I.M., A sugar-containing basal medium for vitamin B_{12}-assay with *Euglena*. Applications to body fluids. J. Protozool. **3,** 101, 1956.

[19] Beck, W.S., Hook, S., Barnett, B.H., The metabolic functions of vitamin B_{12}. I. Distinctive modes of unbalanced growth behavior in *Lactobacillus Leichmanii,* Biochem. Biophys. Acta **55,** 455-469, 1962.

[20] Matthews, D.M., Observations on the estimation of serum vitamin B_{12} using *Lactobacillus Leichmanii,* Clin. Science **22,** 101-111, 1962.

[21] Lau, K.S., Gottlieb, C.W., Wasserman, L.R., Herbert, V., Measurment of serum vitamin B_{12} level using radioisotope dilution and coated charcoal, Blood **26,** 202, 1965.

[22] Linnell, J.C., Hussein, H.A.-A., Matthews, D.M., A two-dimensional chromato-bioautographic method for complete separation of individual plasma cobalamins, J. Clin. Pathol. **23,** 820-821, 1970.

[23] Linnell, J.C., Hoffbrand, A.V., Peters, T.J., Matthews, D.M., Chromatographic and bioautographic estimation of plasma cobalamins in various disturbances of vitamin B_{12} metabolism, Clin. Science **40,** 1-16, 1971.

[24] Gimsing, P., Hippe, E., Nexø, E., Determination of the plasma cobalamins by one-dimensional thin-layer chromatography, in Vitamin B_{12}, Eds. B. Lagalak and W. Friedrich, Walter de Gruyter, Berlin 1979, pp. 665-669.

[25] Nexø, E., Gimsing, P., The pattern of cobalamin in human erythrocytes, in Vitamin B_{12}, Eds. B. Lagalak and W. Friedrich, Walter de Gruyter, Berlin 1979, pp. 967-970.

[26] Begley, J.A., Hall, C.A., Presence of sulfitocobalamin in cell extracts. Resolution and identification by SP-Sephadex C-25 cation exchange chromatography, in Vitamin B_{12}, Eds. B. Lagalak and W. Friedrich, Walter de Gruyter, Berlin 1979, pp. 971-974.

[26a] Omran, N., Neumann, E., Demonstration of vitamin B_{12} deficiency by a simple cytochemical reaction in the peripheral blood smear, Blut **39,** 359-364, 1979.

[26b] Muir, M., Chanarin, J., Separation of cobalamin analogs in human sera binding to intrinsic factor and to R-type vitamin B_{12} binders, Br. J. Haematol. **54,** 613-622, 1983.

[27] Fisher, I.M., Mackay, I.R., Taylor, K.B., Ungar, B., An immunological study of categories of gastrities, Lancet **1,** 176, 1967.

[28] Nicolas, J.P., Heterogeneity of antibodies against B_{12} binders in pernicious anaemia, in Vitamin B_{12}, Eds. B. Lagalak and W. Friedrich, Walter de Gruyter, Berlin 1979, pp. 791-802.

[29] Chen, J.W., silberstein E.B., Maxon, M.R., Volle, C.P., Sohnlein, B.M., Semiautomated system for simultaneous assays of serum vit. B_{12} and folic acid in serum evaluated, Clin. Chem. **28,** 2161-2165, 1982.

[30] Liu, Y.K., Sullivan, L.W., An improved radioisotope dilution assay for serum vitamin B_{12} using hemoglobincoated charcoal, Blood. **39,** 426-432, 1972.

[31] Bain, B., Broom, G.N., Woodside, J., Litwinczuk, R.A., Wickramasinghe, S.N., Assessment of a radioisotopic assay for vitamin B_{12} using an intrinsic factor preparation with R proteins blocked by vitamin B_{12} analogues. J. Clin. Pathol. **35,** 1110-1113, 1982.

[32] Kolhouse, J.F., Kondo, H., Allen, N.C., Podell, E., Allen, R.H., Cobalamin analogues are present in human plasma and can mask cobalamin deficiency because current radioisotope dilution assays are not specific for true cobalamin, N. Engl. J. Med. **299,** 785-792, 1978.

[33] Begley, J.A., Hall, C.A., Forms of vitamin B_{12} in radioisotope dilution assays, J. Clin. Pathol. **34,** 630-636, 1981.

[34] Sourial, N.A., Use of an improved *E. coli* method for the measurement of cobalamin in serum: comparison with the *E. gracilis* assay results, J. Clin. Pathol. **34,** 351-355, 1981.

[35] Wickramasinghe, S.N., Saunders, J.E., A fault in the design of the deoxyuridine suppression test, Clin. Lab. Haematol. **1,** 69-71, 1979.

[36] Cooper, B.A., Whitehead, V.M., Evidence that some patients with pernicious anaemia are not recognised by radio dilution assay for cobalamin in serum, N. Engl. J. Med. **299,** 816-818, 1978.

[37] Chen, I.W., Silberstein, E.B., Maxon, H.R., Sperling, M., Barnes, E., Clinical significance of serum vitamin B_{12} measured by radioassay using pure intrinsic factor, J. Nucl. Med. **22,** 447-451, 1981.

[38] Green, R., Newmark, P.A., Musso, A.M., Mollin, D.L., The use of chicken serum for measurement of serum vitamin B_{12} concentra-

tion by radioisotopic dilution: description of method and comparison with microbiological assay results, Br. J. Haematol. **27**, 507-526, 1974.

[39] Zacharakis, R. E. A., Muir, M., Chanarin, I., Comparison of serum vitamin B_{12} estimation by saturation analysis with intrinsic factor and with R-protein as binding agents, J. Clin. Pathol. **34**, 357-360, 1981.

[40] Dawson, D. W., Fish, D. I., Frew, I. D. O., Orton, B., Roome, T., Vitamin B_{12} quality control trials, Clin. Lab. Haematol. **3**, 323-330, 1981.

[41] Houts, T. M., ^{125}I-labelled cobalamin derivative compatible with intrinsic factor, Clin. Chim. Acta, **126**, 315-322, 1982.

[42] Leonard, J. P., Beckers, C., Vitamin B_{12} and folic acid studies by saturation analysis techniques. A review, Int. J. Nucl. Med. Biol. **2**, 89-96, 1975.

[43] Endres, D. B., Painter, K., Niswender, G. D., A solid-phase radioimmunoassay for vitamin B_{12}, Clin. Chem. **24**, 460-465, 1978.

[44] Houts, T. M., Carney, J. A., Radioassay for cobalamin (vitamin B_{12}) requiring no pretreatment of serum, Clin. Chem. **27**, 263-267, 1981.

[45] Willard, H. F., Rosenberg, L. E., Rapid prenatal and postnatal detection of inborn errors of propionate, methylmalonate and cobalamin metabolism: A sensitive assay using cultured cells, Hum. Genet. **34**, 277-283, 1976.

[46] Fenton, W. A., Rosenberg, L. E., Genetic and biochemical analysis of human cobalamin mutants in cell culture, Annu. Rev. Genet. **12**, 223-248, 1978.

[47] Giorgio, A. J., Plaut, G. W. E., A method for the colorimetric determination of urinary methylmalonic acid in pernicious anemia, J. Lab. Clin. Med. **66**, 667-676, 1965.

[48] Dreyfus, P. M., Dube, V. E., The rapid detection of methylmalonic acid in urine: a sensitive index of vitamin B_{12} deficiency, Clin. Chim. Acta. **15**, 525-528, 1967.

[49] Jacob, E., Wong, K. T. J., Limitations of protein-coated charcoal in the separation of free from bound vit. B_{12} in serum, J. Clin. Pathol. (Lond) **36**, 1022-1027, 1983.

[50] Cardarelli, J. A., Burrows, B. A., Miller, A., Slingerland, D. W., Correction of attenuation in whole-body determination of cobalt-57-labeled vitamin B_{12} absorption, J. Nucl. Med. **24**, 949-951, 1983.

[28] Chen, J. W., Silberstein, F. B., Maxon, H. R., Volle, C. P., Sohnlein, B. H., Semiautomated system for simultaneous assays of serum vit. B_{12} and folic acid in serum evaluated, Clin. Chem. **28**, 2161-2165, 1982.

9.5 Some pharmaceutical preparations of vitamin B_{12}

AdoCbl preparations:
Anabasi (Pierrel, Italy): microtablets – 1 mg, vials – 1 mg
Cobaforte (Maestretti, Italy): capsules 1 mg and 2.5 mg, vials 1 mg and 5 mg
Cobalion (Roussel, France): vials: 50 g emulsion
Cobanzyme (Bouchara, France): caps. 1 mg, vial 2 mg
Dibencozan (Houdé, Italy, Holland): vial 5 mg
Dibencozan fort: vial. 20 mg
Dolonevran (Dausse, France): vial. 20 mg
Heraclene (Spret-Mauchant, France): caps. 1 mg
Indusil (Recordati, Italy): caps. 2 mg; vials 2,5 mg and 30 mg
Indusil T (Diamant, France): caps. 3 mg; vials 10 mg
Nevrizide (Amour-Montagn, France): vials 10 mg
Ripresil (Erba, Italy): caps. 1 mg

Vitamin B_{12} coenzyme (Sefton AG, Switzerland):
Typical Analysis

Appearance	Dark reddish-brown crystalline powder

Assay U. V.	100.3% w/w	
Loss at 105 °C in vacuo	2.2% w/w	
Absorption ration	260 nm/375 nm	3.2
	525 nm/375 nm	0.75
Coloured impurities	1.05% w/w	
Manufacturer	Glaxo Laboratories Greenford, Middlesex, England.	
Form	Supplied as non-sterile powders.	
Specifications	These products are not described in the British Pharmacopoeia or U.S. Pharmacopoeia. Copies of their specifications are available on request.	
Stability	The products are stable for at least 2 years when stored under normal conditions.	
Packing	Glass vials holding 1 g, 5 g, 10 g or 25 g Aluminium containers holding 50 g, 100 g, 500 g or 1 kg.	
Xobaline	(Albert-Roussel, D): caps. 1 mg; amp. 1 mg	

MeCbl preparations:

Methycobal (Eisai, Japan): vial 5 mg

Methylcobalamin (Sefton AG, Switzerland):

Typical Analysis		
Appearance	Bright-red odourless crystalline powder	
Total cobalamins U.V.	100.1% w/w	
Loss at 100 °C in vacuo	7.2% w/w	
Absorption ratio	267 nm/343 nm	1.44
	522 nm/343 nm	0.65
Acidic impurities	0.6%	
Basic impurities	0.48%	
Coloured impurities	2.18%	
Manufacturer	Glaxo Laboratories	
Form	Supplied as non-sterile powders.	
Specifications	These products are not described in the British Pharmacopoeia or U.S. Pharmacopoeia. Copies of their specifications are available on request.	
Stability	The products are stable for at least 2 years when stored under normal conditions.	
Packing	Glass vials holding 1 g, 5 g, 10 g or 25 g Aluminium containers holding 50 g, 100 g, 500 g or 1 kg.	

Methylcobaz (Labaz, France): caps. 3 mg, vial 24 g (0.125%) powder

OH-Cbl preparations:

Alpha-Ruvite (Savage, USA): vial 10 ml (1 ml = 1 mg)

Alphamin (Vortech, USA): vial 10 ml and 30 ml

Alpha-Redisol (Merck Sharp, Dohme, USA): vial 10 ml

AQUO-CYTOBION (Merck, FRG): vavial 0,5 mg/ml

B_{12}-COMPOSITUM (Siegfried, Switzerland): tabl. 0,005 mg
B_{12} DEPOT (Siegfried, FRD): vavial 1 mg/ml
B_{12} DEPOT VICOTRAT (Heyl, FRG): vavial 0,1 mg and 0,5 mg/ml
B_{12} DEPOT (Siegfried, Switzerland): vavial 0,5 mg, 1 mg/ml
BEHEPAN (Kabi, Sweden): vavial 1 mg/ml
BERUBI (Redal, FRG): vial 20 ml
BERUBI LONG (Redal, FRG): vavial 0.5 mg/ml
BETRIMAX-FORTE (Lipha, FRG): vavial 1 mg/ml
COBALPAREN (Saarstickstoff-Fatol, FRG): vavial 1 ml; vial 5 ml, 0,1 mg/ml
COBALPAREN-DEPOT (Saarstickstoff-Fatol, FRG): vavial 0,5 mg/ml
DEPOGAMMA (Cooper, FRG): vavial 0,5 mg/ml
DOCIVIT DEPOT (Robisch, FRG): vavial 1 mg/ml
DODE CAVIT (Aron, France): vavial 10 mg/2 ml
DROXOMIN (Tutag, USA): vial. 10 ml (mg/ml)
DROXOVITE (Kay, USA): vial 10 ml
Hepagon Novum (Astra, Sweden): amp. 1 ml (1 mg), vial 10 ml (10 mg)
HYDROBEXAN (Keene, USA): vial 10 mg
HYDROX-B_{12} (Rocky, USA): vial 10 ml
HYDROXO 5000 (Arphar-Rolland): vavial 5 mg/2 ml
HYDROXOBASE (Lipha, FRG): vavial 5 mg/2 ml
HYDROXOCOBALAMIN (Richter, H): vavial 0,1 mg and 0,25 mg/ml

Hydroxocobalamin acetate (Sefton AG, Switzerland):

Typical Analysis	
Appearance	Dark-red, odourless, crystalline powder
Assay	As hydroxocobalamin acetate 98.8% w/w
Loss at 105 °C in vacuo	9.9% w/w
Coloured impurities	1.76% w/w
Other cobalamins	0.4% w/w
Acidic impurities	0.5% w/w
Manufacturer	Glaxo Laboratories
Form	Supplied as a non-sterile powder.
Specifications	The product is not described in the BP or U.S.P. Copies of the products specifications are available on request.
Stability	Hydroxocobalamin acetate is an unstable product and we recommend storage at −20 °C, where it is stable for at least 16 months. If stored below 0 °C it is stable for at least 1 year.
Packing	Glass vials holding 1 g, 5 g, 10 g or 25 g Aluminium containers holding 50 g, 100 g, 500 g or 1 kg.

HYDROXOCOBALAMINE (Anphar-Rolland, France): 1 mg/2 ml

Hydroxocobalamin hydrochloride (Sefton AG, Switzerland):

Typical Analysis	
Appearance	Dark red-brown, odourless, crystalline powder

Assay	As hydroxocobalamin hydrochloride 98.3% w/w
Loss at 105 °C in vacuo	10.0% w/w
Coloured impurities	1.3% w/w
Other cobalamins	0.49% w/w
Acidic impurities	0.46% w/w
Manufacturer	Glaxo Laboratories,
Form	Supplied as a non-sterile powder.
Specifications	Our minimum specifications for hydroxocobalamin hydrochloride are defined in the latest edition of the British Pharmacopoeia.
Stability	The products is stable for at least 2 years when stored under normal conditions.
Packing	Glass vials holding 1 g, 5 g, 10 g or 25 g Aluminium containers holding 50 g, 100 g, 500 g or 1 kg.

IDROBAMINA (Tiber, Italy): vial 1 mg/2 ml
LOPHAKOMP-B_{12} (Lomapharm, FRG): vavial 1,6 mg
LOPHAKOMP-B_{12} DEPOT: vavial 0,5 mg i 1 mg/2 ml

Neo-Cytamen (Glaxo, England)

Presentation	Ampoules containing a clear red solution which provides either 250 micrograms (Neo-Cytamen "250") or 1000 micrograms (Neo-Cytamen "1000") of hydroxocobalamin (vitamin B_{12}b) per ml. Neo-Cytamen injection complies with the specification for Hydroxocobalamin Injection BP, BNF.
Uses	Addisonian pernicious anaemia. Prophylaxis and treatment of other macrocytic anaemias associated with vitamin B_{12} deficiency. Tobacco amblyopia and Leber's optic atrophy.
Dosage and administration	The following dosage schemes are suitable for adults and children. Addisonian pernicious anaemia and other macrocytic anaemias without neurological involvement: Initially: 250 to 1000 micrograms intramuscularly on alternate days for one to two weeks, then 250 micrograms weekly until the blood count is normal. Maintenance: 1000 micrograms every two or three months. Addisonian pernicious anaemia and other macrocytic anaemias with neurological involvement: Initially: 1000 micrograms on alternate days as long as improvement is occurring. Maintenance: 1000 micrograms every two months.

	Prophylaxis of microcytic anaemia associated with vitamin B_{12} deficiency resulting from gastrectomy, some malabsorption syndromes, and strict vegetarianism: 1000 micrograms every two or three months. Tobacco amblyopia and Leber's optic atrophy: Initially: 1000 micrograms or more daily by intramuscular injection for two weeks, then twice weekly as long as improvement is occurring. Maintenance: 1000 micrograms monthly.
Contraindications	Hypersensitivity to hydroxocobalamin.
Precautions	The dosage schemes given above are usually satisfactory, but regular examination of the blood is advisable. If megaloblastic anaemia fails to respond to Neo-Cytamen, folate metabolism should be investigated.
Side effects	Sensitisation to hydroxocobalamin is rare, but may manifest itself as itching exanthema, and exceptionally, anaphylaxis.
Pharmaceutical precautions	Protect from light.
Legal category	POM
Packs	1 ml ampoules in boxes of 5
Further information	An intramuscular injection of hydroxocobalamin produces higher serum levels than the same dose of cyanocobalamin, and these levels are well-maintained.
Product licence numbers	Neo-Cytamen "250" PLR/0004/5025 Neo-Cytamen "1000" PLR/0004/5026
Product licence holder	Glaxo Laboratories Ltd

NOVIDROXIN (Saarstickstoff Fatol, FRG): vavial 10 mg/2 ml
NOVOBEDOUZE (Bouchara, France): vavial 1 mg and 10 mg/2 ml
OHB_{12} (Galenika, YU): vavial 2,5 mg/2 ml

Cyanocobalamin (Sefton AG, Switzerland):
Typical Analysis

Appearance	Dark-red, odourless, crystalline powder	
Assay U.V.	99.1% w/w	
Absorption ratio	361 nm/278 nm	1.82
	361 nm/550 nm	3.26
Foreign pigments	0.88% w/w	
Loss at 105 °C	1.10% w/w	
Pseudo cobalamins	Passes USP test	
Manufacturer	Glaxo Laboratories	

Form	Supplied as a non-sterile powder also available with higher moisture contents on request.
Specifications	Our minimum specifications for cyanocobalamin are defined in the latest edition of the European Pharmacopoeia, additionally the standard product has a maximum moisture content of 3 percent.
Stability	The product is stable for at least 5 years when stored under normal conditions.
Packing	Glass vials holding 1 g, 5 g, 10 g, 25 g or 50 g Aluminium containers holding 100 g, 500 g or 1 kg.

Cytamen (Glaxo, England):

Presentation	Cytamen "250" is a clear, red solution containing 250 µm cyanocobalamin per millilitre. Cytamen "1000" is containing 1000 µm cyanocobalamin per millilitre. Cytamen complies with the specification for Cyanocobalamin Injection, BP.
Uses	Addisonian pernicious anaemia. Orophylaxis and treatment of other macrocytic anaemias associated with vitamin B_{12}deficiency. Schilling test. Not indicated for treatment of toxic amblyopias – use Neo-Cytamen.
Dosage and administration	Addisonian pernicious anaemia and other macrocytic anaemias without neurological involvement: Initially: 250–1000 µm intramuscularly on alternate days for one to two weeks, then 250 µm weekly until the blood count is normal. Maintenance: 1000 µg monthly. Addisonian pernicious anaemia and other macrocytic anaemias with neurological complications: Initially: 1000 µg intramuscularly on alternate days as long as improvement is occurring. Maintenance: 1000 µg monthly. Prophylaxis of macrocytic anaemia associated with vitamin B_{12} deficiency resulting from gastrectomy, some malabsorption syndromes and strict vegetarianism: 250–1000 µg monthly. Schilling test: An intramuscular injection of 1000 µg cyanocobalamin is an essential part of this test.
Contraindications	Hypersensitivity to cyanocobalamin.
Precautions	Doses in excess of 10 µg daily may produce a haematological response in patients with folate deficiency. Indiscriminate administration may mask the true diagnosis.

Side effects	Sensitisation to vitamin B_{12} is rare, but it may present as an itching exanthema, and, exceptionally, as anaphylactic shock.
Pharmaceutical precautions	Protect from light.
Legal category	POM
Packs	Ampoules of 1 ml in boxes of 5
Product licence numbers	Cytamen 250: 0004/5016 Cytamen 1000: 0004/5017

Rubramin PC Cyanocobalamin Injection U. S. P. (E. R. Squibb and Sons, Inc., USA):

Description	Rubramin PC (Vitamin B_{12}) is a clear, sterile, aqueous solution with a characteristic red color. The cobalt content of the 100 μg/ml and 1000 μg/ml Rubramin PC is 4 μg/ml, and 40 μg/ml, respectively. The preparation contains 1% (w/v) benzylalcohol as a preservative and sodium chloride for isotonicity; the pH has been adjusted with sodium hydroxide or hydrochloric acid.
Actions	Vitamin B_{12} is essential to growth, cell reproduction, hematopoiesis, nucleoprotein and myelin synthesis. Within 48 hours after injection of 100 to 1000 μg of vitamin B_{12}, 50 to 98% of the injected dose may appear in the urine. The major portion is excreted within the first 8 hours.
Indications	Rubramin PC is indicated for vitamin B_{12} deficiency due to malabsorption occurring in Addisonian (pernicious) anaemia, gastrointestinal pathology, dysfunction or surgery, fish tapeworm infestation, gluten enteropathy, sprue, and concomitant folic acid deficiency. It is also indicated in pregnancy, thyrotoxicosis, hemolytic anaemia, hemorrhage, malignancy, hepatic and renal disease because of increased requirements for vitamin B_{12}. Rubramin PC is also suitable for use as the flushing dose in the Schilling (vitamin B_{12} absorption) Test for pernicious anaemia.
Contraindications	This preparation is contraindicated in patients who are sensitive to cobalt or vitamin B_{12}.
Warnings	The preparation should be protected from sunlight. If a vitamin B_{12} deficiency is allowed to progress for more than 3 months, permanent degenerative spinal cord lesions may occur; such lesions have been observed when folic acid is used as the sole hematopoietic agent.

Patients who have early Leber's disease (hereditary optic nerve atrophy) have been found to suffer severe and swift optic atrophy when treated with vitamin B_{12}. Hypokalemia and sudden death may occur when severe megaloblastic anaemia is treated intensively. Lack of therapeutic response may be due to infection, uremia, concomitant treatment with chloramphenicol or misdiagnosis.
The preparation should be not administered intravenously.

Precautions

Before administering vitamin B_{12}, an intradermal test dose is recommended for patients known to be sensitive to cobalamines. Most antibiotics, methotrexate, and pyrimethamine invalidate folic acid and vitamin B_{12} diagnostic microbiological blood assays.
Colchicine, para-aminosalicylic acid or excessive alcohol intake longer than 2 weeks may produce malabsorption of vitamin B_{12}. Doses of vitamin B_{12} exceeding 10 μg daily may produce a hematologic response in patients who have a folate deficiency. Indiscriminate administration of vitamin B_{12} may mask the true diagnosis of pernicious anaemia. A dietary deficiency of only vitamin B_{12} is rare. Multiple vitamin deficiency is expected in any dietary deficiency.

Adverse reactions

Mild transient diarrhea, polycythemia vera, peripheral vascular thrombosis, itching transitory exanthema, feeling of swelling of entire body, pulmonary edema and congestive heart failure early in treatment, anaphylactic shock and death.

Dosage and administration

A. General Rules
In patients with Addisonian (pernicious) anaemia, parenteral therapy with vitamin B_{12} is the recommended method of treatment and will be required for the remainder of the patient's life. Oral therapy is not dependable. Serum potassium must be watched closely the first 48 hours; and potassium should be replaced if necessary. Reticulocyte plasma count, vitamin B_{12} and folic acid levels must be obtained prior to treatment and between the fifth and seventh day of therapy.

In patients with other types of vitamin B_{12} deficiency due to malabsorption, the malabsorption should be corrected. In all patients a well balanced dietary intake should be prescribed and poor dietary habits should be corrected.

B. Treatment of Vitamin B_{12} Deficiency: Thirty micrograms daily for 5 to 10 days followed by 100 μg monthly injected intramuscularly or deep subcutaneously. Folic acid should be administered concomitantly early in the treatment unless folic acid levels are adequate.
In patients having normal intestinal absorption where the vitamin B_{12} deficiency is due to increased requirements, an oral therapeutic multivitamin preparation containing 15 μg vitamin B_{12} should be given daily for 1 month.
C. Schilling Test:
The flushing dose is 1000 μg.

How supplied — Rubramin PC Cyanocobalamin Injection U.S.P. is available in multiple-dose vials providing potencies of 100 and 1000 μg cyanocobalamin per ml.
Rubramin PC is also available in 1 ml Unimatic single dose preassembled syringes and 1 ml cartridge-needle units providing potencies of 100 μg and 1000 μg of cyanocobalamin (see package insert accompanying the product for complete information).

Vitamin B_{12} (Polfa, Poland):
amp. 0.02 mg, 0.05 mg and 0.1 mg in 1 ml, amp. 1 mg in 2 ml.

IF-12 (Polfa, Poland):
tabl. 0.01 mg CN-Cbl, 10 I.U. intrinsic factor (s. section 8.1), 8 mg liver extract and 1 mg $CoCl_3$.

Liver Injection (Lederle, USA): (20 μg for intramuscular injection)

Description — Each ml of LIVER INJECTION, 20 μg, contains Vitamin B_{12} activity equivalent to 20 micrograms of cyanocobalamin. Inactive ingredient consists of not more than 0.1% sodium bisulfite with 0.45% phenol as a preservative. Sodium hydroxide or hydrochloric acid is used to adjust the pH.

Indications — Liver Injection is used in the treatment of pernicious anaemia and tropical and non-tropical sprue, as well as certain cases of megaloblastic anaemia of infancy, achrestic anaemia, and nutritional macrocytic anaemia.

Precautions — Do not inject this liver extract intravenously. The intramuscular administration of liver extract should be surrounded by every known precaution for the prevention and arrest of allergic and other untoward reactions. A careful history should review possible sensitivity.

Administration and dosage

Initial Therapy

The intramuscular injection of a total of 2 to 3 ml of Liver Injection, 20 µg, is suggested as initial therapy. These totals (equivalent in Vitamin B_{12} activity to 40 to 60 micrograms of cyanocobalamin) may be divided into aliquots and administered intramuscularly over a period of 2 or 3 successive days. Clinical and hematological improvement usually occurs.

Intermediate Therapy

Beginning about 10 days after the completion of initial therapy, it is recommended that intramuscular injections of 0.5 to 1.0 ml of Liver Injection, 20 µg, (equivalent in Vitamin B_{12} activity, in the instance, is 10 to 20 micrograms of cyanocobalamin) be administered at weekly intervals until a normal hematologic status has been achieved. Maintenance therapy can then be instituted.

Maintenance Therapy

The intramuscular injection, at intervals of 10 to 15 days or more, of from 0.5 to 1.0 ml of Liver Injection, 20 µg (equivalent in Vitamin B_{12} activity, in the instance, is 10 to 20 micrograms of cyanocobalamin) is suggested for adult maintenance therapy. The specific requirement of each patient should be determined by observation and should be sufficient to maintain clinical and hematological remission, in the presence of infection. Vitamin B_{12} requirements are increased, so the maintenance dose should also be increased. Patients exhibiting extensive neurological involvement and those advanced in years, require higher than average doses. In such instances, an alternative to higher doses could be average doses administered at shorter than usual intervals.

Treatment of Relapse

In cases where relapse has occurred in patients being treated with other types of antianaemia medication, it is suggested that Liver Injection be administered in doses similar to those recommended for initial therapy. Subsequent treatment can be adjusted according to the directions given in intermediate therapy and maintenance therapy.

Adjuvant Therapy

Many patients suffering from megaloblastic anaemias require, in addition to liver extract, large amounts of the Vitamin B complex, including folic acid.

Index of Species

Subject Index

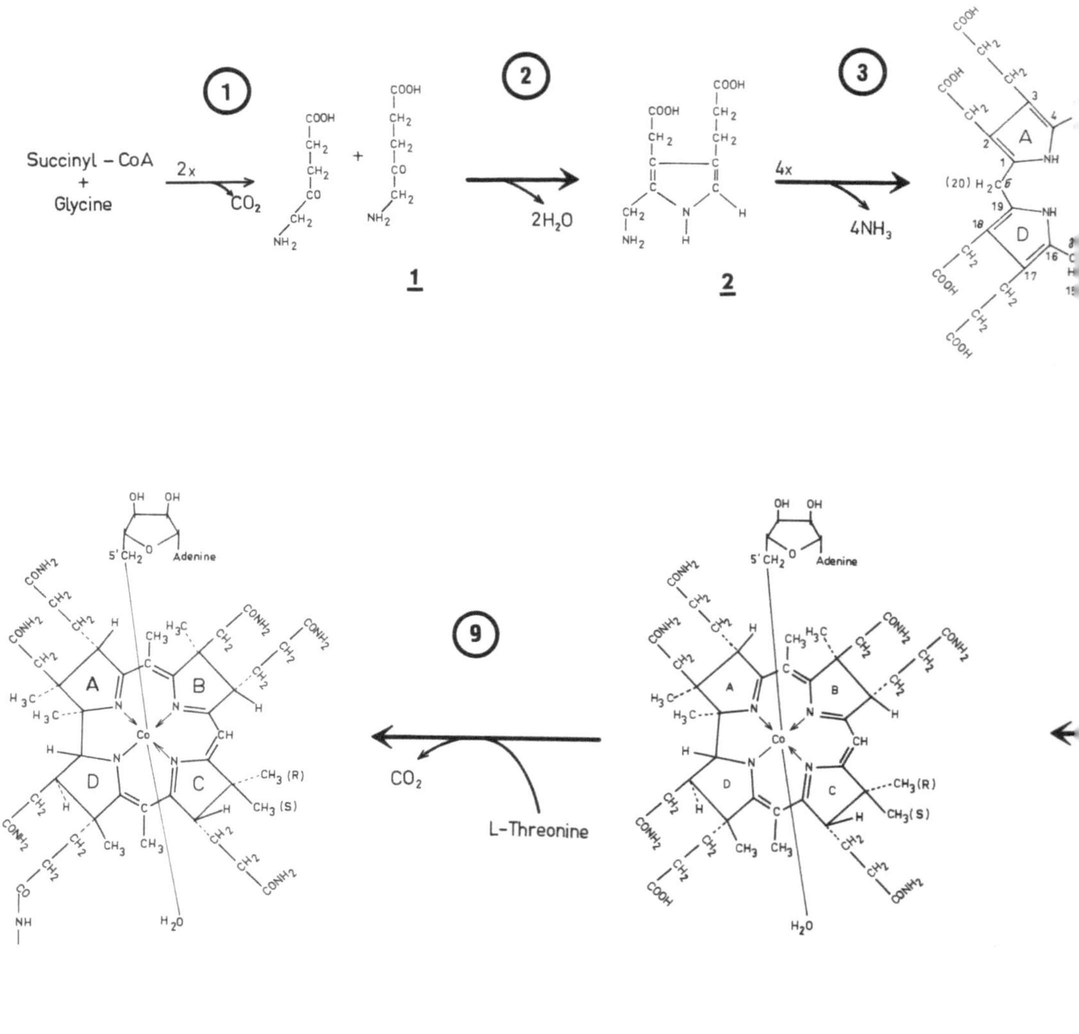

1
2
3
9
Succinyl – CoA
+
Glycine
2x
CO2
2H2O
4x
4NH3
1
2
L-Threonine
CO2
Adenine
H2O

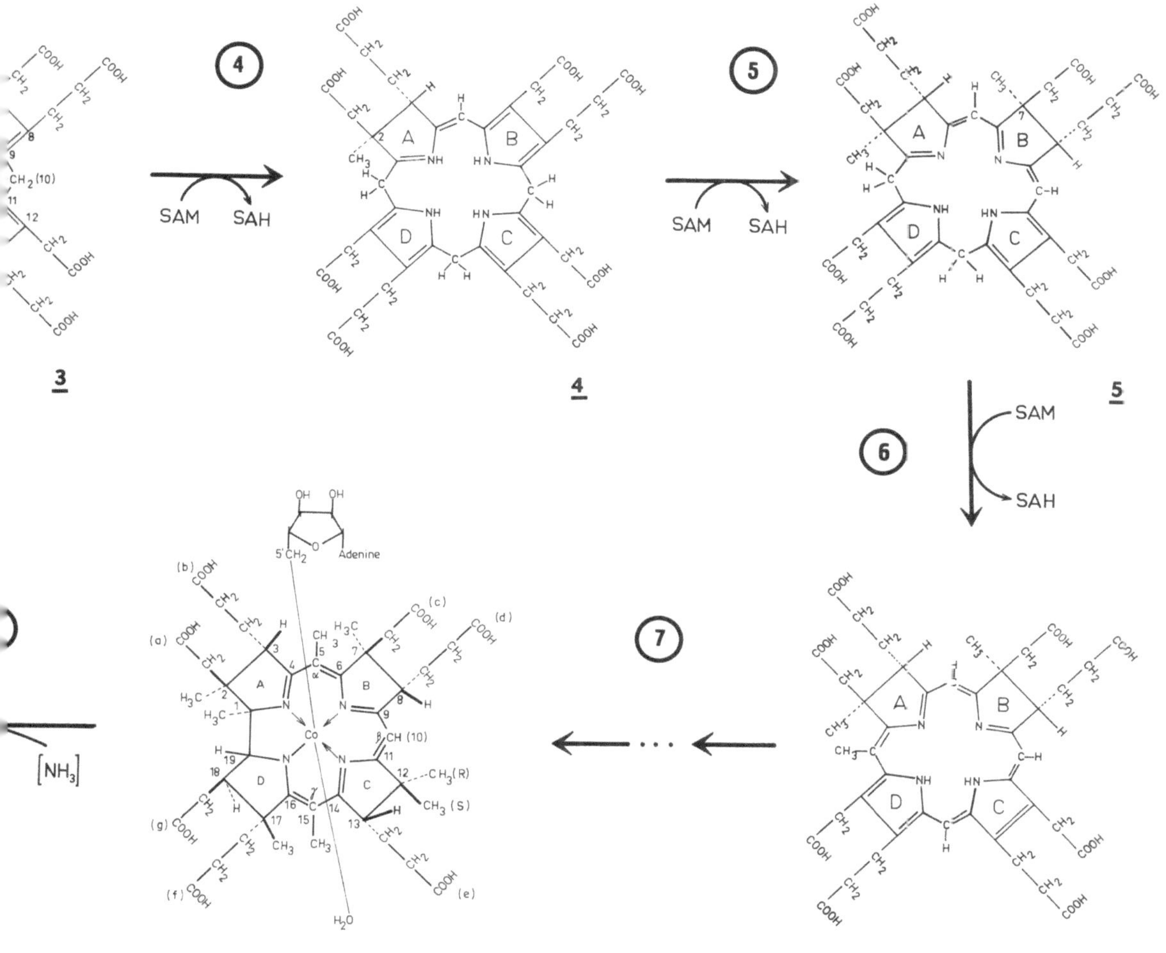

4
SAM
SAH
3
4
5
SAM
SAH
5
6
SAM
SAH
7
[NH3]
Adenine
H2O
Co

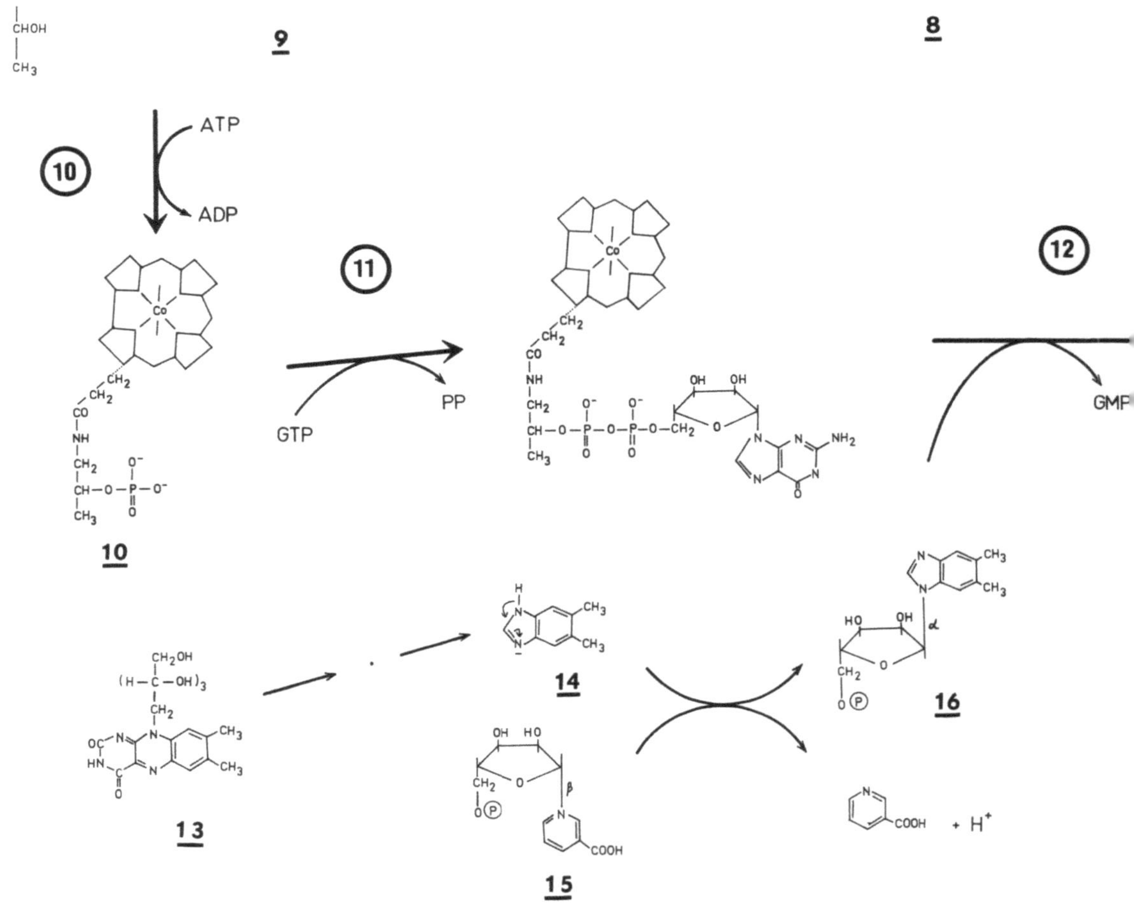

Fig. 4-2 Pathway of Vitamin B_{12} Biosynthesis (see text pp. 99-103)

7 **6**

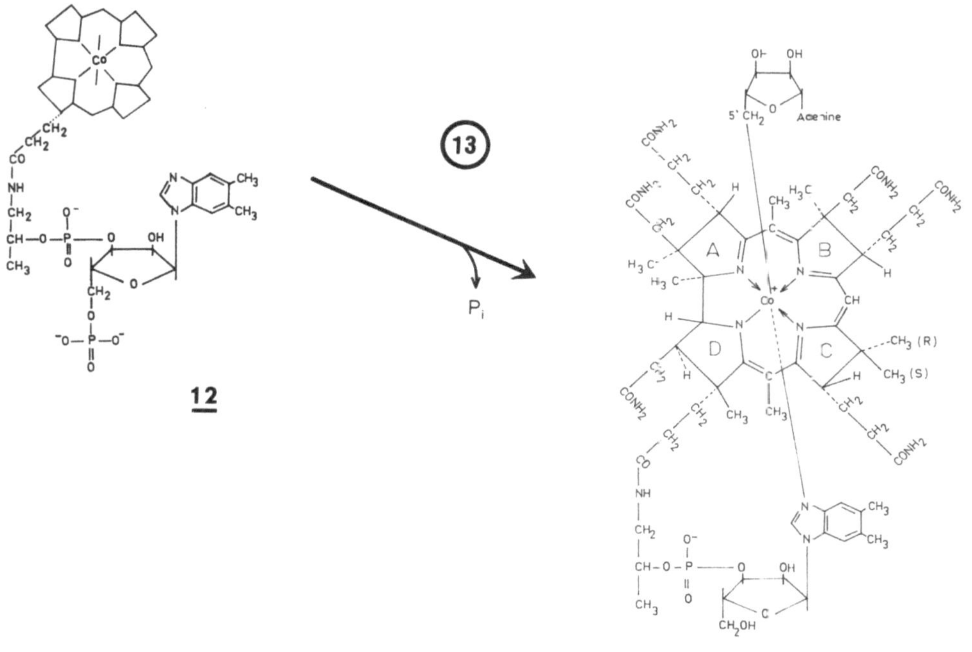

Co
CH2
CO
NH
CH-O-P-O
CH3
OH
N
CH3
12
13
Pi
OH OH
5' CH2
Adenine
CONH2
A
B
C
D
Co+
CH3 (R)
CH3 (S)
17

Bei Fragen zur Produktsicherheit wenden Sie sich bitte an:
If you have any questions regarding product safety,
please contact:

Walter de Gruyter GmbH
Genthiner Straße 13
10785 Berlin
productsafety@degruyterbrill.com